For further information on TI optoelectronic devices, contact the following:

Texas Instruments Incorporated
Optoelectronics Marketing
P.O. Box 225012, MS 12
Dallas, Texas 75265
Phone: (214) 995-3821

AF784271

Optoelectronics Data Book

1983-84

Infrared, Imaging, and Visible Products

Texas Instruments

IMPORTANT NOTICES

Texas Instruments reserves the right to make changes at any time in order to improve design and to supply the best product possible.

Texas Instruments assumes no responsibility for infringement of patents or rights of others based on Texas Instruments applications assistance or product specifications, since TI does not possess full access to data concerning the use or applications of customer's products. TI also assumes no responsibility for customer product designs.

Information contained herein supercedes previously published data on optoelectronics and electro-optical devices from TI, including data book CC-4231.

ISBN 0-89512-115-8
Library of Congress No. 83-070442

Copyright © 1983 Texas Instruments Incorporated

THE OPTOELECTRONICS DATA BOOK

Few people in the electronics industry realize that optoelectronics technology has a history that precedes the invention of the integrated circuit. It is also a relatively unknown fact that Texas Instruments was a pioneer in the development and manufacture of some of the first optoelectronic components, viz infrared detectors and photovoltaic solar cells, back in 1957.

During the past 26 years TI has continued to develop and build optoelectronic devices and assemblies for end application in the space, military, computer, industrial, and consumer industries. TI opto devices have helped to revolutionize the industry and to make it easier for the design engineer to accomplish his job.

In addition, TI offers the broadest line of opto products in the industry. This ensures that design engineers can obtain more answers to questions involving circuitry and operating conditions by contacting TI.

To complete the service aspect, TI has a worldwide distributor network that stocks almost 225 standard opto devices and assemblies. This means that customers can obtain fast delivery on small quantities required for initial circuit evaluation and purchasing departments can be assured of a local source of supply for production quantities.

It is the purpose of this data book to better acquaint our customers with TI opto products and capabilities. It offers the user a categorized listing of optoelectronic data sheets, application reports, and other information for more than 250 standard devices including 116 new types not included in the fifth edition of this data book. Each product section has a quick reference guide that lists the key electrical parameters and features for products in that section. The table of contents and alphanumeric index identify the new devices in this data book in bold type. A handy replacement guide for obsolete devices is also included.

To further assist the user, there is an interchangeability guide that lists more than 600 optoelectronics devices built by other manufacturers, along with the nearest TI equivalent devices. There is also a glossary of optoelectronic terminology to answer questions on optoelectronic terms and phrases.

This data book's new format will make the designer's job easier to use optoelectronics devices in his new and existing products or applications.

General Information

- **Table of Contents**
  New Data Sheets and Applications are Highlighted
- **Alphanumeric Index**
  Devices Added Since 5th Edition are Highlighted
- **Deleted Part Numbers — Replacement Guide**

Contents

1 GENERAL INFORMATION

Pages

ALPHANUMERIC INDEX

The bold type identifies data sheets added since the Fifth Edition of *The Optoelectronics Data Book*.

TYPE

TYPE

TYPE

TEXAS INSTRUMENTS
INCORPORATED
POST OFFICE BOX 225012 • DALLAS, TEXAS 75265

The bold type identifies data sheets added since the Fifth Edition of *The Optoelectronics Data Book.*

TI DELETED PART NUMBERS (REPLACEMENT GUIDE)

The following part numbers have been deleted from the Texas Instruments Optoelectronics product line since the Fifth Edition of *The Optoelectroncs Data Book.* This list indicates the nearest replacement devices available from TI Optoelectronics Department. It is our hope that these replacement devices will allow you to fulfill your Opto requirements.

Obsolete TI Part Number	TI Nearest Replacement
LS400	TIL601 thru TIL604, TIL81
TIED55	TIED56
TIED80	TIED89
TIED82	None
TIED83 thru TIED86	TIED87, TIED88
TIED90 thru TIED98	None
TIED451	TIED88
TIED452	None
TIES12	TIES13, TIES13A
TIES16B, TIES16C	TIES16A
TIES36	TIES35
TIES471	TIES35, TIES494, TIES495, TIES496
TIES472	None
TIL23HR thru TIL25HR	TIL24HR2
TIL26	TIL31B
TIL31	TIL31B
TIL33	TIL33B
TIL34	TIL34B
TIL31A	TIL31B
TIL33A	TIL33B
TIL34A	TIL34B
TIL41 thru TIL50	Contact Opto Marketing
TIL63 thru TIL67	TIL81
TIL107, TIL108	4N47, 4N48, 4N49, 4N22, 4N23, 4N24
TIL131, TIL132, TIL133	Use TIL23, TIL601 series devices
TIL134, TIL135, TIL136	Use TIL23, TIL601 series devices
TIL141, TIL142	Contact Opto Marketing
TIL147	TIL147A
TIL148	TIL148A
TIL227	TIL224
TIL231	TIL228, TIL221

TEXAS INSTRUMENTS
INCORPORATED
POST OFFICE BOX 225012 • DALLAS, TEXAS 75265

Obsolete TI Part Number	TI Nearest Replacement
TIL236	TIL234
TIL261 thru TIL270	Contact Opto Marketing
TIL271 thru TIL280	Contact Opto Marketing
TIL281 thru TIL290	Contact Opto Marketing
TIL360	TIL393
TIL401 thru TIL406	TIL601 thru TIL604, TIL81
TIL501	4N41 (new JEDEC number, same device)
TIL504	TIL507, TIL305
TIL505	4N56 (new JEDEC number, same device)
TIL506	4N57 (new JEDEC number, same device)
TIL507	4N58 (new JEDEC number, same device)
TIL560	4N58
TIL601HR thru TIL604HR	TIL604HR2
TIL605 thru TIL608	TIL601 thru TIL604
TIL621 thru TIL630	Call Opto Marketing
TIL807, TIL808	TIL312, TIL313
TIL829 thru TIL834	TIL321A, 322A, TIL729, 730
TIL835, TIL836	TIL321A, 322A, 330A, TIL729, 730
TIL837, TIL838	TIL312, TIL313
TIL839 thru TIL842	TIL321A, TIL322A, TIL729, TIL730
TILM2	TILM4
TILM3C, TILM3R, TILM3Y, TILM3G	None
TXED453	TIED459
TXED454 thru TXES457 series	TIED459
TXEF402 series	None
TXES37	TIES16A
TXES475, TXES476	TIES494, TIES495, TIES496
TXES478 thru TXES483 series	TIES494, TIES495, TIES496
TXES485, TXES486	TIES494, TIES495, TIES496
TXES488 thru TXES493 series	TIES494, TIES495, TIES496

CCD Image Sensors

(Charged-Coupled Devices)

- Quick Reference Guide
- Virtual Phase Technology Breakthrough
- Linear Arrays
  - Evaluation Boards
  - Evaluation Kits Available
- Area Arrays

LINEAR ARRAYS
QUICK REFERENCE GUIDE

DEVICE	PIXELS	PIXEL SIZE	SENSITIVITY	PACKAGE
TC101	1728 × 1	12.7 μm × 12.7 μm	3.5 V/μJ/cm^2	24-pin CDIP (0.600 inch)
TC102	128 × 1	12.7 μm × 12.7 μm	3.5 V/μJ/cm^2	10-pin CDIP (0.300 inch)
TC103	2048 × 1	12.7 μm × 12.7 μm	3.5 V/μJ/cm^2	24-pin CDIP (0.600 inch)
TC104	3456 × 1	10.7 μm × 10.7 μm	2.0 V/μJ/cm^2	24-pin CDIP (0.600 inch)

EVALUATION BOARDS
QUICK REFERENCE GUIDE

PART NO.	DEVICE EVALUATED	REMARKS
PC401	TC101,TC103,TC104	Device socket fits TC101, TC103 or TC104 (See TCK101, TCK103, TCK104 below)
PC402	TC102	Device socket fits only TC102 (See TCK102 below)

EVALUATION KITS
QUICK REFERENCE GUIDE

PART NO.	CONTENTS	REMARKS
TCK101	TC101 plus PC401	Includes complete instructions to evaluate TC101
TCK102	TC102 plus PC402	Includes complete instructions to evaluate TC102
TCK103	TC103 plus PC401	Includes complete instructions to evaluate TC103
TCK104	TC104 plus PC401	Includes complete instructions to evaluate TC104

AREA ARRAYS
QUICK REFERENCE GUIDE

DEVICE	PIXELS	PIXEL SIZE	SENSITIVITY	PACKAGE
TC201*	328 × 490	24.4 μm × 24.4 μm	0.48 A/W	20-pin CDIP (0.800 inch)
TC202*	390 × 584	22 μm × 22 μm	0.40 A/W	20-pin CDIP (0.800 inch)

*Availability of these devices is scheduled for 3rd quarter 1983.

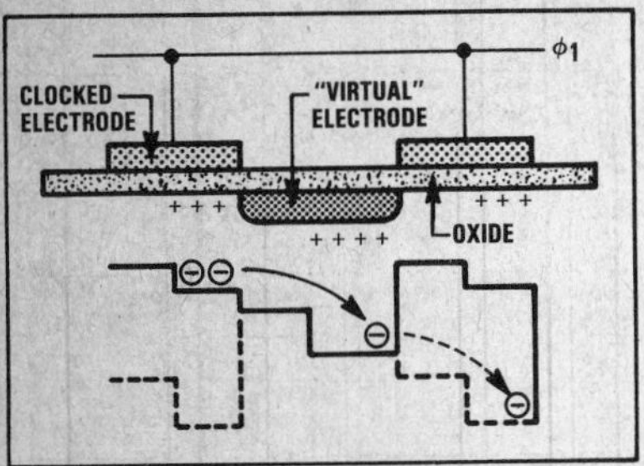

Fig. 1 TI's Patented Virtual Phase Design

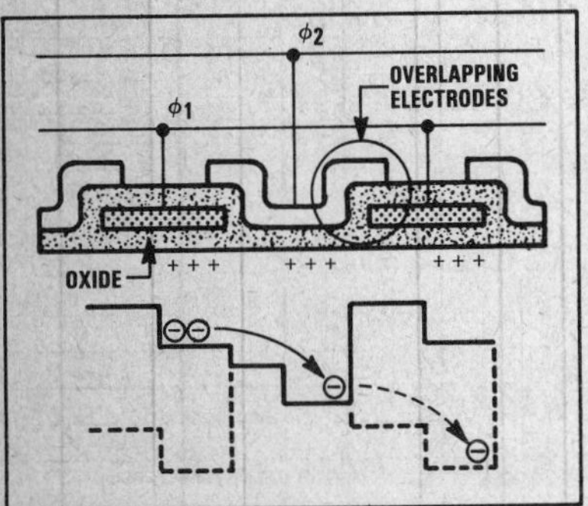

Fig. 2 Standard 2 Phase Design

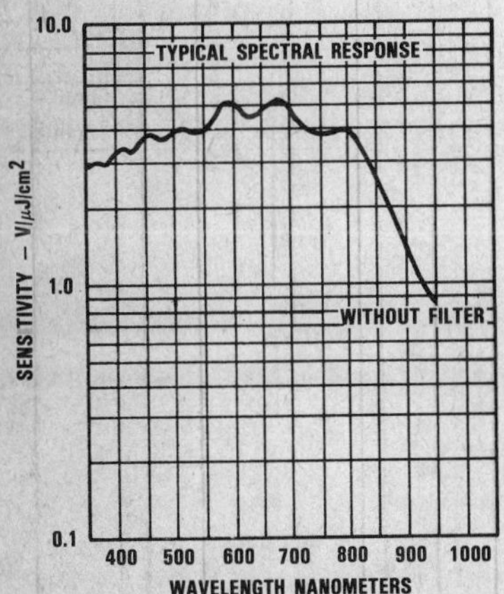

Fig. 3. Typical Sensitivity vs Wavelength

The CCD (Charge Coupled Device) approach to linear image sensing will become the leading edge among industry methods because of process and performance advantages.

Multiple-clock-electrode CCD processing methods have remained complex and difficult to implement in the manufacturing environment with any measure of cost/performance effectiveness ... until now.

The breakthrough: Now, Texas Instruments announces a breakthrough in CCD image sensor processing technology ... Virtual Phase (VP).

This giant technological stride greatly simplifies the processing techniques by **reducing the number of clock electrodes on the device surface to one** (Fig. 1). Other techniques require anywhere from two to three levels (Fig. 2). Additional benefits of this milestone process include simplified device operation and enhanced device quality.

Now, with just one level, the possibility of surface damage and shorts, common to the multilevel approach, is inherently reduced. So, the new Virtual Phase technology can boast the same degree of reliability as standard MOS technology.

The benefits of this TI-patented Virtual Phase technology are:

- Simplified clocking
- Lower noise/Higher dynamic range
- Greater sensitivity to light
- Ease of processing and use
- Greater stability
- Lower dark current
- Improved spectral response in the lower wave length (blue) regions (Fig. 3).

Features:

- Virtual Phase N-Channel silicon MOS technology
- High spectral responsivity ... particularly in the blue region
- Approximately 1-V peak-to-peak output signal
- Dynamic range typically 1000:1
- End-of-scan signal
- Internal dark and white references
- Blemish-free uniformity of image
- Simple, stable operation

TYPE TC101
1728 X 1 CCD LINEAR IMAGE SENSOR

D2663, FEBRUARY 1982

TOP VIEW

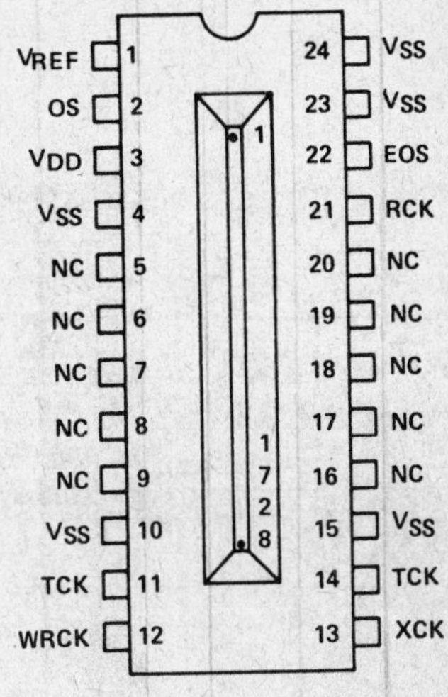

NC – No internal connection

- 1728 x 1 Sensor Element Organization
- Virtual-Phase N-Channel Silicon MOS Technology
- High Quantum Efficiency
- Enhanced Blue Response
- Output Signal Approximately 1 Volt Peak-to-Peak
- Dynamic Range Relative to Peak-to-Peak Noise Typically 1000:1
- End-of-Scan Signal
- Internal Black and White References
- Simple and Stable Operation

description

The TC101, a 1728-element CCD line image sensor, functions in high-resolution image scanning applications such as facsimile and optical character recognition. The 1728 sensor elements provide a 200-points-per-inch resolution across 8.5 inches. The TC101 incorporates virtual-phase MOS technology, which provides simplified operation and high reliability.

This device is supplied in a 24-pin dual-in-line ceramic side-braze package designed for insertion in mounting-hole rows on 15.2-mm (0.600-inch) centers. The glass window may be cleaned by wiping with a cotton swab soaked in alcohol.

Caution. These devices have limited built-in gate protection. The leads should be shorted together or the device placed in conductive foam during storage or handling to prevent electrostatic damage to the MOS gates. Avoid shorting either OS or EOS to V_{SS} during operation to prevent damage to the output amplifiers.

virtual phase technology

This patented design results in simplified clocking circuits, reduced noise, and greater light sensitivity. The virtual phase utilizes a junction-gate region at the substrate dc potential. This accomplishes the same gating and transport function as a separate gate electrode requiring multiple layers and multiple process steps common in other device designs. The resulting simplicity of process and ease of operation will increase performance and reliability for the user.

Copyright © 1982 by Texas Instruments Incorporated

TEXAS INSTRUMENTS
INCORPORATED
POST OFFICE BOX 225012 • DALLAS, TEXAS 75265

functional block diagram

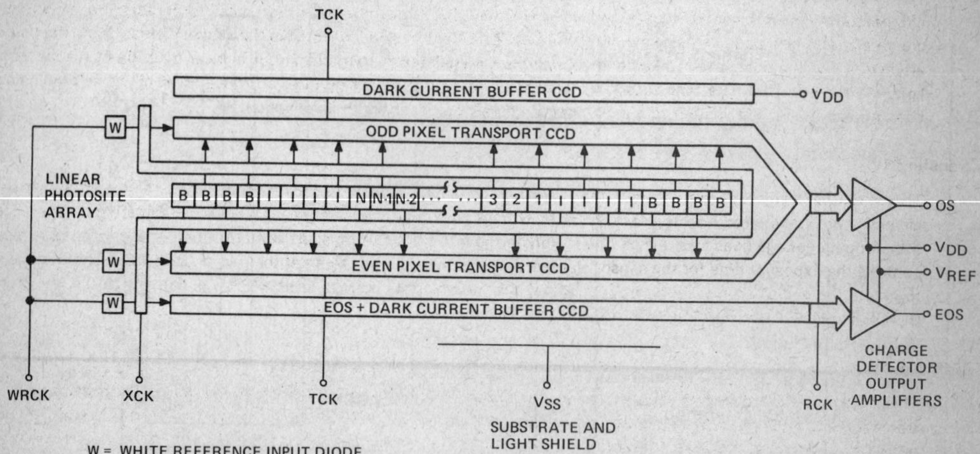

W = WHITE REFERENCE INPUT DIODE
B = BLACK REFERENCE ELEMENT
I = ISOLATION ELEMENT
N = 1728 SENSOR ELEMENTS

PIN FUNCTIONAL DESCRIPTION

PIN NUMBER	SIGNATURE	NAME	DESCRIPTION
1	V_{REF}	Reference Voltage	Bias input for the output amplifiers.
2	OS	Output Signal	Video output from a cascaded source-follower MOS amplifier.
3	V_{DD}	Supply Voltage	Output amplifier supply voltage.
11,14	TCK	Transport Clock	Drives the CCD transport registers.
12	WRCK	White Reference Clock	Injects a controlled charge into the white reference CCD shift register elements to become white-reference and end-of-scan pulses.
13	XCK	Transfer Clock	Controls the transfer of charge packets from sensor elements to shift registers. The interval between pulses of the transfer clock determines the exposure time.
21	RCK	Reset Clock	Controls recharging of the charge-detection diodes in the output amplifiers, and clocks the output shift registers where the odd and even signals have been merged.
22	EOS	End-of-Scan Pulse	Indicates that all charge packets have been shifted out of the transport registers.
4,10,15,23,24	V_{SS}	Substrate	All voltages are referenced to the substrate.

functional description

image sensor elements

The line of sensor elements (also called photosites or pixels) consists of 1728 photo-sensitive areas, 12.7 micrometers (0.5 milliinches) square and approximately 12.7 micrometers from center to center. Image photons create electron-hole pairs in the single-crystal silicon. The electrons are collected in the sensor elements and the holes are swept into the substrate. The amount of charge accumulated in each element is a linear function of the incident light and the exposure time. The output signal charge will vary in an analog manner from a thermally generated noise background at zero illumination to a maximum at saturation under bright illumination.

transfer gate

This structure is adjacent to the line of image sensor elements. The charge packets accumulated in the image sensor elements are transferred into the transfer gate storage well when the transfer gate voltage goes high. When the transfer gate voltage goes low, the charge is transferred into the CCD transport shift registers. The transfer gate also controls the exposure time for the sensor elements and permits charges to enter the end-of-scan (EOS) shift registers to create the end-of-scan waveform. In addition, the transfer gate permits entry of charge packets to the transport CCD shift register to create the white reference signals.

shift registers

There are two CCD transport registers, one on each side of the line of image sensor elements and outside of the transfer gate. Alternate charge packets are transferred to the CCD transport shift registers and moved serially to the output amplifier. The phase relationship of the reset clock and the transport clock and the geometric layout of the paths provide for alternate delivery of charge packets to re-establish the original sequence of the linear image data. The two outer buffer CCD shift registers protect the signal charges in the inner transport CCD shift registers from peripherally generated dark current noise.

black and white reference elements

Four additional sensor elements at each end of the sensor element array (labelled "B" in the block diagram) are covered by opaque metallization. They provide a black (no illumination) signal reference that is delivered at each end of the linear image output signal. Also included on the transport CCD shift register, at the opposite end from the amplifier, is an input diode that provides two white reference pulses in the output signal. The reference pulses are useful as inputs to external dc restoration and/or automatic exposure control circuitry. The white reference pulse amplitude is approximately 70% of the maximum output signal amplitude.

output signal amplifier

The charge packets are transported to a precharge diode whose potential changes linearly in response to the amount of the signal charge delivered. This potential is applied to the input gate of an N-channel MOS double-source-follower amplifier to produce an output signal (OS). A reset transistor, driven by the reset clock (RCK), recharges the charge-detector-diode capacitance before the arrival of each new signal charge packet from the CCD shift registers. A reference voltage (V_{REF}) is applied to the drain of the reset transistor and acts to bias the OS and EOS amplifiers. A current sink is used as an on-chip load for the amplifier output. No external current sink is needed. The output signal is a series of negative-going pulses on a dc level.

resolution

The modulation transfer function decreases at longer wavelengths. (See Figures 7 and 8.) If optimum resolution is required with a light source that has a significant infrared component, then the designer must use appropriate filters to restrict the optical pass band to shorter wavelengths.

end-of-scan amplifier

The EOS amplifier is similar to the OS amplifier. XCK transfers charge from the input diode into the EOS register where it is transported at the TCK clock frequency to the EOS amplifier. This EOS pulse is coincident with the first of the two white reference pulses that pass through the odd and even transport CCDs, respectively. The EOS output can be used to alert the external circuitry that the linear image data readout has been completed.

clocks

The transfer clock (XCK) pulse controls the exposure time of the sensor elements. The minimum exposure time is the time required to shift the entire contents of the transport registers to the output signal amplifier and equals 1761 multiplied by the RCK period. The maximum exposure time is determined by the tolerable level of dark signal.

The transport clock (TCK) transports the linear image signal charge from the sensor element region to the output amplifier.

The reset clock (RCK) operates at twice the transport clock frequency so as to recombine the signal charge in the original sequence and present the charge to the output amplifier. The data rate is equal to the reset clock frequency.

The white reference clock (WRCK) runs at the transfer clock frequency and generates the white reference and the end-of-scan pulses. These pulses can be eliminated by connecting WRCK to V_{DD}. Transients on WRCK going below zero volts will cause charge injection resulting in an increase in apparent dark signal.

Figure 3 presents a suggested circuit for generating the clock waveforms. The RCK clock generator runs continuously. A binary divider halves the frequency to create TCK. After all signal charges have been transported to the output amplifier, TCK continues to run to keep thermally generated charges from accumulating in the transport registers.

The XCK and WRCK clock frequencies are submultiples of the TCK frequency. Figure 2 details the timing relationships among the different clock pulses.

TEXAS INSTRUMENTS
INCORPORATED
POST OFFICE BOX 225012 • DALLAS, TEXAS 75265

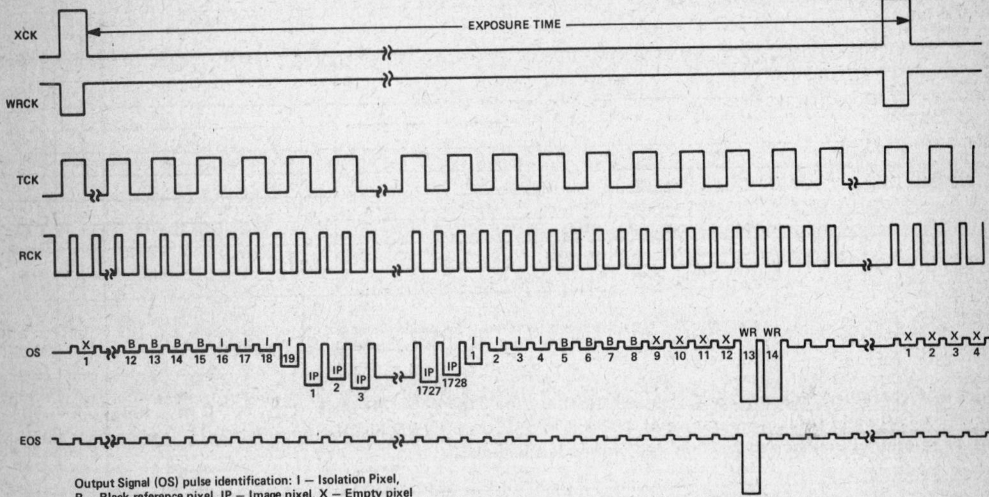

Output Signal (OS) pulse identification: I — Isolation Pixel, B — Black reference pixel, IP — Image pixel, X — Empty pixel See Figure 3.

FIGURE 1 – OPERATING INPUT AND OUTPUT VOLTAGE WAVEFORMS

absolute maximum ratings over operating free-air temperature range (unless otherwise noted) (see note 1)

Amplifier drain voltage (V_{DD})	−0.3 V to 30 V
Amplifier reference voltage (V_{REF})	−0.3 V to 30 V
Transfer clock (XCK) voltage	−25 V to 5 V
Transport clock (TCK) voltage	−25 V to 5 V
Reset clock (RCK) voltage	−25 V to 5 V
White reference clock (WRCK) voltage	−0.3 V to 30 V
Storage temperature	−25 °C to 125 °C
Operating free-air temperature	−25 °C to 70 °C

NOTE 1: Voltage values are with respect to V_{SS}.

recommended operating conditions at $T_A = 25°C$

		MIN	NOM	MAX	UNIT
V_{DD}	Supply voltage	15	16	20	V
V_{REF}	Amplifier reference voltage	6	7	8	V
$V_{IH(X)}$	Transfer clock high-level input voltage	1	2	3	V
$V_{IL(X)}$	Transfer clock low-level input voltage	−17¶	−16	−15	V
$V_{IH(T)}$	Transport clock high-level input voltage	1	2	3	V
$V_{IL(T)}$	Transport clock low-level input voltage	−17¶	−16	−15	V
$V_{IH(R)}$	Reset clock high-level input voltage	1	2	3	V
$V_{IL(R)}$	Reset clock low-level input voltage	−17¶	−16	−15	V
$V_{IH(WR)}$	White reference clock high-level input voltage	15	16	20	V
$V_{IL(WR)}$	White reference clock low-level input voltage	6	7	8	V
f_{RCK}	Reset clock frequency (output data rate)		2	10	MHz

¶The algebraic convention, where the most negative limit is designated as minimum, is used in this data sheet for clock voltage levels only.

electrical characteristics at 25°C free-air temperature, f_{RCK} = 0.5 MHz, t_{exp} = 10 ms, tungsten light source operating at color temperature of 2854 K with 2.0-mm-thick Fish-Schurman HA-11 IR-absorbing filter, and all operating voltages at nominal recommended values

PARAMETER		MIN	TYP	MAX	UNIT
Dark-signal amplitude	Average		0.5	10	mV
	Low-frequency component		0.5	5	
	Nonuniformity relative to average of adjacent pixels		1	20	
Sensitivity		2	3.5	5	V/(μJ/cm²)
Output amplitude variation (PRNU)‡	Peak-to-peak		50	100	mV
	Adjacent pixels from alternate registers (imbalance)		10		
Peak-to-peak noise			1		mV
Equivalent exposure§ of peak-to-peak noise			0.35		nJ/cm²
Saturation exposure§			350		nJ/cm²
Saturation output amplitude		700	1000	1400	mV
Dynamic range relative to peak-to-peak noise		500:1	1000:1		
Charge transfer efficiency (CTE)			0.99999		
White reference amplitude		500	700		mV
End-of-scan amplitude		300	500		mV
Output offset (dc) voltage			10		V
Output impedance			1		kΩ
Resistance to V_{SS}	Transfer gate		170		kΩ
	Transport gate		120		
	Reset gate		260		
Capacitance to V_{SS}	Transfer gate		260		pF
	Transport gate		580		
	Reset gate		16		
I_{REF}	Amplifier reference current		100		nA
I_{DD}	Supply current		6.3	9.4	mA
Power dissipation			100		mW

‡Measured at 700 mV output amplitude with an f/2.8 lens.

§Exposure = intensity x time

timing recommendations

		MIN	NOM	MAX	UNITS
t_{THXH}	Time delay from the transport clock rising edge to the transfer clock rising edge	0		100	ns
t_{THWL}	Time delay from the transport clock rising edge to the white reference clock falling edge	0		100	ns
t_{THRH}	Time delay from the transport clock rising edge to the reset clock rising edge	0			ns
$t_{w(RH)}$	Pulse duration of the high state for the reset clock	40			ns
t_{TLXL}	Time delay from the transport clock falling edge to the transfer clock falling edge	50			ns
t_{TLWH}	Time delay from the transport clock falling edge to the white reference clock rising edge	0		100	ns
t_{XLTH}	Time delay from the transfer clock falling edge to the rising edge of the next transport clock pulse	50			ns
t_r	rise time (all clocks)	15			ns
t_f	fall time (all clocks)	5			ns

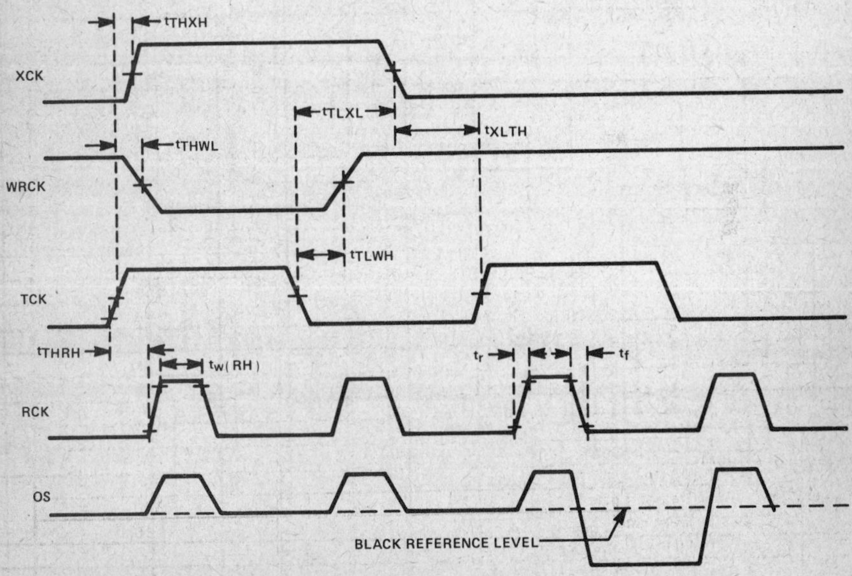

FIGURE 2 – DEVICE TIMING REQUIREMENTS

PARAMETER MEASUREMENT INFORMATION

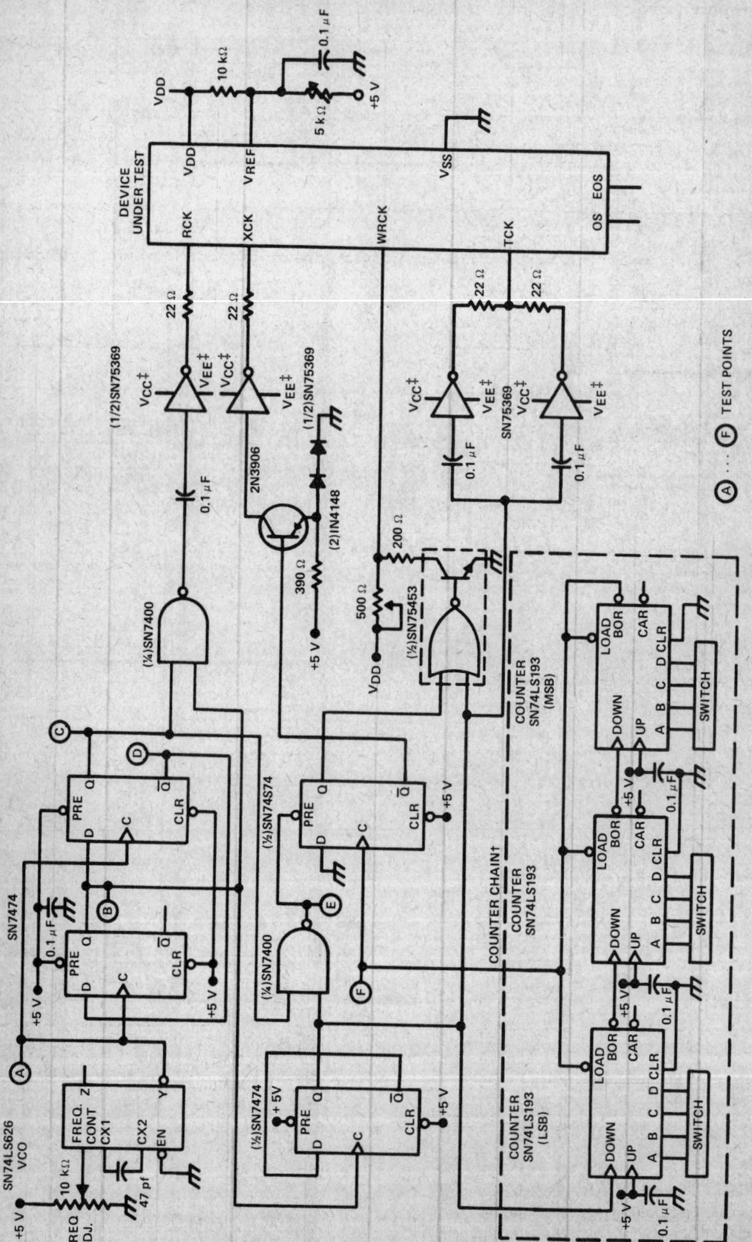

†This counter chain counts transport clock periods to generate the exposure time interval. The data rate is twice the count rate.
‡V_{CC} and V_{EE} are the voltages that will produce the desired values of V_{IH} and V_{IL}, respectively, at the RCK, XCK, and TCK inputs.

FIGURE 3 — DRIVER CIRCUIT FOR TESTING LINE IMAGE SENSOR

TEXAS INSTRUMENTS
INCORPORATED
POST OFFICE BOX 225012 • DALLAS, TEXAS 75265

FIGURE 4 – WAVEFORMS IN DRIVER CIRCUIT

TYPICAL CHARACTERISTICS

(In the circuit of Figure 3 with $T_A = 25°C$, $f_{RCK} = 0.5$ MHz, $t_{exp} = 10$ ms, and all operating voltages at nominal recommended values, unless otherwise noted)

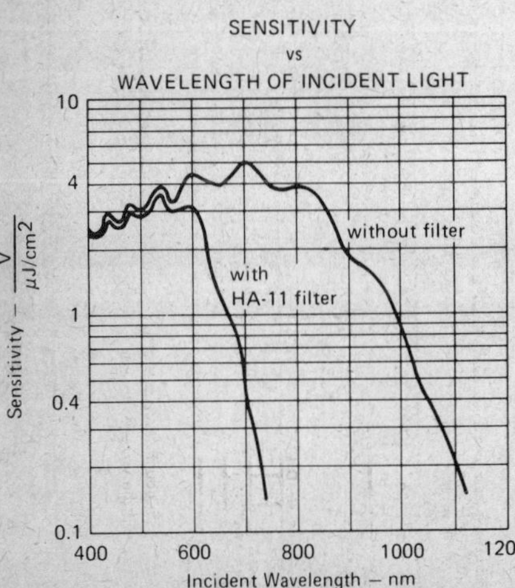

FIGURE 5

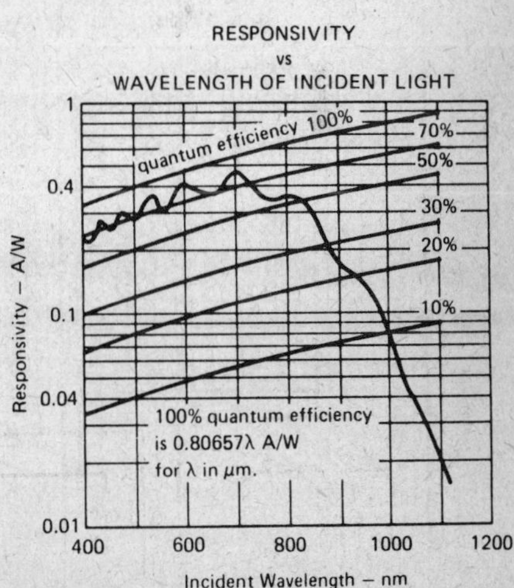

FIGURE 6

TEXAS INSTRUMENTS
INCORPORATED
POST OFFICE BOX 225012 • DALLAS, TEXAS 75265

TYPICAL CHARACTERISTICS

(In the circuit of Figure 3 with $T_A = 25°C$, $f_{RCK} = 0.5$ MHz, $t_{exp} = 10$ ms, and all operating voltages at nominal recommended values, unless otherwise noted)

MODULATION TRANSFER FUNCTION
vs
SPATIAL FREQUENCY

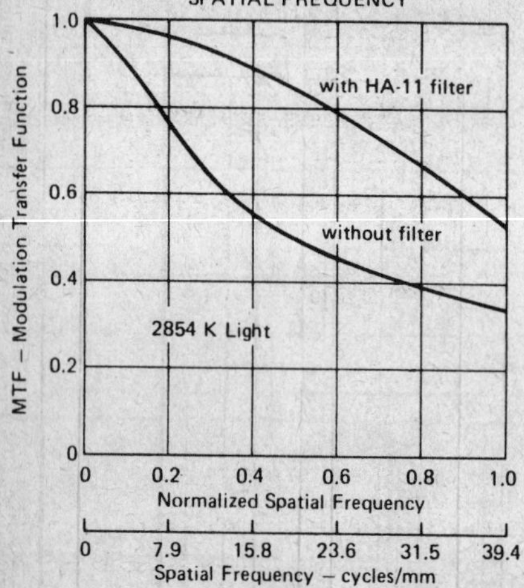

FIGURE 7

MODULATION TRANSFER FUNCTION
vs
SPATIAL FREQUENCY

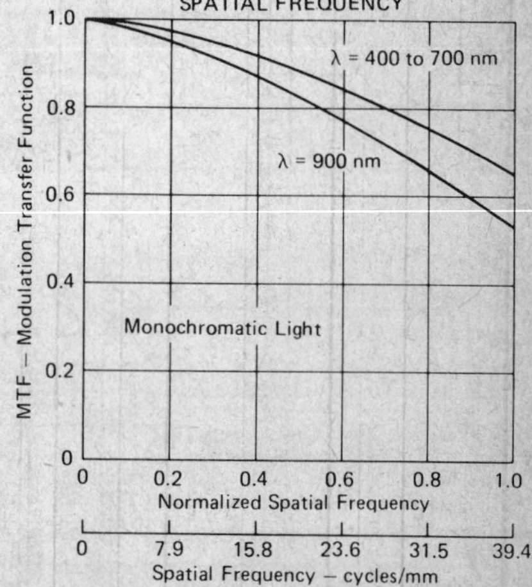

FIGURE 8

AVERAGE AND LOW-FREQUENCY
DARK SIGNAL
vs
EXPOSURE TIME

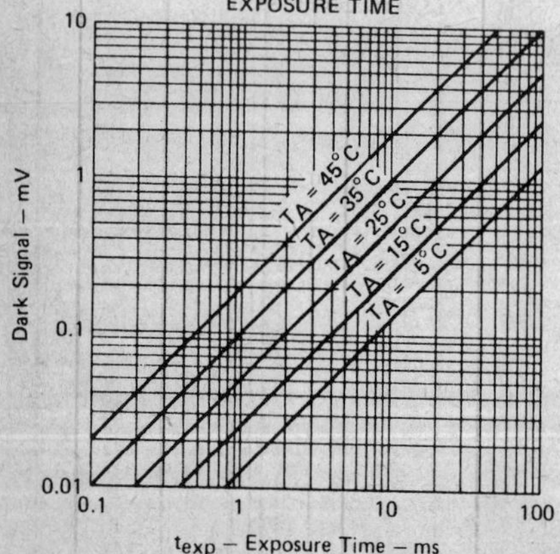

FIGURE 9

OUTPUT SIGNAL VOLTAGE RELATIVE TO
SATURATED OUTPUT VOLTAGE
vs
EXPOSURE TIME

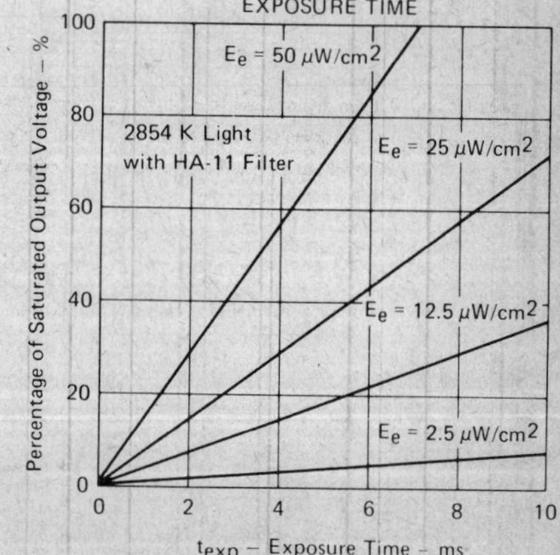

FIGURE 10

TEXAS INSTRUMENTS
INCORPORATED
POST OFFICE BOX 225012 • DALLAS, TEXAS 75265

MECHANICAL DATA

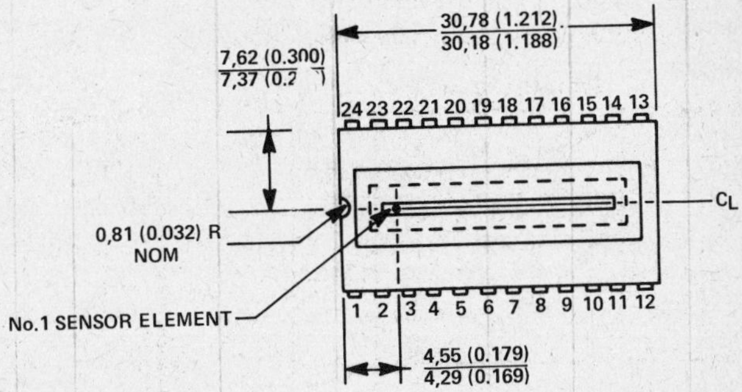

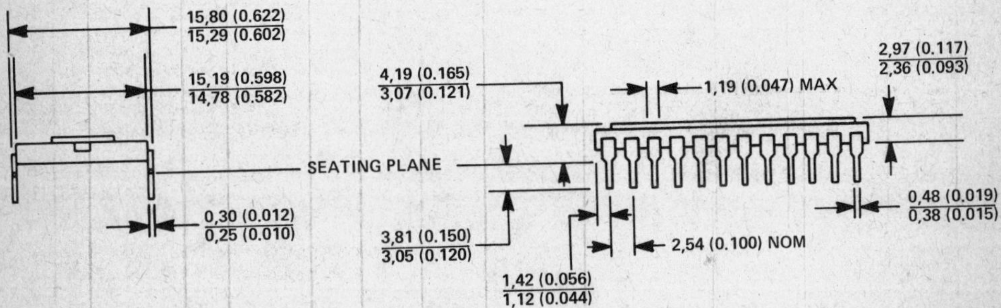

NOTES: 1. All dimensions are in millimeters and parenthetically in inches.

2. The distance between the top surface of the window and the surface of the sensor is nominally 0, 89 (0.035). This is determined by observing the vertical motion of a microscope focused first at one plane, then at the other.

TEXAS INSTRUMENTS
INCORPORATED
POST OFFICE BOX 225012 • DALLAS, TEXAS 75265

128 × 1 CCD LINEAR IMAGE SENSOR

D2664, APRIL 1982

- 128 x 1 Sensor Element Organization
- Virtual-Phase N-Channel Silicon MOS Technology
- High Quantum Efficiency
- Enhanced Blue Response
- Output Signal Approximately 1 Volt Peak-to-Peak
- Dynamic Range Relative to Peak-to-Peak Noise Typically 1000:1
- End-of-Scan Signal
- Internal Black and White References
- Simple and Stable Operation

CERAMIC DUAL-IN-LINE PACKAGE
(TOP VIEW)

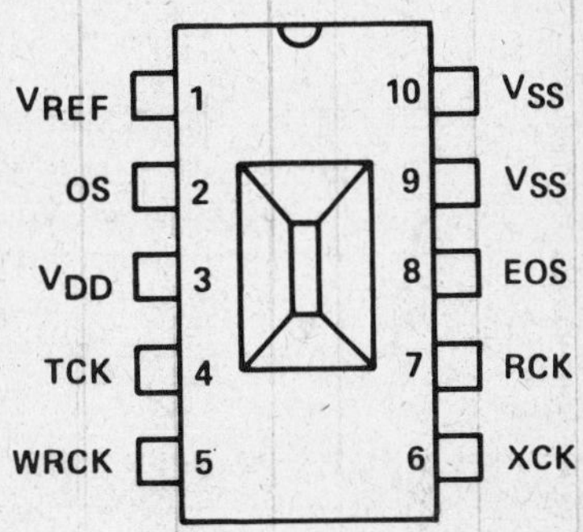

Caution. These devices have limited built-in gate protection. The leads should be shorted together or the device placed in conductive foam during storage or handling to prevent electrostatic damage to the MOS gates. Avoid shorting either OS or EOS to V_{SS} during operation to prevent damage to the output amplifiers.

description

The TC102, a 128-element CCD line image sensor, functions in high-resolution image scanning applications such as document reading and optical character recognition. The TC102 incorporates virtual-phase MOS technology, which provides simplified operation and high reliability.

This device is supplied in a 10-pin dual-in-line ceramic side-braze package designed for insertion in mounting-hole rows on 7.6-mm (0.300-inch) centers. The glass window may be cleaned by wiping with a cotton swab soaked in alcohol.

virtual-phase technology

This patented design results in simplified clocking circuits, reduced noise, and greater light sensitivity. Virtual-phase technology utilizes a junction-gate region at the substrate dc potential. This accomplishes the same gating and transport function as a separate gate electrode requiring multiple layers and multiple process steps common in other device designs. The resulting simplicity of process and ease of operation will increase performance and reliability for the user.

Copyright © 1982 by Texas Instruments Incorporated

TEXAS INSTRUMENTS
INCORPORATED
POST OFFICE BOX 225012 • DALLAS, TEXAS 75265

functional block diagram

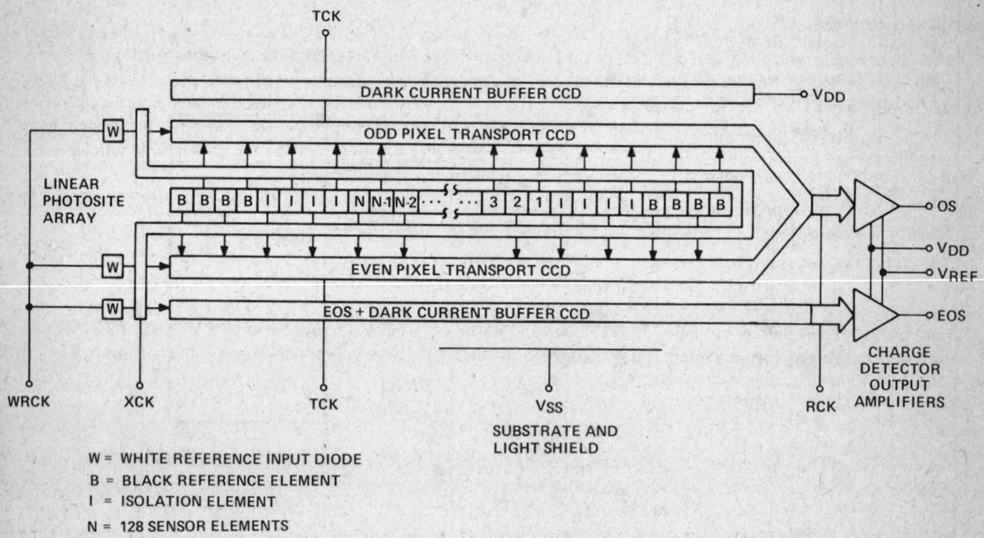

PIN FUNCTIONAL DESCRIPTION

PIN NUMBER	SIGNATURE	NAME	DESCRIPTION
1	V_{REF}	Reference Voltage	Bias input for the output amplifiers.
2	OS	Output Signal	Video output from a cascaded source-follower MOS amplifier.
3	V_{DD}	Supply Voltage	Output amplifier supply voltage.
4	TCK	Transport Clock	Drives the CCD transport registers.
5	WRCK	White Reference Clock	Injects a controlled charge into the white reference CCD shift register elements to become white-reference and end-of-scan pulses.
6	XCK	Transfer Clock	Controls the transfer of charge packets from sensor elements to shift registers. The interval between pulses of the transfer clock determines the exposure time.
7	RCK	Reset Clock	Controls recharging of the charge-detection diodes in the output amplifiers, and clocks the output shift registers where the odd and even signals have been merged.
8	EOS	End-of-Scan Pulse	Indicates that all charge packets have been shifted out of the transport registers.
9, 10	V_{SS}	Substrate	All voltages are referenced to the substrate.

TEXAS INSTRUMENTS
INCORPORATED
POST OFFICE BOX 225012 • DALLAS, TEXAS 75265

functional description

image sensor elements

The line of sensor elements (also called photosites or pixels) consists of 128 photo-sensitive areas, 12.7 micrometers (0.5 milliinches) square and approximately 12.7 micrometers from center to center. Image photons create electron-hole pairs in the single-crystal silicon. The electrons are collected in the sensor elements and the holes are swept into the substrate. The amount of charge accumulated in each element is a linear function of the incident light and the exposure time. The output signal charge will vary in an analog manner from a thermally generated noise background at zero illumination to a maximum at saturation under bright illumination.

transfer gate

This structure is adjacent to the line of image sensor elements. The charge packets accumulated in the image sensor elements are transferred into the transfer gate storage well when the transfer gate voltage goes high. When the transfer gate voltage goes low, the charge is transferred into the CCD transport shift registers. The transfer gate also controls the exposure time for the sensor elements and permits charges to enter the end-of-scan (EOS) shift registers to create the end-of-scan waveform. In addition, the transfer gate permits entry of charge packets to the transport CCD shift register to create the white reference signals.

shift registers

There are two CCD transport registers, one on each side of the line of image sensor elements and outside of the transfer gate. Alternate charge packets are transferred to the CCD transport shift registers and moved serially to the output amplifier. The phase relationship of the reset clock and the transport clock and the geometric layout of the paths provide for alternate delivery of charge packets to re-establish the original sequence of the linear image data. The two outer buffer CCD shift registers protect the signal charges in the inner transport CCD shift registers from peripherally generated dark current noise.

black and white reference elements

Four additional sensor elements at each end of the sensor element array (labelled ''B'' in the block diagram) are covered by opaque metallization. They provide a black (no illumination) signal reference that is delivered at each end of the linear image output signal. Also included on the transport CCD shift register, at the opposite end from the amplifier, is an input diode that provides two white reference pulses in the output signal. The reference pulses are useful as inputs to external dc restoration and/or automatic exposure control circuitry. The white reference pulse amplitude is approximately 70% of the maximum output signal amplitude.

output signal amplifier

The charge packets are transported to a precharge diode whose potential changes linearly in response to the amount of the signal charge delivered. This potential is applied to the input gate of an N-channel MOS double-source-follower amplifier to produce an output signal (OS). A reset transistor, driven by the reset clock (RCK), recharges the charge-detector-diode capacitance before the arrival of each new signal charge packet from the CCD shift registers. A reference voltage (V_{REF}) is applied to the drain of the reset transistor and acts to bias the OS and EOS amplifiers. A current sink is used as an on-chip load for the amplifier output. No external current sink is needed. The output signal is a series of negative-going pulses on a dc level.

resolution

The modulation transfer function decreases at longer wavelengths (see Figures 7 and 8). If optimum resolution is required with a light source that has a significant infrared component, then the designer must use appropriate filters to restrict the optical pass band to shorter wavelengths.

end-of-scan amplifier

The EOS amplifier is similar to the OS amplifier. XCK transfers charge from the input diode into the EOS register where it is transported at the TCK clock frequency to the EOS amplifier. This EOS pulse is coincident with the first of the two white reference pulses that pass through the odd and even transport CCDs, respectively. The EOS output can be used to alert the external circuitry that the linear image data readout has been completed.

clocks

The transfer clock (XCK) pulse controls the exposure time of the sensor elements. The minimum exposure time is the time required to shift the entire contents of the transport registers to the output signal amplifier and equals 161 multiplied by the RCK period. The maximum exposure time is determined by the tolerable level of dark signal.

The transport clock (TCK) transports the linear image signal charge from the sensor element region to the output amplifier.

The reset clock (RCK) operates at twice the transport clock frequency so as to recombine the signal charge in the original sequence and present the charge to the output amplifier. The data rate is equal to the reset clock frequency.

The white reference clock (WRCK) runs at the transfer clock frequency and generates the white reference and the end-of-scan pulses. These pulses can be eliminated by connecting WRCK to V_{DD}. Transients on WRCK going below zero volts will cause charge injection resulting in an increase in apparent dark signal.

Figure 3 presents a suggested circuit for generating the clock waveforms. The RCK clock generator runs continuously. A binary divider halves the frequency to create TCK. After all signal charges have been transported to the output amplifier, TCK continues to run to keep thermally generated charges from accumulating in the transport registers.

The XCK and WRCK clock frequencies are submultiples of the TCK frequency. Figure 2 details the timing relationships among the different clock pulses.

TEXAS INSTRUMENTS
INCORPORATED
POST OFFICE BOX 225012 • DALLAS, TEXAS 75265

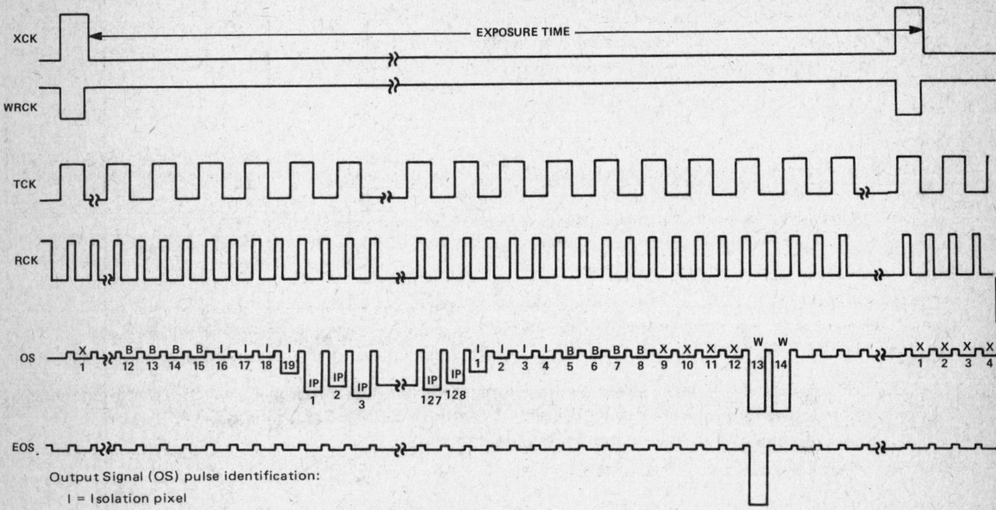

Output Signal (OS) pulse identification:
- I = Isolation pixel
- IP = Image pixel
- B = Black reference pixel
- W = White reference pixel
- X = Empty pixel

FIGURE 1 – OPERATING INPUT AND OUTPUT VOLTAGE WAVEFORMS

absolute maximum ratings over operating free-air temperature range (unless otherwise noted) (see Note 1)

Amplifier drain voltage (V_{DD})	−0.3 V to 30 V
Amplifier reference voltage (V_{REF})	−0.3 V to 30 V
Transfer clock (XCK) voltage	−25 V to 5 V
Transport clock (TCK) voltage	−25 V to 5 V
Reset clock (RCK) voltage	−25 V to 5 V
White reference clock (WRCK) voltage	−0.3 V to 30 V
Storage temperature	−25 °C to 125 °C
Operating free-air temperature	−25 °C to 70 °C

NOTE 1: Voltage values are with respect to V_{SS}.

recommended operating conditions at T_A = 25°C

		MIN	NOM	MAX	UNIT
V_{DD}	Supply voltage	15	16	20	V
V_{REF}	Amplifier reference voltage	6	7	8	V
$V_{IH(X)}$	Transfer clock high-level input voltage	1	2	3	V
$V_{IL(X)}$	Transfer clock low-level input voltage	−17¶	−16	−15	V
$V_{IH(T)}$	Transport clock high-level input voltage	1	2	3	V
$V_{IL(T)}$	Transport clock low-level input voltage	−17¶	−16	−15	V
$V_{IH(R)}$	Reset clock high-level input voltage	1	2	3	V
$V_{IL(R)}$	Reset clock low-level input voltage	−17¶	−16	−15	V
$V_{IH(WR)}$	White reference clock high-level input voltage	15	16	20	V
$V_{IL(WR)}$	White reference clock low-level input voltage	6	7	8	V
f_{RCK}	Reset clock frequency (output data rate)		2	10	MHz

¶The algebraic convention, where the most negative limit is designated as minimum, is used in this data sheet for clock voltage levels only.

electrical characteristics at 25°C free-air temperature, f_{RCK} = 0.5 MHz, t_{exp} = 10 ms, tungsten light source operating at color temperature of 2854 K with 2.0-mm-thick Fish-Schurman HA-11 IR-absorbing filter, and all operating voltages at nominal recommended values.

PARAMETER		MIN	TYP	MAX	UNIT
Dark-signal amplitude	Average		0.5	10	mV
	Low frequency component		0.5	5	
	Nonuniformity relative to average of adjacent pixels		1	20	
Sensitivity		2	3.5	5	V/(μJ/cm²)
Output amplitude variation (PRNU)‡	Peak-to-peak		50	100	mV
	Adjacent pixels from alternate registers (imbalance)		10		
Peak-to-peak noise			1		mV
Equivalent exposure§ of peak-to-peak noise			0.35		nJ/cm²
Saturation exposure§			350		nJ/cm²
Saturation output amplitude		700	1000	1400	mV
Dynamic range relative to peak-to-peak noise		500:1	1000:1		
Charge transfer efficiency			0.99999		
White reference amplitude		500	700		mV
End-of-scan amplitude		300	500		mV
Output offset (dc) voltage			10		V
Output impedance			1		kΩ
Resistance to V_{SS}	Transfer gate		45		kΩ
	Transport gate		45		
	Reset gate		45		
Capacitance to V_{SS}	Transfer gate		26		pF
	Transport gate		57		
	Reset gate		7		
I_{REF}	Amplifier reference current		3		nA
I_{DD}	Supply current		6.3	9.4	mA
Power dissipation			100		mW

‡Measured at 700 mV output amplitude with an f/2.8 lens.
§Exposure = intensity x time

TEXAS INSTRUMENTS
INCORPORATED
POST OFFICE BOX 225012 • DALLAS, TEXAS 75265

timing requirements

		MIN	NOM	MAX	UNIT
t_{THXH}	Time delay from the transport clock rising edge to the transfer clock rising edge	0		100	ns
t_{THWL}	Time delay from the transport clock rising edge to the white reference clock falling edge	0		100	ns
t_{THRH}	Time delay from the transport clock rising edge to the reset clock rising edge	0			ns
$t_{w(RH)}$	Pulse duration of the high state for the reset clock	40			ns
t_{TLXL}	Time delay from the transport clock falling edge to the transfer clock falling edge	50			ns
t_{TLWH}	Time delay from the transport clock falling edge to the white reference clock rising edge	0		100	ns
t_{XLTH}	Time delay from the transfer clock falling edge to the rising edge of the next transport clock pulse	50			ns
t_r	rise time (all clocks)	15			ns
t_f	fall time (all clocks)	5			ns

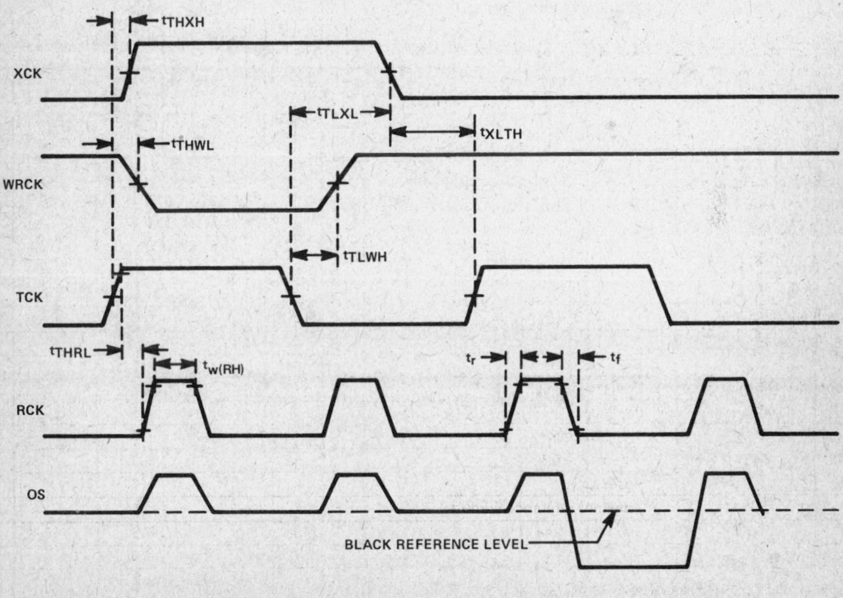

FIGURE 2 – DEVICE TIMING REQUIREMENTS

PARAMETER MEASUREMENT INFORMATION

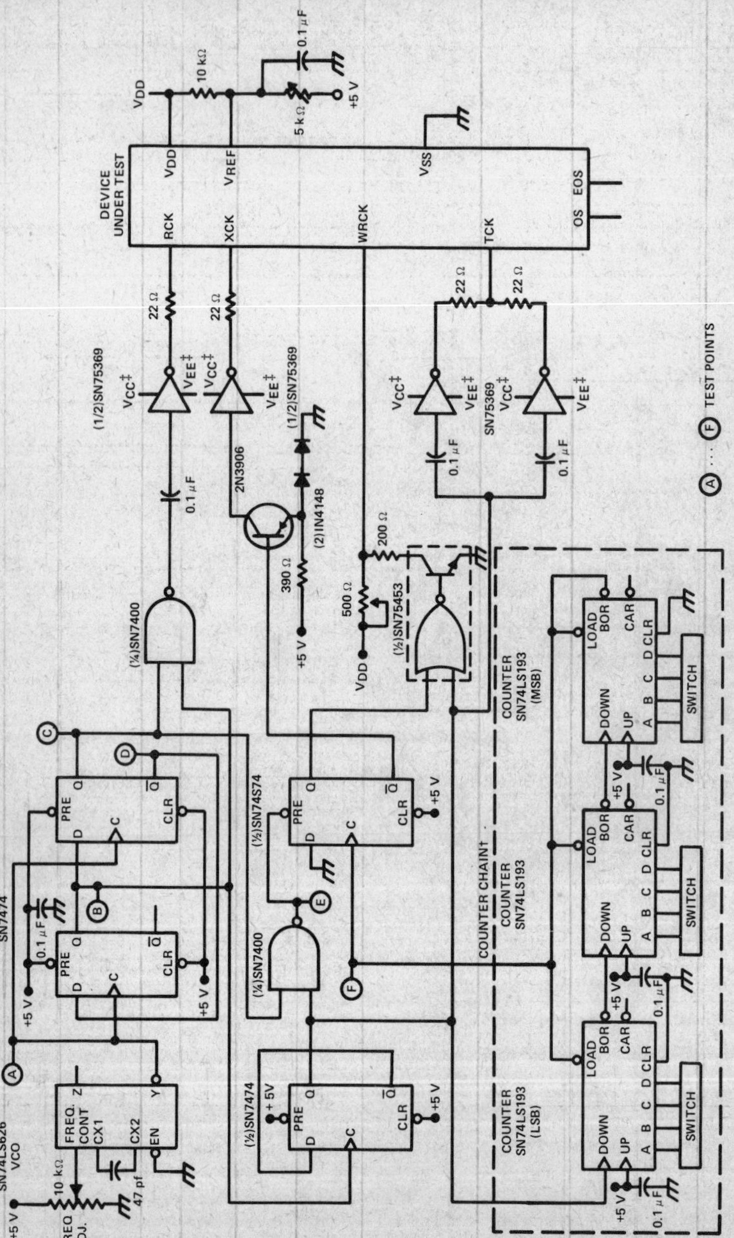

†This counter chain counts transport clock periods to generate the exposure time interval. The data rate is twice the count rate.

‡V_{CC} and V_{EE} are the voltages that will produce the desired values of V_{IH} and V_{IL}, respectively, at the RCK, XCK, and TCK inputs.

FIGURE 3 – DRIVER CIRCUIT FOR TESTING LINE IMAGE SENSOR

FIGURE 4 – WAVEFORMS IN DRIVER CIRCUIT

TYPICAL CHARACTERISTICS

(In the circuit of Figure 3 with T_A = 25°C, f_{RCK} = 0.5 MHz, t_{exp} = 10 ms, and all operating voltages at nominal recommended values, unless otherwise noted)

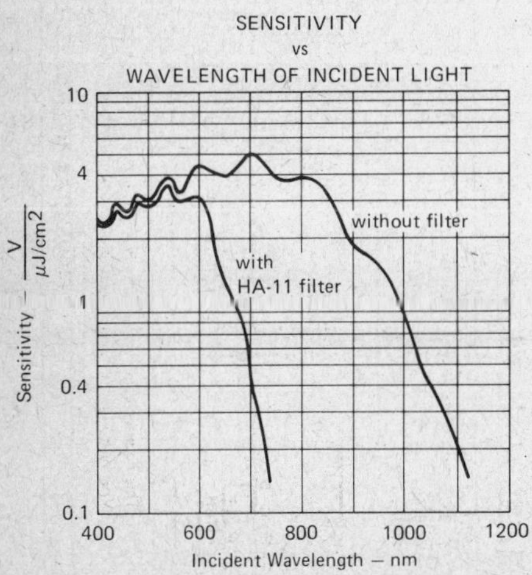

FIGURE 5

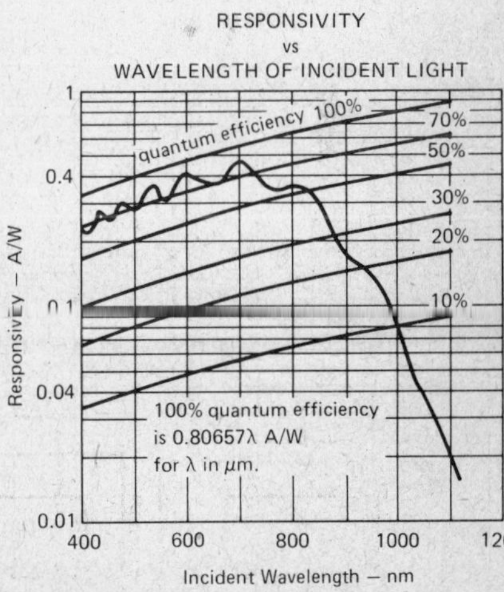

FIGURE 6

TEXAS INSTRUMENTS
INCORPORATED
POST OFFICE BOX 225012 • DALLAS, TEXAS 75265

TYPICAL CHARACTERISTICS

(In the circuit of Figure 3 with $T_A = 25°C$, $f_{RCK} = 0.5$ MHz, $t_{exp} = 10$ ms, and all operating voltages at nominal recommended values, unless otherwise noted)

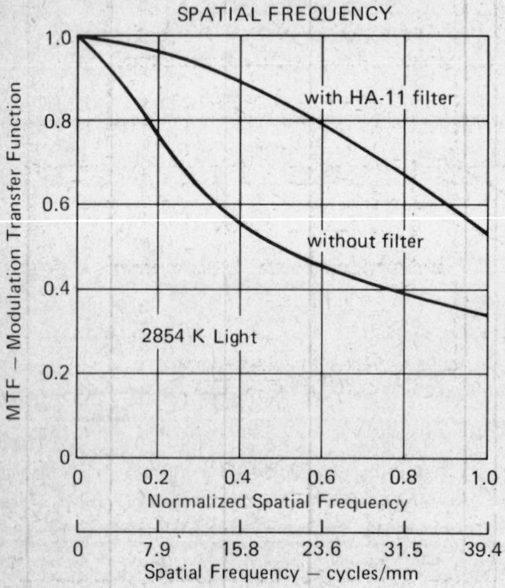

FIGURE 7

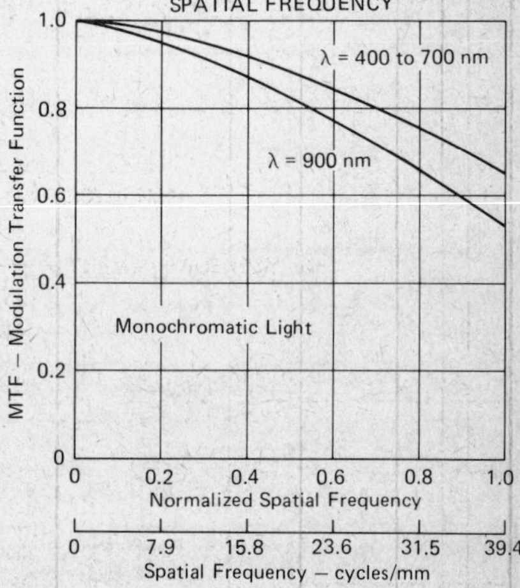

FIGURE 8

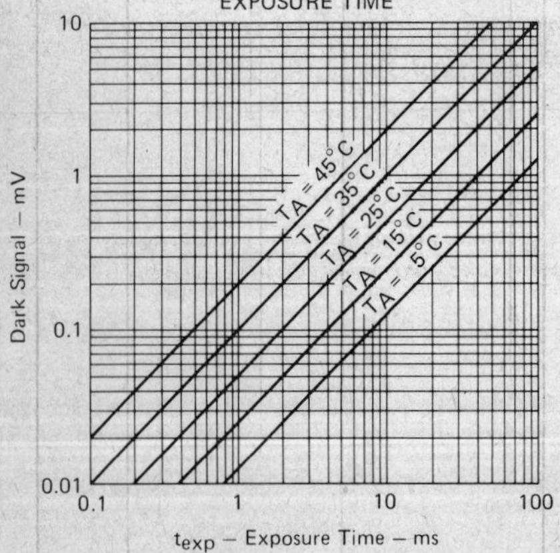

FIGURE 9

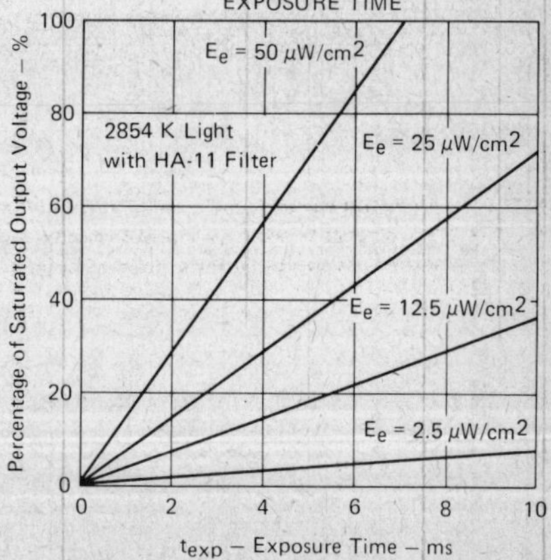

FIGURE 10

TEXAS INSTRUMENTS
INCORPORATED
POST OFFICE BOX 225012 • DALLAS, TEXAS 75265

MECHANICAL DATA

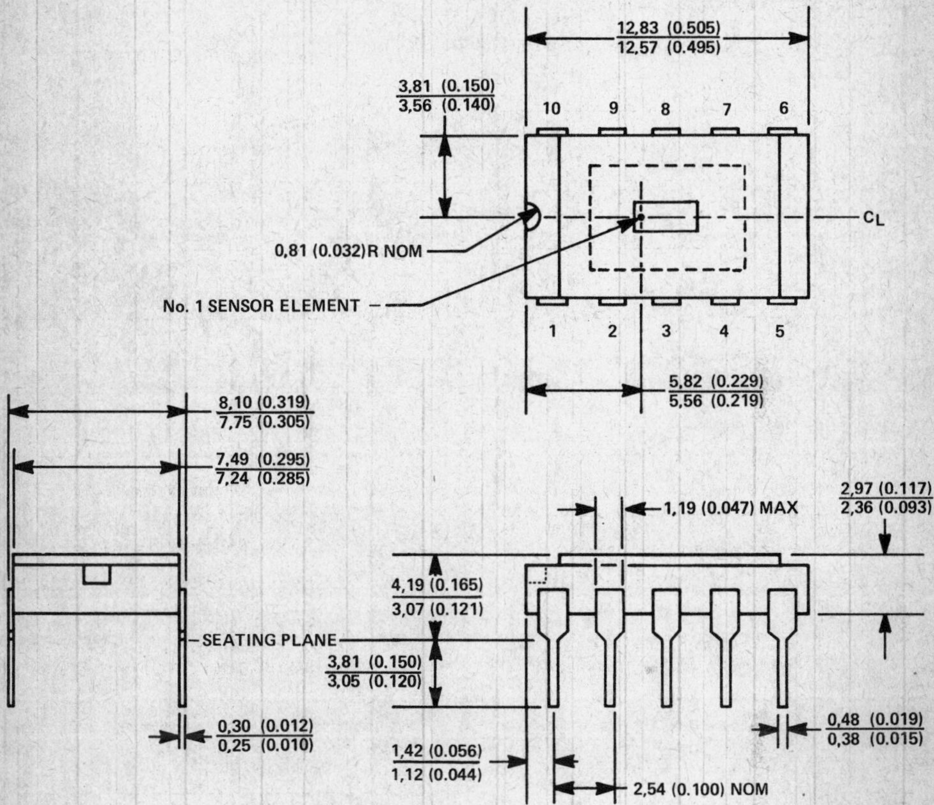

NOTES: 1. All dimensions are in millimeters and parenthetically in inches.

2. The distance between the top surface of the window and the surface of the sensor is nominally 0,89 (0.035). This is determined by observing the vertical motion of a microscope focused first at one plane, then at the other.

TEXAS INSTRUMENTS
INCORPORATED
POST OFFICE BOX 225012 • DALLAS, TEXAS 75265

D2686, FEBRUARY 1983

TYPE TC103
2048 x 1 CCD LINEAR IMAGE SENSOR

- 2048 x 1 Sensor Element Organization
- Virtual-Phase N-Channel Silicon MOS Technology
- High Quantum Efficiency
- Enhanced Blue Response
- Output Signal Approximately 1.0 Volt Peak-to-Peak
- Dynamic Range Relative to Peak-to-Peak Noise Typically 1000:1
- End-of-Scan Signal
- Internal Black and White References
- Simple and Stable Operation
- OPTIONAL FEATURE:
  Internal Reference Voltage

TC103 . . . DUAL-IN-LINE PACKAGE
(TOP VIEW)

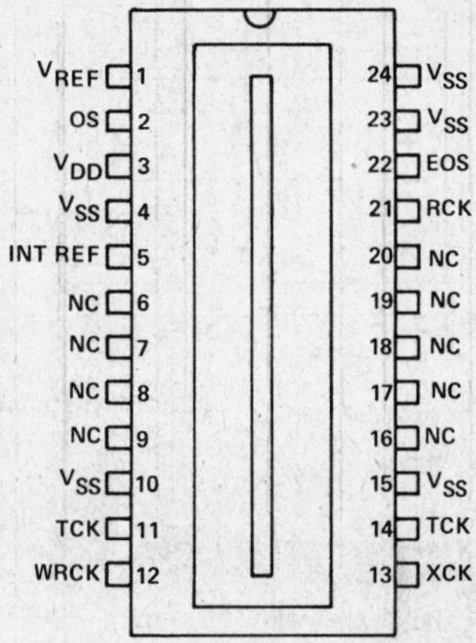

NC — No internal connection

description

The TC103, a 2048-element CCD line image sensor, functions in high-resolution image scanning applications such as facsimile and optical character recognition. The TC103 incorporates virtual-phase MOS technology, which provides simplified operation and high reliability. The 2048 sensor elements provide 8 points-per-millimeter resolution across 256 millimeters.

This device is supplied in a 24-pin dual-in-line ceramic side-braze package designed for insertion in mounting-hole rows on 15,2-mm (0,600-inch) centers. The glass window may be cleaned by wiping with a cotton swab soaked in alcohol.

Caution. These devices have limited built-in gate protection. The leads should be shorted together or the device placed in conductive foam during storage or handling to prevent electrostatic damage to the MOS gates. Avoid shorting either OS or EOS to V_{SS} during operation to prevent damage to the amplifiers.

virtual phase technology

This patented design results in simplified clocking circuits, reduced noise, and greater light sensitivity. The virtual phase utilizes a junction-gate region at the substrate dc potential. This accomplishes the same gating and transport function as a separate gate electrode requiring multiple layers and multiple process steps common in other device designs. The resulting simplicity of process and ease of operation will increase performance and reliability for the user.

Copyright © 1983 by Texas Instruments Incorporated

functional block diagram

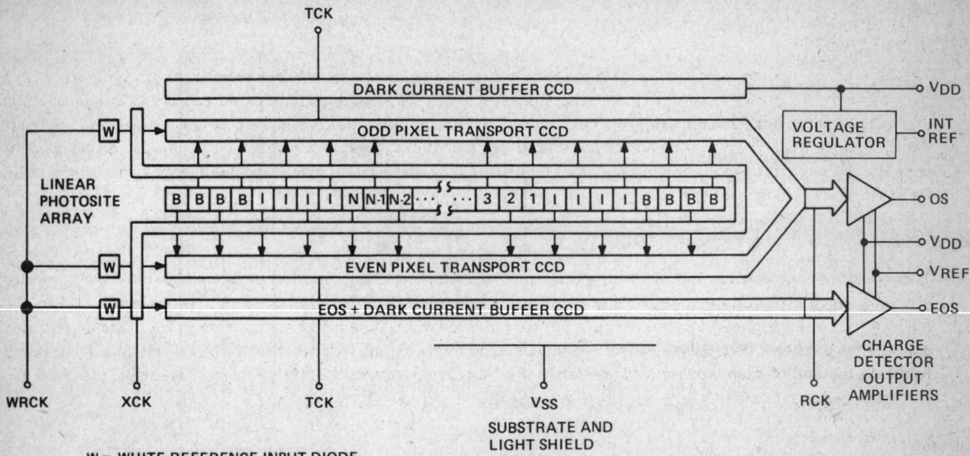

W = WHITE REFERENCE INPUT DIODE
B = BLACK REFERENCE ELEMENT
I = ISOLATION ELEMENT
N = 2048 SENSOR ELEMENTS

PIN FUNCTIONAL DESCRIPTION

PIN NUMBER	SIGNATURE	NAME	DESCRIPTION
1	V_{REF}	Reference Voltage	Bias input for the output amplifiers and internal reference.
2	OS	Output Signal	Video output from a cascaded source-follower MOS amplifier.
3	V_{DD}	Supply Voltage	Output amplifier supply voltage.
4, 10, 15, 23, 24	V_{SS}	Substrate	All voltages are referenced to the substrate.
5	INT REF	Internal Reference	Potential derived internally for operational reference voltage.
6, 7, 8, 9, 16 17, 18, 19, 20	NC		No internal connection.
11, 14	TCK	Transport Clock	Drives the CCD transport registers.
12	WRCK	White Reference Clock	Injects a controlled charge into the white reference CCD shift register elements to become white-reference and end-of-scan pulses.
13	XCK	Transfer Clock	Controls the transfer of charge packets from sensor elements to shift registers. The interval between pulses of the transfer clock determines the exposure time.
21	RCK	Reset Clock	Controls recharging of the charge-detection diodes in the output amplifiers, and clocks the output shift registers where the odd and even signals have been merged.
22	EOS	End-of-Scan Pulse	Indicates that all charge packets have been shifted out of the transport registers.

TEXAS INSTRUMENTS
INCORPORATED
POST OFFICE BOX 225012 • DALLAS, TEXAS 75265

functional description

image sensor elements

The line of sensor elements (also called photosites or pixels) consists of 2048 photo-sensitive areas, 12,7 micrometers (0.5 milliinches) square and approximately 12,7 micrometers from center to center. Image photons create electron/hole pairs in the single-crystal silicon. The electrons are collected in the sensor elements, and the holes are swept into the substrate. The amount of charge accumulated in each element is a linear function of the incident light and the exposure time. The output signal charge will vary in an analog manner from a thermally generated noise background at zero illumination to a maximum at saturation under bright illumination.

transfer gate

This structure is adjacent to the line of image sensor elements. The charge packets accumulated in the image sensor elements are transferred into the transfer gate storage well when the transfer gate voltage goes high. When the transfer gate voltage goes low, the charge is transferred into the CCD transport shift registers. The transfer gate also controls the exposure time for the sensor elements and permits charges to enter the end-of-scan (EOS) shift registers to create the end-of-scan waveform. In addition, the transfer gate permits entry of charge packets to the transport CCD shift register to create the white reference signals.

shift registers

There are two CCD transport registers, one on each side of the line of image sensor elements and outside of the transfer gate. Alternate charge packets are transferred to the CCD transport shift registers and moved serially to the output amplifier. The phase relationship of the reset clock and the transport clock and the geometric layout of the paths provide for alternate delivery of charge packets to re-establish the original sequence of the linear image data. The two outer buffer CCD shift registers protect the signal charges in the inner transport CCD shift registers from peripherally generated dark current noise.

black and white reference elements

Four additional sensor elements at each end of the sensor element array (labelled "B" in the block diagram) are covered by opaque metallization. They provide a black (no illumination) signal reference that is delivered at each end of the linear image output signal. Also included on the transport CCD shift register, at the opposite end from the amplifier, is an input diode that provides two white reference pulses in the output signal. The reference pulses are useful as inputs to external dc restoration and/or automatic exposure control circuitry. The white reference pulse amplitude is approximately 70% of the maximum output signal amplitude.

output signal amplifier

The charge packets are transported to a precharge diode whose potential changes linearly in response to the amount of the signal charge delivered. This potential is applied to the input gate of an N-channel MOS double-source-follower amplifier to produce an output signal (OS). A reset transistor, driven by the reset clock (RCK), recharges the charge-detector-diode capacitance before the arrival of each new signal charge packet from the CCD shift registers. Reference voltage (V_{REF}) is applied to the drain of the reset transistor and acts to bias the OS and EOS amplifiers. A current sink is used as an on-chip load for the amplifier output, so no external current sink is needed. The output signal on pin 2 is a series of negative-going pulses on a dc level.

internal reference voltage

An internal reference voltage (INT REF) is available on the chip to provide the V_{REF} voltage. The required connections appear in Figure 3. If the internal reference voltage is not used, an external voltage is connected directly to pin 1. Pin 5 is then left unconnected.

TEXAS INSTRUMENTS
INCORPORATED
POST OFFICE BOX 225012 • DALLAS, TEXAS 75265

resolution

The modulation transfer function decreases at longer wavelengths. (See Figures 7 and 8.) If optimum resolution is required with a light source that has a significant infrared component, then the designer must use appropriate filters to restrict the optical pass band to shorter wavelengths.

end-of-scan amplifier

The EOS amplifier is similar to the OS amplifier. XCK transfers charge from the input diode into the EOS register where it is transported at the TCK clock frequency to the EOS amplifier. This EOS pulse is coincident with the first of the two white reference pulses that pass through the odd and even transport CCDs, respectively. The EOS output can be used to alert the external circuitry that the linear image data readout has been completed.

clocks

The transfer clock (XCK) pulse controls the exposure time of the sensor elements. The minimum exposure time is the time required to shift the entire contents of the transport registers to the output signal amplifier and equals 2081 multiplied by the RCK period. The maximum exposure time is determined by the tolerable level of dark signal.

The transport clock (TCK) transports the linear image signal charge from the sensor element region to the output amplifier.

The reset clock (RCK) operates at twice the transport clock frequency so as to recombine the signal charge in the original sequence and present the charge to the output amplifier. The data rate is equal to the reset clock frequency.

The white reference clock (WRCK) runs at the transfer clock frequency and generates the white reference and the end-of-scan pulses. These pulses can be eliminated by connecting WRCK to V_{DD}. Transients on WRCK going below zero volts will cause charge injection resulting in an increase in apparent dark signal.

Figure 3 presents a suggested circuit for generating the clock waveforms. The RCK clock generator runs continuously. A binary divider halves the frequency to create TCK. After all signal charges have been transported to the output amplifier, TCK continues to run to keep thermally generated charges from accumulating in the transport registers.

The XCK and WRCK clock frequencies are submultiples of the TCK frequency. Figure 2 details the timing relationships among the different clock pulses.

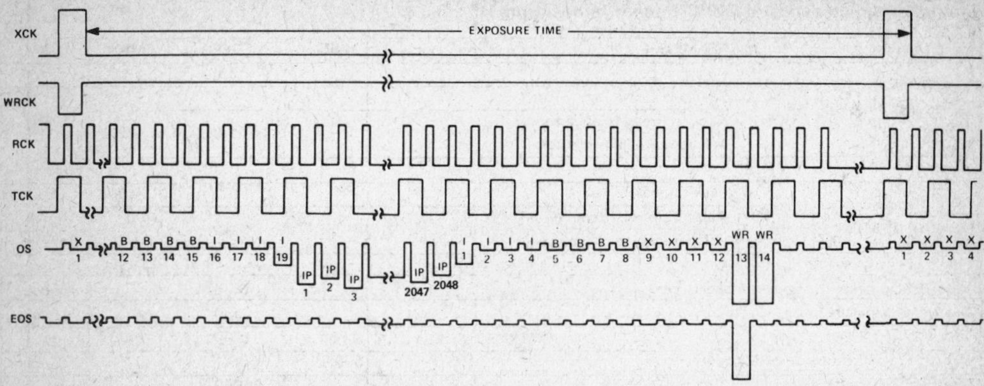

Output Signal (OS) pulse identification: I = isolation pixel, IP = Image pixel, B = Black reference pixel, WR = White reference pixel, X = empty pixel.

FIGURE 1 — OPERATING INPUT AND OUTPUT VOLTAGE WAVEFORMS

absolute maximum ratings over operating free-air temperature range (unless otherwise noted) (see Note 1)

Amplifier drain voltage (V_{DD1})	−0.3 V to 30 V
Transfer clock (XCK) voltage	−25 V to 5 V
Transport clock (TCK) voltage	−25 V to 5 V
Reset clock (RCK) voltage	−25 V to 5 V
White reference clock (WRCK) voltage	−0.3 V to 30 V
Storage temperature	−25°C to 125°C
Operating free-air temperature	−25°C to 70°C

recommended operating conditions at T_A = 25°C (see Note 1)

		MIN	NOM	MAX	UNIT
V_{DD}	Amplifier supply voltage	13	14	15	V
$V_{IH(X)}$	Transfer clock high-level input voltage	3	4	5	V
$V_{IL(X)}$	Transfer clock low-level input voltage	−15†	−14	−13	V
$V_{IH(T)}$	Transport clock high-level input voltage	3	4	5	V
$V_{IL(T)}$	Transport clock low-level input voltage	−15†	−14	−13	V
$V_{IH(R)}$	Reset clock high-level input voltage	3	4	5	V
$V_{IL(R)}$	Reset clock low-level input voltage	−15†	−14	−13	V
$V_{IH(WR)}$	White reference clock high-level input voltage	13	14	15	V
$V_{IL(WR)}$	White reference clock low-level input voltage	6	7	8	V
f_{RCK}	Reset clock frequency (output data rate)			10	MHz

†The algebraic convention, where the most negative limit is designated as minimum, is used in this data sheet for clock voltage levels only.
NOTE 1: Voltage values are with respect to V_{SS}.

electrical characteristics at 25°C free-air temperature¶

PARAMETER		MIN	TYP	MAX	UNIT
Dark-signal amplitude	Average		0.5	10	mV
	Low frequency component		0.5	5	mV
	Nonuniformity relative to average of adjacent pixels		1.0	20	mV
Sensitivity		2	3.5	5	V/(μJ/cm^2)
Output amplitude variation (PRNU)‡	Peak-to-peak		50	100	mV
	Adjacent pixels from alternate registers (imbalance)		10		mV
Peak-to-peak noise			1		mV
Equivalent exposure§ of peak-to-peak noise			0.35		nJ/cm^2
Saturation exposure§			350		nJ/cm^2
Saturation output amplitude		700	1000	1400	mV
Dynamic range relative to peak-to-peak noise		500:1	1000:1		
Charge transfer efficiency			0.99999		
White reference amplitude		500	700		mV
End-of-scan amplitude		300	500		mV
Output offset (dc) voltage			10		V
Output impedance			1		kΩ
Resistance to V_{SS}	Transfer gate		150		kΩ
	Transport gate		500		kΩ
	Reset gate		500		kΩ
Amplifier reference voltage, V_{REF}			7		V
Capacitance to V_{SS}	Transfer gate		250		pF
	Transport gate		600		pF
	Reset gate		16		pF
Amplifier supply current			8	12	mA
Total power dissipation			110		mW

‡Measured at 700 mV output amplitude with an f/2.8 lens.

§Exposure = intensity x time

¶Test conditions are f_{RCK} = 0.5 MHz, t_{exp} = 10 ms, tungsten light source operating at color temperature of 2854 K with 2.0-mm-thick Fish-Schurman HA-11 IR-absorbing filter, and all operating voltages at nominal recommended values. The internal reference voltage is used.

timing requirements

		MIN	NOM	MAX	UNITS
t_{THXH}	Time delay from the transport clock rising edge to the transfer clock rising edge.	0		50	ns
t_{THWL}	Time delay from the transport clock rising edge to the white reference clock falling edge.	0		50	ns
t_{THRH}	Time delay from the transport clock rising edge to the reset clock rising edge.	0			ns
$t_{w(RH)}$	Pulse duration of the high state for the reset clock.	40			ns
t_{TLXL}	Time delay from the transport clock falling edge to the transfer clock falling edge.	50			ns
t_{TLWH}	Time delay from the transport clock falling edge to the white reference clock rising edge.	0		50	ns
t_{XLTH}	Time delay from the transfer clock falling edge to the rising edge of the next transport clock pulse.	50			ns
t_r	rise time (all clocks)	15			ns
t_f	fall time (all clocks)	5			ns

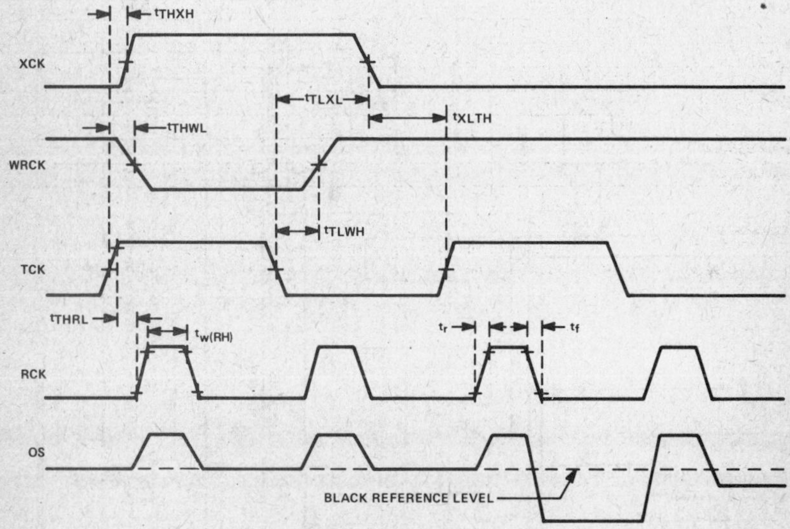

FIGURE 2—DEVICE TIMING REQUIREMENTS

PARAMETER MEASUREMENT INFORMATION

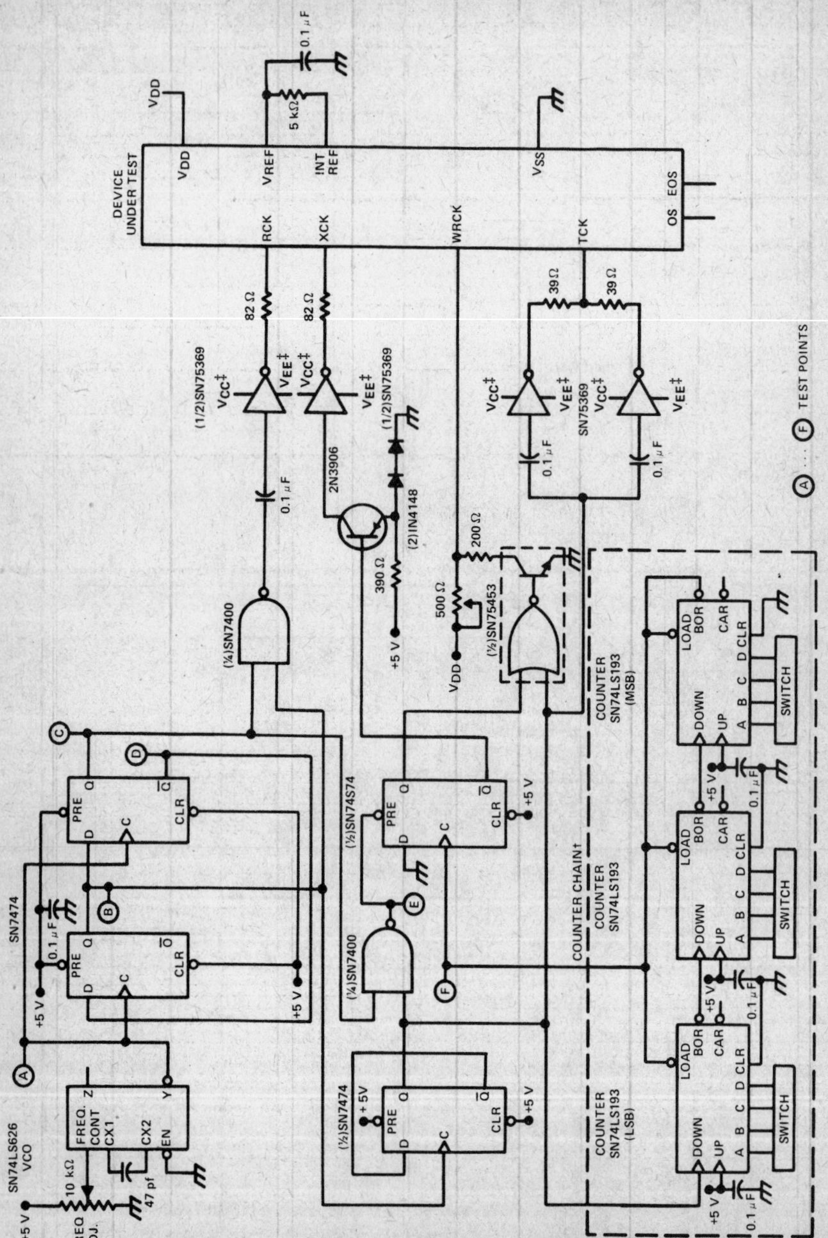

†This counter chain counts transport clock periods to generate the exposure time interval. The data rate is twice the count rate.

‡V_{CC} and V_{EE} are the voltages that will produce the desired values of V_{IH} and V_{IL}, respectively, at the RCK, XCK, and TCK inputs.

FIGURE 3—DRIVER CIRCUIT FOR TESTING IMAGE SENSOR

FIGURE 4 — WAVEFORMS IN DRIVER CIRCUIT

TYPICAL CHARACTERISTICS

(In the circuit of Figure 3 with T_A = 25 °C, f_{RCK} = 0.5 MHz, t_{exp} = 10 ms, and all operating voltages at nominal recommended values, unless otherwise noted)

SENSITIVITY
vs
WAVELENGTH OF INCIDENT LIGHT

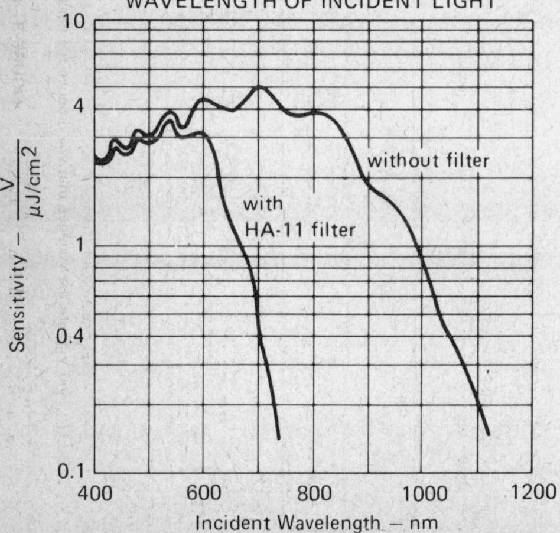

FIGURE 5

RESPONSIVITY
vs
WAVELENGTH OF INCIDENT LIGHT

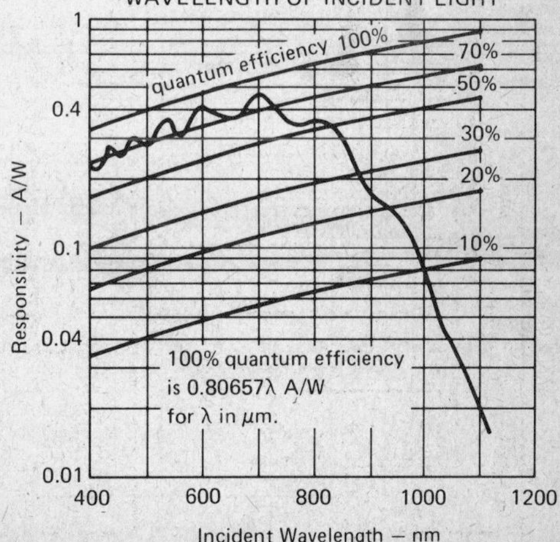

FIGURE 6

TYPICAL CHARACTERISTICS

(In the circuit of Figure 3 with T_A = 25 °C, f_{RCK} = 0.5 MHz, t_{exp} = 10 ms, and all operating voltages at nominal recommended values, unless otherwise noted)

MODULATION TRANSFER FUNCTION
vs
SPATIAL FREQUENCY

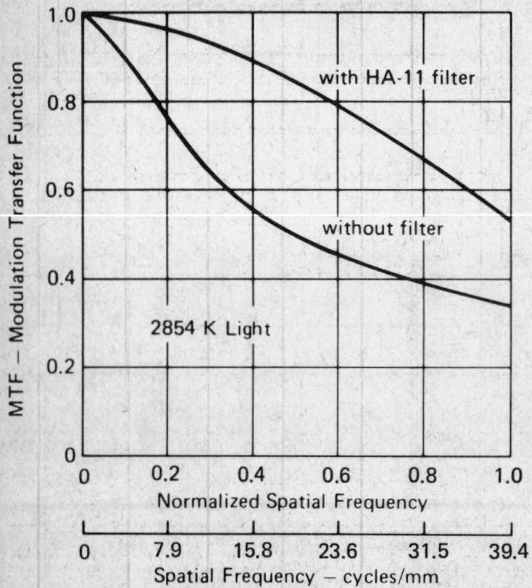

FIGURE 7

MODULATION TRANSFER FUNCTION
vs
SPATIAL FREQUENCY

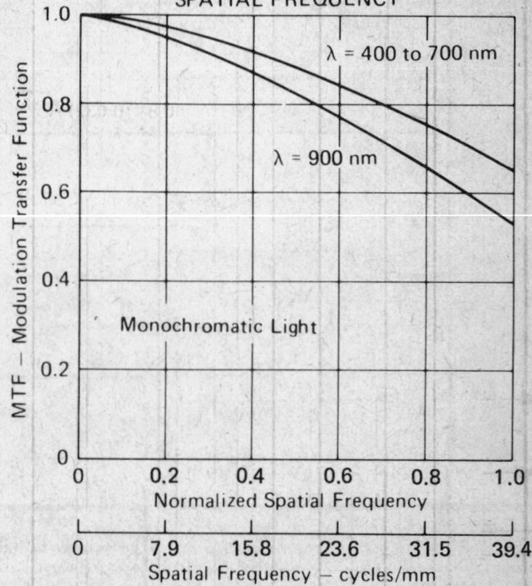

FIGURE 8

AVERAGE AND LOW FREQUENCY
DARK SIGNAL
vs
EXPOSURE TIME

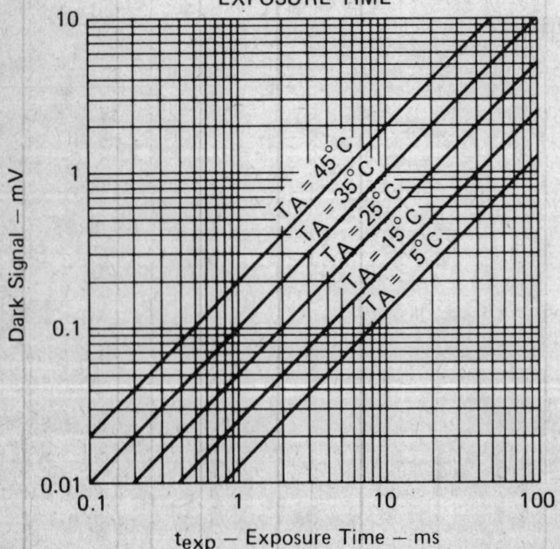

FIGURE 9

OUTPUT SIGNAL VOLTAGE RELATIVE TO
SATURATED OUTPUT VOLTAGE
vs
EXPOSURE TIME

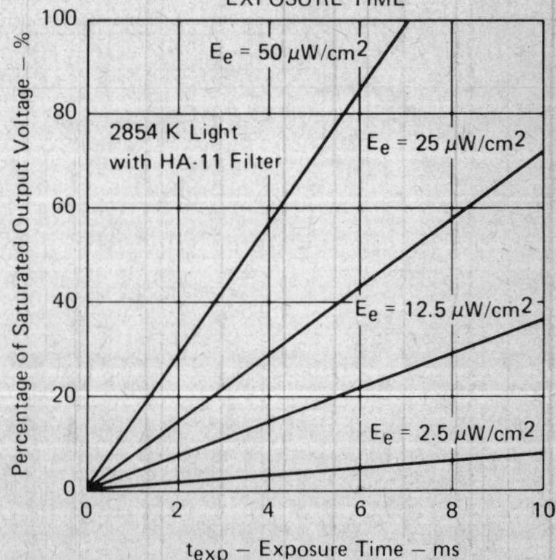

FIGURE 10

TEXAS INSTRUMENTS
INCORPORATED
POST OFFICE BOX 225012 • DALLAS, TEXAS 75265

MECHANICAL DATA

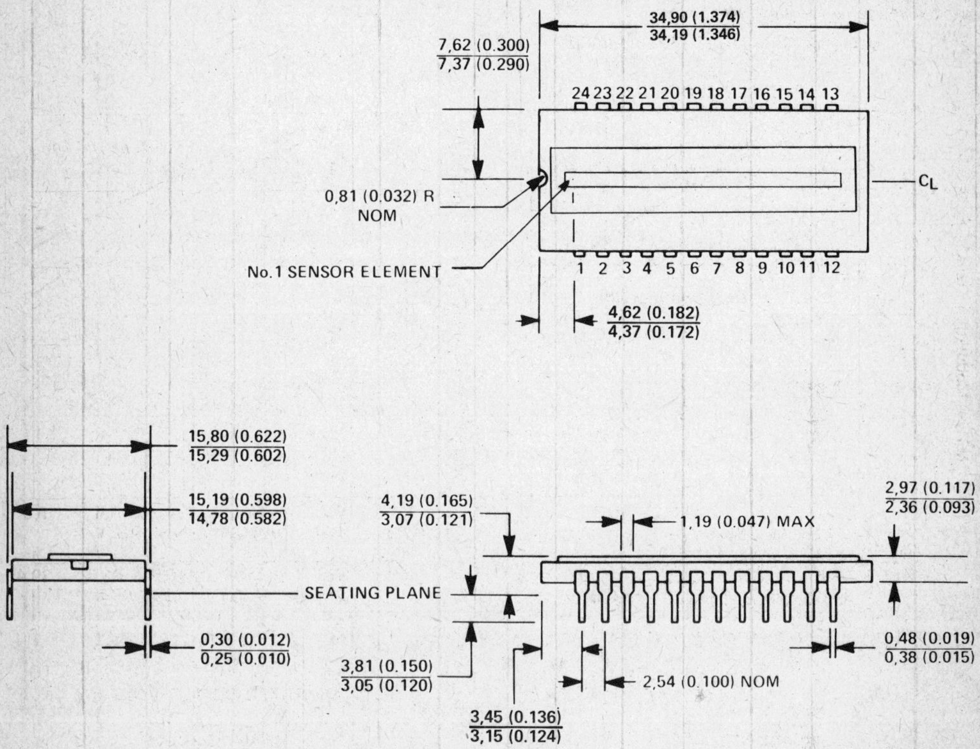

ALL DIMENSIONS ARE IN MILLIMETERS AND PARENTHETICALLY IN INCHES.

NOTE 1: The distance between the top surface of the window and the surface of the sensor is nominally 0,89 (0.035). This is determined by observing the vertical motion of a microscope focused first at one plane, then at the other.

TEXAS INSTRUMENTS
INCORPORATED
POST OFFICE BOX 225012 • DALLAS, TEXAS 75265

TYPE TC104
3456 x 1 CCD LINEAR IMAGE SENSOR

- **3456 x 1 Sensor Element Organization**
- **Virtual-Phase N-Channel Silicon MOS Technology**
- **High Quantum Efficiency**
- **Enhanced Blue Response**
- **Output Signal Approximately 0.6 Volt Peak-to-Peak**
- **Dynamic Range Relative to Peak-to-Peak Noise Typically 1000:1**
- **End-of-Scan Signal**
- **Internal Black and White References**
- **Simple and Stable Operation**
- **OPTIONAL FEATURE: Internal Reference Voltage**

TC104 . . . DUAL-IN-LINE PACKAGE
(TOP VIEW)

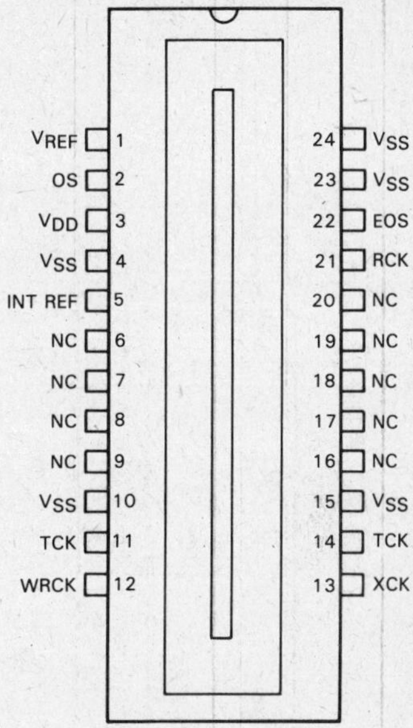

NC — No internal connection.

description

The TC104, a 3456-element CCD line image sensor, functions in high-resolution image scanning applications such as document reading and optical character recognition. The TC104 incorporates virtual-phase MOS technology, which provides simplified operation and high reliability. The 3456 sensor elements provide 400 points-per-inch resolution across 8.5 inches.

This device is supplied in a 24-pin dual-in-line ceramic side-braze package designed for insertion in mounting-hole rows on 15.2-mm (0,600-inch) centers. The glass window may be cleaned by wiping with a cotton swab soaked in alcohol.

Caution. These devices have limited built-in gate protection. The leads should be shorted together or the device placed in conductive foam during storage or handling to prevent electrostatic damage to the MOS gates. Avoid shorting either OS or EOS to V_{SS} during operation to prevent damage to the amplifiers.

virtual phase technology

This patented design results in simplified clocking circuits, reduced noise, and greater light sensitivity. The virtual phase utilizes a junction-gate region at the substrate dc potential. This accomplishes the same gating and transport function as a separate gate electrode requiring multiple layers and multiple process steps common in other device designs. The resulting simplicity of process and ease of operation will increase performance and reliability for the user.

Copyright © 1983 by Texas Instruments Incorporated

functional block diagram

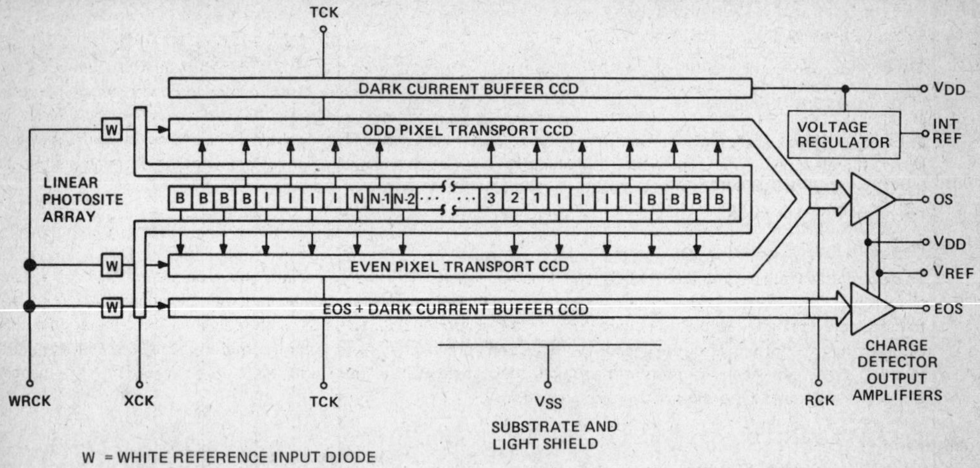

W = WHITE REFERENCE INPUT DIODE
B = BLACK REFERENCE ELEMENT
I = ISOLATION ELEMENT
N = 3456 SENSOR ELEMENTS

PIN FUNCTIONAL DESCRIPTION

PIN NUMBER	SIGNATURE	NAME	DESCRIPTION
1	V_{REF}	Reference Voltage	Bias input for the output amplifiers and internal reference.
2	OS	Output Signal	Video output from a cascaded source-follower MOS amplifier.
3	V_{DD}	Supply Voltage	Output amplifier supply voltage.
4, 10, 15, 23, 24	V_{SS}	Substrate	All voltages are referenced to the substrate.
5	INT REF	Internal Reference	Potential derived internally for operational reference voltage.
6, 7, 8, 9, 16 17, 18, 19, 20	NC		No internal connection.
11, 14	TCK	Transport Clock	Drives the CCD transport registers.
12	WRCK	White Reference Clock	Injects a controlled charge into the white reference CCD shift register elements to become white-reference and end-of-scan pulses.
13	XCK	Transfer Clock	Controls the transfer of charge packets from sensor elements to shift registers. The interval between pulses of the transfer clock determines the exposure time.
21	RCK	Reset Clock	Controls recharging of the charge-detection diodes in the output amplifiers, and clocks the output shift registers where the odd and even signals have been merged.
22	EOS	End-of-Scan Pulse	Indicates that all charge packets have been shifted out of the transport registers.

functional description

image sensor elements

The line of sensor elements (also called photosites or pixels) consists of 3456 photo-sensitive areas, 10,7 micrometers (0.42 milliinches) square and approximately 10,7 micrometers from center to center. Image photons create electron/hole pairs in the single-crystal silicon. The electrons are collected in the sensor elements, and the holes are swept into the substrate. The amount of charge accumulated in each element is a linear function of the incident light and the exposure time. The output signal charge will vary in an analog manner from a thermally generated noise background at zero illumination to a maximum at saturation under bright illumination.

transfer gate

This structure is adjacent to the line of image sensor elements. The charge packets accumulated in the image sensor elements are transferred into the transfer gate storage well when the transfer gate voltage goes high. When the transfer gate voltage goes low, the charge is transferred into the CCD transport shift registers. The transfer gate also controls the exposure time for the sensor elements and permits charges to enter the end-of-scan (EOS) shift registers to create the end-of-scan waveform. In addition, the transfer gate permits entry of charge packets to the transport CCD shift register to create the white reference signals.

shift registers

There are two CCD transport registers, one on each side of the line of image sensor elements and outside of the transfer gate. Alternate charge packets are transferred to the CCD transport shift registers and moved serially to the output amplifier. The phase relationship of the reset clock and the transport clock and the geometric layout of the paths provide for alternate delivery of charge packets to re-establish the original sequence of the linear image data. The two outer buffer CCD shift registers protect the signal charges in the inner transport CCD shift registers from peripherally generated dark current noise.

black and white reference elements

Four additional sensor elements at each end of the sensor element array (labelled "B" in the block diagram) are covered by opaque metallization. They provide a black (no illumination) signal reference that is delivered at each end of the linear image output signal. Also included on the transport CCD shift register, at the opposite end from the amplifier, is an input diode that provides two white reference pulses in the output signal. The reference pulses are useful as inputs to external dc restoration and/or automatic exposure control circuitry. The white reference pulse amplitude is approximately 100% of the maximum output signal amplitude.

output signal amplifier

The charge packets are transported to a precharge diode whose potential changes linearly in response to the amount of the signal charge delivered. This potential is applied to the input gate of an N-channel MOS double-source-follower amplifier to produce an output signal (OS). A reset transistor, driven by the reset clock (RCK), recharges the charge-detector-diode capacitance before the arrival of each new signal charge packet from the CCD shift registers. Reference voltage (V_{REF}) is applied to the drain of the reset transistor and acts to bias the OS and EOS amplifiers. A current sink is used as an on-chip load for the amplifier output, so no external current sink is needed. The output signal on pin 2 is a series of negative-going pulses on a dc level.

internal reference voltage

An internal reference voltage (INT REF) is available on the chip to provide the V_{REF} voltage. The required connections appear in Figure 3. If the internal reference voltage is not used, an external voltage is connected directly to pin 1. Pin 5 is then left unconnected.

resolution

The modulation transfer function decreases at longer wavelengths. (See Figures 7 and 8.) If optimum resolution is required with a light source that has a significant infrared component, then the designer must use appropriate filters to restrict the optical pass band to shorter wavelengths.

end-of-scan amplifier

The EOS amplifier is similar to the OS amplifier. XCK transfers charge from the input diode into the EOS register where it is transported at the TCK clock frequency to the EOS amplifier. This EOS pulse is coincident with the first of the two white reference pulses that pass through the odd and even transport CCDs, respectively. The EOS output can be used to alert the external circuitry that the linear image data readout has been completed.

clocks

The transfer clock (XCK) pulse controls the exposure time of the sensor elements. The minimum exposure time is the time required to shift the entire contents of the transport registers to the output signal amplifier and equals 3489 multiplied by the RCK period. The maximum exposure time is determined by the tolerable level of dark signal.

The transport clock (TCK) transports the linear image signal charge from the sensor element region to the output amplifier.

The reset clock (RCK) operates at twice the transport clock frequency so as to recombine the signal charge in the original sequence and present the charge to the output amplifier. The data rate is equal to the reset clock frequency.

The white reference clock (WRCK) runs at the transfer clock frequency and generates the white reference and the end-of-scan pulses. These pulses can be eliminated by connecting WRCK to V_{DD}. Transients on WRCK going below zero volts will cause charge injection resulting in an increase in apparent dark signal.

Figure 3 presents a suggested circuit for generating the clock waveforms. The RCK clock generator runs continuously. A binary divider halves the frequency to create TCK. After all signal charges have been transported to the output amplifier, TCK continues to run to keep thermally generated charges from accumulating in the transport registers.

The XCK and WRCK clock frequencies are submultiples of the TCK frequency. Figure 2 details the timing relationships among the different clock pulses.

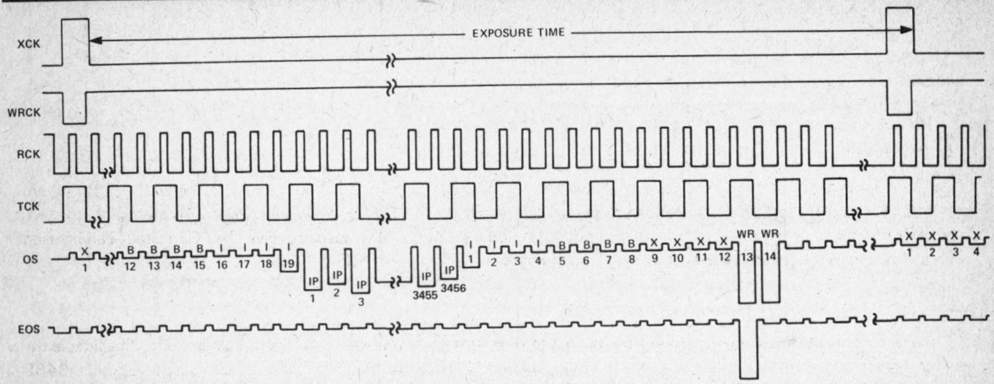

Output Signal (OS) pulse identification: I = Isolation pixel, IP = Image pixel, B = Black reference pixel, WR = White reference pixel, X = empty pixel.

FIGURE 1—OPERATING INPUT AND OUTPUT VOLTAGE WAVEFORMS

absolute maximum ratings over operating free-air temperature range (unless otherwise noted) (see Note 1)

Amplifier drain voltage (V_{DD})	−0.3 V to 30 V
Transfer clock (XCK) voltage	−25 V to 5 V
Transport clock (TCK) voltage	−25 V to 5 V
Reset clock (RCK) voltage	−25 V to 5 V
White reference clock (WRCK) voltage	−0.3 V to 30 V
Storage temperature	−25°C to 125°C
Operating free-air temperature	−25°C to 70°C

NOTE 1: Voltage values are with respect to V_{SS}.

recommended operating conditions at T_A = 25°C (see Note 1)

		MIN	NOM	MAX	UNIT
V_{DD}	Amplifier supply voltage	13	14	15	V
$V_{IH(X)}$	Transfer clock high-level input voltage	3	4	5	V
$V_{IL(X)}$	Transfer clock low-level input voltage	−15†	−14	−13	V
$V_{IH(T)}$	Transport clock high-level input voltage	3	4	5	V
$V_{IL(T)}$	Transport clock low-level input voltage	−15†	−14	−13	V
$V_{IH(R)}$	Reset clock high-level input voltage	3	4	5	V
$V_{IL(R)}$	Reset clock low-level input voltage	−15†	−14	−13	V
$V_{IH(WR)}$	White reference clock high-level input voltage	13	14	15	V
$V_{IL(WR)}$	White reference clock low-level input voltage	6	7	8	V
f_{RCK}	Reset clock frequency (output data rate)			8	MHz

†The algebraic convention, where the most negative limit is designated as minimum, is used in this data sheet for clock voltage levels only.

NOTE 1: Voltage values are with respect to V_{SS}.

electrical characteristics at 25 °C free-air temperature ¶

PARAMETER		MIN	TYP	MAX	UNIT
Dark-signal amplitude	Average		0.5	10	mV
	Low frequency component		0.5	5	mV
	Nonuniformity relative to average of adjacent pixels		4	20	
Sensitivity		1.4	2	3.5	V/(μJ/cm^2)
Output amplitude variation (PRNU) ‡	Peak-to-peak		30	60	mV
	Adjacent pixels from alternate registers (imbalance)		10		
Peak-to-peak noise			0.6		mV
Equivalent exposure § of peak-to-peak noise			0.3		nJ/cm^2
Saturation exposure §			300		nJ/cm^2
Saturation output amplitude		400	600	800	mV
Dynamic range relative to peak-to-peak noise		500:1	1000:1		
Charge transfer efficiency			0.99999		
White reference amplitude		400	600		mV
End-of-scan amplitude		200	350		mV
Output offset (dc) voltage			6		V
Output impedance			1		kΩ
Resistance to V_{SS}	Transfer gate		150		kΩ
	Transport gate		700		
	Reset gate		700		
Amplifier reference voltage, V_{REF}			7		V
Capacitance to V_{SS}	Transfer gate		400		pF
	Transport gate		900		
	Reset gate		16		
Amplifier supply current			8	12	mA
Total power dissipation			112		mW

‡Measured at 400 mV output amplitude with an f/2.8 lens.

§Exposure = intensity x time

¶ Test conditions are f_{RCK} = 0.5 MHz, t_{exp} = 10 ms, tungsten light source operating at color temperature of 2854 K with 2.0-mm-thick Fish-Schurman HA-11 IR-absorbing filter, and all operating voltages at nominal recommended values using the internal reference voltage.

TEXAS INSTRUMENTS
INCORPORATED
POST OFFICE BOX 225012 • DALLAS, TEXAS 75265

timing requirements

		MIN	NOM	MAX	UNITS
t_{THXH}	Time delay from the transport clock rising edge to the transfer clock rising edge.	0		50	ns
t_{THWL}	Time delay from the transport clock rising edge to the white reference clock falling edge.	0		50	ns
t_{THRH}	Time delay from the transport clock rising edge to the reset clock rising edge.	0			ns
$t_{w(RH)}$	Pulse duration of the high state for the reset clock.	40			ns
t_{TLXL}	Time delay from the transport clock falling edge to the transfer clock falling edge.	50			ns
t_{TLWH}	Time delay from the transport clock falling edge to the white reference clock rising edge.	0		50	ns
t_{XLTH}	Time delay from the transfer clock falling edge to the rising edge of the next transport clock pulse.	50			ns
t_r	Rise time (all clocks)	15			ns
t_f	Fall time (all clocks)	5			ns

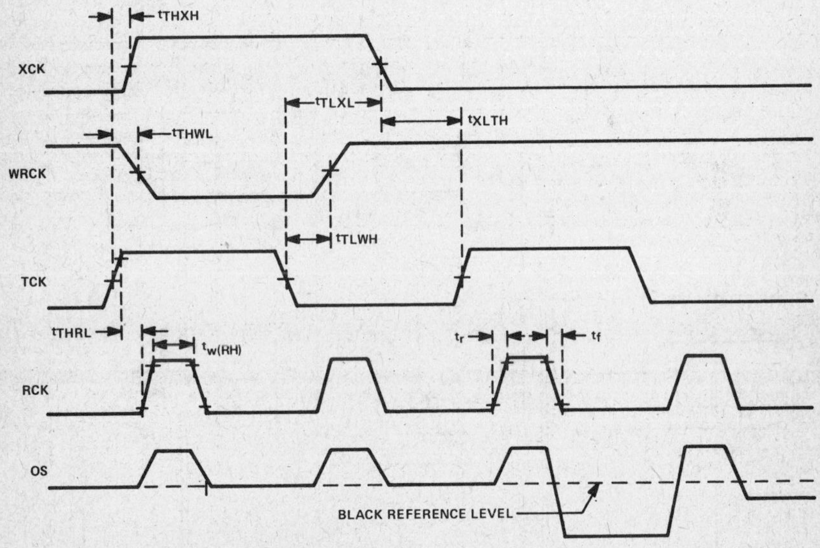

FIGURE 2—DEVICE TIMING REQUIREMENTS

PARAMETER MEASUREMENT INFORMATION

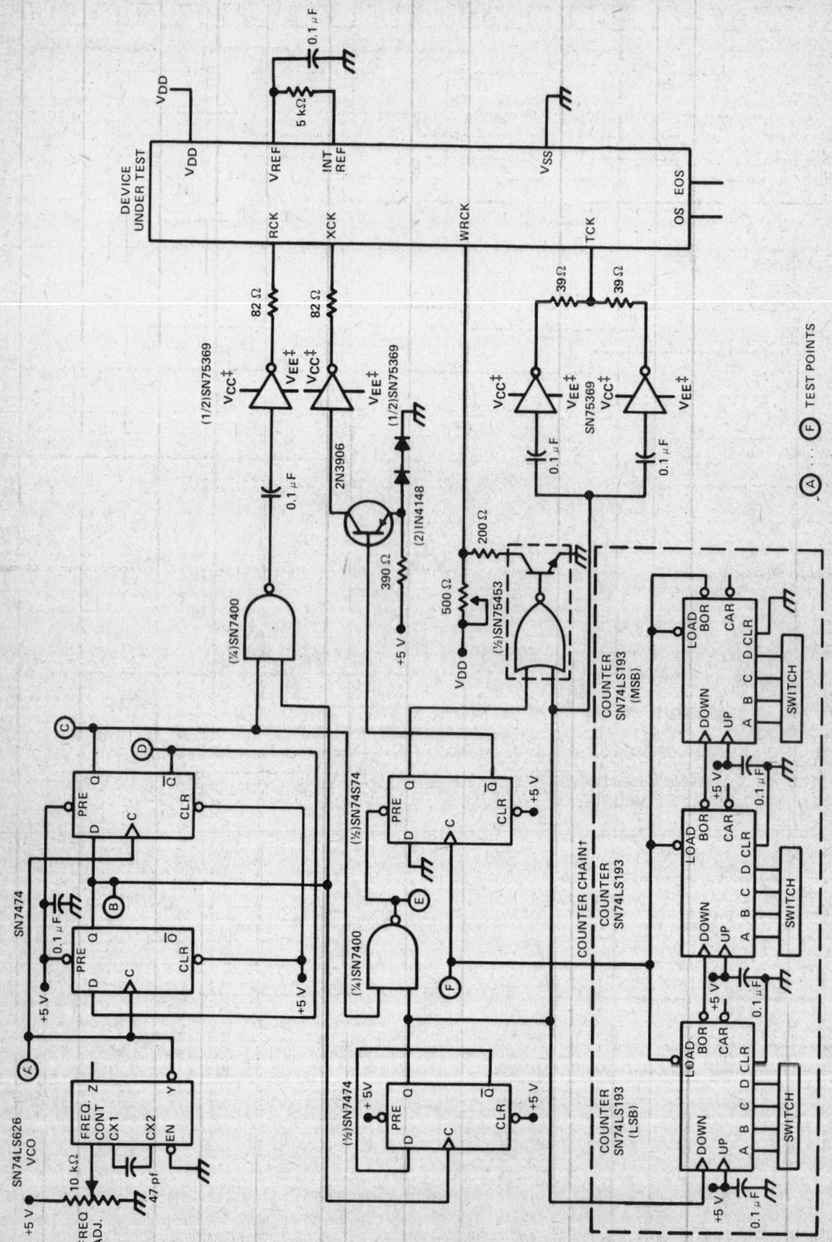

†This counter chain counts transport clock periods to generate the exposure time interval. The data rate is twice the count rate.

‡V_{CC} and V_{EE} are the voltages that will produce the desired values of V_{IH} and V_{IL}, respectively, at the RCK, XCK, and TCK inputs.

FIGURE 3—DRIVER CIRCUIT FOR TESTING IMAGE SENSOR

TEXAS INSTRUMENTS
INCORPORATED
POST OFFICE BOX 225012 • DALLAS, TEXAS 75265

FIGURE 4 — WAVEFORMS IN DRIVER CIRCUIT

TYPICAL CHARACTERISTICS

(In the circuit of Figure 3 with T_A = 25°C, f_{RCK} = 0.5 MHz, t_{exp} = 10 ms, and all operating voltages at nominal recommended values, unless otherwise noted)

SENSITIVITY
vs
WAVELENGTH OF INCIDENT LIGHT

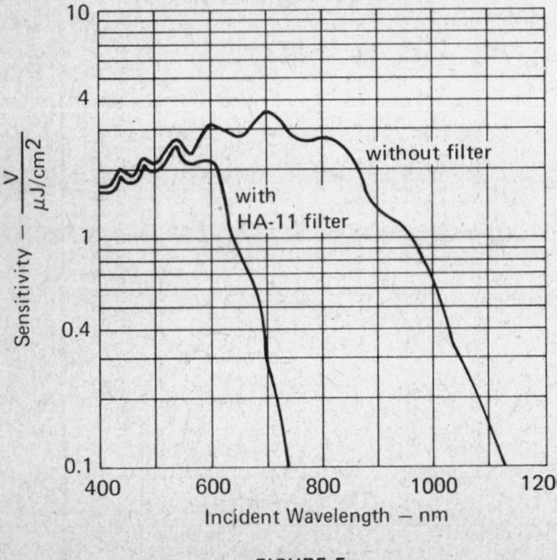

FIGURE 5

RESPONSIVITY
vs
WAVELENGTH OF INCIDENT LIGHT

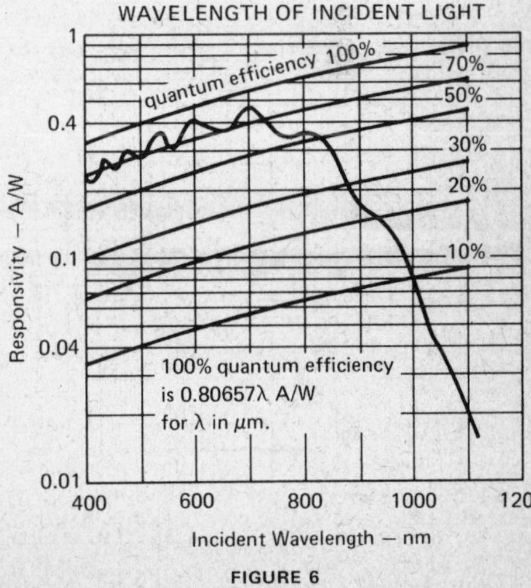

FIGURE 6

TYPICAL CHARACTERISTICS

(In the circuit of Figure 3 with T_A = 25°C, f_{RCK} = 0.5 MHz, t_{exp} = 10 ms, and all operating voltages at nominal recommended values, unless otherwise noted)

MODULATION TRANSFER FUNCTION
vs
SPATIAL FREQUENCY

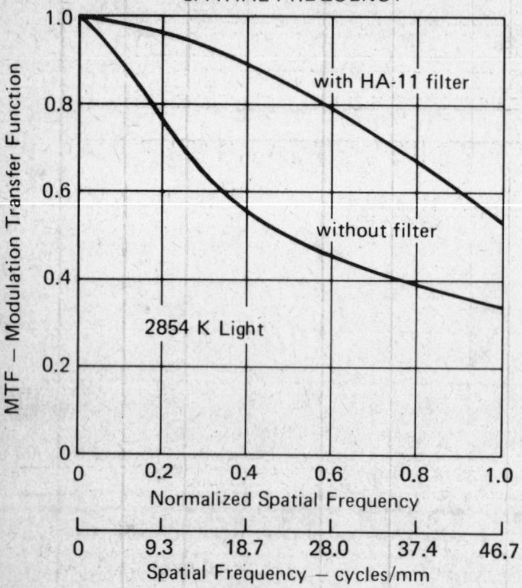

FIGURE 7

MODULATION TRANSFER FUNCTION
vs
SPATIAL FREQUENCY

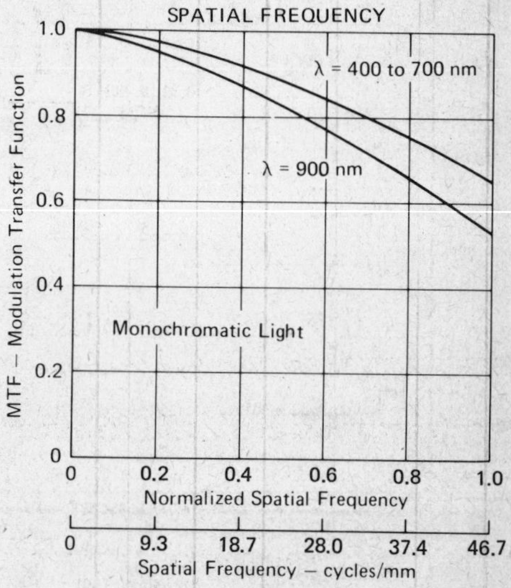

FIGURE 8

AVERAGE AND LOW FREQUENCY
DARK SIGNAL
vs
EXPOSURE TIME

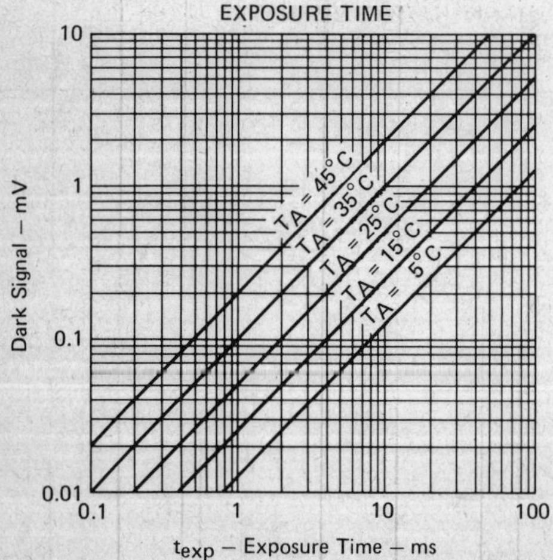

FIGURE 9

OUTPUT SIGNAL VOLTAGE RELATIVE TO
SATURATED OUTPUT VOLTAGE
vs
EXPOSURE TIME

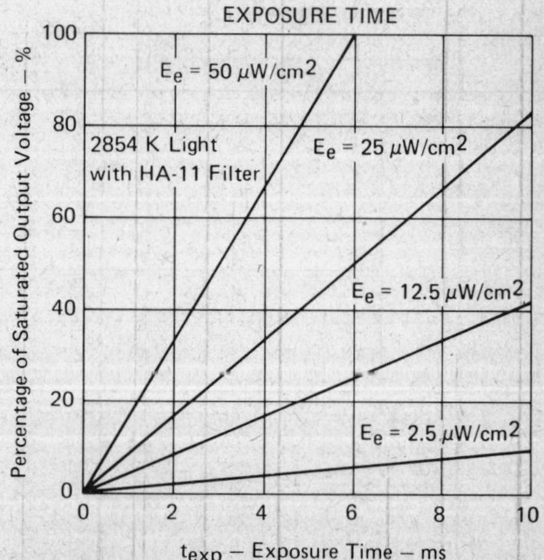

FIGURE 10

TEXAS INSTRUMENTS
INCORPORATED
POST OFFICE BOX 225012 • DALLAS, TEXAS 75265

MECHANICAL DATA

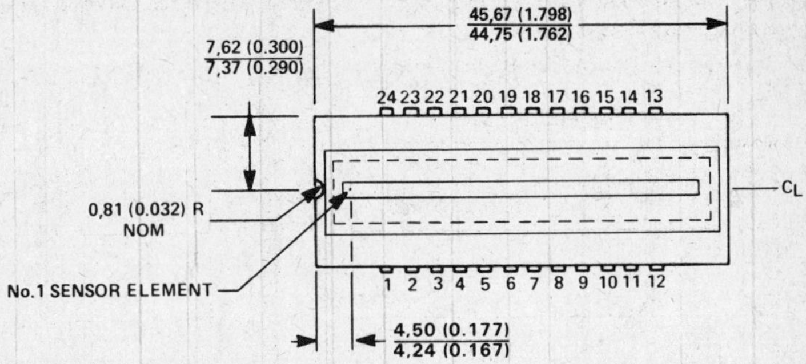

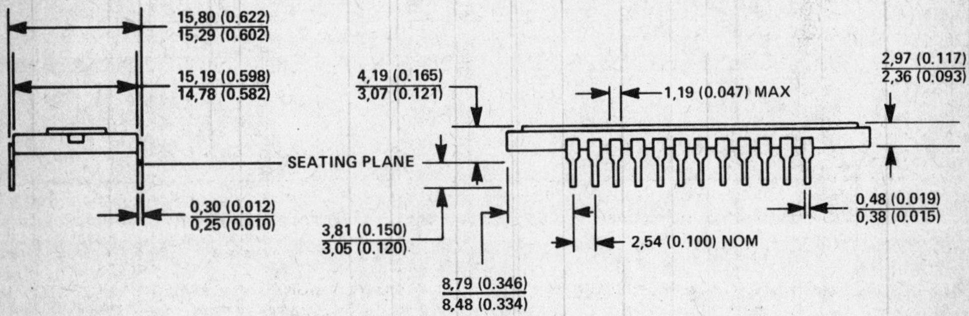

ALL DIMENSIONS ARE IN MILLIMETERS AND PARENTHETICALLY IN INCHES.

NOTE 1: The distance between the top surface of the window and the surface of the sensor is nominally 0,89 (0.035). This is determined by observing the vertical motion of a microscope focused first at one plane, then at the other.

2

CCD IMAGE SENSORS

- 328- X 490-Pixel Format for Frame-Store or Full-Frame Mode Operation
- Virtual Phase (VP), Front Side Illuminated
- Buried Channel Registers
- High Charge Transfer Efficiency
- High Resolution
- Interlaced 525-line TV Output (Frame-Store Mode)
- No Residual Imaging
- No Microphonics
- Small Size, 11-mm Image Diagonal (Frame-Store Mode)
- No Image Burn-in
- High Uniformity

CERAMIC PACKAGE
(TOP VIEW)

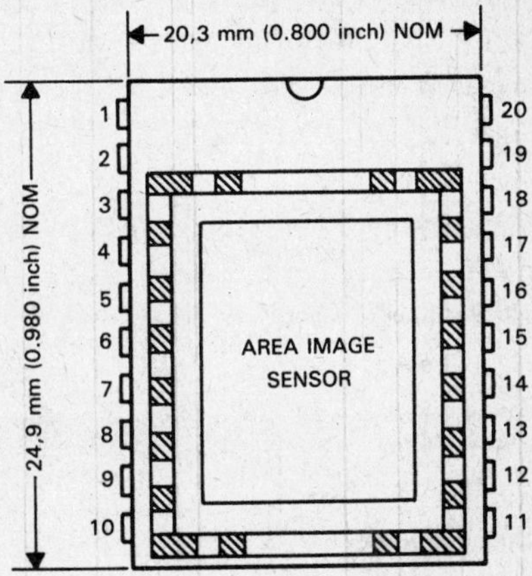

description

This 328- X 490-Pixel area sensor is designed to operate in the frame-store mode as a 328H X 245V imager for 525-line US TV applications. The device can also be used in the full-frame mode for long-integration, single-frame applications. Charge packets are transported to a precharged diode whose potential changes in response to the quantity of the signal charge delivered. This potential is applied to an on-chip floating diffusion amplifier to produce a signal voltage at the video output pin. A reset transistor is used to recharge the charge-detector-diode capacitance before the arrival of each new signal charge packet from the serial transport register. The imager is fabricated using virtual phase MOS, technology, which provides greater reliability at lower cost than the conventional 2-phase MOS technologies.

This device is supplied in a 20-pin dual-in-line ceramic side-braze package designed for insertion in mounting-hole rows on 20,3-mm (0.800-inch) centers. The pins in each row are located on 2,54-mm (0.100-inch) centers. Optical quality glass lids cover the image area.

Availability of this device is scheduled for 3rd quarter 1983

PRODUCT PREVIEW
This document contains information on a product under development. Texas Instruments reserves the right to change or discontinue this product without notice.

Copyright © 1983 by Texas Instruments Incorporated

390- X 584-PIXEL CCD AREA IMAGE SENSOR

D2732, MARCH 1983

- 390- X 584-Pixel Format for Frame-Store or Full-Frame Mode Operation
- Virtual Phase (VP), Front Side Illuminated
- Buried Channel Registers
- High Charge Transfer Efficiency
- High Resolution
- Interlaced 625-line TV Output (Frame-Store Mode)
- No Residual Imaging
- No Microphonics
- Small Size, 11-mm Image Diagonal (Frame-Store Mode)
- No Image Burn-in
- High Uniformity

CERAMIC PACKAGE
(TOP VIEW)

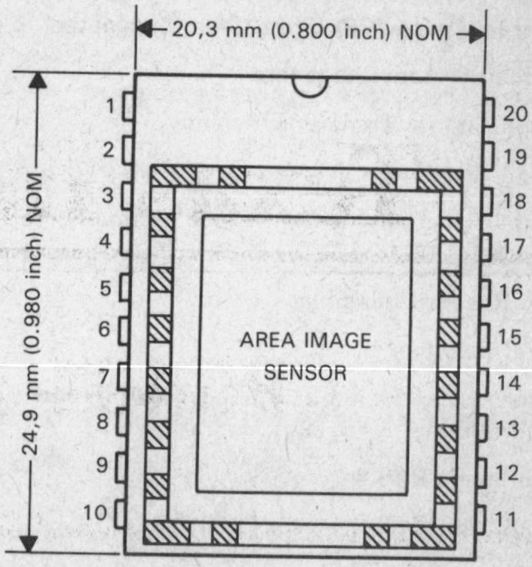

description

This 390- X 584-pixel area sensor is designed to operate in the frame-store mode as a 390H X 292V imager for 625-line European TV applications. The device can also be used in the full-frame mode for long-integration, single-frame applications. Charge packets are transported to a precharged diode whose potential changes in response to the quantity of the signal charge delivered. This potential is applied to an on-chip floating diffusion amplifier to produce a signal voltage at the video output pin. A reset transistor is used to recharge the charge-detector-diode capacitance before the arrival of each new signal charge packet from the serial transport register. The imager is fabricated using virtual phase MOS technology, which provides greater reliability at lower cost than the conventional 2-phase MOS technologies.

The device is supplied in a 20-pin dual-in-line ceramic side-braze package designed for insertion in mounting-hole rows on 20,3-mm (0.800-inch) centers. The pins in each row are located on 2,54-mm (0.100-inch) centers. Optical quality glass lids cover the image area.

Availablity of this device is scheduled for 3rd quarter 1983

PRODUCT PREVIEW
This document contains information on a product under development. Texas Instruments reserves the right to change or discontinue this product without notice.

Copyright © 1983 by Texas Instruments Incorporated

TEXAS INSTRUMENTS
INCORPORATED
POST OFFICE BOX 225012 • DALLAS, TEXAS 75265

Texas Instruments

PC401 and PC402

EVALUATION BOARDS

for operating
Virtual Phase
Linear CCD Sensors
TC101 (1728 X 1)
TC102 (128 X 1)
TC103 (2048 X 1)
TC104 (3456 X 1)

TEXAS INSTRUMENTS
OPTOELECTRONICS
P.O. Box 225012, MS12
(214) 995-3821
Dallas, Texas 75265

EVALUATION BOARD SCHEMATIC

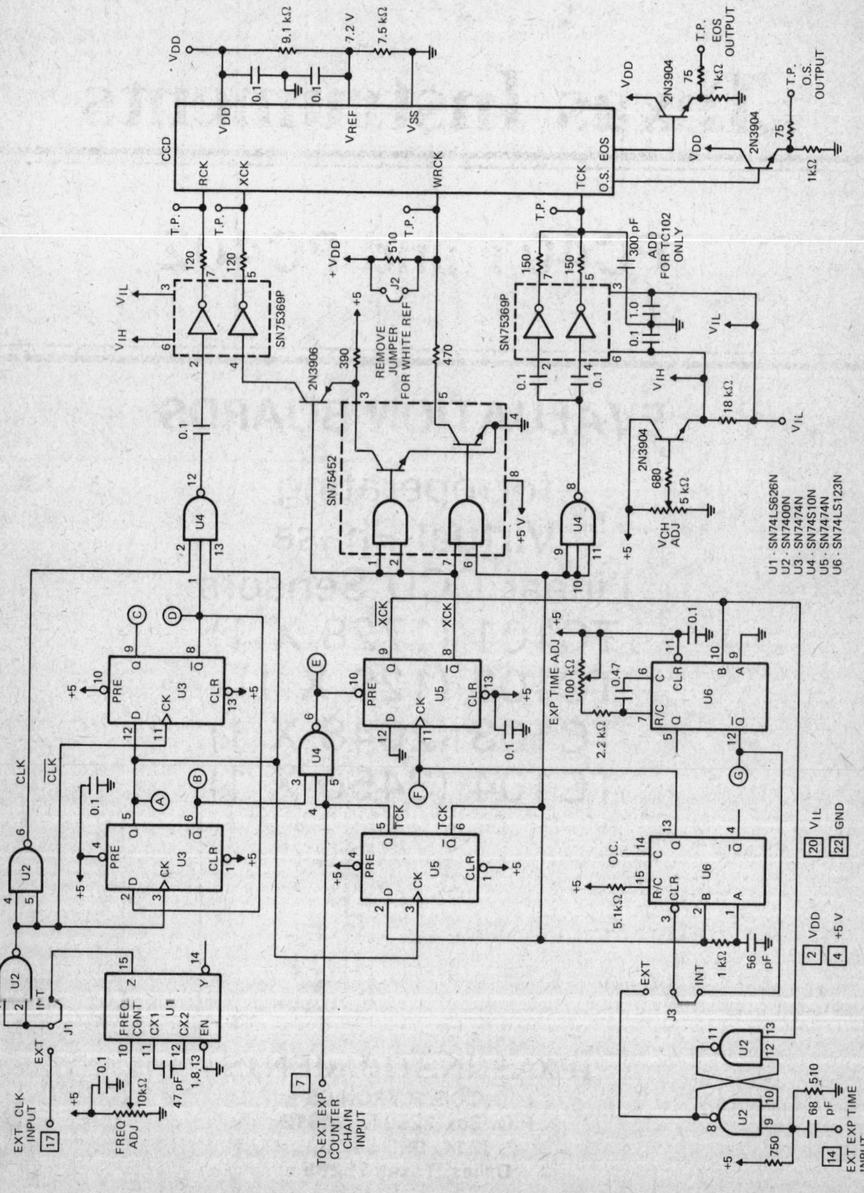

LOGIC TIMING WAVEFORM

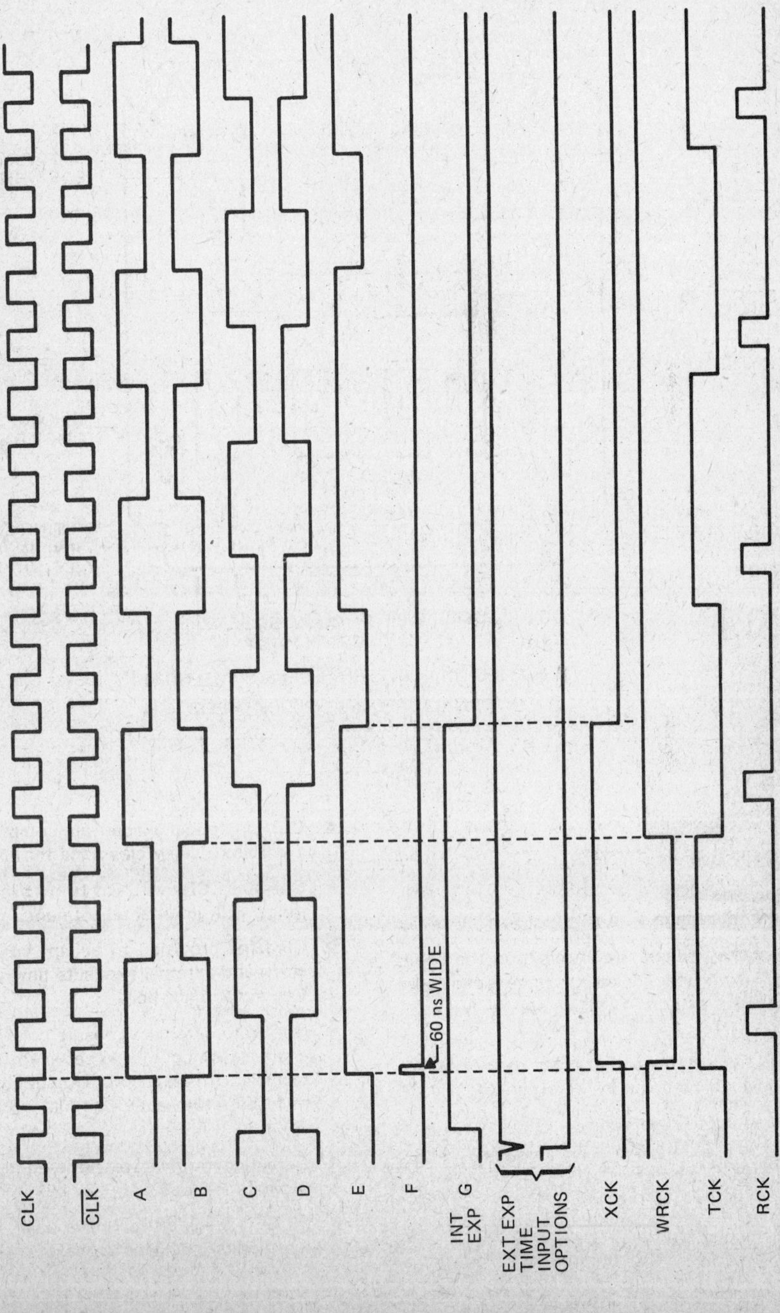

LINEAR CCD EVALUATION BOARD

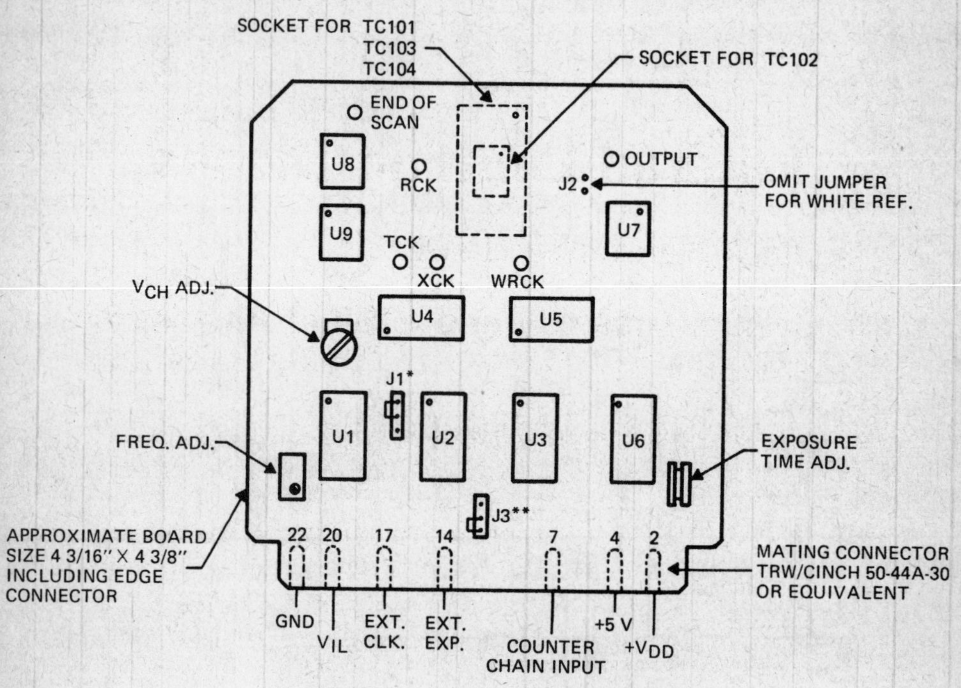

(J1 & J3 SHOWN IN INTERNAL SELECT POSITION)

* J1 = INTERNAL/EXTERNAL CLOCK SELECT

** J3 = INTERNAL/EXTERNAL EXPOSURE TIME SELECT

FEATURES:

- PC401 operates TC101, TC103, and TC104.
- PC402 operates TC102.
- Operates CCD over a 0.2 MHz to 2.0 MHz data rate range.
- HI clock voltage is controlled on the board (V_{CH}) while the LO clock voltage is controlled from the external negative supply (V_{IL}).
- White reference control allows evaluation of this clock's injection stability as well as the signal's elimination in the output.
- Three supply operation:
  +5 V @ 165 mA
  V_{DD} @ 15 mA*
  V_{IL}(R,T,X) nom @ 40 mA*

  *Values from DATA SHEETS
- CCD on opposite side from components and controls to allow clearance for optics.
- Operates CCD over 2- to 16-ms exposure time range with internal adjustment.
- Contains provision to accept both external clock and external exposure time with jitter-free synchronization.
- Output to external exposure control counter chain allows counter to set exposure time in multiples of TCK periods. Output from counter must be returned to external exposure time input.
- Output emitter follower buffered to drive high capacitance load.
- All CCD clock signals have easily accessed test points for scope probes.

Infrared-Emitting Diodes

- Quick Reference Guide
- Gallium Arsenide and Gallium Aluminum Arsenide
- Low-Cost Plastic Packages
  - T-1
  - T-1¾
  - Sidelookers
- Hermetically Sealed Packages
  - Pill
  - TO-18
- High-Reliability Devices (HR2)
  - Pill
  - TO-18
- Measuring the Output of IREDs and LEDs

See Section 4 for Special Function Infrared-Emitting Diodes.

3 IR EMITTERS

INFRARED EMITTERS
QUICK REFERENCE GUIDE

DEVICE	POWER OUTPUT MIN @ mW	I_F mA	θ_{HI}	V_F MAX @ V	I_F mA	λ_p TYP nm	FEATURES
TIL23	0.4	50	35°	1.5	50	940	Pill package for mounting on double-sided printed circuit boards. Compatible with TIL601 Series
TIL24†	1	50	35°	1.5	50	940	
TIL25	0.75	50	35°	1.5	50	940	
TIL31B†	3.3	100	10°	1.75	100	940	Hermetically sealed TO-18 package
TIL32	0.5	20	35°	1.6	20	940	Low-cost plastic package
TIL33B	2.5	100	80°	1.75	100	940	Hermetically sealed TO-18 package
TIL34B	1.6	100	10°	1.75	100	940	Hermetically sealed TO-18 package
TIL38	6	100	50°	1.75	100	940	Low-cost plastic package T 1¾ package
TIL39	6	100	20°	1.75	100	940	Low-cost plastic T 1¾ package
TIL40	0.05	20	30°	1.6	20	940	Low-cost plastic sidelooker package
TIL902-1	1.5	20	35°	1.6	20	880	Low-cost plastic T 1 package
TIL902-2	2.5	20	35°	1.6	20	880	Mechanically similar to TIL32
TIL903-1	6	100	10°	2.1	100	880	Hermetically sealed TO-18 package
TIL903-2	9	100	10°	2.1	100	880	Mechanically similar to TIL31B
TIL904-1	5	100	80°	2.1	100	880	Hermetically sealed TO-18 package
TIL904-2	9	100	80°	2.1	100	880	Mechanically similar to TIL33B
TIL905-1	1.5	20	50°	1.6	20	880	Low-cost plastic T 1¾ package
TIL905-2	2.5	20	50°	1.6	20	880	Mechanically similar to TIL38
TIL906-1	1.5	20	20°	1.6	20	880	Low-cost plastic T 1¾ package
TIL906-2	2.5	20	20°	1.6	20	880	Mechanically similar to TIL39

†High-reliability versions (TIL24HR2 and TIL31BHR2) are also available.

D2132, FEBRUARY 1970—REVISED FEBRUARY 1983

DESIGNED TO EMIT NEAR-INFRARED RADIATION WHEN FORWARD BIASED

- Output Spectrally Compatible with Silicon Sensors
- High Power Efficiency
- High Power Output
- Small Size Permits Matrix Assembly Directly into Printed Circuit Boards
- High Radiant Intensity
- TIL24HR2* Includes High-Reliability Processing and Lot Acceptance (See page 3-7 for Summary of Processing

mechanical data

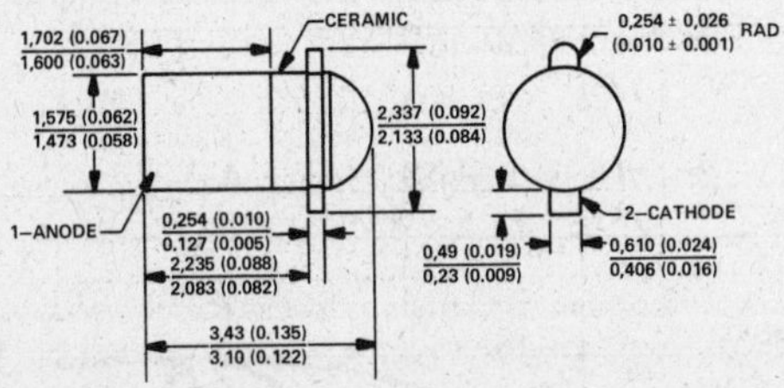

4 TIMES
ACTUAL SIZE

ALL LINEAR DIMENSIONS ARE IN MILLIMETERS AND PARENTHETICALLY IN INCHES

absolute maximum ratings

Reverse Voltage at 25°C Case Temperature	2 V
Continuous Forward Current at 25°C Case Temperature (See Note 1)	100 mA
Operating Case Temperature Range	−65°C to 125°C
Storage Temperature Range	−65°C to 150°C
Soldering Temperature (10 seconds)	240°C

*All electrical and mechanical specifications for the TIL24 also apply for TIL24HR2.

NOTE 1: Derate linearly to 125°C case temperature at the rate of 1 mA/°C. For pulsed operation at higher currents, see Figures 8 and 9.

Copyright © 1983 by Texas Instruments Incorporated

operating characteristics at 25°C case temperature

PARAMETER		TEST CONDITIONS	TIL23			TIL24			TIL25			UNIT
			MIN	TYP	MAX	MIN	TYP	MAX	MIN	TYP	MAX	
P_O	Radiant Power Output	I_F = 50 mA	0.4			1			0.75			mW
λ_p	Wavelength at Peak Emission		915	940	975	915	940	975	915	940	975	nm
$\Delta\lambda$	Spectral Bandwidth			50	75		50	75		50	75	nm
θ_{HI}	Half-Intensity Beam Angle			35°			35°			35°		
V_F	Static Forward Voltage			1.25	1.5			1.5			1.5	V

TYPICAL CHARACTERISTICS

RELATIVE SPECTRAL CHARACTERISTICS

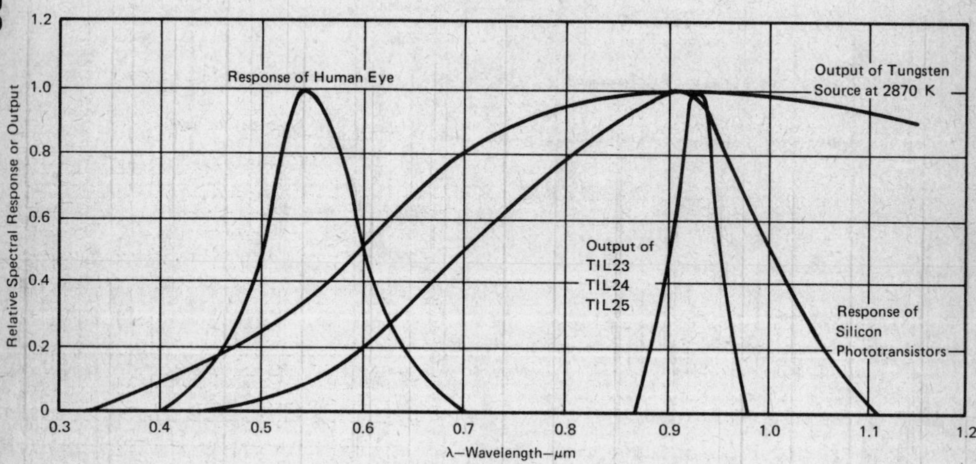

FIGURE 1

TYPICAL CHARACTERISTICS

RELATIVE PHOTON INTENSITY
vs
ANGULAR DISPLACEMENT

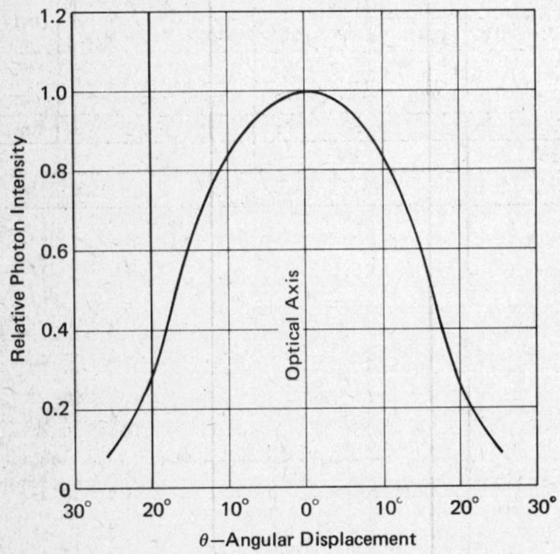

FIGURE 2

RELATIVE POWER OUTPUT
vs
CASE TEMPERATURE

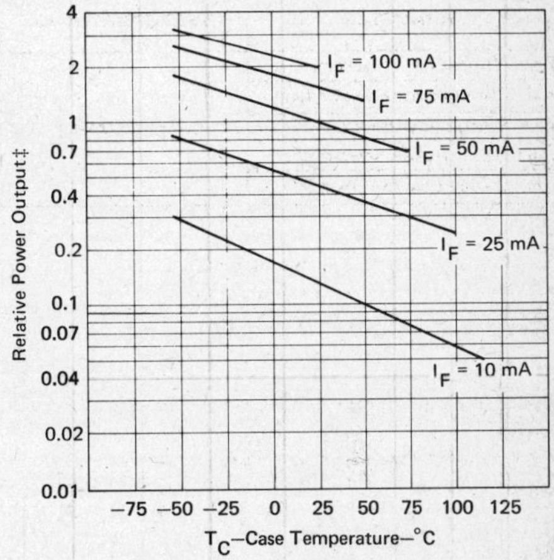

FIGURE 3

3
IR EMITTERS

CHANGE IN WAVELENGTH OF PEAK INTENSITY
vs
CASE TEMPERATURE

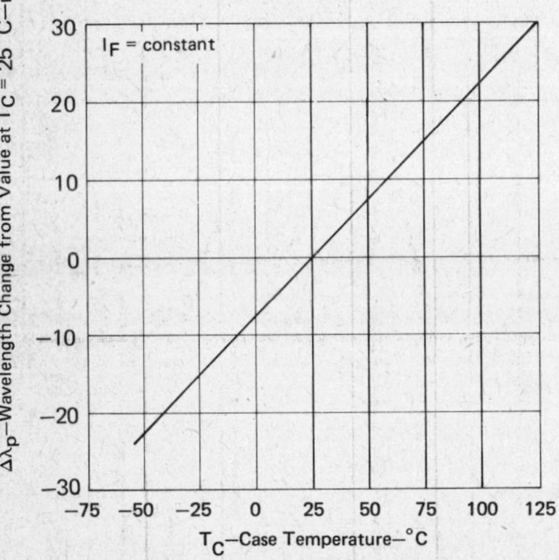

FIGURE 4

RELATIVE POWER OUTPUT
vs
FORWARD CURRENT

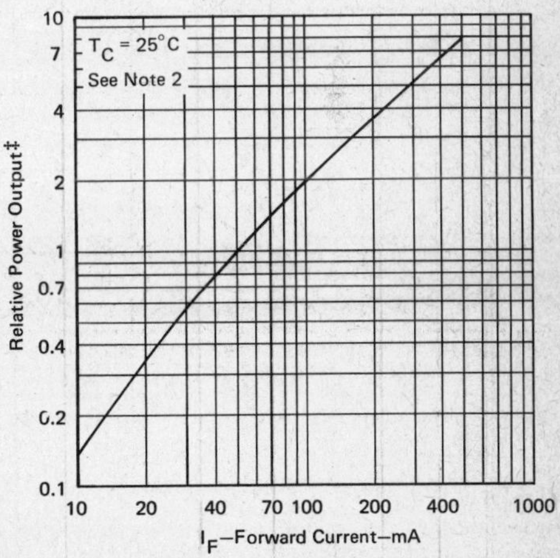

FIGURE 5

NOTE 2: These parameters must be measured using pulse techniques: t_w = 0.04 ms, duty cycle ≤ 10%.
‡Normalized to output at I_F = 50 mA, T_C = 25°C.

TYPICAL CHARACTERISTICS

TIL23

FORWARD CONDUCTION CHARACTERISTICS

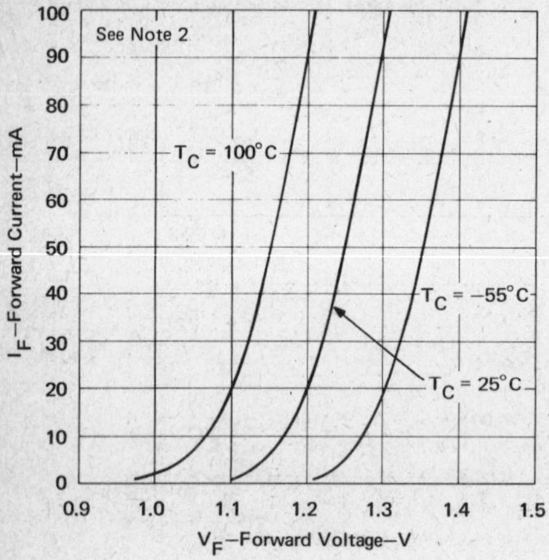

FIGURE 6

COUPLING CHARACTERISTICS OF TIL23, TIL24, AND TIL25 WITH TIL602

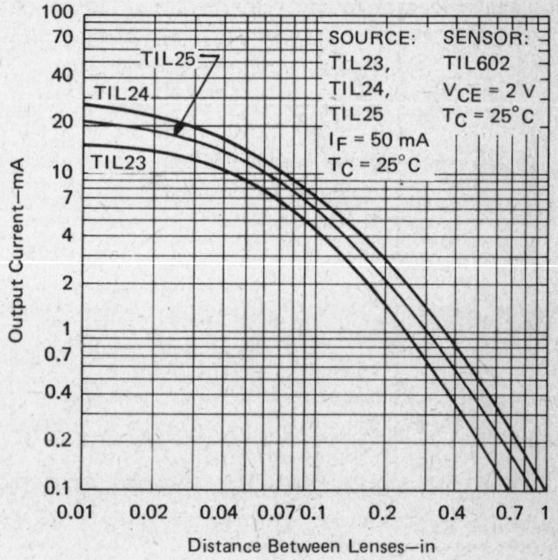

FIGURE 7

NOTE 2: These parameters must be measured using pulse techniques: t_w = 0.04 ms, duty cycle ⩽ 10%.

THERMAL CHARACTERISTICS

MAXIMUM PULSE WIDTH
vs
FORWARD CURRENT

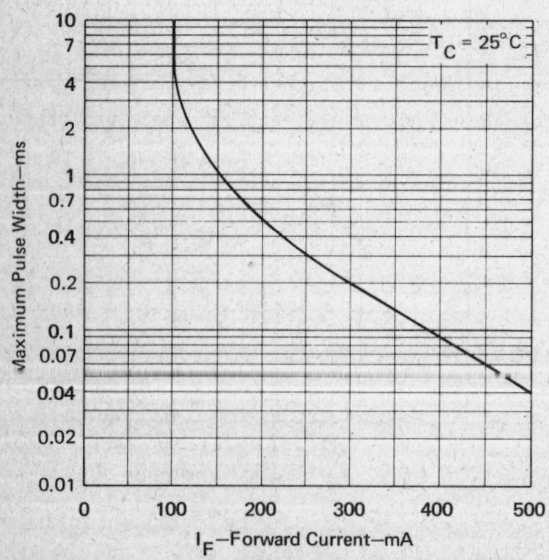

FIGURE 8

MAXIMUM DUTY CYCLE
vs
FORWARD CURRENT

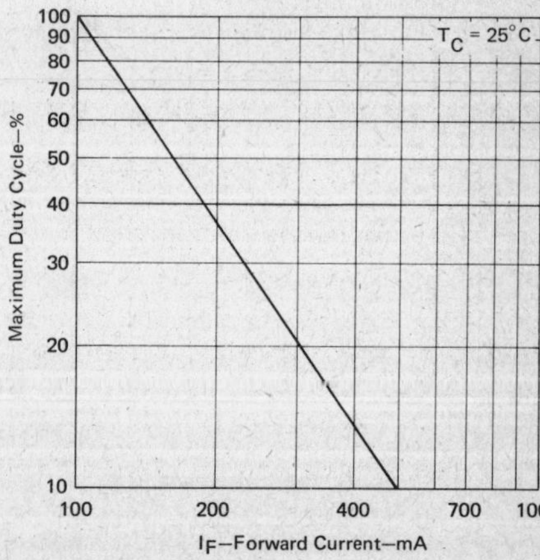

FIGURE 9

- **This processing applies only to devices ordered under the part number TIL24HR2**
- **For electrical and mechanical specifications, refer to page 3-3**

This processing and lot acceptance follows the sequence of tests in MIL-S-19500 for JANTX types. This is not to be construed to be a JANTX-qualified part. A detail specification is available upon request through your TI Field Sales Office or Optoelectronics Marketing at the following address:

Texas Instruments Incorporated
Optoelectronics Marketing
P.O. Box 225012, MS 12
Dallas, Texas 75265
Phone: (214) 995-3821

TEST	MIL-STD-750 TEST METHOD
100% Processing	
Storage: T_A = 125°C, t = 24 h	1032
Temperature Cycle: −55°C to 125°C, 10 cycles	1051
Constant Acceleration: 20,000 G, Y_1 axis	2006
Power Burn-in: I_F = 50 mA, t = 168 h	1039
Hermetic Seal, Fine	1071 Cond. G or H
Hermetic Seal, Gross	1071 Cond. C or D
External Visual	2071
Product Acceptance	
Group A: LTPD = 5	
External Visual	2071
Electrical: T_A = 25°C	per detail spec
Group B-1: LTPD = 15	
Solderability	2026
Group B-2: LTPD = 10	
Thermal Shock	1051 Cond. B-1
Hermetic Seal, Fine	1071 Cond. G or H
Hermetic Seal, Gross	1071 Cond. C or D
Group B-3: LTPD = 5	
Steady-State Operating Life: t = 340 h	1027
Group B-4:	
Decap, Internal Visual; Design Verification	
1 Device/0 Failure	2075
Bond Strength LTPD = 20 (C = 0)	2037 Cond. A
Group B-5: Not Applicable	
Group B-6: LTPD = 7	
High-Temperature Life (Nonoperating) t = 340 h	1032

TEST	MIL-STD-750 TEST METHOD
(Group C Tests are run on one lot every six months)	
Group C-1: LTPD = 15	
Physical Dimensions	2066
Group C-2: LTPD = 10	
Thermal Shock (Glass Strain)	1056 Cond. A
Hermetic Seal, Fine	1071 Cond. G or H
Hermetic Seal, Gross	1071 Cond. C or D
Moisture Resistance	1021
External Visual	2071
Group C-3: LTPD = 10	
Shock: 1500 G	2016
Vibration: 50 G	2056
Acceleration: 20000 G	2006
Group C-4: LTPD = 15	
Salt Atmosphere	1041
Group C-5: Not Applicable	
Group C-6: λ = 10	
Steady-State Operating Life: t = 1000 h	1026

D1934, NOVEMBER 1974—REVISED FEBRUARY 1983

DESIGNED TO EMIT NEAR-INFRARED RADIATION WHEN FORWARD BIASED

- "B" Versions Especially Designed for Low Degradation and are Direct Replacements for the "A" Versions
- Spectrally and Mechanically Compatible with TIL81 and TIL99 Phototransistors
- Typical Applications Include Card Readers, Encoders, Intrusion Alarms, Sector Sensors, Level Indicators, and Beginning-of-Tape/End-of-Tape Indicators
- TIL31HR2* Includes High-Reliability Processing and Lot Acceptance (See Page 3-11 for Summary of Processing)

mechanical data

Each device is in a hermetically sealed welded case similar to JEDEC TO-18 with window. The TIL31B and TIL34B have convex lenses while that of the TIL33B is essentially flat. A coin header is used to increase dissipation capability. All TO-18 registration notes also apply to this outline. Approximate weight is 0.35 gram.

THE CATHODE IS IN ELECTRICAL CONTACT WITH THE CASE

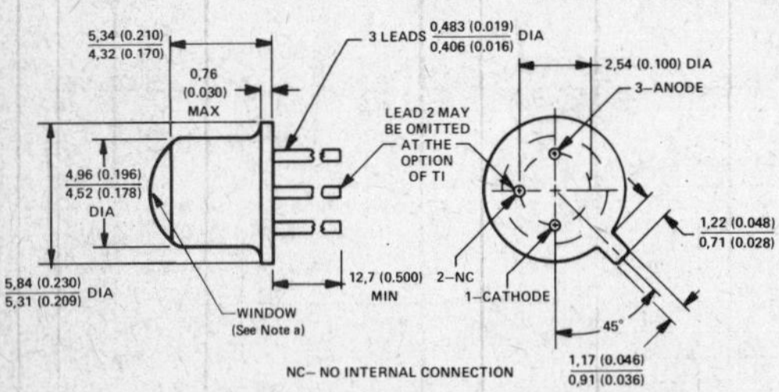

NOTE:
a. The window is Corning 7052 glass or equivalent.
Window dimensions are:
Minimum diameter . . 3,5 (0.140)
Maximum extension from top of case
TIL31B and TIL34B . . . 1,02 (0.040)
TIL33B 0,26 (0.010)

ALL LINEAR DIMENSIONS ARE IN MILLIMETERS AND PARENTHETICALLY IN INCHES

CL-18

3 IR EMITTERS

*On the original TIL31, TIL33, and TIL34, the anode was in electrical contact with the case. Lead 2, which had no internal connection, is omitted on the B-suffix versions.

absolute maximum ratings

Reverse Voltage at 25°C Case Temperature . . . . . 5 V
Continuous Forward Current at 25°C Case Temperature (See Note 1) . . . . . 200 mA
Operating Case Temperature Range . . . . . −65°C to 150°C
Storage Temperature Range . . . . . −65°C to 150°C
Lead Temperature 1,6 mm (1/16 Inch) from Case for 10 Seconds . . . . . 240°C

operating characteristics at 25°C case temperature

PARAMETER		TEST CONDITIONS	TIL31B			TIL33B			TIL34B			UNIT
			MIN	TYP	MAX	MIN	TYP	MAX	MIN	TYP	MAX	
P_O	Radiant Power Output	I_F = 100 mA	3.3	6		2.5	5		2	3		mW
λ_p	Wavelength at Peak Emission		915	940	975	915	940	975	915	940	975	nm
$\Delta\lambda$	Spectral Bandwidth			50	75		50	75		50	75	nm
θ_{HI}	Half-Intensity Beam Angle			10°			80°			10°		
V_F	Static Forward Voltage			1.4	1.75		1.4	1.75		1.4	1.75	V
t_r	Radiant Pulse Rise Time†	I_{FM} = 100 mA, $t_w \geq 5\ \mu s$		600			600			600		ns
t_f	Radiant Pulse Fall Time†			350			350			350		

*All electrical and mechanical specifications for the TIL24 also apply for TIL24HR2.
†Radiant pulse rise time is the time required for a change in radiant intensity from 10% to % of its peak value for a step change in current; radiant pulse fall time is the time required for a change in radiant intensity from 90% to 10% of its peak value for a step change in current.
NOTE 1: Derate linearly to 150°C case temperature at the rate of 1.6 mA/°C.

Copyright © 1983 by Texas Instruments Incorporated

TYPICAL CHARACTERISTICS

RELATIVE SPECTRAL CHARACTERISTICS

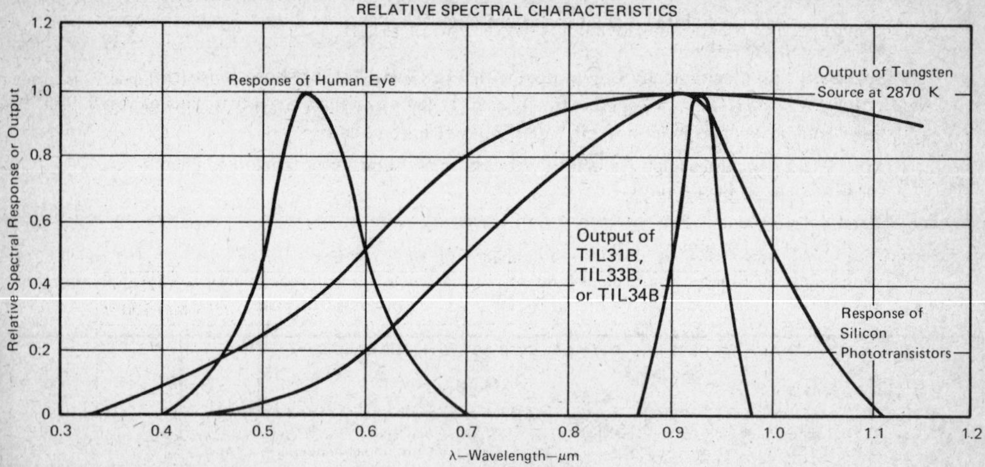

FIGURE 1

RELATIVE POWER OUTPUT vs CASE TEMPERATURE

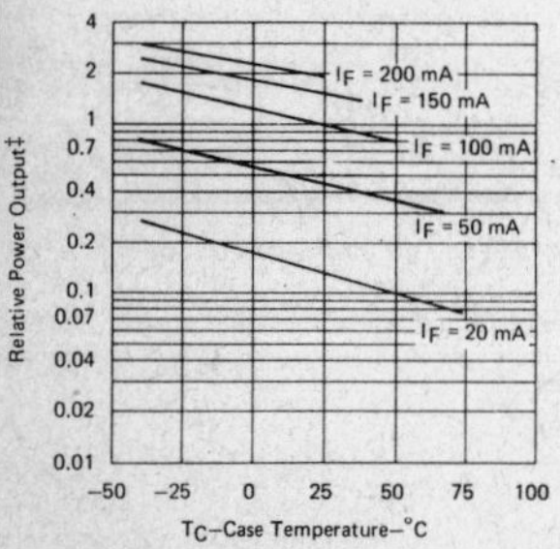

FIGURE 2

RELATIVE POWER OUTPUT vs FORWARD CURRENT

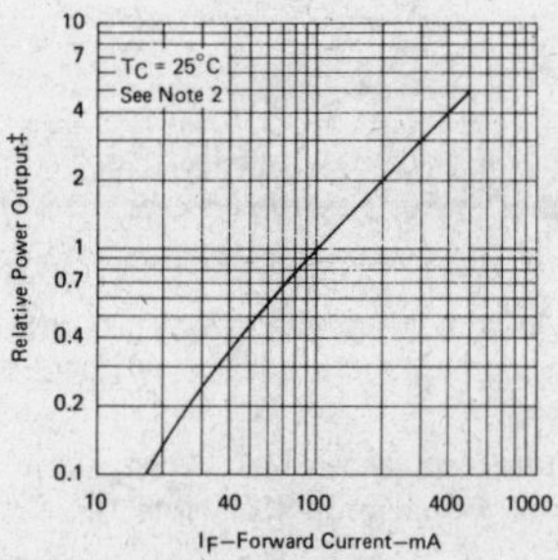

FIGURE 3

CHANGE IN WAVELENGTH AT PEAK EMISSION vs CASE TEMPERATURE

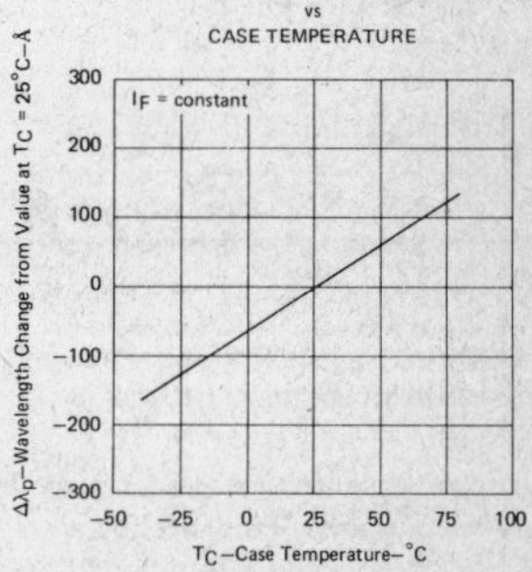

FIGURE 4

TIL31B, TIL34B
RELATIVE RADIANT INTENSITY vs ANGULAR DISPLACEMENT

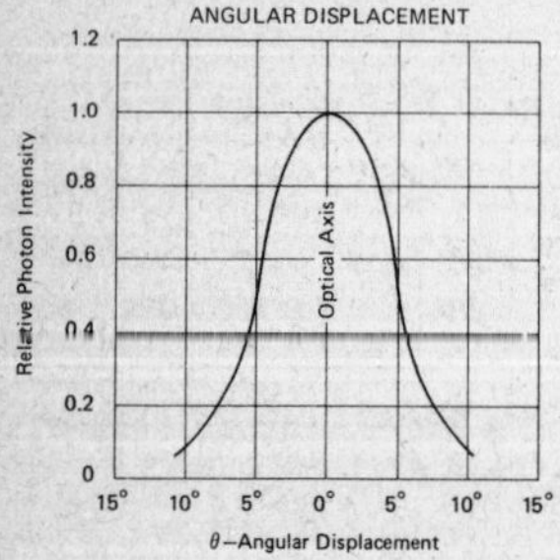

FIGURE 5

TIL33B
RELATIVE RADIANT INTENSITY vs ANGULAR DISPLACEMENT

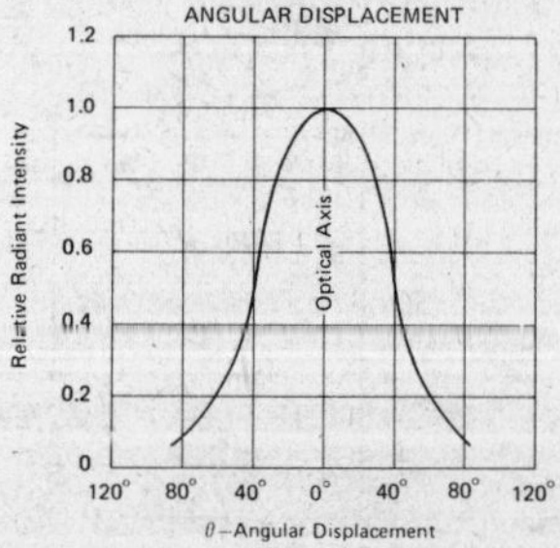

FIGURE 6

FORWARD CONDUCTION CHARACTERISTICS

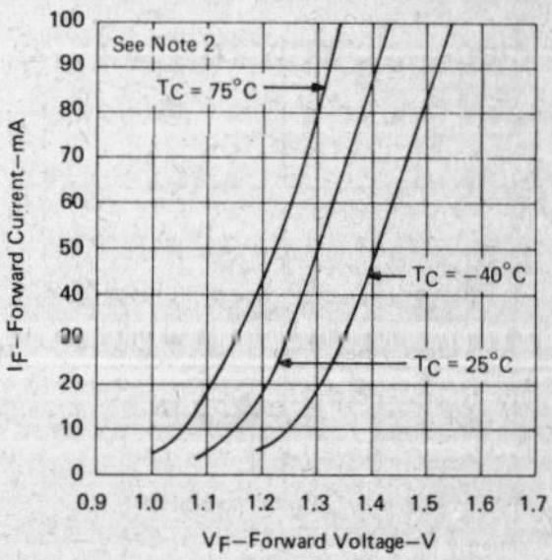

FIGURE 7

NOTE 2: This parameter must be measured using pulse techniques. t_w = 0.04 ms, duty cycle ⩽ 10%.

‡Normalized to output at I_F = 10 mA, T_C = 25°C.

- **This processing applies only to devices ordered under the part number TIL31BHR2**
- **For electrical and mechanical specifications, refer to page 3-9**

This processing and lot acceptance follows the sequence of tests in MIL-S-19500 for JANTX types. This is not to be construed to be a JANTX-qualified part. A detail specification is available upon request through your TI Field Sales Office or Optoelectronics Marketing at the following address:

Texas Instruments Incorporated
Optoelectronics Marketing
P.O. Box 225012, MS 12
Dallas, Texas 75265
Phone: (214) 995-3821

TEST	MIL-STD-750 TEST METHOD
100% Processing	
Storage: T_A = 125°C, t = 24 h	1032
Temperature Cycle: −55°C to 125°C, 10 cycles	1051
Constant Acceleration: 20,000 G, Y_1 axis	2006
Power Burn-in: I_F = 100 mA, t = 168 h	1039
Hermetic Seal, Fine	1071 Cond. G or H
Hermetic Seal, Gross	1071 Cond. C or D
External Visual	2071
Product Acceptance	
Group A: LTPD = 5	
External Visual	2071
Electrical: T_A = 25°C	per detail spec
Group B-1: LTPD = 15	
Solderability	2026
Resistance to Solvents	1022
Group B-2: LTPD = 10	
Thermal Shock	1051 Cond. B-1
Hermetic Seal, Fine	1071 Cond. G or H
Hermetic Seal, Gross	1071 Cond. C or D
Group B-3: LTPD = 5	
Steady-State Operating Life: t = 340 h	1027
Group B-4:	
Decap, Internal Visual; Design Verification 1 Device/0 Failure	2075
Bond Strength LTPD = 20 (C = 0)	2037 Cond. A
Group B-5: Not Applicable	
Group B-6: LTPD = 7	
High-Temperature Life (Nonoperating) t = 340 h	1032

TEST	MIL-STD-750 TEST METHOD
(Group C Tests are run on one lot every six months)	
Group C-1: LTPD = 15	
Physical Dimensions	2066
Group C-2: LTPD = 10	
Thermal Shock (Glass Strain)	1056 Cond. A
Terminal Strength	2036 Cond. E
Hermetic Seal, Fine	1071 Cond. G or H
Hermetic Seal, Gross	1071 Cond. C or D
Moisture Resistance	1021
External Visual	2071
Group C-3: LTPD = 10	
Shock: 1500 G	2016
Vibration: 50 G	2056
Acceleration: 20000 G	2006
Group C-4: LTPD = 15	
Salt Atmosphere	1041
Group C-5: Not Applicable	
Group C-6: λ = 10	
Steady-State Operating Life: t = 1000 h	1026

TEXAS INSTRUMENTS
INCORPORATED
POST OFFICE BOX 225012 • DALLAS, TEXAS 75265

P-N GALLIUM ARSENIDE INFRARED-EMITTING DIODE

D1855, SEPTEMBER 1971—REVISED APRIL 1983

DESIGNED TO EMIT NEAR-INFRARED RADIATION WHEN FORWARD BIASED

- Output Spectrally Compatible with Silicon Sensors (e.g., TIL78)
- High Power Efficiency
- High Power Output
- High Radiant Intensity
- Plastic Package with Two Leads for Ease of Handling

mechanical data

This device has a tinted molded plastic body similar in size to lamp style T-1.

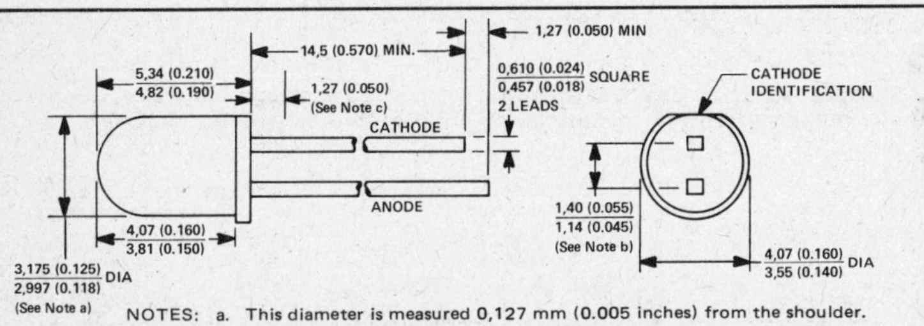

NOTES: a. This diameter is measured 0,127 mm (0.005 inches) from the shoulder.
b. Lead spacing is measured where the leads emerge from the package.

ALL DIMENSIONS ARE IN MILLIMETERS AND PARENTHETICALLY IN INCHES

CL-9

absolute maximum ratings

Reverse Voltage at 25°C Free-Air Temperature . . . . 2 V
Continuous Forward Current at (or below) 25°C Free-Air Temperature (See Note 1) . . . . 40 mA
Operating Free-Air Temperature Range . . . . −40°C to 100°C
Storage Temperature Range . . . . −40°C to 125°C
Lead Temperature 1,6 mm (1/16 Inch) from Case for 5 Seconds . . . . 240°C

operating characteristics at 25°C free-air temperature

	PARAMETER	TEST CONDITIONS	MIN	TYP	MAX	UNIT
P_O	Radiant Power Output	I_F = 20 mA	0.5	1.2		mW
λ_p	Wavelength at Peak Emission		915	940	975	nm
$\Delta\lambda$	Spectral Bandwidth			50	75	nm
θ_{HI}	Half-Intensity Beam Angle			35°		
V_F	Static Forward Voltage			1.2	1.6	V
t_r	Radiant Pulse Rise Time†	I_{FM} = 40 mA, $t_w \geq 5$ μs		600		ns
t_f	Radiant Pulse Fall Time†			350		

†Radiant pulse rise time is the time required for a change in radiant power output from 10% to 90% of its peak value for a step change in current; radiant pulse fall time is the time required for a change in radiant power output from 90% to 10% of its peak value for a step change in current.

NOTE 1: Derate linearly to 100°C free-air temperature at the rate of 0.53 mA/°C.

Copyright © 1983 by Texas Instruments Incorporated

483

TYPICAL CHARACTERISTICS

RELATIVE SPECTRAL CHARACTERISTICS

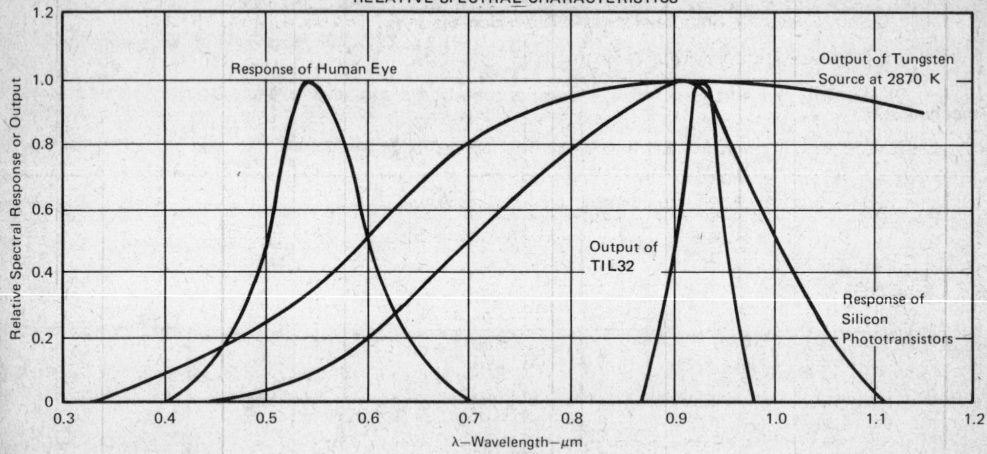

FIGURE 1

RELATIVE POWER OUTPUT vs FREE-AIR TEMPERATURE

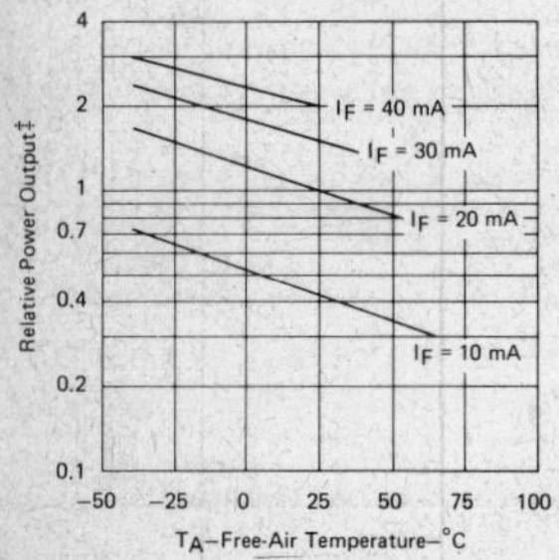

FIGURE 2

RELATIVE POWER OUTPUT vs FORWARD CURRENT

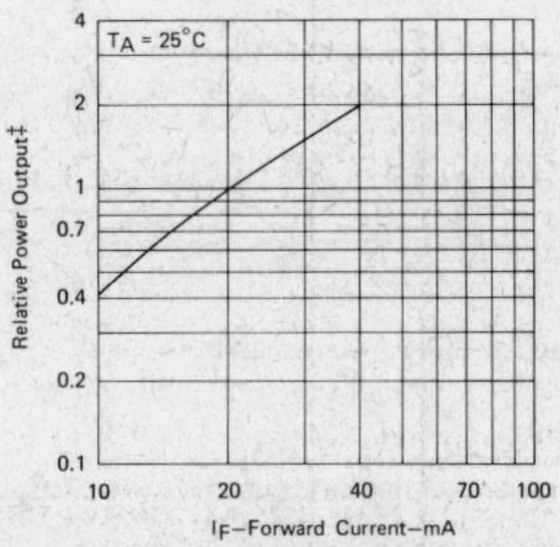

FIGURE 3

CHANGE IN WAVELENGTH OF PEAK INTENSITY vs FREE-AIR TEMPERATURE

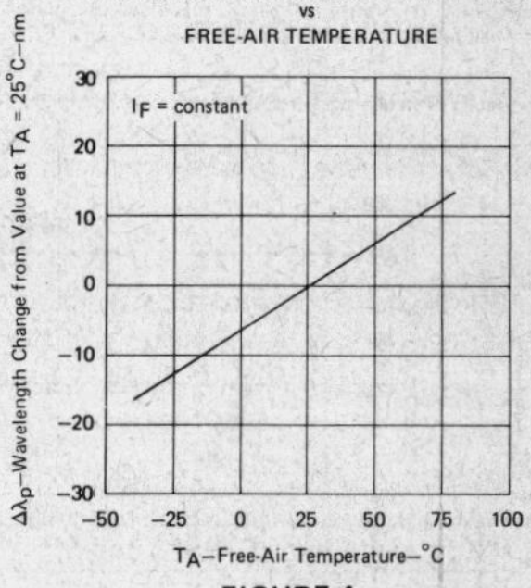

FIGURE 4

RELATIVE PHOTON INTENSITY vs ANGULAR DISPLACEMENT

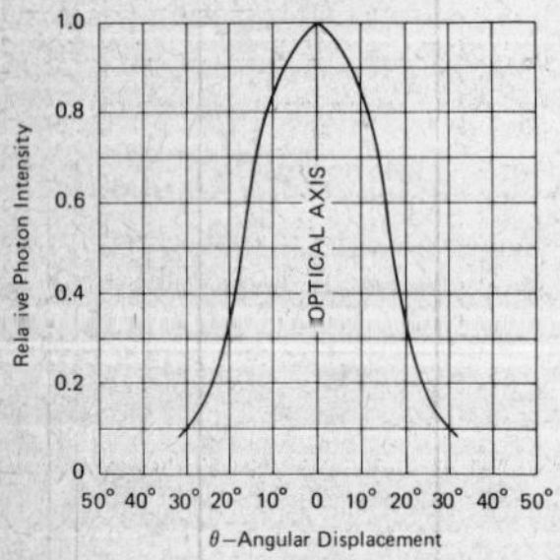

FIGURE 5

COUPLING CHARACTERISTICS OF TIL32 WITH TIL78

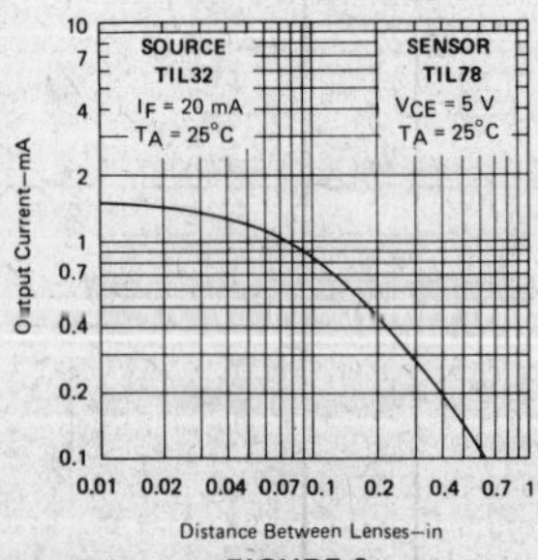

FIGURE 6

FORWARD CONDUCTION CHARACTERISTICS

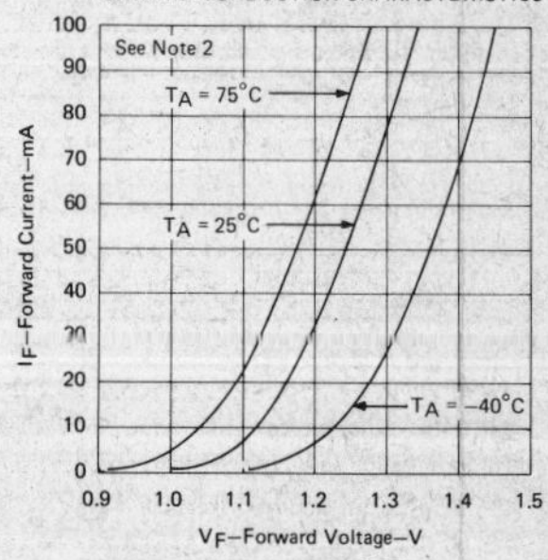

FIGURE 7

NOTE 2: This parameter must be measured using pulse techniques: t_w = 0.04 ms, duty cycle ≤ 10%.

‡Normalized to Output at I_F = 20 mA, T_A = 25°C.

D2594, JULY 1980—REVISED APRIL 1983

DESIGNED TO EMIT NEAR-INFRARED RADIATION

- **Output Spectrally Compatible with Silicon Sensors (e.g., TIL100, TIL413, TIL414)**
- **High Power Output with a Beam Angle of 50°**

mechanical data

This device has a tinted molded plastic body similar in size to lamp style T-1¾.

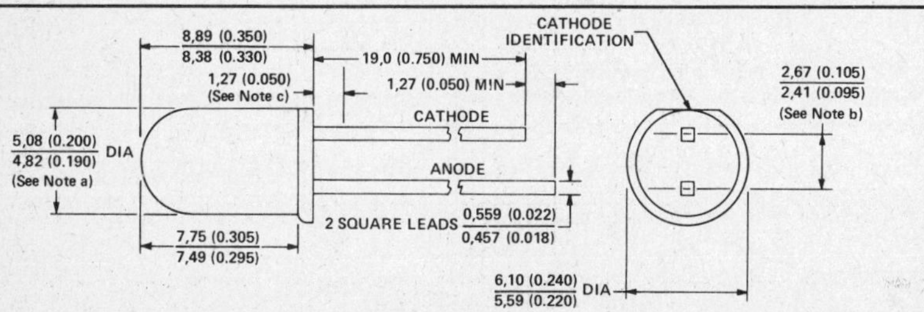

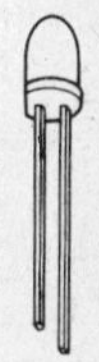

NOTES: a. This diameter is measured 0,127 mm (0.005 inch) from the package shoulder.
b. Lead spacing is measured where the leads emerge from the package.
c. Within this zone the package contour is not controlled.

ALL DIMENSIONS ARE IN MILLIMETERS AND PARENTHETICALLY IN INCHES

CL-10

absolute maximum ratings at 25°C free-air temperature (unless otherwise noted)

Reverse Voltage	5 V
Continuous Forward Current at (or below) 25°C Free-Air Temperature (See Note 1)	100 mA
Peak Forward Current (See Note 2)	2 A
Operating Free-Air Temperature Range	−40°C to 80°C
Storage Temperature	−40°C to 100°C
Lead Temperature 1,6 mm (1/16 inch) from Case for 5 Seconds	240°C

operating characteristics at 25°C free-air temperature

PARAMETER		TEST CONDITIONS	MIN	TYP	MAX	UNIT
P_O	Radiant Power Output	I_F = 100 mA, See Note 3	6	8		mW
I_e	Axial Radiant Intensity†	I_F = 100 mA, See Note 3		15		mW/sr
λ_p	Wavelength at Peak Emission	I_F = 20 mA	915	940	975	nm
$\Delta\lambda$	Spectral Bandwidth Between Half-Power Points	I_F = 20 mA		50	75	nm
θ_{HI}	Emission Beam Angle Between Half-Intensity Points	I_F = 20 mA		50°		
V_F	Static Forward Voltage	I_F = 100 mA		1.4	1.75	V
V_F	Static Forward Voltage	I_F = 1 A, t_w = 10 μs, duty cycle ≤ 1%		2.55		V
C	Capacitance	V_F = 0, f = 1 MHz		25		pF
t_r	Radiant Pulse Rise Time‡	I_{FM} = 100 mA, t_w ≥ 5 μs		600		ns
t_f	Radiant Pulse Fall Time‡	I_{FM} = 100 mA, t_w ≥ 5 μs		350		ns

†Axial radiant intensity is measured over 0.1 steradian on the mechanical axis. One steradian is the solid angle at the center of a sphere subtended by a portion of the surface area equal to the square of the radius of the sphere. There are 4 π steradians in a complete sphere.

‡Radiant pulse rise time is the time required for a change in radiant power output from 10% to 90% of its peak value for a step change in current; radiant pulse fall time is the time required for a change in radiant power output from 90% to 10% of its peak value for a step change in current.

NOTES: 1. Derate linearly to 80°C free-air temperature at the rate of 1.82 mA/°C.
2. This value applies for t_w ≤ 10 μs, f ≤ 1 kHz. See Figure 1.
3. These parameters must be measured using pulse techniques, t_w = 10 ms, duty cycle ≤ 1%.

Copyright © 1983 by Texas Instruments Incorporated

TEXAS INSTRUMENTS
INCORPORATED
POST OFFICE BOX 225012 • DALLAS, TEXAS 75265

ABSOLUTE MAXIMUM RATINGS

PEAK FORWARD CURRENT
vs
PULSE WIDTH

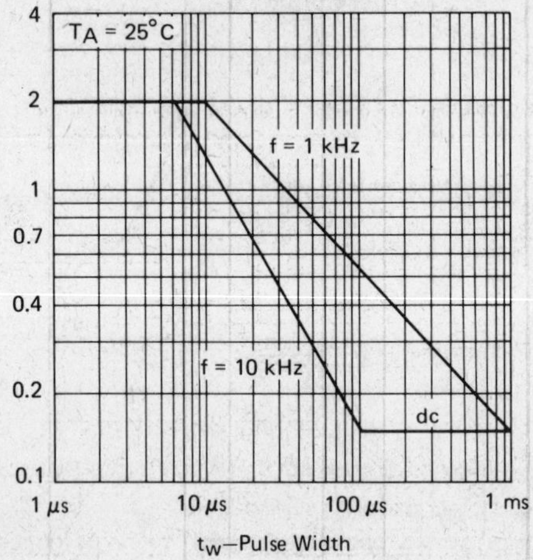

FIGURE 1

TYPICAL CHARACTERISTICS

FORWARD CONDUCTION CHARACTERISTICS

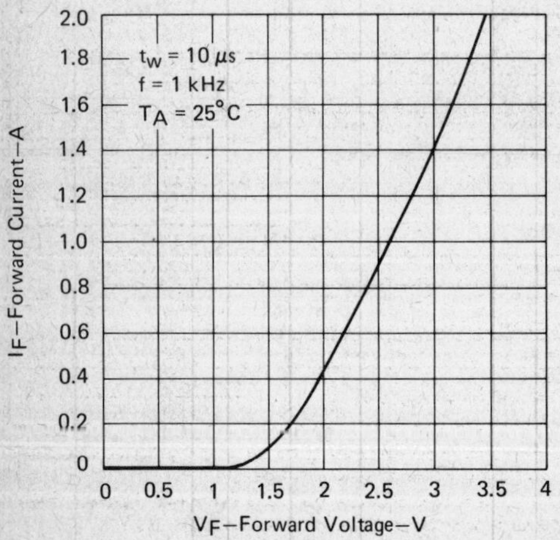

FIGURE 2

AXIAL RADIANT INTENSITY
vs
PEAK FORWARD CURRENT

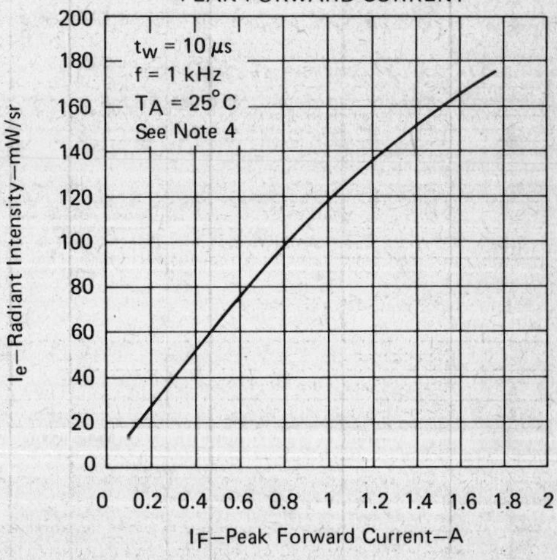

FIGURE 3

NOTE 4: Axial radiant intensity is measured over 0.01 steradian on the mechanical axis.

TEXAS INSTRUMENTS
INCORPORATED
POST OFFICE BOX 225012 • DALLAS, TEXAS 75265

D2594, JULY 1980—REVISED APRIL 1983

DESIGNED TO EMIT NEAR-INFRARED RADIATION

- **Output Spectrally Compatible with Silicon Sensors (e.g., TIL100, TIL413, TIL414)**
- **High Power Output with a Beam Angle of 20°**

mechanical data

This device has a tinted molded plastic body similar in size to lamp style T-1¾

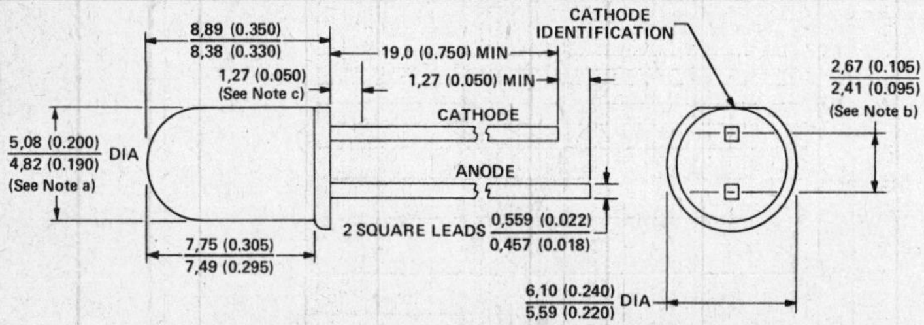

NOTES: a. This diameter is measured 0,127 mm (0.005 inch) from the package shoulder.
b. Lead spacing is measured where the leads emerge from the package.
c. Within this zone the package contour is not controlled.

ALL DIMENSIONS ARE IN MILLIMETERS AND PARENTHETICALLY IN INCHES

CL-10

absolute maximum ratings at 25°C free-air temperature (unless otherwise noted)

Reverse Voltage . . . . 5 V
Continuous Forward Current at (or below) 25°C Free-Air Temperature (See Note 1) . . . . . . 100 mA
Peak Forward Current (See Note 2) . . . . 2 A
Operating Free-Air Temperature Range . . . . −40°C to 80°C
Storage Temperature . . . . −40°C to 100°C
Lead Temperature 1,6 mm (1/16 inch) from Case for 5 Seconds . . . . 240°C

operating characteristics at 25°C free-air temperature

	PARAMETER	TEST CONDITIONS	MIN	TYP	MAX	UNIT
P_O	Radiant Power Output	I_F = 100 mA, See Note 3	6	8		mW
	Axial Power Output into a 10° Cone	I_F = 20 mA, See Note 4		150		μW
I_e	Axial Radiant Intensity†	I_F = 100 mA, See Note 3		35		mW/sr
λ_p	Wavelength at Peak Emission	I_F = 20 mA	915	940	975	nm
$\Delta\lambda$	Spectral Bandwidth Between Half-Power Points	I_F = 20 mA		50	75	nm
θ_{HI}	Emission Beam Angle Between Half-Intensity Points	I_F = 20 mA		20°		
V_F	Static Forward Voltage	I_F = 100 mA		1.4	1.75	V
		I_F = 1 A, t_w = 10 μs, duty cycle ≤ 1%		2.55		V
C	Capacitance	V_F = 0, f = 1 MHz		25		pF
t_r	Radiant Pulse Rise Time‡	I_{FM} = 100 mA, t_w ≥ 5 μs		600		ns
t_f	Radiant Pulse Fall Time‡	I_{FM} = 100 mA, t_w ≥ 5 μs		350		ns

†Axial radiant intensity is measured over 0.1 steradian on the mechanical axis. One steradian is the solid angle at the center of a sphere subtended by a portion of the surface area equal to the square of the radius of the sphere. There are 4 π steradians in a complete sphere.

‡Radiant pulse rise time is the time required for a change in radiant power output from 10% to 90% of its peak value for a step change in current; radiant pulse fall time is the time required for a change in radiant power output from 90% to 10% of its peak value for a step change in current.

NOTES: 1. Derate linearly to 80°C free-air temperature at the rate of 1.82 mA/°C.
2. This value applies for t_w ≤ 10 μs, f ≤ 1 kHz. See Figure 1.
3. These parameters must be measured using pulse techniques, t_w = 10 ms, duty cycle ≤ 1%.
4. The nominal 10° cone is defined by an aperture that has a diameter of 6,76 mm (0.266 inch) and is located 38.6 mm (1.52 inch) from the lens side of the flange.

Copyright © 1983 by Texas Instruments Incorporated

ABSOLUTE MAXIMUM RATINGS

PEAK FORWARD CURRENT
vs
PULSE WIDTH

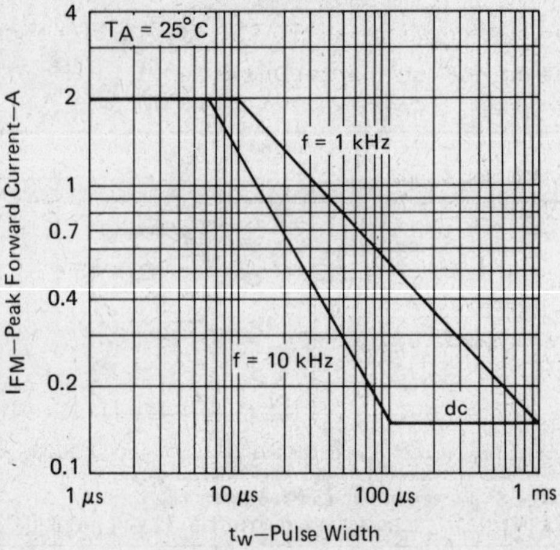

FIGURE 1

TYPICAL CHARACTERISTICS

FORWARD CONDUCTION CHARACTERISTICS

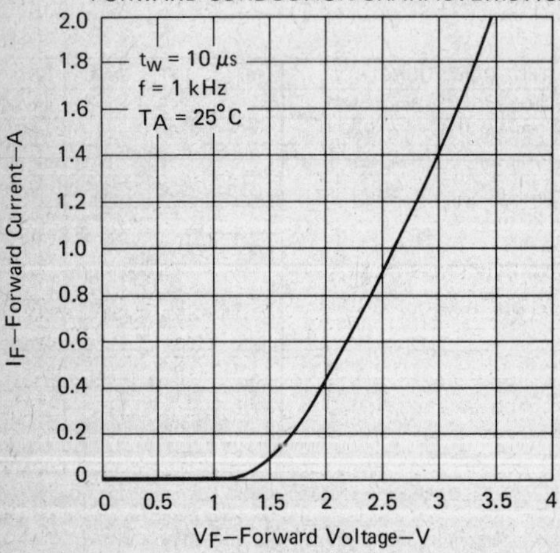

FIGURE 2

AXIAL RADIANT INTENSITY
vs
PEAK FORWARD CURRENT

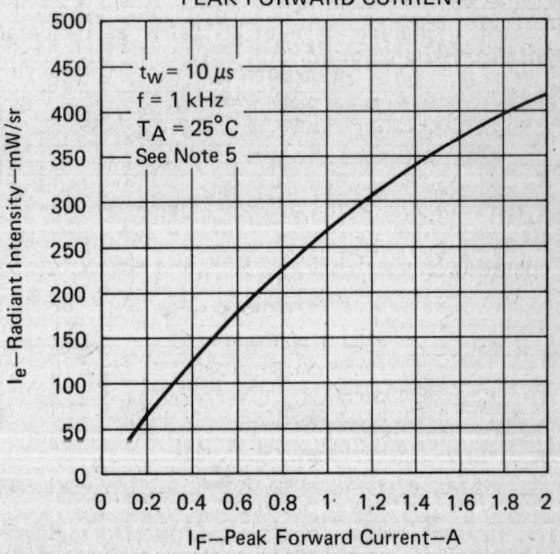

FIGURE 3

NOTE 5: Radiant intensity is measured over 0.01 steradian on the mechanical axis.

P-N GALLIUM ARSENIDE INFRARED-EMITTING DIODE

D2558, JULY 1980—REVISED APRIL 1983

DESIGNED TO EMIT NEAR-INFRARED RADIATION WHEN FORWARD BIASED

- **Recommended for Applications Requiring Low-Cost Discrete Infrared Emitters**
- **Spectrally and Mechanically Compatible with TIL411, TIL412, TIL415, and TIL416.**
- **Designed for use in Housings or Printed Circuit Boards**

mechanical data

This device has a gray-tinted molded plastic body.

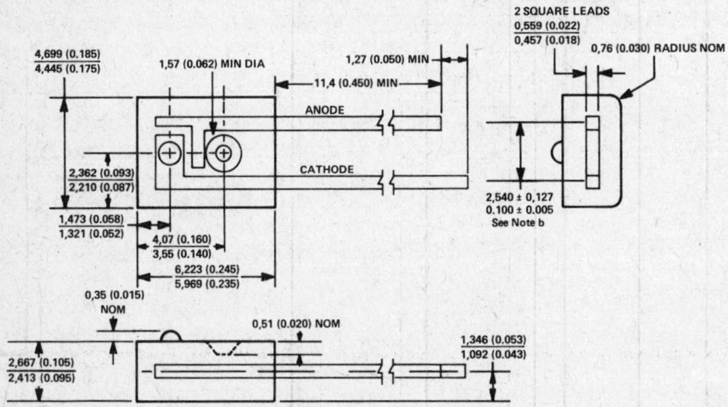

NOTES: a. All dimensions are in millimeters and parenthetically in inches.
b. Lead spacing is measured where the leads emerge from the package.

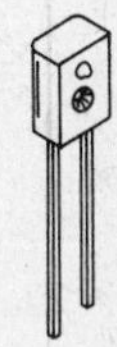

absolute maximum ratings

Reverse Voltage at 25°C Free-Air Temperature . . . . 2 V
Continuous Forward Current at (or below) 25°C Free-Air Temperature (See Note 1) . . . . 40 mA
Operating Free-Air Temperature Range . . . . −40°C to 80°C
Storage Temperature Range . . . . −40°C to 100°C
Lead Temperature 1,6 mm (1/16 Inch) from Case for 5 Seconds . . . . 240°C

operating characteristics at 25°C free-air temperature (unless otherwise noted)

PARAMETER		TEST CONDITIONS	MIN	TYP	MAX	UNIT
P_O	Radiant Power Output	$I_F = 20$ mA	50	100		μW
	Axial Power Output into a 10° Cone			10		μW
λ_p	Wavelength at Peak Emission			940		nm
$\Delta\lambda$	Spectral Bandwidth			50	75	nm
θ_{HI}	Half-Intensity Beam Angle			30°		
V_F	Static Forward Voltage			1.2	1.6	V
t_r	Radiant Pulse Rise Time†	$I_{FM} = 40$ mA, $t_w \geq 5$ μs		600		ns
t_f	Radiant Pulse Fall Time†			350		

†Radiant pulse rise time is the time for a change in radiant power output from 10% to 90% of its peak value for a step change in current; radiant pulse fall time is the time for a change in radiant power output from 90% to 10% of its peak value for a step change in current.

NOTE 1: Derate linearly to 80°C free-air temperature at the rate of 0.73 mA/°C.

Copyright © 1983 by Texas Instruments Incorporated

TYPICAL CHARACTERISTICS

RELATIVE SPECTRAL CHARACTERISTICS

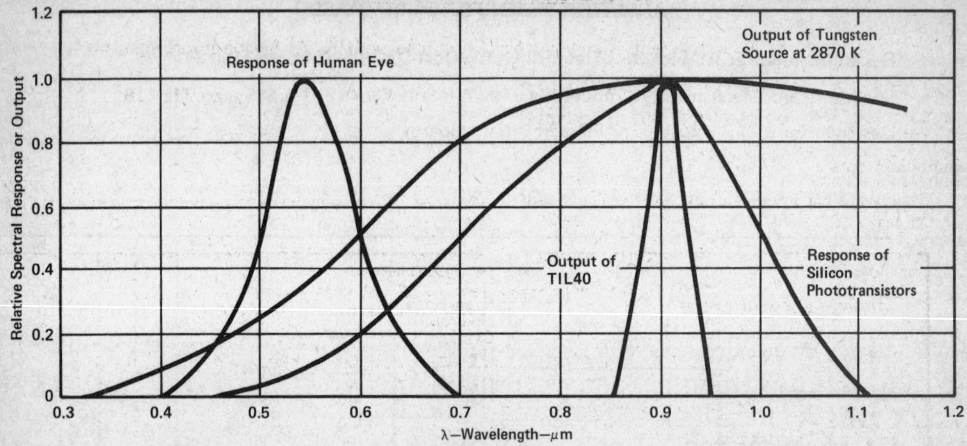

FIGURE 1

RELATIVE POWER OUTPUT
vs
FREE-AIR TEMPERATURE

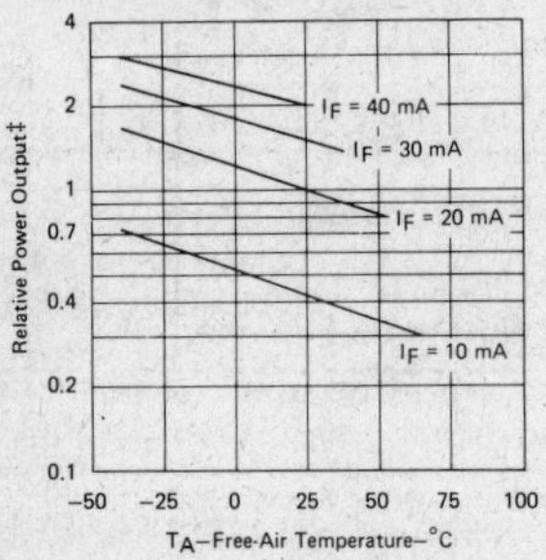

FIGURE 2

RELATIVE POWER OUTPUT
vs
FORWARD CURRENT

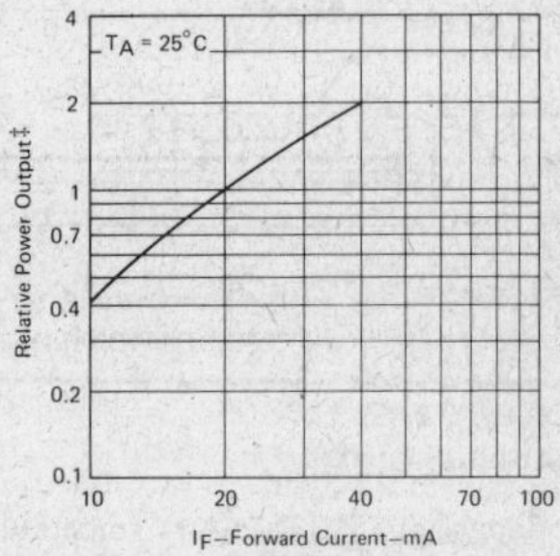

FIGURE 3

RELATIVE PHOTON INTENSITY
vs
ANGULAR DISPLACEMENT

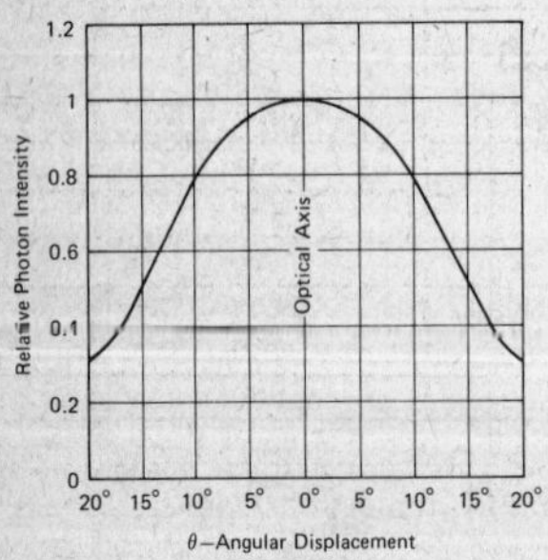

FIGURE 4

COUPLING CHARACTERISTICS
OF TIL40 WITH TIL411

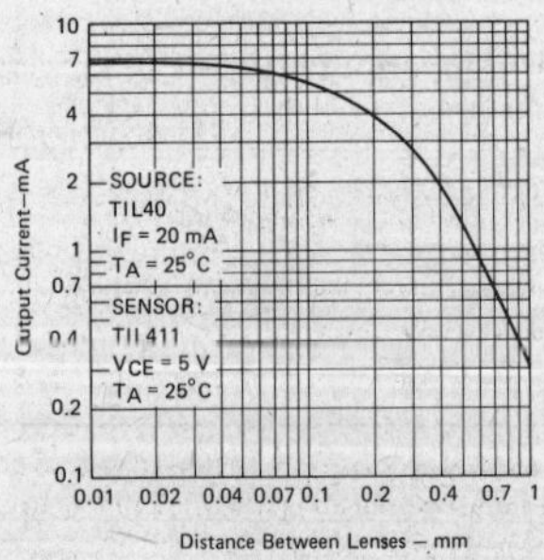

FIGURE 5

FORWARD CONDUCTION CHARACTERISTICS

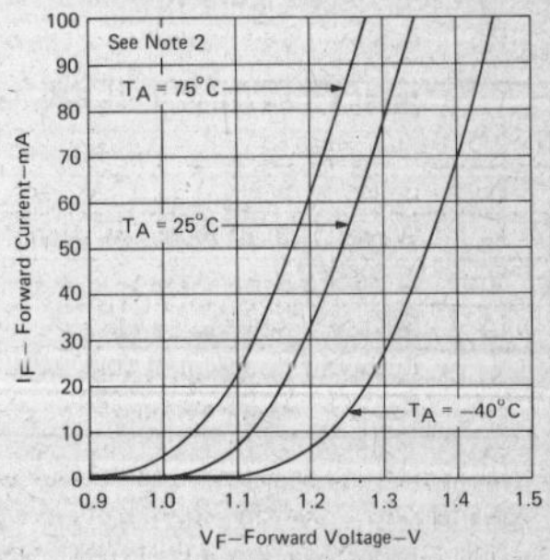

FIGURE 6

NOTE 2: This parameter must be measured using pulse techniques: t_w = 0.04 ms, duty cycle ≤ 10%.
‡Normalized to Output at I_F = 20 mA, T_A = 25°C.

TEXAS INSTRUMENTS
INCORPORATED
POST OFFICE BOX 225012 • DALLAS, TEXAS 75265

P-N GALLIUM ALUMINUM ARSENIDE INFRARED-EMITTING DIODE

D2699, MARCH 1983

DESIGNED TO EMIT NEAR-INFRARED RADIATION WHEN FORWARD BIASED

- **Spectrally and Mechanically Compatible with TIL78 Silicon Phototransistor**
- **High Power Output**
- **Low-Cost Plastic Package**

mechanical data

This device has a tinted plastic body similar in size to lamp style T-1 and may be panel mounted using mounting clip TILM1.

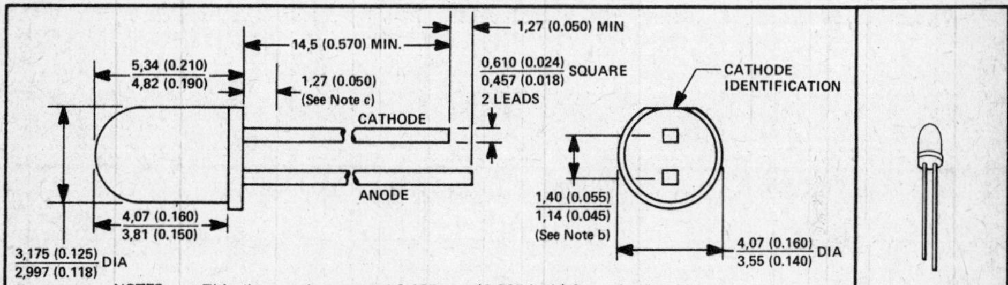

NOTES: a. This diameter is measured 0,127 mm (0.005 inch) from the shoulder.
b. Lead spacing is measured where the leads emerge from the package.
c. Within this zone package contour is not controlled.

ALL DIMENSIONS ARE IN MILLIMETERS AND PARENTHETICALLY IN INCHES

CL-9

absolute maximum ratings at 25°C free-air temperature (unless otherwise noted)

Reverse Voltage	3 V
Continuous Forward Current at (or below) 25°C Free-Air Temperature (See Note 1)	40 mA
Peak Forward Current (See Note 2)	2 A
Operating Free-Air Temperature Range	−40°C to 80°C
Storage Temperature Range	−40°C to 100°C
Lead Temperature 1,6 mm (1/16 inch) from Case for 5 Seconds	240°C

operating characteristics at 25°C free-air temperature

PARAMETER		TEST CONDITIONS		MIN	TYP	MAX	UNIT
P_O	Radiant Power Output	I_F = 20 mA	TIL902-1	1.5			mW
			TIL902-2	2.5			
P_A	Radiant Power Output into an Aperture (see Note 3)		TIL902-1		0.4		mW
			TIL902-2		0.7		
θ_{HI}	Emission Beam Angle Between Half-Intensity Points	I_F = 20 mA			30°		
λ_p	Wavelength at Peak Emission				880		nm
V_F	Static Forward Voltage				1.25	1.75	V
I_R	Reverse Current	V_R = 3 V				100	μA
t_r	Radiant Pulse Rise Time‡	I_{FM} = 20 mA, $t_w \geq 5\ \mu s$			600		ns
t_f	Radiant Pulse Fall Time‡				350		

‡ Radiant pulse rise time is the time required for a change in radiant power output from 10% to 90% of its peak value for a step change in current; radiant pulse fall time is the time required for a change in radiant power output from 90% to 10% of its peak value for a step change in current.

NOTES: 1. Derate linearly to 80°C free-air temperature at the rate of 0.73 mA/°C.
2. This value applies for $t_w \leq 10\ \mu s$, $f \leq 1$ kHz. See Figure 1.
3. The parameter is measured with an aperture angle of 30°. The symmetry line of the cone is the mechanical axis of the device under test. The apex of the cone coincides with the package shoulder.

Copyright © 1983 by Texas Instruments Incorporated

TEXAS INSTRUMENTS
INCORPORATED
POST OFFICE BOX 225012 • DALLAS, TEXAS 75265

ABSOLUTE MAXIMUM RATINGS

PEAK FORWARD CURRENT vs PULSE WIDTH

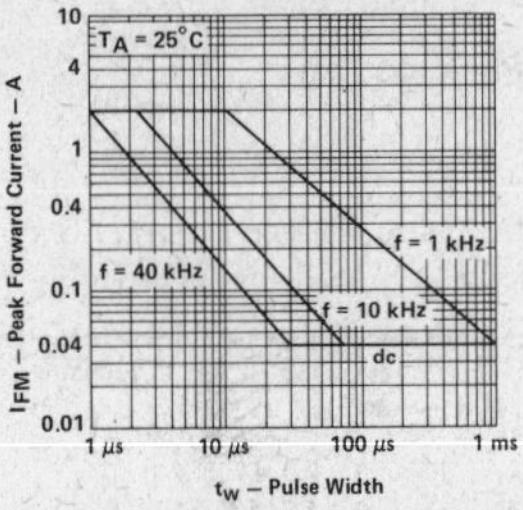

FIGURE 1

TYPICAL CHARACTERISTICS

RELATIVE POWER OUTPUT vs FORWARD CURRENT

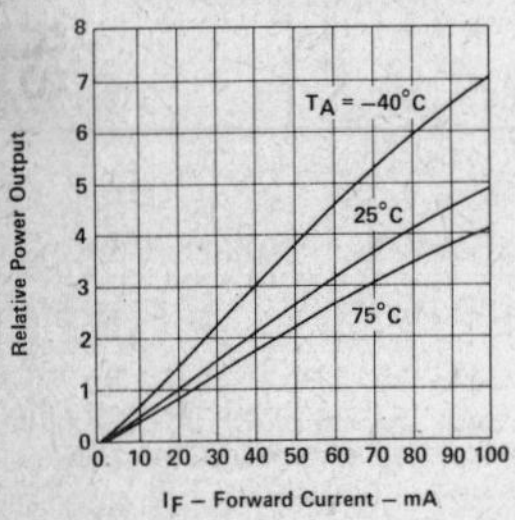

FIGURE 2

RADIANT POWER OUTPUT vs FORWARD CURRENT

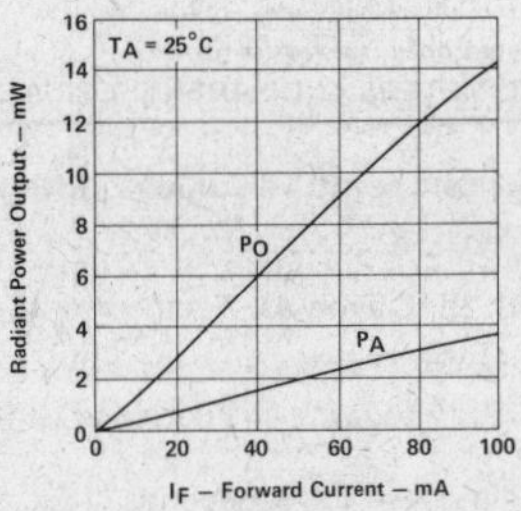

FIGURE 3

RELATIVE RADIANT INTENSITY vs ANGULAR DISPLACEMENT

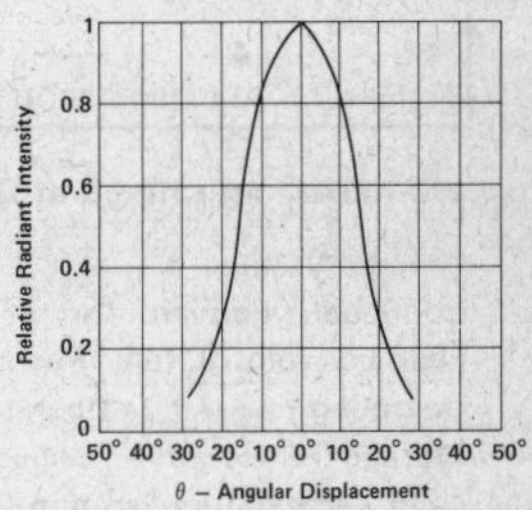

FIGURE 4

COUPLING CHARACTERISTICS OF TIL902 WITH TIL78

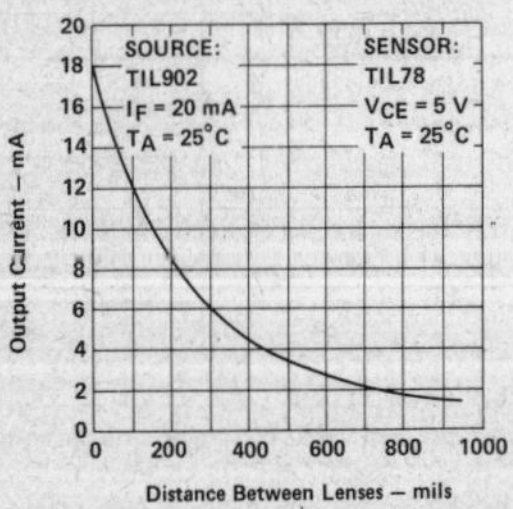

FIGURE 5

FORWARD CONDUCTION CHARACTERISTICS

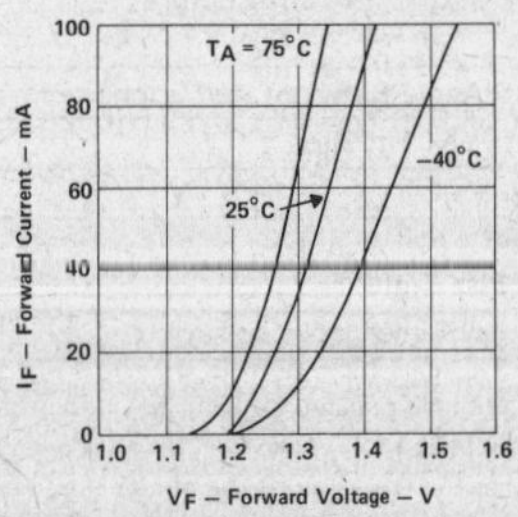

FIGURE 6

TEXAS INSTRUMENTS
INCORPORATED
POST OFFICE BOX 225012 • DALLAS, TEXAS 75265

TYPES TIL903, TIL904 GALLIUM ALUMINUM ARSENIDE INFRARED-EMITTING DIODES

D2719, FEBRUARY 1983

DESIGNED TO EMIT NEAR-INFRARED RADIATION WHEN FORWARD BIASED

- **Output Spectrally Compatible with Silicon Sensors**
- **Mechanically Compatible with TIL81 and TIL99**
- **Typical Applications Include Card Readers, Encoders, Intrusion Alarms, Sector Sensors, Level Indicators, and Beginning-of-Tape/End-of-Tape Indicators**

mechanical data

Each device is in a hermetically sealed welded case similar to JEDEC TO-18 with window. The TIL903 has a convex lens while that of the TIL904 is essentially flat.

All TO-18 registration notes also apply to this outline. Approximate weight is 0.35 gram.

THE ANODE IS IN ELECTRICAL CONTACT WITH THE CASE

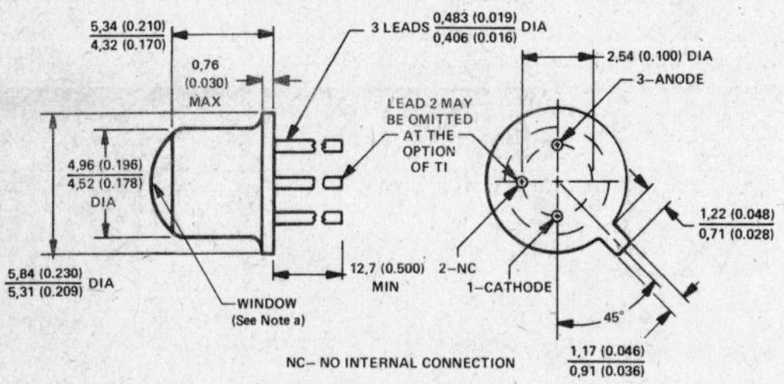

NOTE:

a. The window is Corning 7052 glass or equivalent.
   Window dimensions are:
   Minimum diameter ... 3,5 (0.140)
   Maximum extension from top of case
   TIL903 .... 1,02 (0.040)
   TIL904 .... 0,25 (0.010)

ALL LINEAR DIMENSIONS ARE IN MILLIMETERS AND PARENTHETICALLY IN INCHES

CL-18

absolute maximum ratings

Reverse Voltage at 25 °C Case Temperature .... 3 V
Continuous Forward Current at 25 °C Case Temperature (See Note 1) .... 200 mA
Operating Case Temperature Range .... −65 °C to 125 °C
Storage Temperature Range .... −65 °C to 150 °C
Lead Temperature 1,6 mm (1/16 inch) from Case for 5 Seconds .... 260 °C

operating characteristics at 25 °C case temperature

PARAMETER		TEST CONDITIONS		TIL903			TIL904			UNIT
				MIN	TYP	MAX	MIN	TYP	MAX	
P_O	Radiant Power Output	I_F = 100 mA	TIL903-1	6						mW
			TIL903-2	9						
			TIL904-1				5			
			TIL904-2				10			
λ_p	Wavelength at Peak Emission	I_F = 100 mA			880			880		nm
θ_{HI}	Half-Intensity Beam Angle				10°			80°		
V_F	Static Forward Voltage				1.5	2.1		1.5	2.1	V
I_R	Reverse Current	V_R = 3 V				100			100	μA
t_r	Radiant Pulse Rise Time†	I_{FM} = 100 mA, $t_w \geq 5\ \mu s$			600			600		ns
t_f	Radiant Pulse Fall Time†				350			350		

† Radiant pulse rise time is the time required for a change in radiant power output from 10% to 90% of its peak value for a step change in current; radiant pulse fall time is the time required for a change in radiant power output from 90% to 10% of its peak value for a step change in current.

NOTE 1: Derate linearly to 125 °C case temperature at the rate of 2.0 mA/°C.

Copyright © 1983 by Texas Instruments Incorporated

ABSOLUTE MAXIMUM RATINGS

PEAK FORWARD CURRENT
vs
PULSE WIDTH

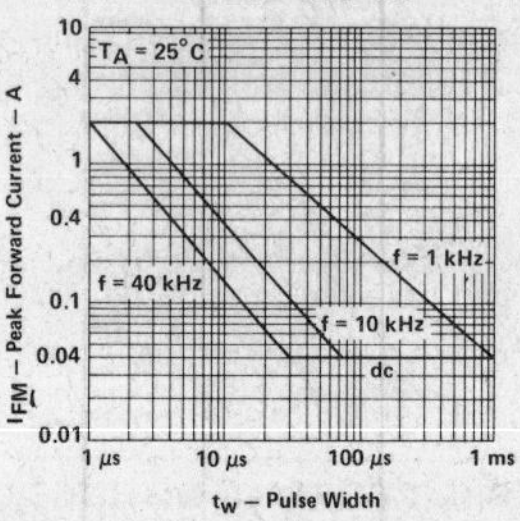

FIGURE 1

TYPICAL CHARACTERISTICS

RELATIVE POWER OUTPUT
vs
FORWARD CURRENT

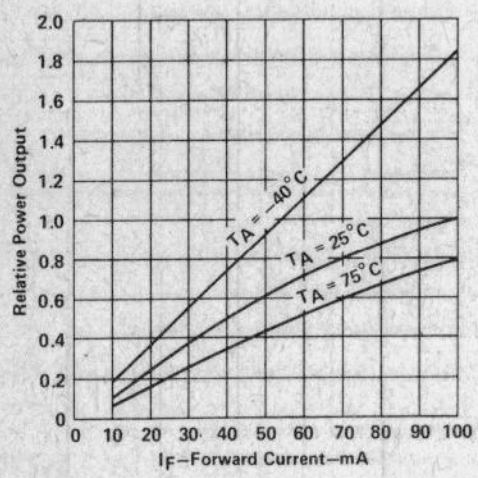

FIGURE 2

TIL903
RELATIVE RADIANT INTENSITY
vs
ANGULAR DISPLACEMENT

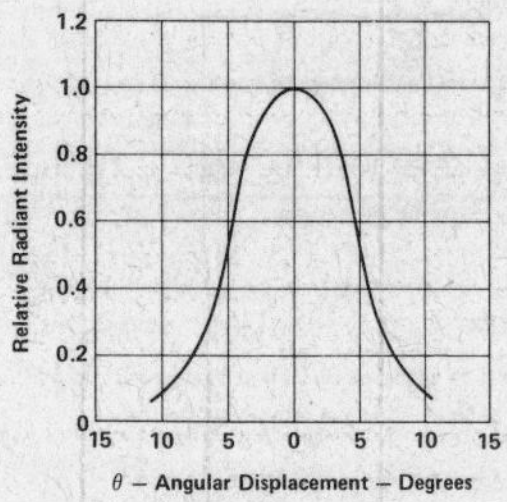

FIGURE 3

TIL904
RELATIVE RADIANT INTENSITY
vs
ANGULAR DISPLACEMENT

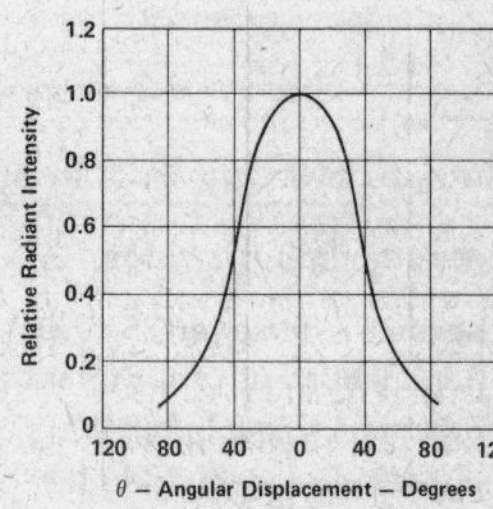

FIGURE 4

COUPLING CHARACTERISTICS
OF TIL81 WITH TIL903 AND TIL904

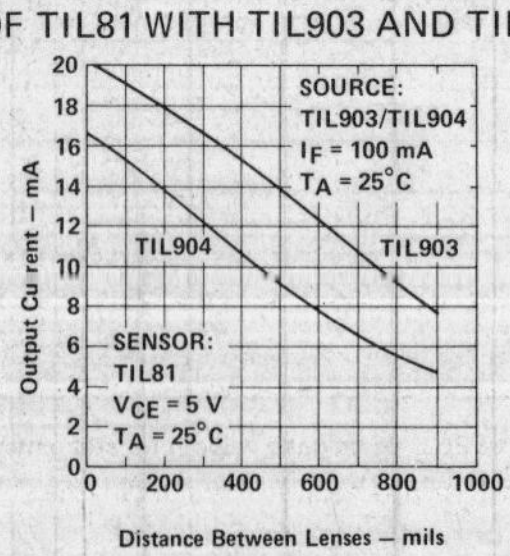

FIGURE 5

FORWARD CONDUCTION
CHARACTERISTICS

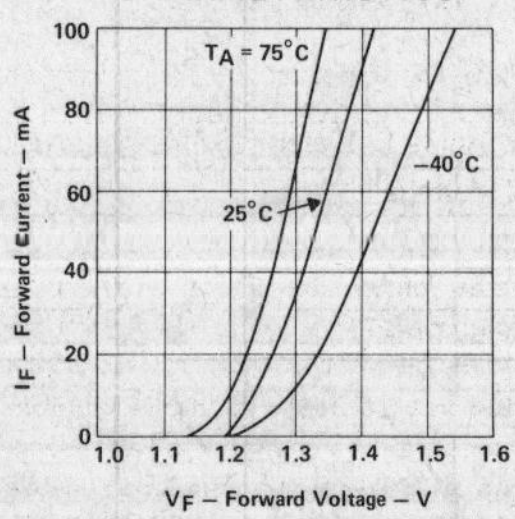

FIGURE 6

TEXAS INSTRUMENTS
INCORPORATED
POST OFFICE BOX 225012 • DALLAS, TEXAS 75265

TYPE TIL905
P-N GALLIUM ALUMINUM ARSENIDE INFRARED-EMITTING DIODE

D2682, FEBRUARY 1983

DESIGNED TO EMIT NEAR-INFRARED RADIATION WHEN FORWARD BIASED

- **Output Spectrally Compatible with Silicon Sensors (e.g., TIL100, TIL413, TIL414)**
- **High Power Output with a 50° Beam Angle**
- **Low-Cost Plastic Package**

mechanical data

This device has a tinted molded plastic body similar in size to lamp style T-1¾ and may be panel mounted using mounting clip TILM4 (formerly TILM2).

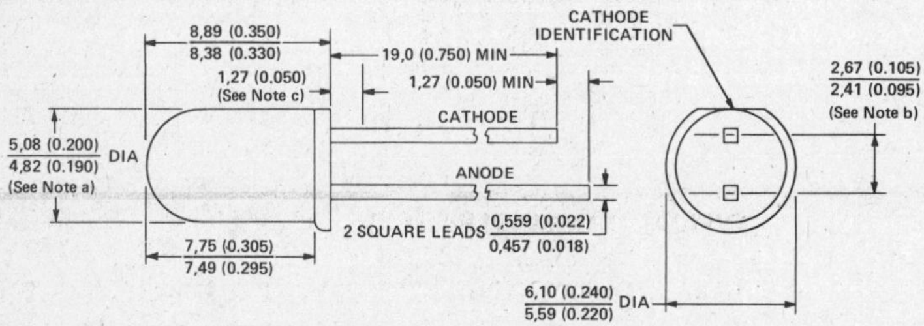

NOTES: a. This diameter is measured 0,127 mm (0.005 inch) from the package shoulder.
b. Lead spacing is measured where the leads emerge from the package.
c. Within this zone the package contour is not controlled.

ALL DIMENSIONS ARE IN MILLIMETERS AND PARENTHETICALLY IN INCHES

3 IR EMITTERS

absolute maximum ratings at 25°C free-air temperature (unless otherwise noted)

Reverse Voltage	3 V
Continuous Forward Current at (or below) 25°C Free-Air Temperature (See Note 1)	100 mA
Peak Forward Current (See Note 2)	2 A
Operating Free-Air Temperature Range	−40°C to 80°C
Storage Temperature Range	−40°C to 100°C
Lead Temperature 1,6 mm (1/16 inch) from Case for 5 Seconds	240°C

operating characteristics at 25°C free-air temperature

	PARAMETER	TEST CONDITIONS		MIN	TYP	MAX	UNIT
P_O	Radiant Power Output	I_F = 20 mA	TIL905-1	1.5			mW
			TIL905-2	2.5			
P_A	Radiant Power Output into an Aperture (see Note 3)		TIL905-1		0.6		mW
			TIL905-2		1.0		
λ_p	Wavelength at Peak Emission	I_F = 20 mA			880		nm
θ_{HI}	Emission Beam Angle Between Half-Intensity Point				50°		
V_F	Static Forward Voltage				1.25	1.75	V
I_R	Reverse Current	V_R = 3 V				100	μA
t_r	Radiant Pulse Rise Time†	I_{FM} = 20 mA, $t_w \geq 5\ \mu s$			600		ns
t_f	Radiant Pulse Fall Time†				350		

† Radiant pulse rise time is the time required for a change in radiant power output from 10% to 90% of its peak value for a step change in current; radiant pulse fall time is the time required for a change in radiant power output from 90% to 10% of its peak value for a step change in current.

NOTES: 1. Derate linearly to 80°C free-air temperature at the rate of 1.82 mA/°C.
2. This value applies for $t_w \leq 10\ \mu s$, $f \leq 1$ kHz. See Figure 1.
3. The parameter is measured with an aperture angle of 30°. The symmetry line of the cone is the mechanical axis of the device under test. The apex of the cone coincides with the package shoulder.

Copyright © 1983 by Texas Instruments Incorporated

ABSOLUTE MAXIMUM RATINGS

PEAK FORWARD CURRENT
vs
PULSE WIDTH

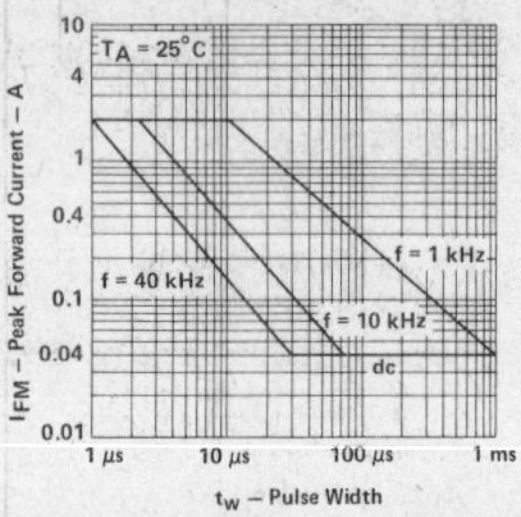

FIGURE 1

TYPICAL CHARACTERISTICS

RELATIVE POWER OUTPUT
vs
FORWARD CURRENT

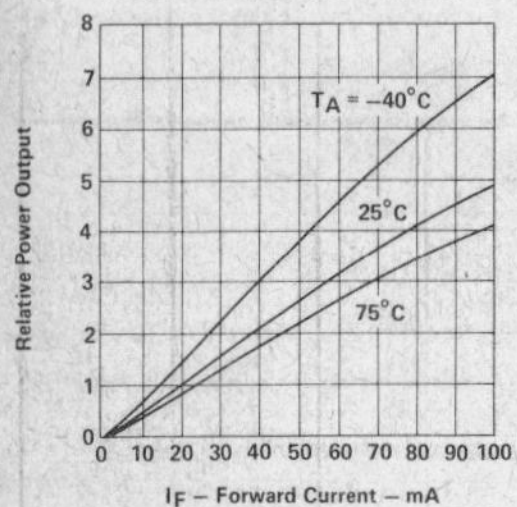

FIGURE 2

RADIANT POWER OUTPUT
vs
FORWARD CURRENT

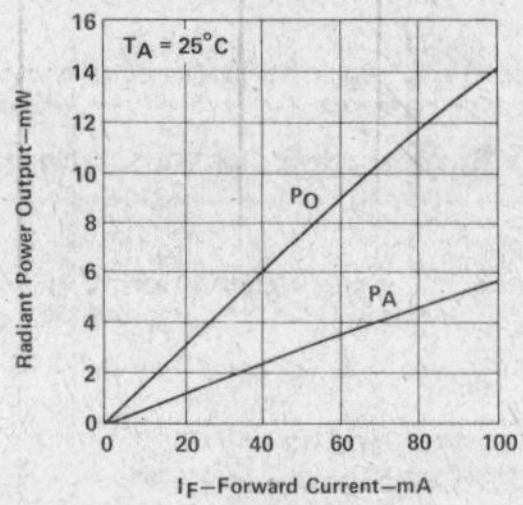

FIGURE 3

RELATIVE RADIANT INTENSITY
vs
ANGULAR DISPLACEMENT

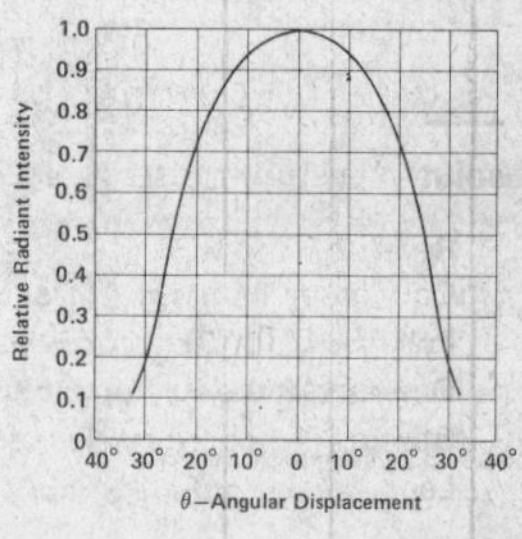

FIGURE 4

COUPLING CHARACTERISTICS
OF TIL905 WITH TIL414

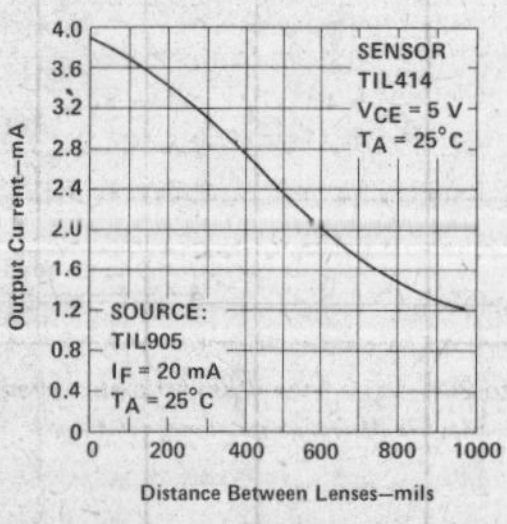

FIGURE 5

FORWARD CONDUCTION
CHARACTERISTICS

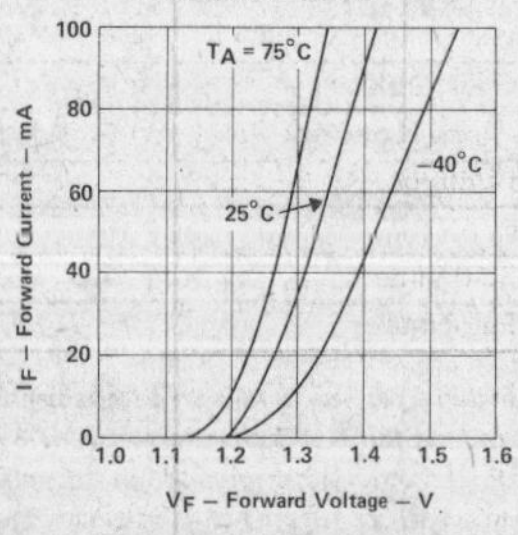

FIGURE 6

P-N GALLIUM ALUMINUM ARSENIDE INFRARED-EMITTING DIODE

D2683, MARCH 1983

DESIGNED TO EMIT NEAR-INFRARED RADIATION WHEN FORWARD BIASED

- **Output Spectrally Compatible with Silicon Sensors (e.g., TIL100, TIL413, TIL414)**
- **High Power Output with a 20° Beam Angle**
- **Low-Cost Plastic Package**

mechanical data

This device has a tinted molded plastic body similar in size to lamp style T-1¾ and may be panel mounted using mounting clip TILM4 (formerly TILM2).

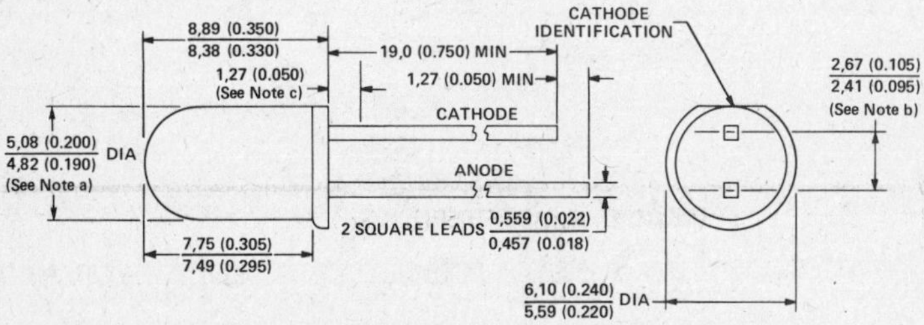

NOTES. a. This diameter is measured 0,127 mm (0.005 inch) from the package shoulder.
b. Lead spacing is measured where the leads emerge from the package.
c. Within this zone the package contour is not controlled.

ALL DIMENSIONS ARE IN MILLIMETERS AND PARENTHETICALLY IN INCHES

CL-10

absolute maximum ratings at 25°C free-air temperature (unless otherwise noted)

Reverse Voltage . . . . 3 V
Continuous Forward Current at (or below) 25°C Free-Air Temperature (See Note 1) . . . . 100 mA
Peak Forward Current (See Note 2) . . . . 2 A
Operating Free-Air Temperature Range . . . . −40°C to 80°C
Storage Temperature Range . . . . −40°C to 100°C
Lead Temperature 1,6 mm (1/16 inch) from Case for 5 Seconds . . . . 240°C

operating characteristics at 25°C case temperature

PARAMETER		TEST CONDITIONS		MIN	TYP	MAX	UNIT
P_O	Radiant Power Output	I_F = 20 mA	TIL906-1	1.5			mW
			TIL906-2	2.5			
P_A	Radiant Power Output into an Aperture (see Note 3)		TIL906-1		0.8		mW
			TIL906-2		1.3		
θ_{HI}	Emission Beam Angle Between Half-Intensity Points	I_F = 20 mA			20°		
λ_p	Wavelength at Peak Emission				880		nm
V_F	Static Forward Voltage				1.25	1.75	V
I_R	Reverse Current	V_R = 3 V				100	µA
t_r	Radiant Pulse Rise Time†	I_{FM} = 20 mA, $t_w \geq 5$ µs			600		ns
t_f	Radiant Pulse Fall Time†				350		

† Radiant pulse rise time is the time required for a change in radiant power output from 10% to 90% of its peak value for a step change in current; radiant pulse fall time is the time required for a change in radiant power output from 90% to 10% of its peak value for a step change in current.

NOTES: 1. Derate linearly to 80°C free-air temperature at the rate of 1.82 mA/°C.
2. This value applies for $t_w \leq 10$ µs, $f \leq 1$ kHz. See Figure 1.
3. This parameter is measured with an aperture angle of 30°. The symmetry line of the cone is the mechanical axis of the device under test. The apex of the cone coincides with the package shoulder.

Copyright © 1983 by Texas Instruments Incorporated

ABSOLUTE MAXIMUM RATINGS

PEAK FORWARD CURRENT
vs
PULSE WIDTH

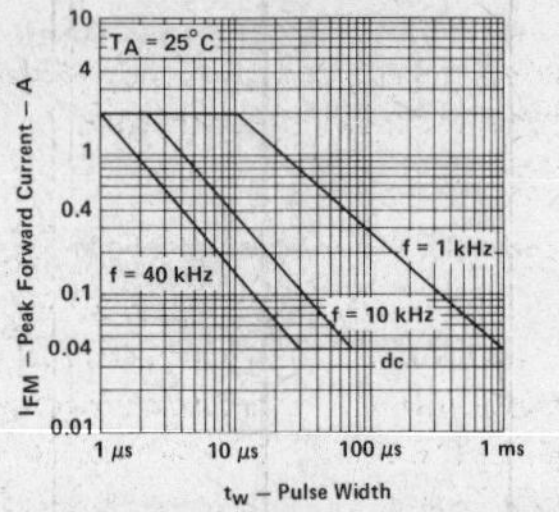

FIGURE 1

TYPICAL CHARACTERISTICS

RELATIVE POWER OUTPUT
vs
FORWARD CURRENT

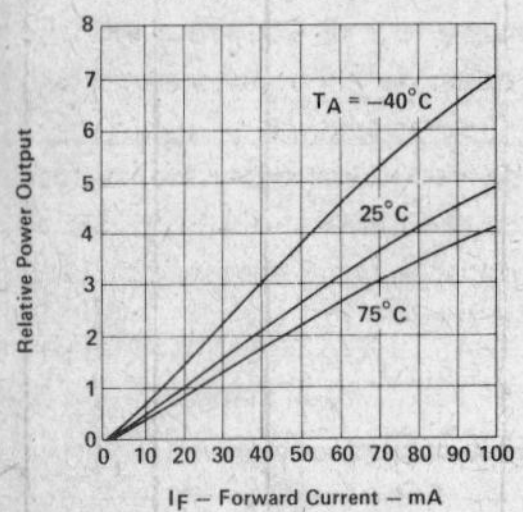

FIGURE 2

RADIANT POWER OUTPUT
vs
FORWARD CURRENT

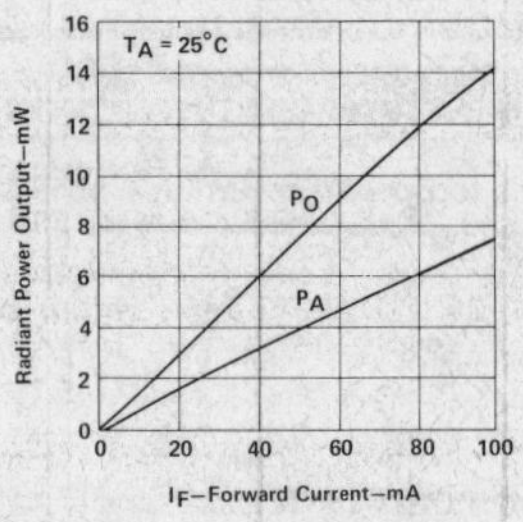

FIGURE 3

RELATIVE RADIANT INTENSITY
vs
ANGULAR DISPLACEMENT

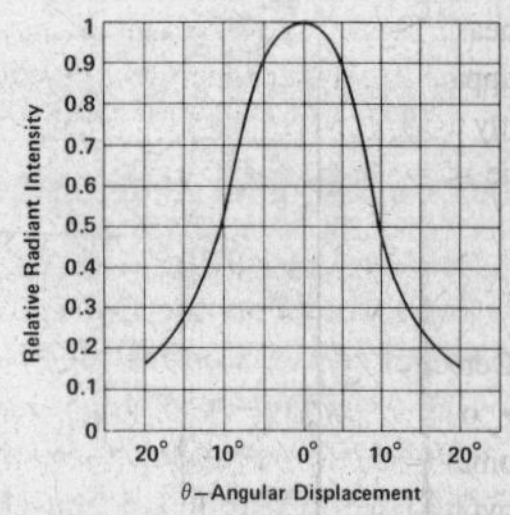

FIGURE 4

COUPLING CHARACTERISTICS
FOR TIL906 WITH TIL414

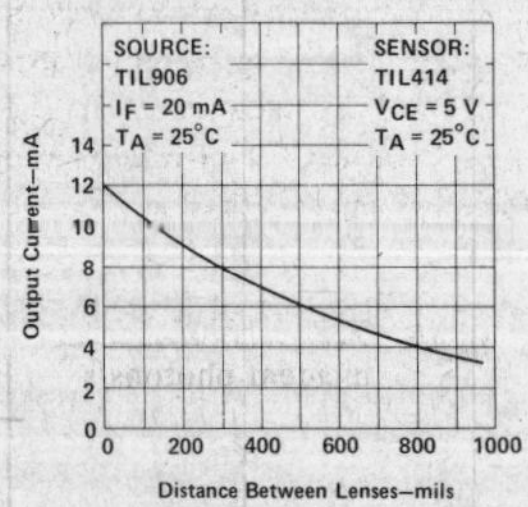

FIGURE 5

FORWARD CONDUCTION
CHARACTERISTICS

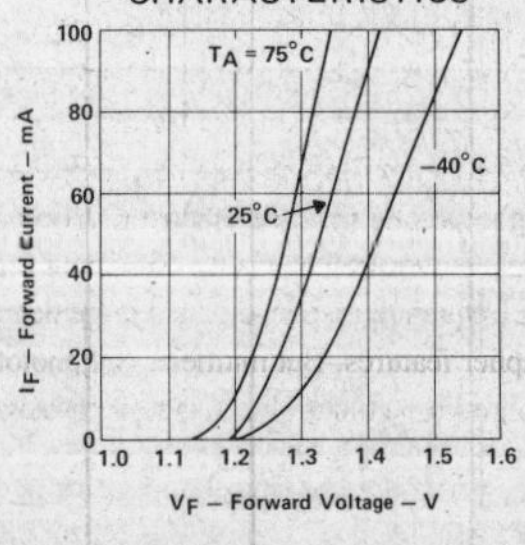

FIGURE 6

TEXAS INSTRUMENTS
INCORPORATED
POST OFFICE BOX 225012 • DALLAS, TEXAS 75265

MEASURING THE OUTPUT OF INFRARED-EMITTING AND LIGHT-EMITTING DIODES

Ronald D. Grotti and Larry D. Major
Optoelectronics Department

Making accurate radiant-energy measurements involves, if not a little black magic, at least a relatively complicated commercial instrument and a skilled operator. However, the increased use of infrared-emitting diodes (IREDs) and light-emitting diodes (LEDs) as a precision system component has necessitated the development of equipment suitable for measuring radiant energy from IREDs and LEDs in the designer's lab, in the quality control lab, and on the production line. This equipment must be easy to use, provide the necessary accuracy, be calibratable, and be inexpensive.

To develop such equipment requires the selection of a suitable photodetector and the development of the proper calibration and operation procedures. This report describes a method that has been used in the Texas Instruments Optoelectronic Device Department for measuring the output of its radiation-emitting diode products. The apparatus consists simply of a photovoltaic detector connected directly to an ammeter, with a special mechanical fixture to prevent escape of radiant energy.

SELECTION OF DETECTOR

Detectors that might be considered for measuring IRED and LED output include thermopiles, photocells, photodiodes, photomultipliers, and photovoltaic cells. To show why the photovoltaic cell was chosen for this application, a review of pertinent detector characteristics is in order.

Thermopiles can be excellent primary detecting devices but are generally unsuitable for most laboratory and quality control types of service. Not only are they difficult to apply properly, but they are costly, lose their calibration when mishandled, and have an inadequate frequency response.

Photodiodes have good frequency capabilities, are reasonably priced, and are being used in pulse and high-frequency applications. However, most IREDs and LEDs are tested under low-frequency conditions, and therefore frequency response is not a critical sensor parameter. Because the photodiode must be electronically biased, a well-regulated bias supply is required to ensure consistent results.

Good sensitivity and frequency response plus a large detection area are some photomultiplier features. But multielement phototubes are expensive, require high-voltage supplies, and since output is a function of supply voltage, stability problems can arise. Also, if improperly applied, photomultipliers can saturate, causing errors and possibly permanent tube damage.

Photovoltaic cells—particularly the solar-cell variety—have a large active area, good long-term stability, and good spectral matching, are easy to use, and are inexpensive. The frequency response from dc to 100 kHz, although less than that of the photomultiplier and photodiode, is satisfactory for this application. These factors, combined with the fact that power or bias supplies are not required, makes the solar cell appear to have the best combination of qualities for this application.

Using the photovoltaic cell to precisely measure the emitter output and determine its quantum efficiency requires detailed knowledge of the cell, the emitter, and how they are optically coupled. Such knowledge depends not only on the mathematical characterization of the two devices, but on an accurate calibration of the photovoltaic cell. Once these steps have been accomplished, the emitter's power output and its quantum efficiency can be calculated using only two measured values — the emitter's input current and the cell's output current.

THE PHOTOVOLTAIC CELL

Before describing how the photovoltaic cell is calibrated, a few comments on the basic characteristics of this semiconductor device are in order. It is not necessary for our purposes to discuss the theory of operation in detail. Suffice it to say that electron-hole pairs are generated within the device as a function of impinging photons. Only those photons that have a quantum energy larger than the band gap between the valence band and the conduction band generate electron-hole pairs. The lower-energy photons simply transmit through the cell and do not cause an output. The ratio of electrons generated to the total number of incident photons is the cell's quantum efficiency, and is defined as

$$\eta_{SC} = \frac{\text{electrons generated/s}}{\text{incident photons/s}}$$

It is necessary to note that the cell's quantum efficiency is a function of the wavelength (Figure 1). This fact is particularly important because the sensor specifications are often based on the device's sensitivity to a particular wavelength. This quantum efficiency curve can be shaped through various means including the deposition of antireflection coatings on the photovoltaic cell's surface.

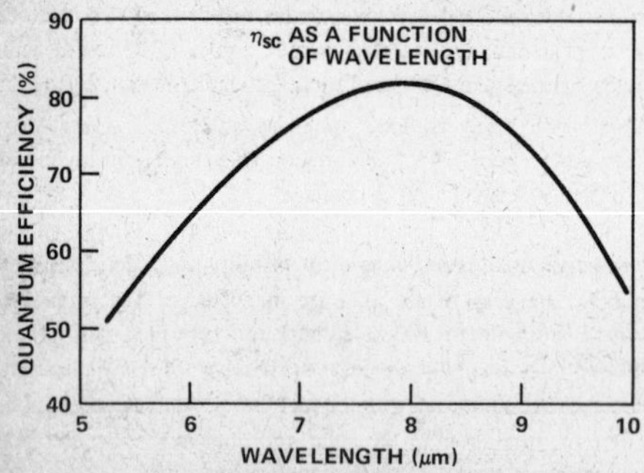

Figure 1. Photodetector Quantum Efficiency Varies as a Function of Wavelength, Thus Making Calibration at a Number of Wavelengths Necessary for General Use

CELL CALIBRATION

Before using the photovoltaic cell to measure the IRED or LED power output, the cell must be calibrated. This calibration is a two-step process, with the first step being the accurate determination of the cell's relative response. This determination is made using a grating monochromator, a tungsten light source, and a thermocouple detector that has a flat response in the spectral region of 500 nm to 1000 nm. Two curves are obtained, one using the thermocouple detector to measure the tungsten source's output and the other using the photovoltaic detector to measure the same output.

By dividing the photovoltaic cell response by the thermocouple response, the relative response of the cell is obtained. The relative response curve allows the measurement of radiation souces with different spectral characteristics to be accurately compared. However, to determine the actual power generated by a particular source using this cell requires another calibration step in which the photovoltaic cell output is determined when illuminated by a radiation source with a known power output. To accomplish this goal, the output of three monochromatic sources (gallium arsenide IRED, helium-neon laser, and argon laser) are measured by the cell being calibrated and by the Eppley thermopile. The quantum efficiency of the cell at the wavelength of each emitter is then found by using the optical power equation:

$$\eta_{SC} = \left(\frac{I_L}{\text{optical power}}\right)\left(\frac{\text{energy}}{\text{photon}}\right)$$

where I_L is the short-circuit current from the photovoltaic cell under test and optical power is the measurement made by the thermopile.

The three quantum efficiencies are then plotted, and a curve is generated that allows the cell to be used to measure accurately any impinging light of known spectral characteristics.

MEASUREMENT PROCEDURE

To employ this calibrated detector in a radiation-emitting diode testing system, it is necessary to develop the relationships that can describe the diode's quantum efficiency.

The diode output is directly proportional to the emitted photon energy and quantity per unit of time. The relation between energy E and wavelength λ is defined as

$$\lambda = \frac{1.24}{E} \text{ or } E = \frac{1.24}{\lambda} \quad \text{(units are } \mu\text{m and eV)}$$

Energy, and therefore wavelength, of any given photon emitted from an IRED or LED source fall within a distribution curve such as that shown in Figure 2 for a GaAs IRED. To be absolutely accurate in calculating the optical power output of a solid-state source requires a time-consuming graphical integration using Figure 1 and Figure 2. Fortunately, all photons emitted by a monochromatic source have the same energy. Since it is a valid assumption to consider the IRED to be monochromatic, the IRED's optical power can be described to a first approximation without any noticeable error.

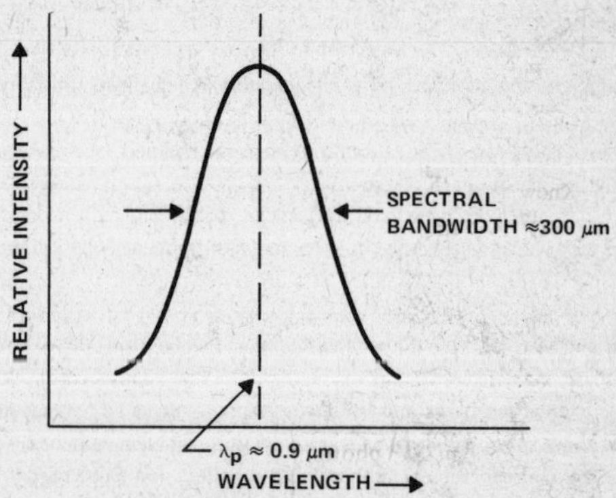

Figure 2. Spectral Characteristics of GaAs Diode Indicate that the Device is Nearly Monochromatic

TEXAS INSTRUMENTS
INCORPORATED
POST OFFICE BOX 225012 • DALLAS, TEXAS 75265

For simplicity of calculation, let us assume that all photons generated by the IRED are collected by the photovoltaic cell. Therefore, considering the ammeter as a load consisting of a calibrated resistor and microvoltmeter (Figure 3), the current I_L (neglecting the internal resistance R_{shunt} of the photocell) is proportional to the number of photons striking the surface:

$$I_L = \left(\frac{\text{electrons}}{\text{s}}\right)(1.602 \times 10^{-19})$$

and

$$I_L = \eta_{sc}\left(\frac{\text{photons}}{\text{s}}\right)(1.602 \times 10^{-19})$$

therefore,

$$\text{photons/s} = \frac{I_L}{\eta_{sc}\,(1.602 \times 10^{-19})}$$

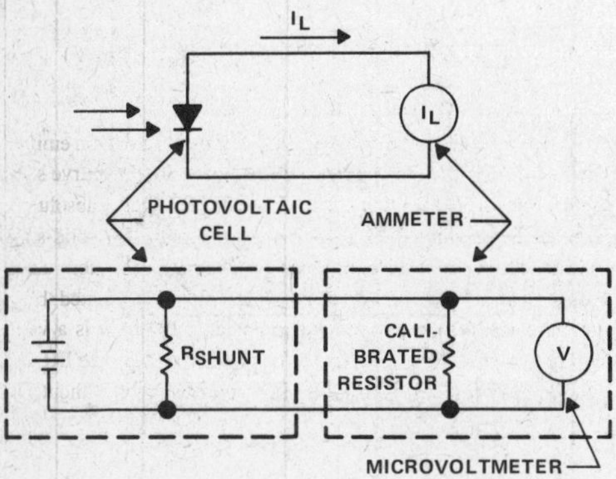

Figure 3. Equivalent Circuit of a Photovoltaic Cell (Silicon Solar Cell) Connected to the Ammeter Used to Measure Short Circuit Current

Knowing I_L, we can now calculate the emitter quantum efficiency η_{em} and optical power P_O:

$$\eta_{em} = \frac{I_L}{\eta_{sc}\, I_D} \quad \text{where } I_D \text{ is IRED current}$$

$$P_O = \left(\frac{I_L}{\eta_{sc}}\right)\left(\frac{\text{energy}}{\text{photon}}\right)$$

Using these equations, we can indeed determine both the quantum efficiency and the optical power generated by the IRED under conditions where all the power emitted is collected by the photovoltaic cells. To ensure the photovoltaic cell receives all emitted photons, it is necessary to build special test fixtures using detectors either singly or in arrays. (See Figures 4 and 5). In either case, the test procedures are the same. However, if such fixtures are not possible, then the percentage of energy emitted that actually reaches the detector must be included in the calculation. This fraction can be determined by dividing the total power emitted by the steradian relationship between the detector and the emitter, the total number of steradians being equal to the aperture area of the detector divided by the square of the distance between the emitter and the detector surface.

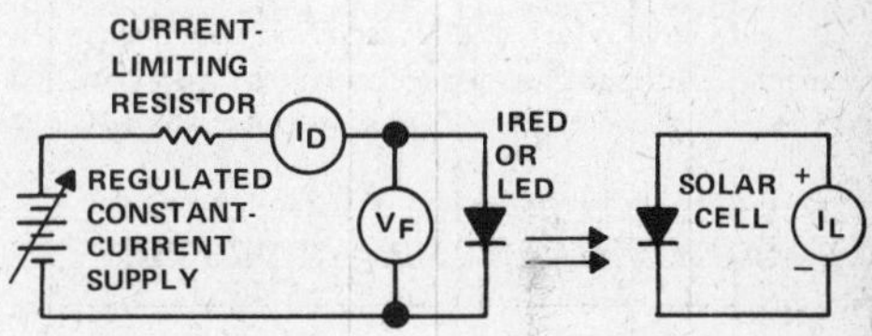

Figure 4. Calibrating the Test Setup

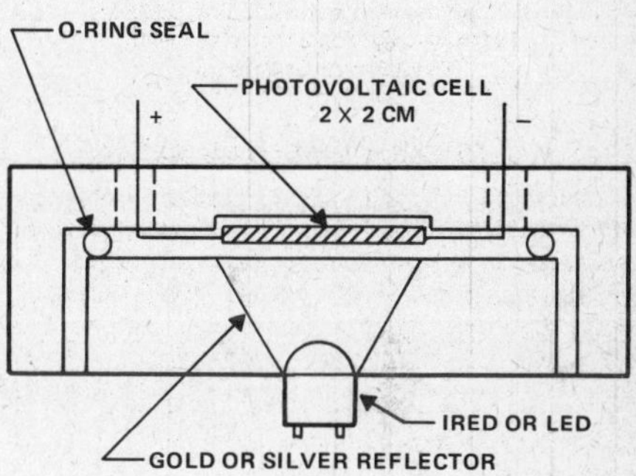

Figure 5. Test Fixture for Capturing the Total Diode Output with a Single Photovoltaic Cell

TESTING PRECAUTIONS

Generally, gallium arsenide and gallium arsenide phosphide (GaAsP) infrared emitters provide an output signal I_L large enough that an ammeter may be used to measure the cell's short-circuit current directly. The measurement of GaAsP visible-light-emitting diodes and tests such as radiant intensity measurements usually produce signal levels that require a calibrated resistor and a microvoltmeter. The important point is that the input impedance of the measuring instrument must be less than 1/10 the value of R_{shunt} to prevent lowering the output of the cell. The exact value of R_{shunt} for photovoltaic cells is difficult to measure, but it is usually in the order of 10 kΩ to 30 kΩ. If the cell has been mistreated, the R_{shunt} may be as low as 1 kΩ or less. Thus, if an electronic ammeter is used in the 3×10^{-6} ampere range, as may be required for testing GaAsP LEDs, the input meter impedance of 300 to 1000 Ω approaches the critical value of the typical solar cell. Thus, these low-level measurements must be made using the resistor-microvoltmeter technique.

The second problem occurs when the photovoltaic cell becomes appreciably self-biased because of the voltage drop developed across the load. Care must be taken to limit this bias to prevent a reduced output signal. As a rule of thumb, this load-voltage drop is kept lower than 50 mV. When measuring high-power emitters, the value of I_L of a 2- by 2-cm photovoltaic cell is capable of reaching the 200-mA level without saturation; therefore at these levels, the input impedance of the ammeter and the value of the calibrated resistor (See Figure 3) must be kept less than 0.25 Ω.

SAMPLE CALCULATION OF DIODE POWER OUTPUT AND QUANTUM EFFICIENCY

Assume the following values:

I_D = emitting diode current = 300 mA

V_F = forward voltage of the emitter = 1.6 volts

I_L = solar cell output signal = 25 mA

λ_p = peak wavelength of the emitter = 0.925 μm

η_{sc} = quantum efficiency of the cell = 0.70 electrons per photon

Then:

η_{em} = emitter quantum efficiency

$$= \left(\frac{I_L}{\eta_{sc}}\right)\left(\frac{I}{I_D}\right) = \left(\frac{25\text{ mA}}{0.7\ \frac{\text{elect}}{\text{photon}}}\right)\left(\frac{1}{300\text{ mA}}\right)$$

$$= 0.119\ \text{photons/electron}$$

$$\eta_{em} = 11.9\%$$

$$\text{Optical Power} = P_O = \left(\frac{I_L}{\eta_{sc}}\right)\left(\frac{\text{energy}}{\text{photon}}\right)$$

$$\text{Where energy} = \frac{1.24}{\lambda_p} = \frac{1.24}{0.925} = 1.341\text{ eV}$$

$$P_O = \left(\frac{25\text{ mA}}{0.7\ \frac{\text{elect}}{\text{photon}}}\right)\left(1.341\ \frac{\text{eV}}{\text{photon}}\right)$$

$$\text{Power efficiency} = \frac{P_O}{\text{Input Power}}$$

$$= \frac{47.9 \times 10^{-3}\text{ W}}{I_D V_F} = \frac{47.9 \times 10^{-3}\text{ W}}{48 \times 10^{-2}\text{ W}}$$

Power efficiency = 0.0998 = 9.98%

This material appeared as an article in *Electro-Optical Systems Design,* Vol. 2 No. 7, July 1970.

Special Function Infrared-Emitting Diodes

- Quick Reference Guide
- High-Efficiency/High-Power
- Hermetically Sealed Packages
- Open Construction on Some Devices
- Article on TIES27 GaAs Noncoherent IR Source

See Section 3 for Standard Infrared-Emitting Diodes.

SPECIAL FUNCTION INFRARED-EMITTING DIODES
QUICK REFERENCE GUIDE

DEVICE	POWER OUTPUT MIN mW	@	I_F mA	θ_{HI} TYP	V_F MAX V	@	I_F mA	λ_p TYP µm	FEATURES
TIES06	0.6		500	115°	2.3		500	0.91	0.19-mm (0.0075-in) dia emitting area
TIES13	20		300	130°	2		300	0.93	0.91-mm (0.036-in) diameter hemispherically shaped chip
TIES13A	30		300	130°	2		300	0.93	
TIES14	60		1000	130°	2		1000	0.93	1.83-mm (0.072-in) diameter hemispherically shaped chip
TIES15	30		1000	130°	2		1000	0.93	
TIES16A	100		2000	150°	2		2000	0.93	
TIES27	15		300	135°	2.2		300	0.93	Stud header with epoxy lens
TIES35	0.9		50	135°	2		50	0.91	0.46-mm (0.018-in) diameter hermispherically shaped chip, 15-ns typical rise time

GALLIUM ARSENIDE INFRARED-EMITTING DIODE

D343, FEBRUARY 1967—REVISED APRIL 1983

DESIGNED TO EMIT NEAR-INFRARED RADIANT ENERGY WHEN FORWARD BIASED

- **Spectrally Matched to Silicon Sensors . . . Peak Emission at 930 nm**
- **Circular, Consistent-Size, Flat Emitting Areas . . . 7.5 Mils Diameter**
- **Recommended for Precision Optical Alignment, Communication, and Photographic Film Annotation**
- **Stud-Mounted Package for Convenient Mounting and Heat-Sinking**

mechanical data

This device is in a hermetically sealed package with a flat glass window in the top of the case. The cathode is in electrical contact with the case and adjacent solder lug. The anode is in electrical contact with the stud, which is insulated from the case by a glass-to-metal seal.

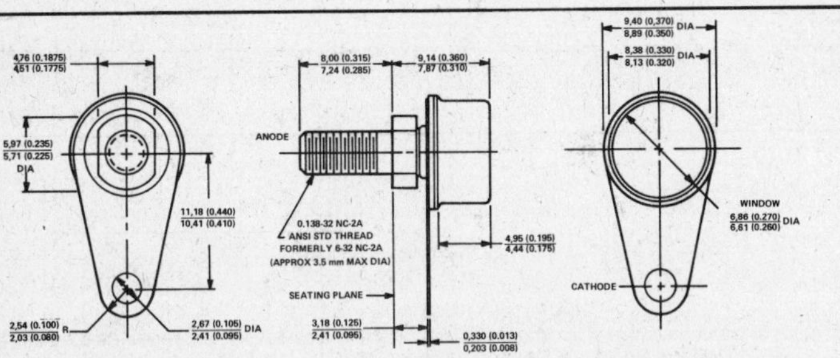

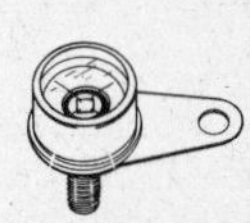

NOTES: a. Torque between stud and can or lug must be avoided. Flats are provided on the stud for tightening to heat sink.

b. The orientation of the lug in relation to the stud flats is not controlled.

ALL DIMENSIONS ARE IN MILLIMETERS AND PARENTHETICALLY IN INCHES

CL-17

absolute maximum ratings

Reverse Voltage at 25°C Stud Temperature . . . . . 2 V
Continuous Forward Current at (or below) 25°C Stud Temperature (See Note 1) . . . . . 500 mA
Peak Forward Current at (or below) 25°C Stud Temperature (See Note 2) . . . . . 700 mA
Storage Temperature Range . . . . . −55°C to 125°C
Solder Lug Temperature for 10 Seconds (See Note 3) . . . . . 240°C

operating characteristics at 25°C stud temperature

PARAMETER		TEST CONDITIONS	MIN	TYP	MAX	UNIT
P_O	Radiant Power Output	I_F = 500 mA	0.6	0.12		mW
λ_P	Wavelength at Peak Emission			930		nm
$\Delta\lambda$	Spectral Bandwidth			25		nm
θ_{HI}	Half-Intensity Beam Angle			120°		
V_F	Static Forward Voltage			1.7	2.3	V
t_r	Radiant Pulse Rise Time	I_{FM} = 100 mA, $t_w \geq$ 100 ns		15		ns
t_f	Radiant Pulse Fall Time			15		

NOTES: 1. Derate linearly to 125°C stud temperature at the rate of 5 mA/°C.

2. This value applies for $t_w \leq 100\ \mu s$, duty cycle ≤ 50%. Derate linearly to 125°C stud temperature at the rate of 7 mA/°C.

3. Soldered connections should not be made directly to the stud because of the low-thermal-resistance path between the stud and emitting element.

Copyright © 1983 by Texas Instruments Incorporated

TYPICAL CHARACTERISTICS

RELATIVE SPECTRAL CHARACTERISTICS

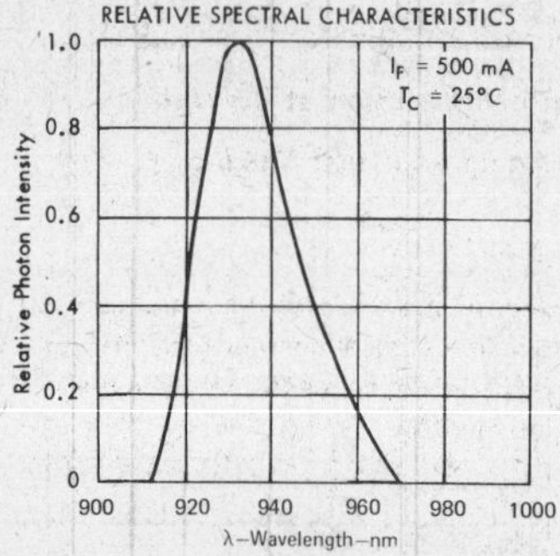

FIGURE 1

RELATIVE PHOTON INTENSITY
vs
ANGULAR DISPLACEMENT

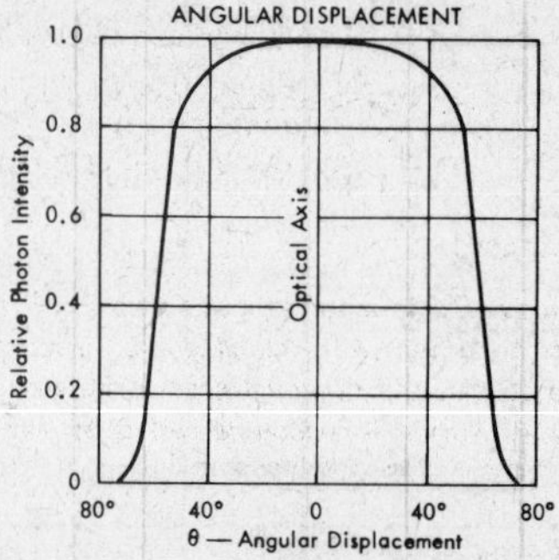

FIGURE 2

RELATIVE PHOTON INTENSITY
vs
CASE TEMPERATURE

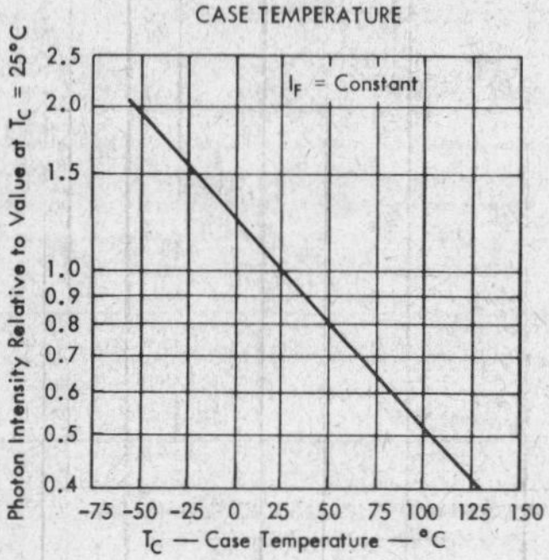

FIGURE 3

CHANGE IN WAVELENGTH OF PEAK INTENSITY
vs
CASE TEMPERATURE

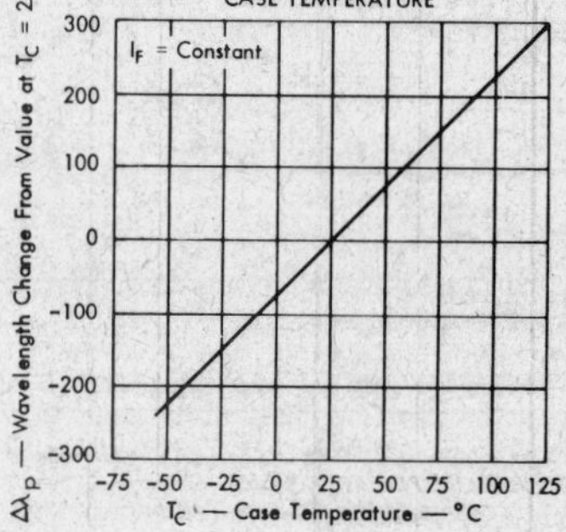

FIGURE 4

RELATIVE PHOTON INTENSITY
vs
FORWARD CURRENT

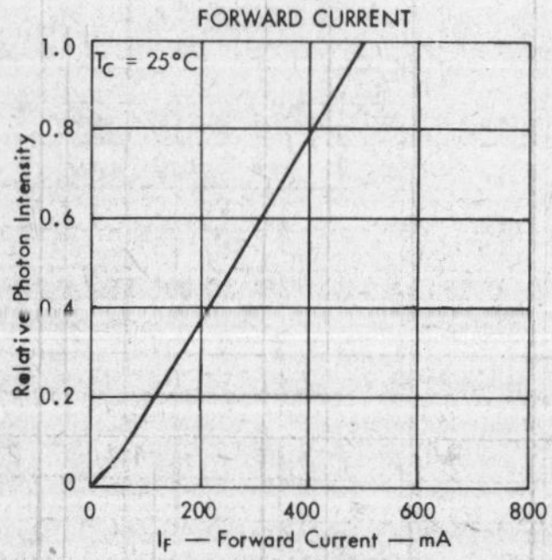

FIGURE 5

FORWARD CONDUCTION CHARACTERISTICS

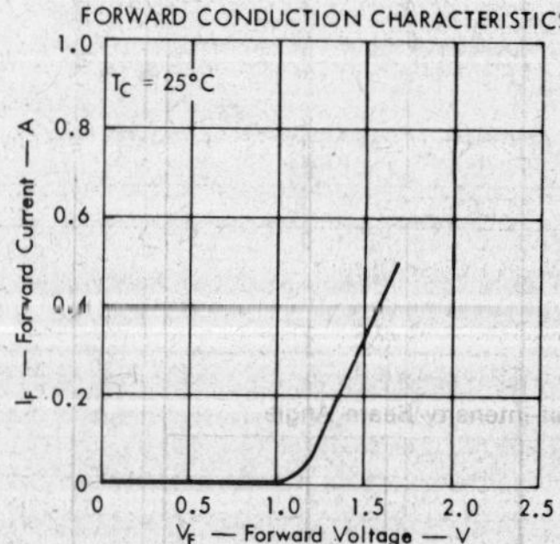

FIGURE 6

TEXAS INSTRUMENTS
INCORPORATED
POST OFFICE BOX 225012 • DALLAS, TEXAS 75265

D2403, MARCH 1969—REVISED APRIL 1983

DESIGNED TO EMIT NEAR-INFRARED RADIANT ENERGY WHEN FORWARD BIASED

- **High Output Efficiency**
- **Hemispherically Shaped Chips with Diameter of 36 Mils**
- **Spectrally Matched to Silicon Sensors . . . Peak Emission at 930 nm**
- **Stud-Mounted Package for Convenient Mounting and Heat Sinking**

mechanical data

Each device is in a hermetically sealed package with a flat glass window in the top of the case. The cathode is in electrical contact with the case and adjacent solder lug. The anode is in electrical contact with the stud, which is insulated from the case by a glass-to-metal seal. Soldered connections should not be made directly to the stud because of the low-thermal-resistance path between the stud and emitting element.

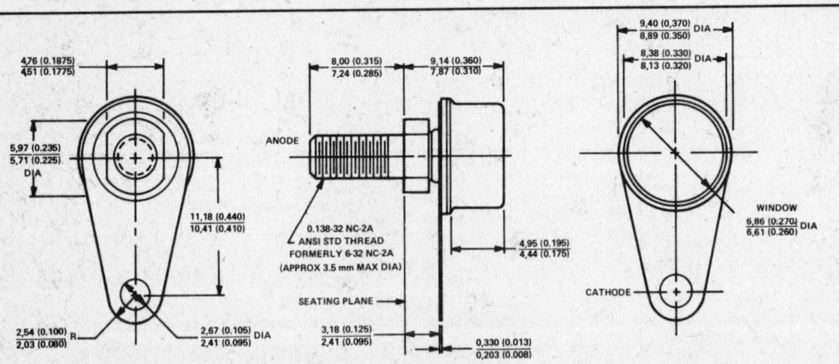

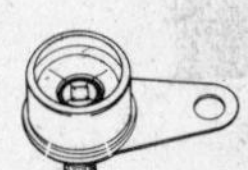

NOTES: a. Torque between stud and can or lug must be avoided. Flats are provided on the stud for tightening to heat sink.

b. The orientation of the lug in relation to the stud flats is not controlled.

ALL DIMENSIONS ARE IN MILLIMETERS AND PARENTHETICALLY IN INCHES

CL-17

absolute maximum ratings

Reverse Voltage at 25 °C Stud Temperature . . . . 2 V
Continuous Forward Current at (or below) 25 °C Stud Temperature (See Note 1) . . . . 300 mA
Peak Forward Current at (or below) 25 °C Stud Temperature (See Note 2) . . . . 500 mA
Storage Temperature Range . . . . −55 °C to 100 °C
Solder Lug Temperature for 10 Seconds . . . . 240 °C

operating characteristics at 25 °C stud temperature

PARAMETER		TEST CONDITIONS	TYPE	MIN	TYP	MAX	UNIT
P_O	Radiant Power Output	I_F = 300 mA	TIES13	20	25		mW
			TIES13A	30	35		
λ_p	Wavelength at Peak Emission		All		930		nm
$\Delta\lambda$	Spectral Bandwidth		All		45		nm
θ_{HI}	Half-Intensity Beam Angle		All		130°		
V_F	Static Forward Voltage		All		1.4	2	V
t_r	Radiant Pulse Rise Time	I_{FM} = 100 mA, $t_w \geq 5\ \mu s$	All		600		ns
t_f	Radiant Pulse Fall Time		All		450		ns

NOTES: 1. Derate linearly to 100 °C stud temperature at the rate of 4 mA/°C.

2. This value applies for $t_w \leq 100\ \mu s$, duty cycle ≤50%. Derate linearly to 100 °C stud temperature at the rate of 6.7 mA/°C.

D2403, MARCH 1969—REVISED APRIL 1983

DESIGNED TO EMIT NEAR-INFRARED RADIANT ENERGY WHEN FORWARD BIASED

- **High Output . . . 60 mW Min at 25°C for the TIES14**
- **Hemispherically Shaped Chips with Diameter of 72 Mils**
- **Spectrally Matched to Silicon Sensors . . . Peak Emission at 930 nm**
- **Stud-Mounted Package for Convenient Mounting and Heat Sinking**

mechanical data

Each device is in a hermetically sealed package with a flat glass window in the top of the case. The cathode is in electrical contact with the case and adjacent solder lug. The anode is in electrical contact with the stud, which is insulated from the case by a glass-to-metal seal. Soldered connections should not be made directly to the stud because of the low-thermal-resistance path between the stud and emitting element.

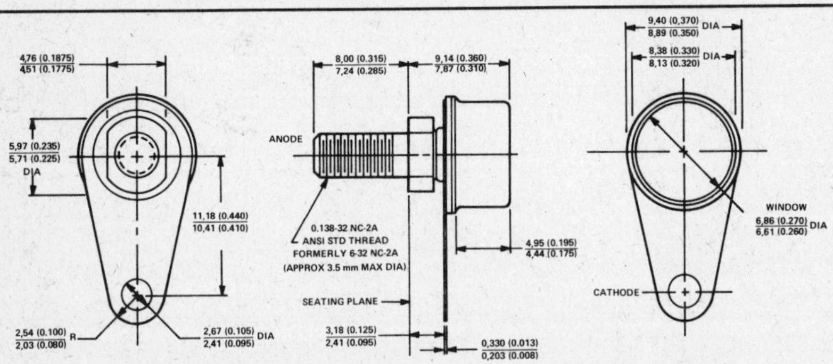

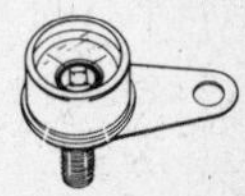

NOTES: a. Torque between stud and can or lug must be avoided. Flats are provided on the stud for tightening to heat sink.
b. The orientation of the lug in relation to the stud flats is not controlled.

ALL DIMENSIONS ARE IN MILLIMETERS AND PARENTHETICALLY IN INCHES

CL-17

absolute maximum ratings

Reverse Voltage at 25°C Stud Temperature . . . . 2 V
Continuous Forward Current at (or below) 25°C Stud Temperature (See Note 1) . . . . 1 A
Peak Forward Current at (or below) 25°C Stud Temperature (See Note 2) . . . . 1.6 A
Storage Temperature Range . . . . −55°C to 100°C
Solder Lug Temperature for 10 Seconds . . . . 240°C

operating characteristics at 25°C stud temperature

	PARAMETER	TEST CONDITIONS	TYPE	MIN	TYP	MAX	UNIT
P_O	Radiant Power Output	$I_F = 1$ A	TIES14	60	75		mW
			TIES15	30	50		
λ_p	Wavelength at Peak Emission		All		930		nm
$\Delta\lambda$	Spectral Bandwidth		All		45		nm
θ_{HI}	Half-Intensity Beam Angle		All		130°		
V_F	Static Forward Voltage		All		1.4	2	V
t_r	Radiant Pulse Rise Time	$I_{FM} = 100$ mA, $t_w \geq 5\ \mu s$	All		600		ns
t_f	Radiant Pulse Fall Time		All		450		ns

NOTES: 1. Derate linearly to 100°C stud temperature at the rate of 13.3 mA/°C.
2. This value applies for $t_w \leq 100\ \mu s$, duty cycle $\leq 50\%$. Derate linearly to 100°C stud temperature at the rate of 21.3 mA/°C.

Copyright © 1983 by Texas Instruments Incorporated.

D1947, NOVEMBER 1972—REVISED FEBRUARY 1983

DESIGNED TO EMIT NEAR-INFRARED RADIENT ENERGY WHEN FORWARD BIASED (FORMERLY TIXL16A)

- **High Output Power . . . 100 mW Min at 25 °C**
- **Hemispherically Shaped 72-Mil-Diameter Chip**
- **Stud Mounting for Convenient Heat Sinking**
- **Open Construction to Allow Flexibility in Optical Design**

mechanical data

This diode is mounted on a copper stud header to provide efficient heat sinking. The anode is in electrical contact with the copper stud. The cathode lead is a varnished 0.01-inch copper wire secured to the stud by a metalized ceramic insulator. Soldered connections should not be made directly to the stud because of the low-thermal-resistance path to the emitting element.

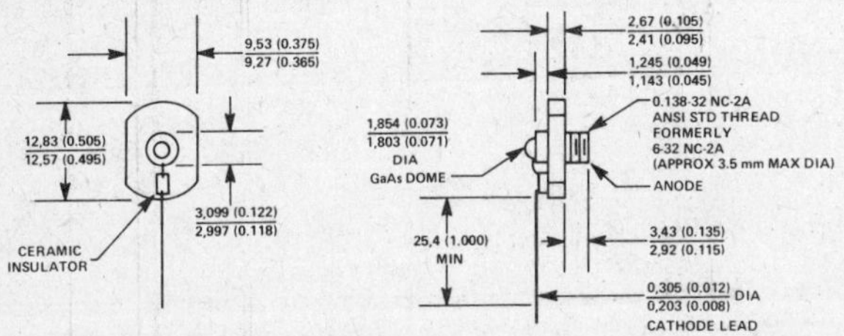

ALL DIMENSIONS ARE IN MILLIMETERS AND PARENTHETICALLY IN INCHES.

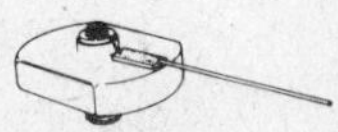

absolute maximum ratings

Reverse Voltage at 25 °C Stud Temperature	2 V
Continuous Forward Current at (or below) 25 °C Stud Temperature (See Note 1)	2 A
Peak Forward Current at (or below) 25 °C Stud Temperature (See Note 2)	3 A
Storage Temperature Range	−55 °C to 100 °C
Lead Temperature 6.4 mm (1/4 Inch) from Ceramic Insulator for 5 Seconds	230 °C

operating characteristics at 25 °C stud temperature

	PARAMETER	TEST CONDITION	MIN	TYP	TYP	UNIT
P_O	Radiant Power Output	I_F = 2 A	100	150		mW
λ_p	Wavelength at Peak Emission			930		nm
$\Delta\lambda$	Spectral Bandwidth			450		Å
θ_{HI}	Half-Intensity Beam Angle			150°		
V_F	Static Forward Voltage			1.6	2	V
t_r	Radiant Pulse Rise Time	I_{FM} = 100 mA, $t_w \geq 5$ µs		600		ns
t_f	Radiant Pulse Fall Time			450		

NOTES: 1. Derate linearly to 100 °C stud temperature at the rate of 26.7 mA/°C.
2. This value applies for $t_w \leq 100$ µs, duty cycle ≤50%. Derate linearly to 100 °C stud temperature at the rate of 40 mA/°C.

Copyright © 1983 by Texas Instruments Incorporated

GALLIUM ARSENIDE INFRARED-EMITTING DIODE

D901, SEPTEMBER 1971–REVISED APRIL 1983

DESIGNED TO EMIT NEAR-INFRARED RADIANT ENERGY WHEN FORWARD BIASED

- High Output Power . . . 15 mW Min at 25°C
- Spectrally Matched to Silicon Sensors . . . Peak Emission at 930 nm
- Stud Mounting for Convenient Heat Sinking
- Recommended for Precision Optical Alignment, Industrial Controls, and Optical Communications

mechanical data

The device is encapsulated and mounted on a stud header. The cathode is in electrical contact with the solder lug. The anode is in electrical contact with the stud, which is insulated from the case by a glass-to-metal seal. Soldered connections should not be made directly to the stud because of the low-thermal-resistance path between stud and emitting element.

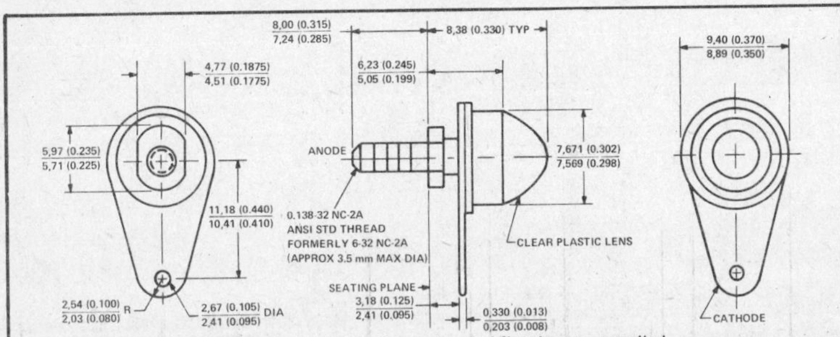

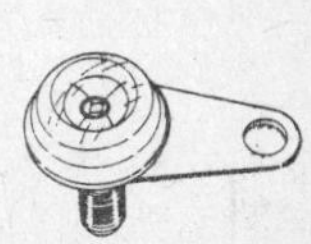

NOTES: a. The orientation of the lug in relation to the stud flats is not controlled.
b. Torque between stud and lug must be avoided. Flats are provided on the stud for tightening to heat sink.

ALL DIMENSIONS ARE IN MILLIMETERS AND PARENTHETICALLY IN INCHES

absolute maximum ratings

Reverse Voltage at 25°C Stud Temperature . . . 2 V
Continuous Forward Current at (or below) 25°C Stud Temperature (See Note 1) . . . 300 mA
Peak Forward Current at (or below) 25°C Stud Temperature (See Note 2) . . . 500 mA
Storage Temperature Range . . . 0°C to 90°C
Solder Lug Temperature for 10 Seconds . . . 240°C

operating characteristics at 25°C stud temperature

PARAMETER		TEST CONDITIONS	MIN	TYP	MAX	UNIT
P_O	Radiant Power Output	I_F = 300 mA	15	20		mW
λ_p	Wavelength at Peak Emission			930		nm
$\Delta\lambda$	Spectral Bandwidth			45		nm
θ_{HI}	Half-Intensity Beam Angle			130°		
V_F	Static Forward Voltage			1.7	2.2	V
t_r	Radiant Pulse Rise Time	I_{FM} = 100 mA, $t_w \geq 5$ μs		600		ns
t_f	Radiant Pulse Fall Time			450		

NOTES: 1. Derate linearly to 70°C stud temperature at the rate of 6.7 mA/°C.
2. This value applies for $t_w \leq 100$ μs, duty cycle ≤50%. Derate linearly to 70°C stud temperature at the rate of 11.1 mA/°C

Copyright © 1983 by Texas Instruments Incorporated

TYPICAL CHARACTERISTICS

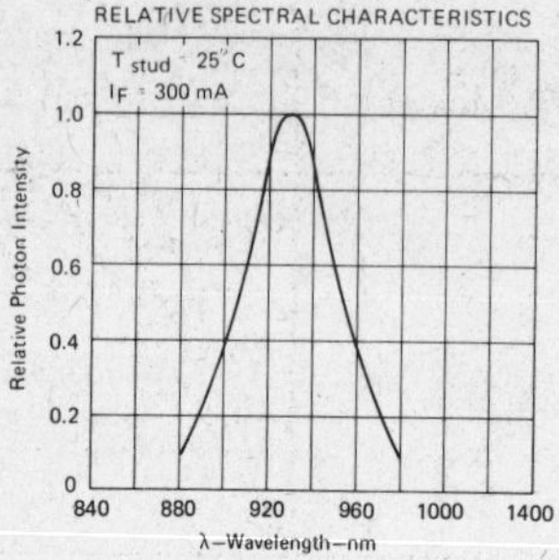

FIGURE 1

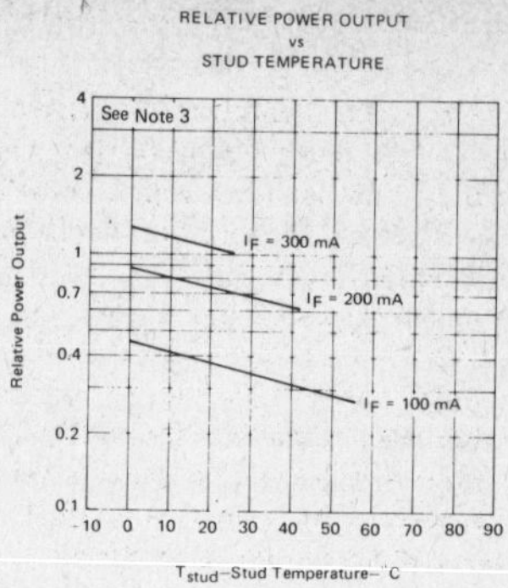

FIGURE 2

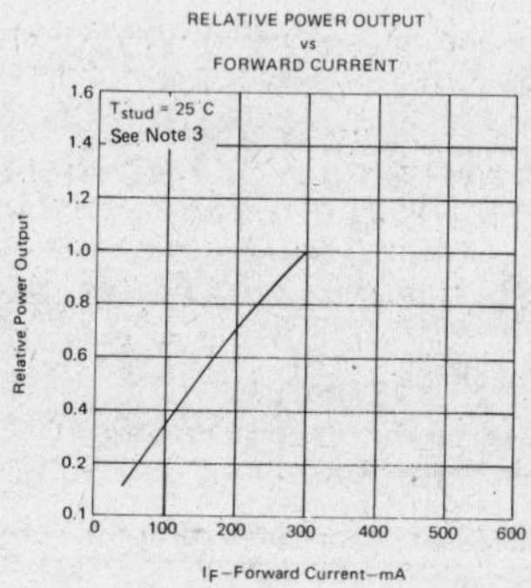

FIGURE 3

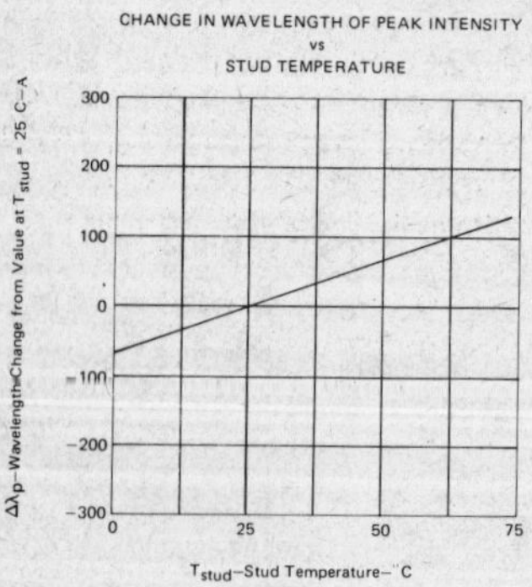

FIGURE 4

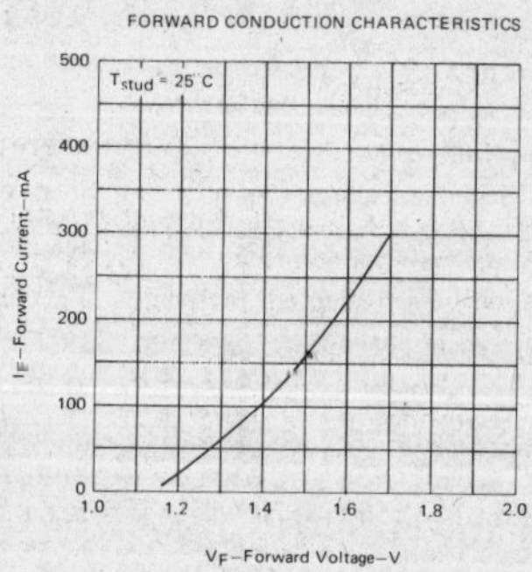

FIGURE 5

NOTE 3: These curves have been normalized to the output at I_F = 300 mA, T_{stud} = 25°C.

TEXAS INSTRUMENTS
INCORPORATED
POST OFFICE BOX 225012 • DALLAS, TEXAS 75265

TIES27 GaAs NONCOHERENT INFRARED SOURCE

The TIES27 GaAs noncoherent infrared source is essentially a solution-grown P-N junction. The output of the device is 15 mW minimum with 20 mW being typical at the rated forward current. The device emits in the near-infrared region.

This report presents basic information necessary to utilize this high-power, low-cost industrial IR source. Included in this discussion are the theory of operation, device performance including forward voltage, optical power, spectral distribution, radiance, radiant intensity, thermal impedance, pulse-mode operation and optical design considerations plus typical mechanical specification and application data.

THEORY OF OPERATION

The TIES27 GaAs noncoherent infrared source is a solution grown P-N junction in the shape of an 18-mil-square chip. The chip is mounted on a stud header and encapsulated in an epoxy dome.

When the P-N junction is forward biased, electrons from the N-region are injected into the P-region and radiant quanta (photons) are generated through recombination. The radiant energy emitted is in the near-infrared region.

A flat-geometry GaAs source emitting into air has a critical angle that can be described by:

$$\text{Sin}\,\theta_c = \frac{N_1}{N_2} \quad (1)$$

where N_1 = index of refraction of air = 1

N_2 = index of refraction of GaAs = 3.6

θ_c = critical angle = 16.1°.

Any radiant energy generated that strikes the surface of the chip at an angle greater than the critical angle will not escape but will be reflected internally. This is shown in Figure 1.

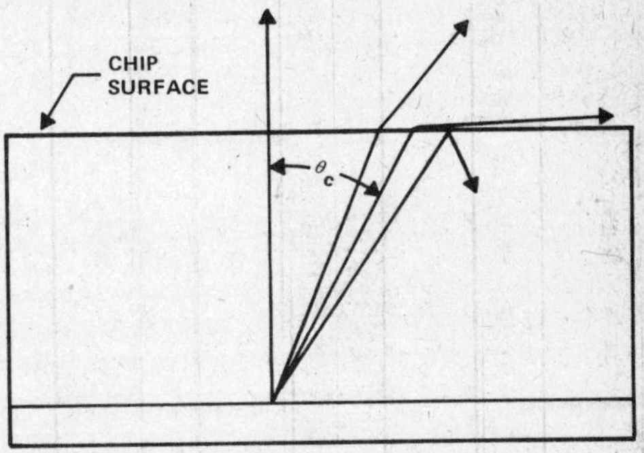

FIGURE 1. Angle of Light Determines if it Escapes or is Reflected Internally

The critical angle of the TIES27 chip has been changed by placing epoxy on the chip. Since the index of refraction of the epoxy is 1.5, the critical angle changes from 16.1° to 24.6°. The improvement factor can be calculated as follows:

$$\alpha = \frac{1 - \text{Cos}\,\theta_2}{1 - \text{Cos}\,\theta_1} = \frac{1 - \text{Cos}\,24.6^\circ}{1 - \text{Cos}\,16.1^\circ} = 2.31 \quad (2)$$

The improvement factor is valid only when the radiant energy that is emitted from the P-N junction can be transmitted through the epoxy and into the air.

The external quantum efficiency of the device can be described as the ratio of optical current output (photons per second) divided by forward input current.

$$\eta_S = \frac{I_\phi}{I_F} \quad (3)$$

DEVICE PERFORMANCE

Forward Voltage

At a constant temperature, the voltage change as a function of current can be predicted from equation (4):

$$\Delta V_F = \frac{nKT}{q} \log e \frac{I_{F1}}{I_{F2}} \qquad (4)$$

where V_F = forward voltage

I_F = forward current

$\frac{KT}{q} \approx 26$ mV

n = constant

The value of n ranges from 1 to 3 for the TIES27 with n being larger at small forward bias currents. Exact values for n may be obtained by characterizing the diode under the operating conditions to which it will be exposed.

The typical distribution of the forward voltage at the rated current of 300 mA will range from 1.3 volts to a maximum of 2.2 volts.

Optical Power

The TIES27 generates an optical output power of 15 mW minimum. The optical output power approximates a linear function of the forward bias current when operated above a few milliamperes and at or below the maximum specified forward current. Figure 2 shows relative optical power versus forward drive current.

The optical output power can be described by Equation (5):

$$P_o = I_\phi E$$

where I_ϕ = optical output current = $\eta_s I_F$ (5)

$$E = \frac{1.24}{\lambda_p}$$

λ_p = peak wavelength in micrometers

The optical power of the TIES27 varies inversely with temperature. A typical curve of optical output power versus temperature is shown in Figure 3.

Spectral Distribution

The distribution of emission wavelengths of the TIES27 is narrow; half-power wavelengths are typically separated by 450 angstroms. The peak wavelength ranges from 9300 to 9450 angstroms when operated at rated forward current (300 mA) at 25°C stud temperature. The peak wavelength (λ_p) is a function of forward bias current and temperature. The change in wavelength of peak intensity versus case (stud) temperature is shown in Figure 4.

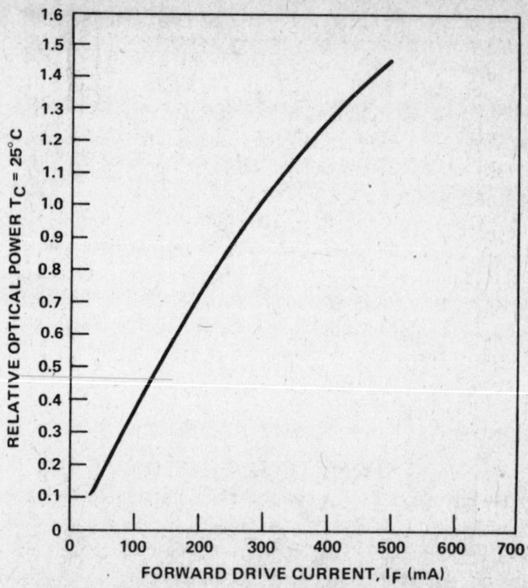

FIGURE 2. Relative Optical Power versus Forward Drive Current for TIES27. T = 25°C.

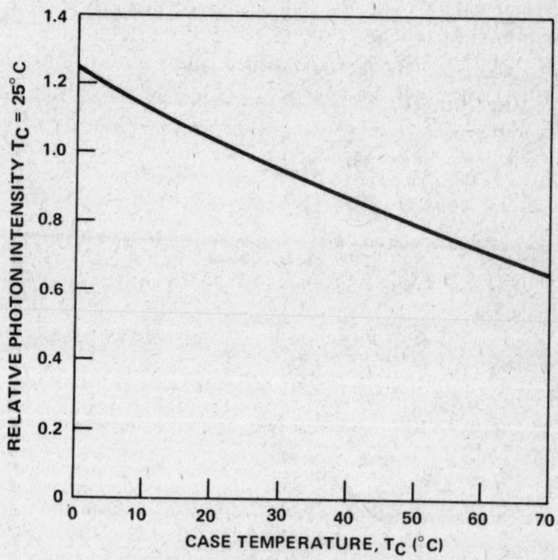

FIGURE 3. Relative Photon Intensity versus Case Temperature

TEXAS INSTRUMENTS
INCORPORATED
POST OFFICE BOX 225012 • DALLAS, TEXAS 75265

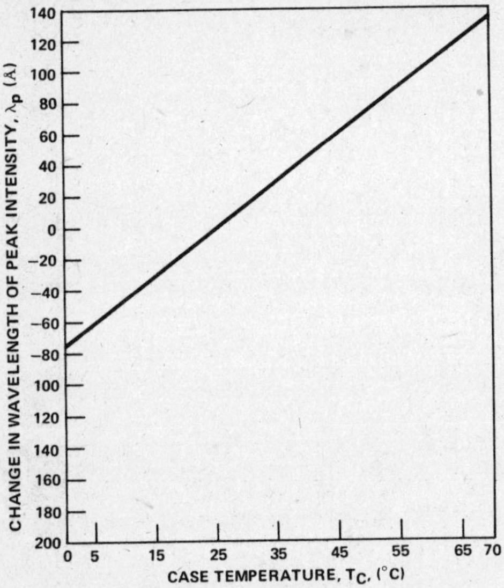

FIGURE 4. Change in Wavelength of Peak Intensity versus Case (Stud) Temperature. I_F = 300 mA.

Radiance

Radiance (L_e) is defined as radiant intensity emitted per unit area. In the case of the TIES27, the radiance can be calculated by using Equations (5) – (8):

$$P_o = I_\phi E \qquad (5)$$

where I_ϕ = optical output current = $\eta_s I_F$

$$E = \frac{1.24}{\lambda_p}$$

λ_p = peak wavelength in micrometers.

$$L_e = \frac{P_o/\Omega}{A} \qquad (6)$$

where P_o = total optical power

Ω = solid angle of emission in steradians

A = Area of active region in cm^2.

For the TIES27 (active area is 18 X 18 mils),

$$A = (0.018 \times 2.54)^2\ cm^2 = 2.09 \times 10^{-3}\ cm^2 \qquad (7)$$

$$L_e = \frac{(15 \times 10^{-3}\ W)/(2\pi\ sr)}{2.09 \times 10^{-3}\ cm^2} \qquad (8)$$

$$= 1.14\ W \cdot sr^{-1}/cm^2$$

It should be pointed out that this is the worst case because the TIES27 does not emit uniformly into 2π steradians but into a solid angle less than 2π.

Radiant Intensity

The radiant intensity of an isotropic radiator is equal in all directions, therefore, the radiant intensity is equal to

$$I_e = \frac{P}{2\pi}. \qquad (9)$$

where I_e = radiant intensity (W/sr)

P = total optical power (W)

However, most GaAs infrared emitters are not perfect isotropic radiators and the radiant intensity is higher on the optical axis or within a few degrees of the optical axis. Figure 5 shows a typical intensity pattern for the TIES27.

Thermal Resistance

The thermal resistance of the TIES27 is typically in the range of 12°C/W. The chip is mounted directly to the stud which when heatsinked properly can be approximated to the first order as an infinite heatsink. It is important to note that the thermal resistance is a very difficult parameter to determine and measured values from different groups of processed material may have a wide distribution.

Pulse Mode Operation

The TIES27 is capable of being pulsed at relatively high peak currents. The limiting factor, as it is in most pulsed mode applications, is the interfaces and not the P-N junction—the power density gets so large in the bonding wire or the contact pad that catastrophic failures occur. For example, a 1-mil gold wire that is 0.5 inches long has a power density of approximately 4200 W/cm^3 with 300 mA flowing through it. However, by increasing the current to 1 amp, the power density increases to approximately 47,000 W/cm^3.

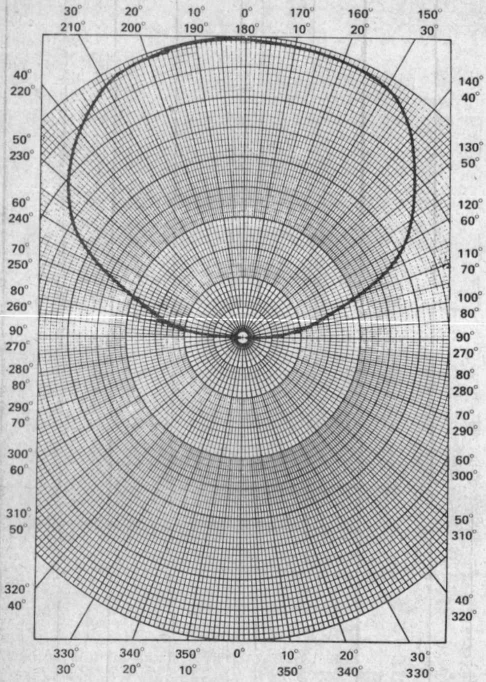

FIGURE 5. A Typical Intensity Pattern for the TIES 27

There are tradeoffs that must be considered when operating in the pulse mode such as duty cycle, repetition rate, and peak current. The peak current can be approximated with reasonable accuracy by using

$$I_{FM} = (I_F \max)/D = I_F \max \left(\frac{T}{t}\right) \qquad (10)$$

where I_{FM} = maximum peak current

I_F max = maximum-rated continuous forward current

D = duty cycle

T = period of frequency

t = diode "on" time

However, careful judgement should be used to ensure that the peak current does not exceed a level that will cause the bonding wires to open. The TIES27 should not be exposed to peak pulses of current greater than 4 amperes with an appropriate duty cycle. Figure 6 shows typical peak power

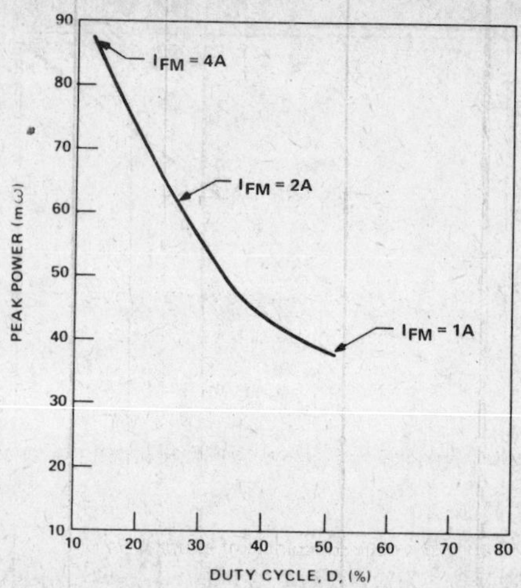

FIGURE 6. Typical Peak Power of TIES 27 at 10 kHz with Various Peak Current Levels at Various Duty Cycles

obtained when the device was operated at a frequency of 10 kHz with current levels of 1 ampere, 2 amperes and 4 amperes at respective duty cycles of 50%, 25%, and 12.5% – higher current pulses than equation (10) defines.

Optical Design Considerations

Since the TIES27 emits into such a large pattern (approximately 2π steradians), it is necessary to use some form of optics to collect and direct that portion of the optical power that will be used.

The amount of optical power collected can be determined quickly once the optics have been defined. The following is an example that illustrates the effect of the f-number of the lens on the power transmitted.

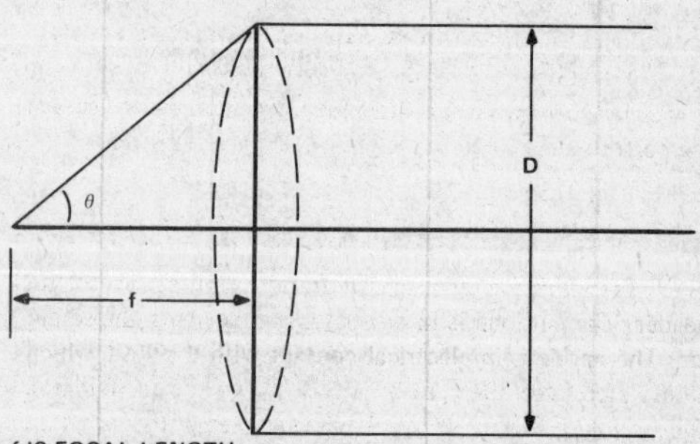

f IS FOCAL LENGTH

FIGURE 7. Typical Optical Collection Configuration

$$P_t = P_o \left(\frac{\Omega_c}{\Omega_e}\right) \eta_t = \qquad (11)$$

$$P_o \left(\frac{2\pi\,(1 - \cos\theta)}{2\pi}\right) \eta_t$$

where P_t = optical power transmitted in the beam of the collection optics.

P_o = the total radiated optical power.

Ω_c = the solid angle of collection in steradians.

Ω_e = the solid angle of emission in steradians.

η_t = the transmission efficiency of the lens.

θ = the half angle of the collection cone.

Table 1

f-number	θ (°)	$1 - \cos\theta$	P_t (mW)
1.0	26.6	0.1	1.5
1.4	19.6	0.06	0.8
2.0	14.0	0.03	0.45
2.8	10.2	0.02	0.3
4.0	7.0	0.01	0.15

TYPICAL APPLICATION DATA

Figure 8 shows an economical approach for modulating a TIES27. This circuit features excellent bandwidth as well as high peak currents. Figures 9, 10, and 11 show the performance data for the circuit shown in Figure 8.

MECHANICAL DATA

The device is encapsulated and mounted on a stud header. The cathode is in electrical contact with the solder lug. The anode is in electrical contact with the stud, which is insulated from the case by a glass-to-metal seal. Soldered connections should not be made directly to the stud because of the low-thermal-resistance path between stud and emitting element.

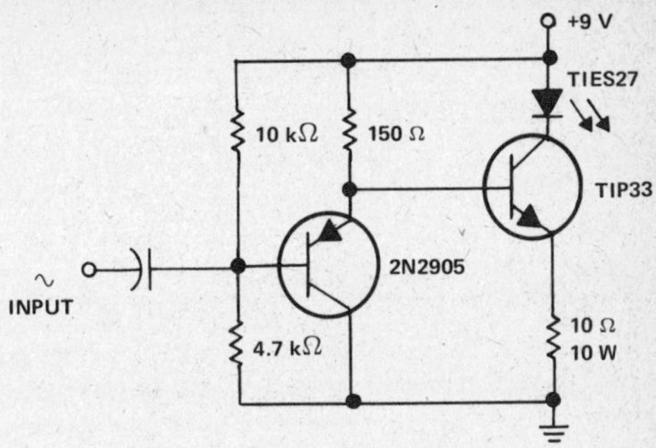

FIGURE 8. An Economical Circuit for Modulating a TIES27

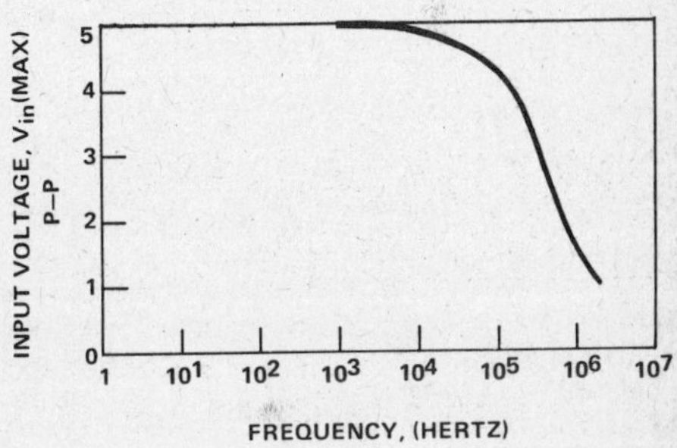

FIGURE 9. Maximum Frequency for Circuit in Figure 8

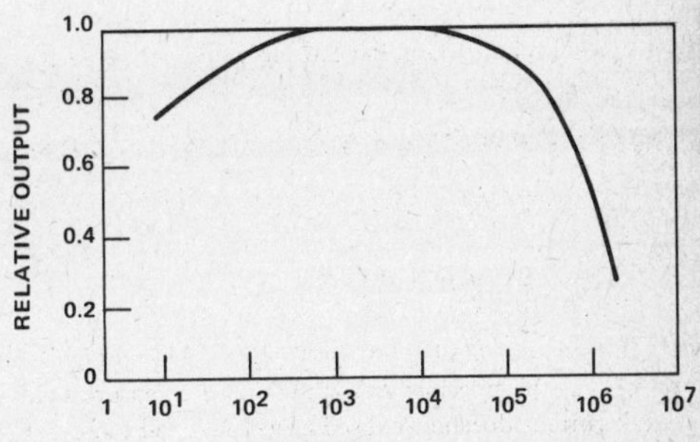

FIGURE 10. Frequency Response of IRED in Circuit of Figure 8

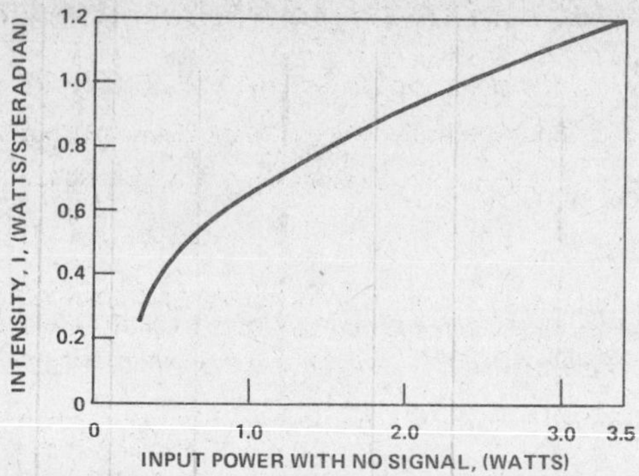

FIGURE 11. Intensity versus Input Power for Circuit in Figure 8 When it is Used with an f/1.6 Lens Which has a 29-Millimeter Diameter

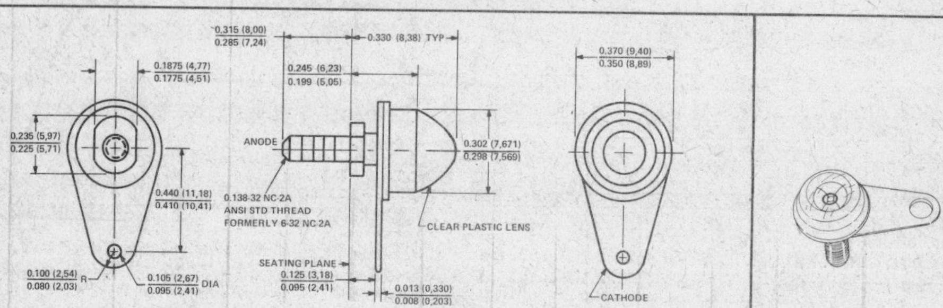

NOTES: a. The orientation of the lug in relation to the stud flats is not controlled.
b. Torque between stud and lug must be avoided. Flats are provided on the stud for tightening to heat sink.
c. All dimensions are in inches and parenthetically in millimeters. Inch dimensions govern.

FIGURE 12. Mechanical Specifications for TIES27

TEXAS INSTRUMENTS
INCORPORATED
POST OFFICE BOX 225012 • DALLAS, TEXAS 75265

D1948, NOVEMBER 1974—REVISED APRIL 1983

DESIGNED TO EMIT NEAR INFRARED RADIANT ENERGY WHEN FORWARD BIASED

- High Speed, High Efficiency
- Hemispherically Shaped 18-Mil-Diameter Chip
- Stud Mounting for Convenient Heat Sinking

mechanical data

The cathode is in electrical contact with the case and adjacent solder lug. The anode is in electrical contact with the stud, which is insulated from the case by a glass-to-metal seal. Soldered connections should not be made directly to the stud because of the low-thermal-resistance path between the stud and emitting element.

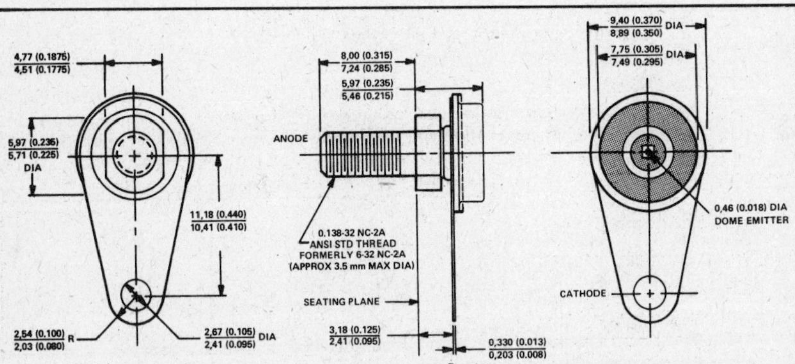

NOTES: a. Torque between stud and lug must be avoided. Flats are provided on the stud for tightening to heat sink.

b. The orientation of the lug in relation to the stud flats is not controlled.

ALL DIMENSIONS ARE IN MILLIMETERS AND PARENTHETICALLY IN INCHES

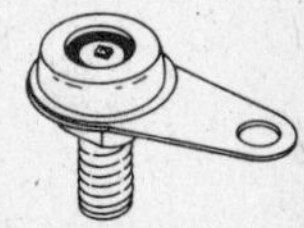

absolute maximum ratings

Reverse Voltage at 25°C Stud Temperature . . . . 2 V
Continuous Forward Current at (or below) 25°C Stud Temperature (See Note 1) . . . . 200 mA
Peak Forward Current at (or below) 25°C Stud Temperature (See Note 2) . . . . 300 mA
Storage Temperature Range . . . . −55°C to 100°C
Solder Lug Temperature for 10 Seconds . . . . 240°C

operating characteristics at 25°C stud temperature

	PARAMETER	TEST CONDITIONS	MIN	TYP	MAX	UNIT
P_O	Radiant Power Output	I_F = 50 mA	900	1200		μW
λ_p	Wavelength at Peak Emission			910		nm
$\Delta\lambda$	Spectral Bandwidth			30		nm
θ_{HI}	Half-Intensity Beam Angle			135°		
V_F	Static Forward Voltage			1.5	2	V
t_r	Radiant Pulse Rise Time	I_{FM} = 50 mA,		15		ns
t_f	Radiant Pulse Fall Time	$t_w \geq 100$ ns		15		

† Radiant pulse rise time is the time required for a change in radiant power output from 10% to 90% of its peak value for a step change in current.

NOTES: 1. Derate linearly to 50 mA at 100°C stud temperature at the rate of 2.0 mA/°C.

2. This value applies for tw ≤100 μs, duty cycle ≤50%. Derate linearly to 100°C stud temperature at the rate of 3.0 mA/°C.

Copyright © 1983 by Texas Instruments Incorporated

TEXAS INSTRUMENTS
INCORPORATED
POST OFFICE BOX 225012 • DALLAS, TEXAS 75265

Photodetectors (Sensors)

- **Quick Reference Guide**
- **Low-Cost Plastic Packages**
  - T-1
  - T-1¾
  - Sidelookers
- **Hermetically Sealed Packages**
  - Pill
  - TO-18
- **High-Reliability Devices (HR2)**
  - Pill
  - TO-18

See Section 6 for Avalanche Photodiodes.

PHOTODETECTORS
QUICK REFERENCE GUIDE

DEVICE	TYPE	LIGHT CURRENT MIN	LIGHT CURRENT MAX @	V	DARK CURRENT MAX @	V	POWER DISS.	FEATURES
1N5722	Phototransistor	0.5 mA	3 mA	5	25 nA	30	50 mW	EIA-Registered versions of TIL601-TIL604
1N5723	Phototransistor	2 mA	5 mA	5	25 nA	30	50 mW	
1N5724	Phototransistor	4 mA	8 mA	5	25 nA	30	50 mW	
1N5725	Phototransistor	7 mA		5	25 nA	30	50 mW	
LS600	Phototransistor	0.8 mA		5	25 nA	30	50 mW	Pill package (See TIL601 Series)
TIL78	Phototransistor	1 mA		5	25 nA	30	50 mW	Low-cost epoxy package compatible with TIL32, TIL902
TIL81†	As Phototransistor	5 mA		5	100 nA	10	250 mW	TO-18 package with narrow field of view. Compatible with TIL31B, TIL33B, TIL34B, TIL902, TIL904
	As Photodiode	170 μA Typ		0-50	10 nA	10	250 mW	
TIL99	As Phototransistor	1 mA		5	100 nA	10	250 mW	Similar to TIL81 except flat lens
	As Photodiode	40 μA Typ		0-50	10 nA	10	250 mW	
TIL100	Photodiode	10 μA		10	50 nA	10	150 mW	Designed for infrared remote-control systems Compatible with TIL38, TIL39, TIL905, and TIL906
TIL411	Phototransistor	100 μA		5	100 nA	5	50 mW	Compatible with TIL40; Reverse pinout of TIL415
TIL412	Photodarlington	500 μA		1	100 nA	5	50 mW	Compatible with TIL40; Reverse pinout of TIL416
TIL413	Photodiode	10 μA		10	50 nA	10	150 mW	Compatible with TIL38, TIL39, TIL905, and TIL906
TIL414	Phototransistor	100 μA		5	50 nA	10	50 mW	
TIL415	Phototransistor	100 μA		5	100 nA	5	50 mW	Compatible with TIL40
TIL416	Photodarlington	500 μA		1	100 nA	5	50 mW	Compatible with TIL40
TIL601	Phototransistor	0.5 mA	3 mA	5	25 nA	30	50 mW	Pill package designed for mounting on double-sided printed board. Compatible with TIL23 series
TIL602	Phototransistor	2 mA	5 mA	5	25 nA	30	50 mW	
TIL603	Phototransistor	4 mA	8 mA	5	25 nA	30	50 mW	
TIL604†	Phototransistor	7 mA		5	25 nA	30	50 mW	

†High-reliability versions (TIL81 HR2 and TIL604 HR2) are also available.

For additional photodetectors, see *Special Electro-optical Components* section of this book.

D974, MARCH 1972–REVISED NOVEMBER 1974

JEDEC-REGISTERED VERSIONS OF TIL601 THRU TIL604

- Recommended for Application in Character Recognition, Tape and Card Readers, Velocity Indicators, and Encoders
- Unique Package Design Allows for Assembly into Printed Circuit Boards

*mechanical data

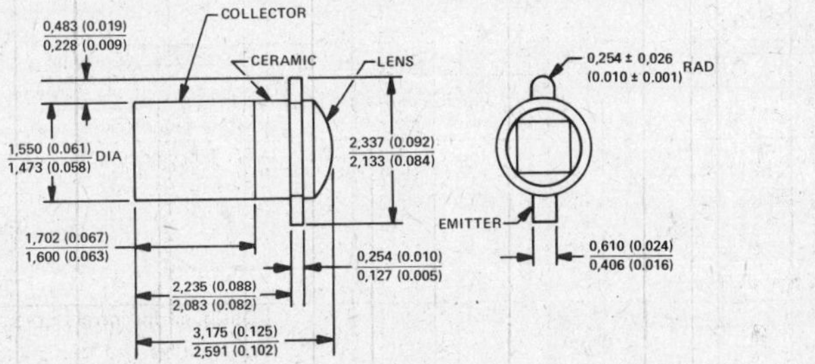

FOUR TIMES ACTUAL SIZE

ALL DIMENSIONS ARE IN MILLIMETERS AND PARENTHETICALLY IN INCHES

*absolute maximum ratings at 25°C case temperature (unless otherwise noted)

Collector-Emitter Voltage	50 V
Emitter-Collector Voltage	7 V
Continuous Device Dissipation at (or below) 25°C Case Temperature (See Note 1)	50 mW
Operating Case Temperature Range	−65°C to 125°C
Storage Temperature Range	−65°C to 150°C
Soldering Temperature (3 minutes)	240°C

*electrical characteristics at 25°C case temperature (unless otherwise noted)

PARAMETER		TEST CONDITIONS	TYPE	MIN	TYP	MAX	UNIT
$V_{(BR)CEO}$	Collector-Emitter Breakdown Voltage	I_C = 100 µA, E_e = 0	ALL	50			V
$V_{(BR)ECO}$	Emitter-Collector Breakdown Voltage	I_E = 100 µA, E_e = 0	ALL	7			V
I_D	Dark Current	V_{CE} = 30 V, E_e = 0	ALL			25	nA
		V_{CE} = 30 V, E_e = 0, T_C = 100°C	ALL		1		µA
I_L	Light Current	V_{CE} = 5 V, E_e = 20 mW/cm², See Note 2	1N5722	0.5		3	mA
			1N5723	2		5	
			1N5724	4		8	
			1N5725	7			
$V_{CE(sat)}$	Collector-Emitter Saturation Voltage	I_C = 0.4 mA, E_e = 20 mW/cm², See Note 2	ALL		0.15		V

NOTES: 1. Derate linearly to 125°C at the rate of 0.5 mW/°C.

2. Irradiance (E_e) is the radiant power per unit area incident upon a surface. For this measurement the source is an unfiltered tungsten linear-filament lamp operating at a color temperature of 2870 K.

*JEDEC registered data. This data sheet contains all applicable JEDEC registered data in effect at the time of publication.

***switching characteristics at 25°C case temperature**

PARAMETER	TEST CONDITIONS	MIN	TYP	MAX	UNIT
t_r Rise Time	V_{CC} = 30 V, I_L = 800 µA,		1.5	2.5	µs
t_f Fall Time	R_L = 1 kΩ, See Figure 1		15	25	

*PARAMETER MEASUREMENT INFORMATION

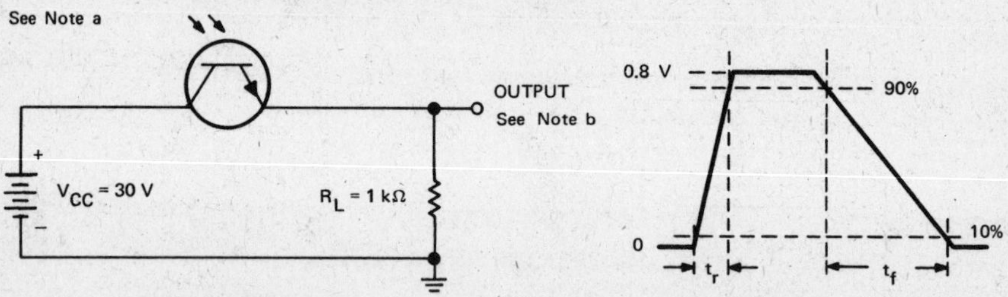

TEST CIRCUIT **OUTPUT VOLTAGE WAVEFORM**

FIGURE 1

NOTES: a. Input irradiance is supplied by a pulsed xenon bulb source. Incident irradiation is adjusted for I_L = 800 µA.

b. Output waveform is monitored on an oscilloscope with the following characteristics: $t_r \leq 25$ ns, $R_{in} \geq 1$ MΩ, $C_{in} \leq 20$ pF.

*JEDEC registered data

TYPICAL CHARACTERISTICS

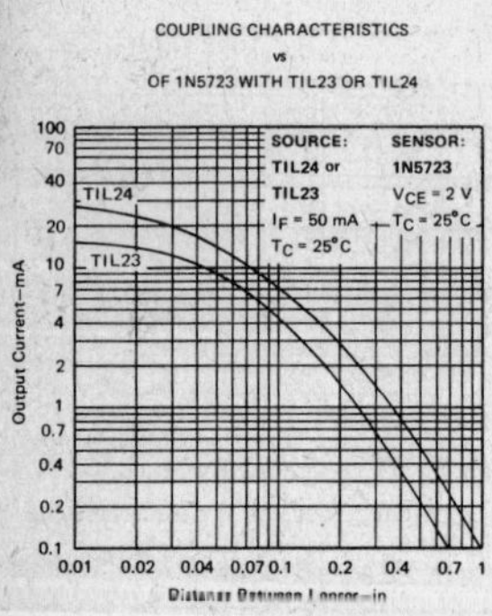

FIGURE 2

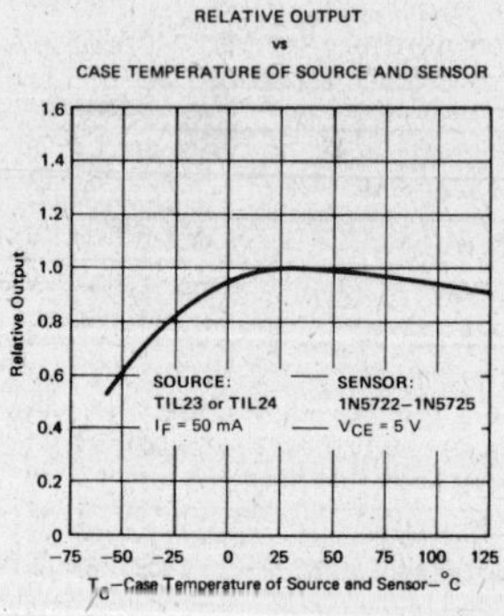

FIGURE 3

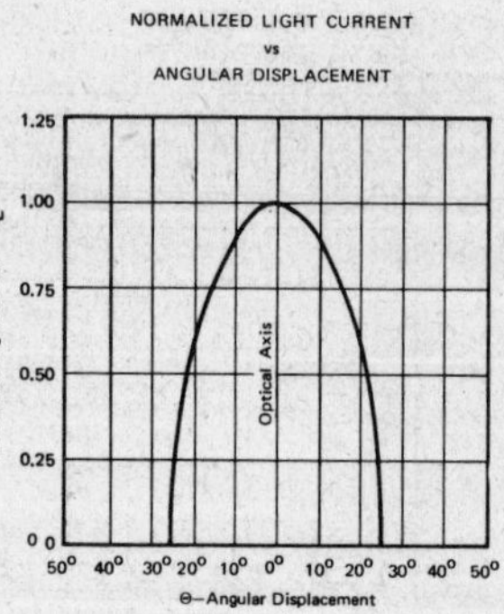

FIGURE 4

TYPICAL CHARACTERISTICS

COLLECTOR CURRENT
vs
COLLECTOR-EMITTER VOLTAGE

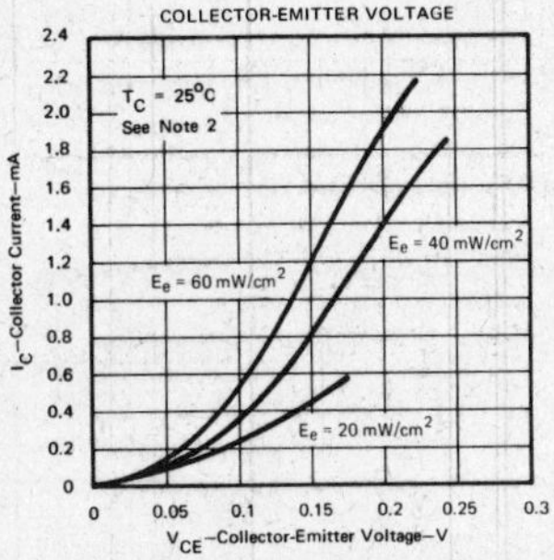

FIGURE 5

1N5723
COLLECTOR CURRENT
vs
COLLECTOR-EMITTER VOLTAGE

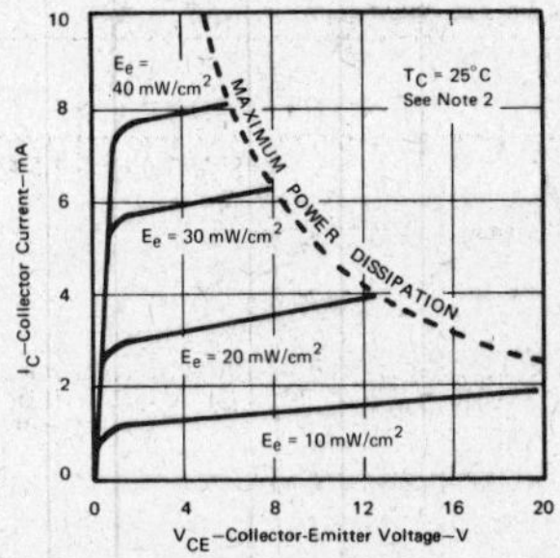

FIGURE 6

COLLECTOR-EMITTER SATURATION VOLTAGE
vs
IRRADIANCE

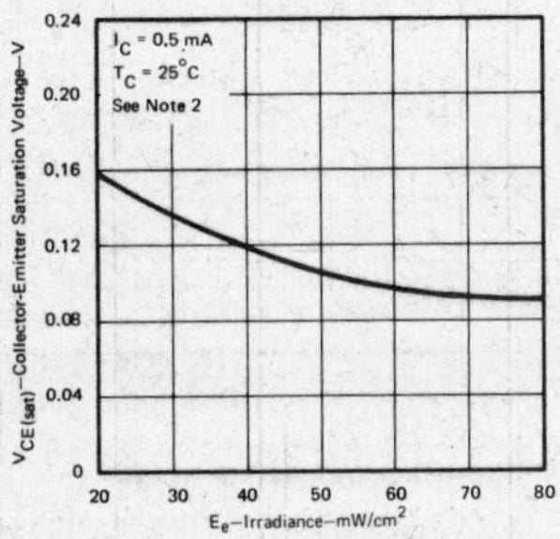

FIGURE 7

1N5723
COLLECTOR CURRENT
vs
IRRADIANCE

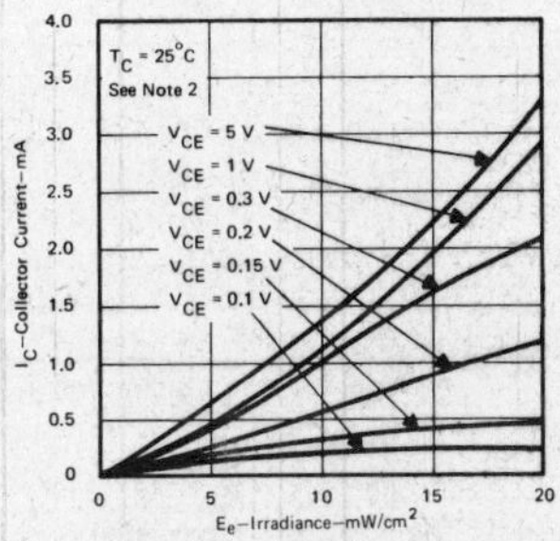

FIGURE 8

DARK CURRENT
vs
CASE TEMPERATURE

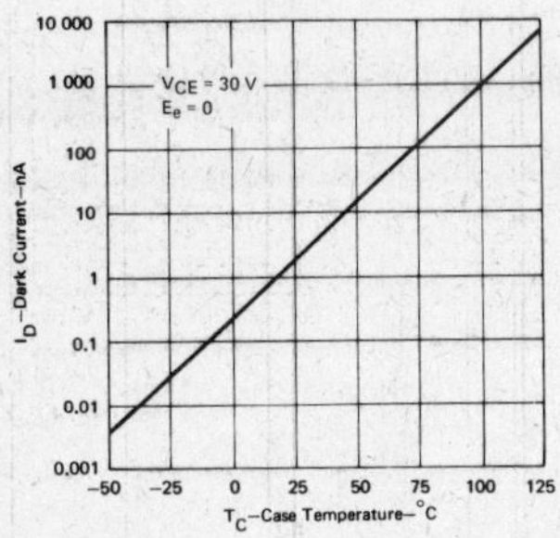

FIGURE 9

RELATIVE OUTPUT
vs
MODULATION FREQUENCY

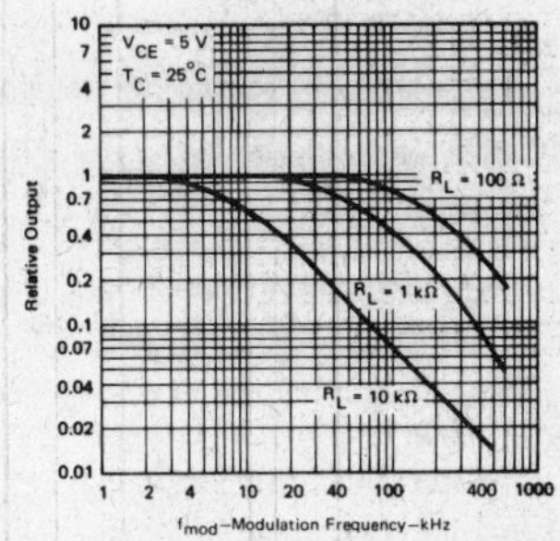

FIGURE 10

NOTE 2: Irradiance (E_e) is the radiant power per unit area incident upon a surface. For this measurement the source is an unfiltered tungsten linear-filament lamp operating at a color temperature of 2870 K.

TYPICAL CHARACTERISTICS

RELATIVE SPECTRAL CHARACTERISTICS

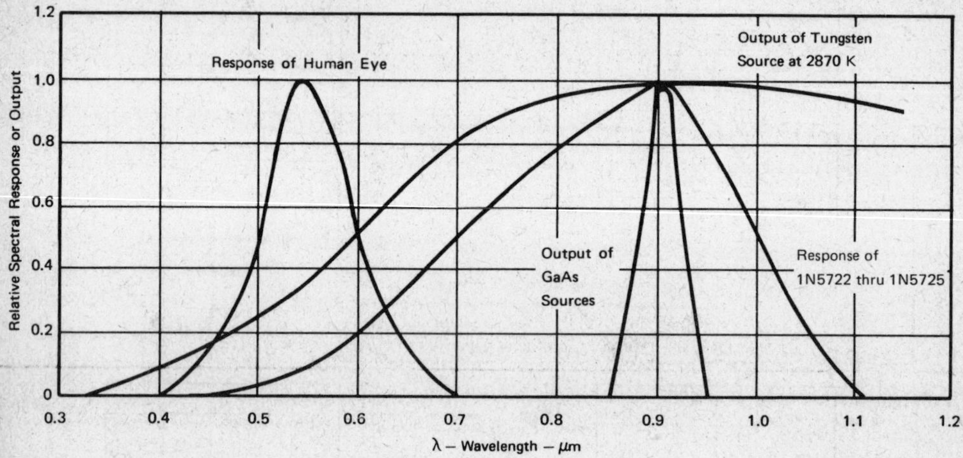

FIGURE 11

TEXAS INSTRUMENTS CUSTOMIZED OPTOELECTRONIC ARRAYS

The 1N5722 through 1N5725 series is available mounted in printed circuit boards for custom-designed array or matrix applications. The array is a complete unit, without the problems associated with small, difficult-to-handle components. These arrays can be designed for punched-card or tape readers, position indicators, pattern and character recognition, shaft encoders, and many other special applications.

Texas Instruments custom-array techniques offer many advantages:

- The arrays are pre-assembled and tested, ready for installation.
- Custom arrays can be manufactured in almost any configuration to allow maximum design flexibility.
- Sensitivity across an entire array will be matched to within 50%.
- GaAs sources can be furnished to give complete solid-state matched sets for specific applications.
- Arrays with components firmly soldered into place on both sides of a printed circuit board are more rugged than individually wired sensing devices.

Specifying optoelectronic arrays is easy; all that is required is a print of the array and the desired specifications.

TI sales engineers will assist in developing specifications for special applications.

TYPE TIL78
N-P-N SILICON PHOTOTRANSISTOR

D1856, SEPTEMBER 1971—REVISED DECEMBER 1982

- **Designed for Automatic or Hand Insertion in Sockets or PC Boards**
- **Recommended for Industrial Applications Requiring Low-Cost Discrete Phototransistors**
- **Spectrally and Mechanically Compatible with TIL32 and TIL902 IR Emitters**

mechanical data

This device has a clear molded epoxy body.

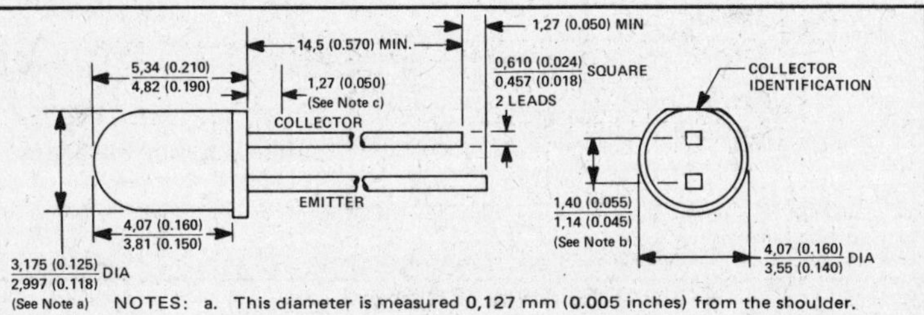

NOTES: a. This diameter is measured 0,127 mm (0.005 inches) from the shoulder.
b. Lead spacing is measured where the leads emerge from the package.
c. Within this zone the package contour is not controlled.

ALL DIMENSIONS ARE IN MILLIMETERS AND PARENTHETICALLY IN INCHES

CL-9

absolute maximum ratings at 25°C free-air temperature (unless otherwise noted)

Collector-Emitter Voltage	50 V
Emitter-Collector Voltage	7 V
Continuous Device Dissipation at (or below) 25°C Free-Air Temperature (See Note 1)	50 mW
Operating Free-Air Temperature Range	−40°C to 100°C
Storage Temperature Range	−40°C to 125°C
Lead Temperature 1,6 mm (1/16 Inch) from Case for 5 Seconds	240°C

electrical characteristics at 25°C free-air temperature (unless otherwise noted)

PARAMETER		TEST CONDITIONS		MIN	TYP	MAX	UNIT
$V_{(BR)CEO}$	Collector-Emitter Breakdown Voltage	I_C = 100 μA, E_e = 0		50			V
$V_{(BR)ECO}$	Emitter-Collector Breakdown Voltage	I_E = 100 μA, E_e = 0		7			V
I_D	Dark Current	V_{CE} = 30 V, E_e = 0				100	nA
		V_{CE} = 30 V, E_e = 0,	T_A = 80°C			1	μA
I_L	Light Current	V_{CE} = 5 V, E_e = 20 mW/cm²	See Note 2	1	7		mA
		V_{CE} = 5 V, E_e = 2 mW/cm²			0.5		
$V_{CE(sat)}$	Collector-Emitter Saturation Voltage	I_C = 2 mA, E_e = 20 mW/cm²,	See Note 2		0.4		V

switching characteristics at 25°C free-air temperature

PARAMETER		TEST CONDITIONS	TYP	UNIT
t_r	Rise Time	V_{CC} = 30 V, I_L = 800 μA,	8	μs
t_f	Fall Time	R_L = 1 kΩ, See Figure 1	6	

NOTES: 1. Derate linearly to 100°C free-air temperature at the rate of 0.67 mA/°C.
2. Irradiance (E_e) is the radiant power per unit area incident upon a surface. For this measurement the source is an unfiltered tungsten linear-filament lamp operating at a color temperature of 2870 K.

Copyright © 1982 by Texas Instruments Incorporated

282

TEXAS INSTRUMENTS
INCORPORATED
POST OFFICE BOX 225012 • DALLAS, TEXAS 75265

PARAMETER MEASUREMENT INFORMATION

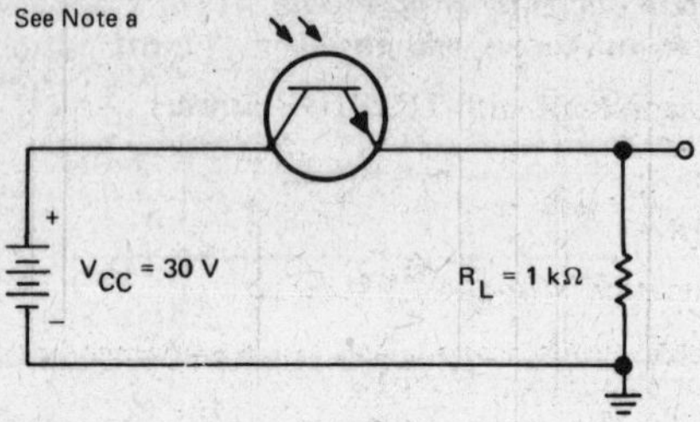

TEST CIRCUIT

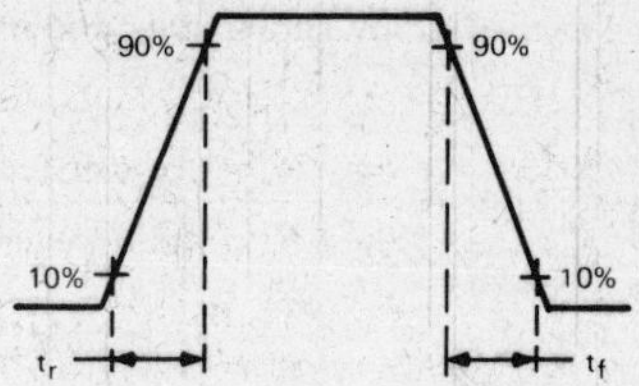

OUTPUT VOLTAGE WAVEFORM

FIGURE 1

NOTES: a. Input irradiance is supplied by a pulsed gallium arsenide infrared emitter with rise and fall times of less than 50 ns. Incident irradiation is adjusted for I_L = 800 μA.

b. Output waveform is monitored on an oscilloscope with the following characteristics: $t_r \leq 25$ ns, $R_{in} \geq 1$ MΩ, $C_{in} \leq 20$ pF.

TYPICAL CHARACTERISTICS

NORMALIZED LIGHT CURRENT
vs
ANGULAR DISPLACEMENT

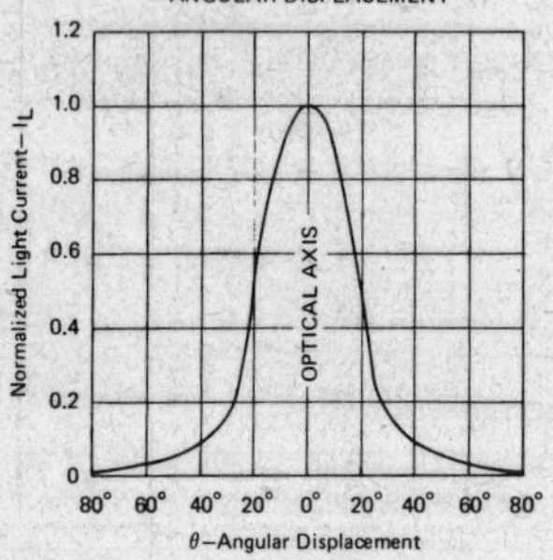

FIGURE 1

DARK CURRENT
vs
FREE-AIR TEMPERATURE

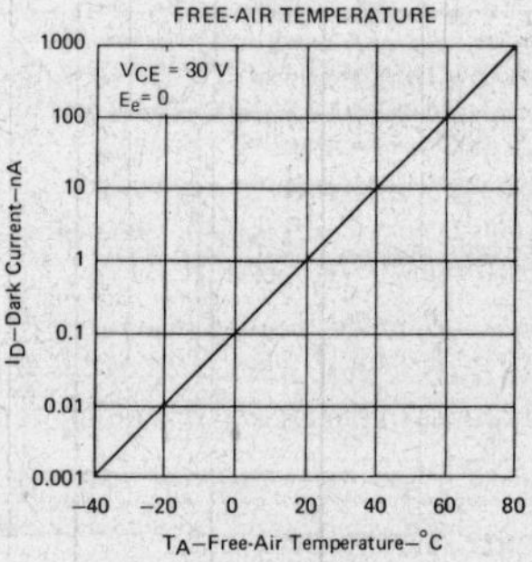

FIGURE 2

COLLECTOR-EMITTER SATURATION VOLTAGE
vs
IRRADIANCE

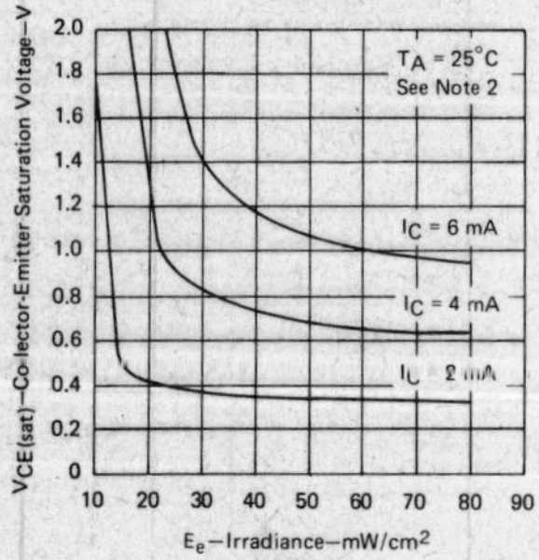

FIGURE 3

COUPLING CHARACTERISTICS
OF
TIL32 WITH TIL78

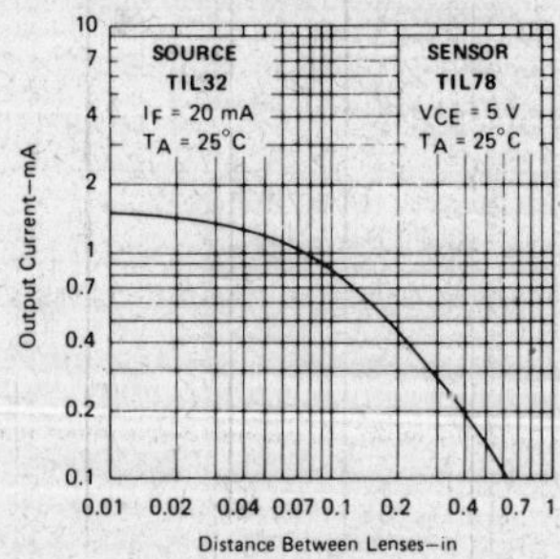

FIGURE 4

NOTE 2: Irradiance (E_e) is the radiant power per unit area incident upon a surface. For this measurement the source is an unfiltered tungsten linear-filament lamp operating at a color temperature of 2870 K.

D1215, MARCH 1972–REVISED DECEMBER 1982

- Recommended for Application in Character Recognition, Tape and Card Readers, Velocity Indicators, and Encoders
- Spectrally and Mechanically Compatible with TIL31B IR Emitter
- Glass-to-Metal-Seal Header
- Base Contact Externally Available
- Saturation Level Directly Compatible with Most TTL/DTL
- TIL81HR2* Includes High-Reliability Processing and Lot Acceptance (See Page 5-13 for Summary of Processing)

mechanical data

The device is in a hermetically sealed package with glass window. The outline of the TIL81 is similar to TO-18 except for the window. All TO-18 registration notes also apply to this outline

THE COLLECTOR IS IN ELECTRICAL CONTACT WITH THE CASE

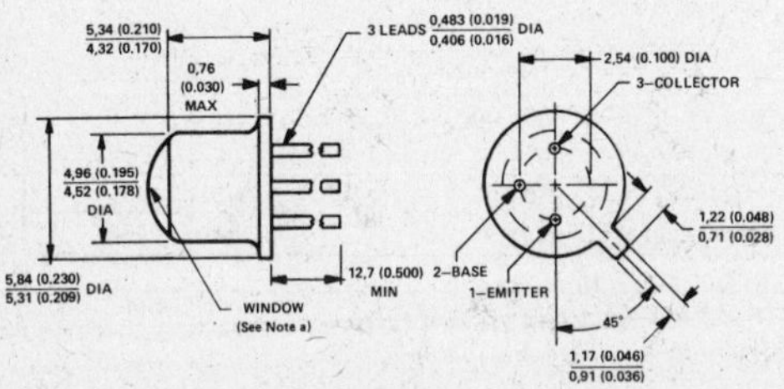

NOTES:

a. The window is Corning 7052 glass or equivalent.
Window dimensions are:
Minimum diameter . . 3,2 (0.125)
Maximum extension from top of case 1,78 (0.070)

ALL DIMENSIONS ARE IN MILLIMETERS AND PARENTHETICALLY IN INCHES

absolute maximum ratings at 25°C free-air temperature (unless otherwise noted)

Collector-Base Voltage	50 V
Collector-Emitter Voltage	30 V
Emitter-Base Voltage	7 V
Emitter-Collector Voltage	7 V
Continuous Collector Current	50 mA
Continuous Device Dissipation at (or below) 25°C Free-Air Temperature (See Note 1)	250 mW
Operating Free-Air Temperature Range	−65°C to 125°C
Storage Temperature Range	−65°C to 150°C
Lead Temperature 1,6 mm (1/16 inch) from Case for 10 Seconds	240°C

NOTE 1: Derate linearly to 125°C free-air temperature at the rate of 2.5 mW/°C.

*All electrical and mechanical specifications for the TIL81 also apply for the TIL81HR2.

Copyright © 1982 by Texas Instruments Incorporated

TEXAS INSTRUMENTS
INCORPORATED
POST OFFICE BOX 225012 • DALLAS, TEXAS 75265

electrical characteristics at 25°C free-air temperature (unless otherwise noted)

PARAMETER			TEST CONDITIONS	MIN	TYP	MAX	UNIT
$V_{(BR)CBO}$	Collector-Base Breakdown Voltage		I_C = 100 μA, I_E = 0, E_e = 0	50			V
$V_{(BR)CEO}$	Collector-Emitter Breakdown Voltage		I_C = 100 μA, I_B = 0, E_e = 0	30			V
$V_{(BR)EBO}$	Emitter-Base Breakdown Voltage		I_E = 100 μA, I_C = 0, E_e = 0	7			V
$V_{(BR)ECO}$	Emitter-Collector Breakdown Voltage		I_E = 100 μA, I_B = 0, E_e = 0	7			V
I_D	Dark Current	Phototransistor Operation	V_{CE} = 10 V, I_B = 0, E_e = 0			0.1	μA
			V_{CE} = 10 V, I_B = 0, E_e = 0, T_A = 100°C		20		
		Photodiode Operation	V_{CB} = 10 V, I_E = 0, E_e = 0			0.01	μA
I_L	Light Current	Phototransistor Operation	V_{CE} = 5 V, I_B = 0, E_e = 5 mW/cm², See Note 2	5	22		mA
		Photodiode Operation	V_{CB} = 0 to 50 V, I_E = 0, E_e = 20 mW/cm², See Note 2		170		μA
h_{FE}	Static Forward Current Transfer Ratio		V_{CE} = 5 V, I_C = 1 mA, E_e = 0		200		
$V_{CE(sat)}$	Collector-Emitter Saturation Voltage		I_C = 2 mA, I_B = 0, E_e = 20 mW/cm², See Note 2		0.2		V

NOTE 2: Irradiance (E_e) is the radiant power per unit area incident upon a surface. For these measurements the source is an unfiltered tungsten linear-filament lamp operating at a color temperature of 2870 K.

switching characteristics at 25°C free-air temperature

PARAMETER			TEST CONDITIONS	TYPICAL	UNIT
t_r	Rise Time	Phototransistor Operation	V_{CC} = 5 V, I_L = 800 μA, R_L = 100 Ω, See Test Circuit A of Figure 1	8	μs
t_f	Fall Time			6	
t_r	Rise Time	Photodiode Operation	V_{CC} = 0 to 50 V, I_L = 60 μA, R_L = 100 Ω, See Test Circuit B of Figure 1	350	ns
t_f	Fall Time			500	

PARAMETER MEASUREMENT INFORMATION

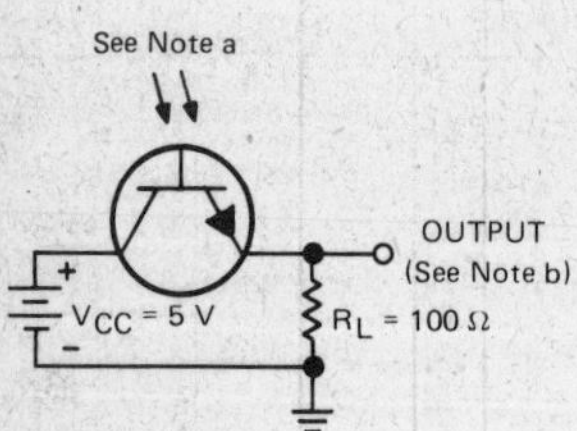

TEST CIRCUIT A
PHOTOTRANSISTOR OPERATION

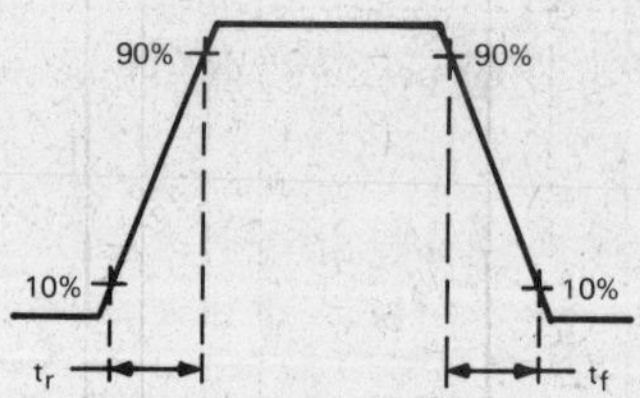

VOLTAGE WAVEFORM

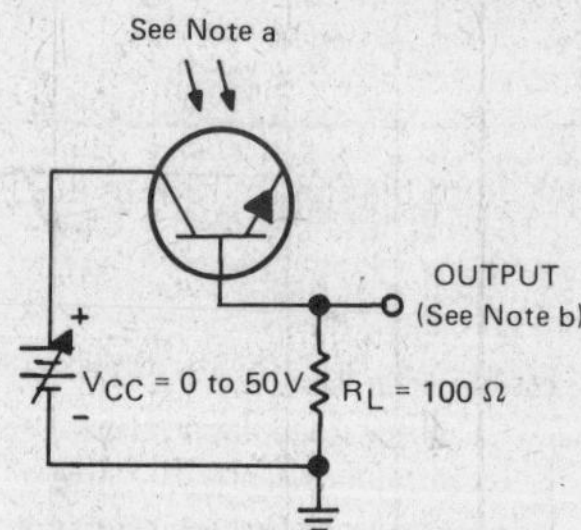

TEST CIRCUIT B
PHOTODIODE OPERATION

NOTES: a. Input irradiance is supplied by a pulsed gallium arsenide infrared emitter with rise and fall times less than 50 ns. Incident irradiation is adjusted for specified I_L.
b. Output waveform is monitored on an oscilloscope with the following characteristics: $t_r \leqslant 25$ ns, $R_{in} \geqslant 1\ M\Omega$, $C_{in} \leqslant 20$ pF.

FIGURE 1

TEXAS INSTRUMENTS
INCORPORATED
POST OFFICE BOX 225012 • DALLAS, TEXAS 75265

TYPICAL CHARACTERISTICS

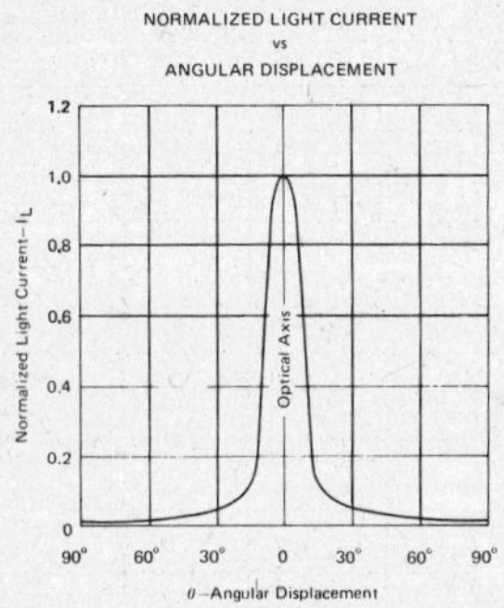

FIGURE 2

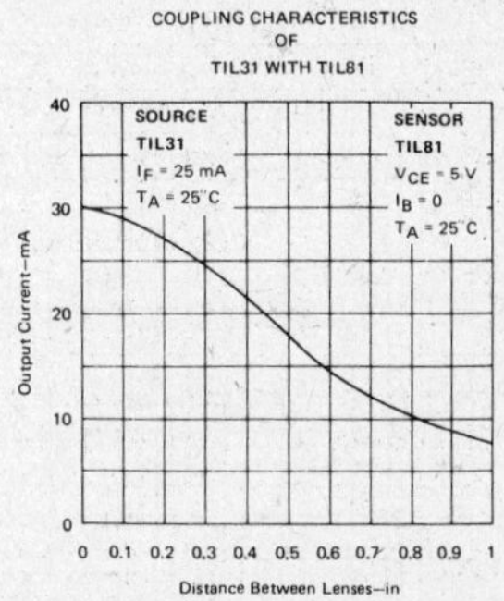

FIGURE 3

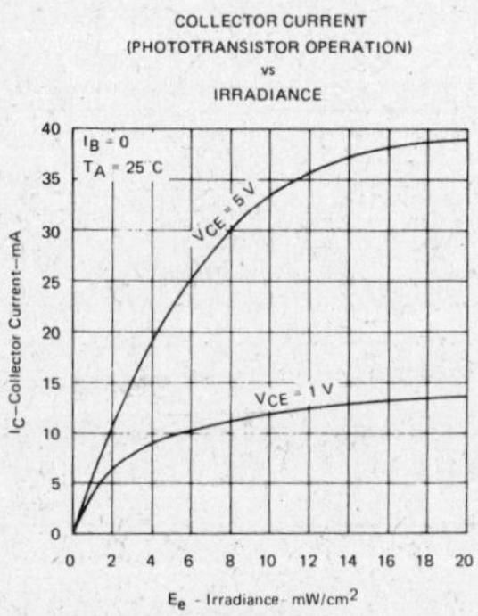

FIGURE 4

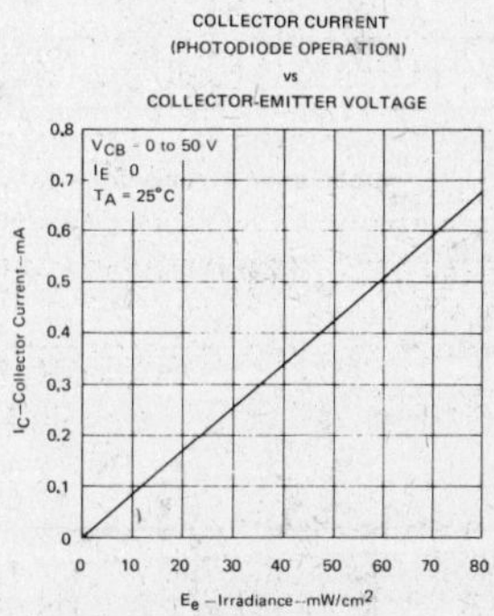

FIGURE 5

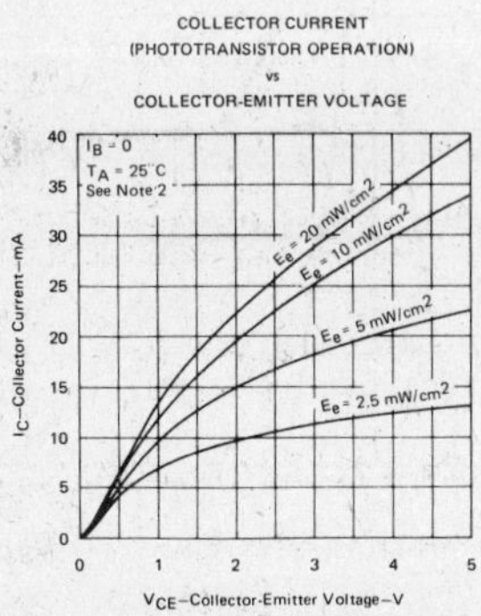

FIGURE 6

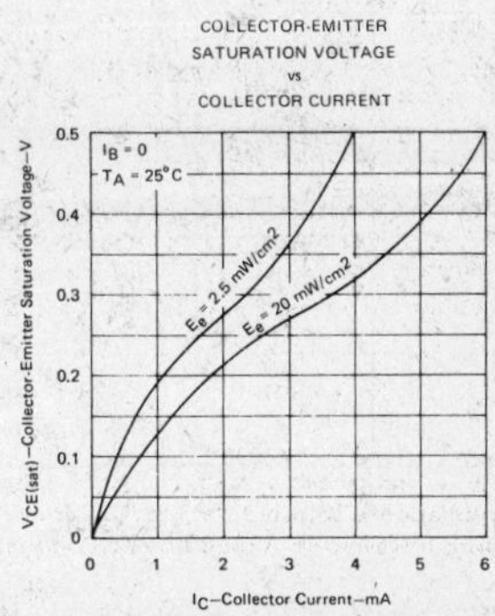

FIGURE 7

NOTE 2: Irradiance (E_e) is the radiant power per unit area incident upon a surface. For these measurements the source is an unfiltered tungsten linear-filament lamp operating at a color temperature of 2870 K.

TEXAS INSTRUMENTS
INCORPORATED
POST OFFICE BOX 225012 • DALLAS, TEXAS 75265

TYPICAL CHARACTERISTICS

RELATIVE OUTPUT
(PHOTOTRANSISTOR OPERATION)
vs
MODULATION FREQUENCY

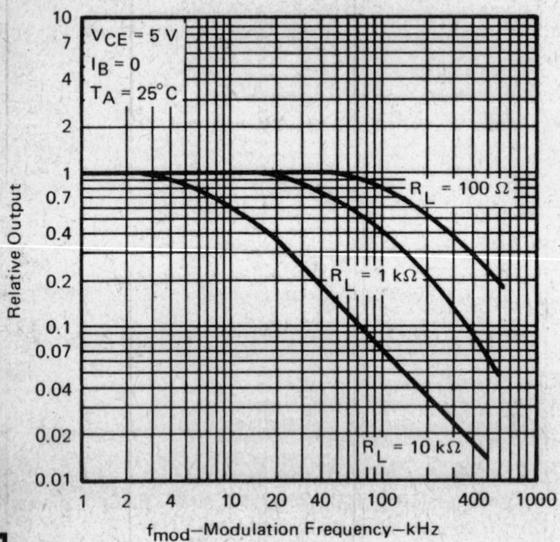

FIGURE 8

NORMALIZED DARK CURRENT
vs
FREE-AIR TEMPERATURE

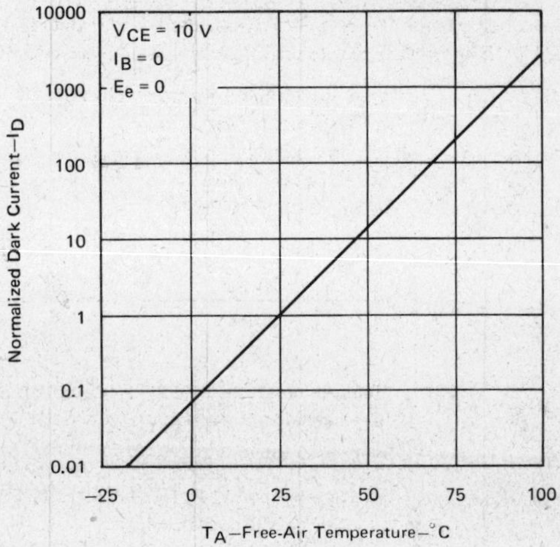

FIGURE 9

RELATIVE SPECTRAL CHARACTERISTICS

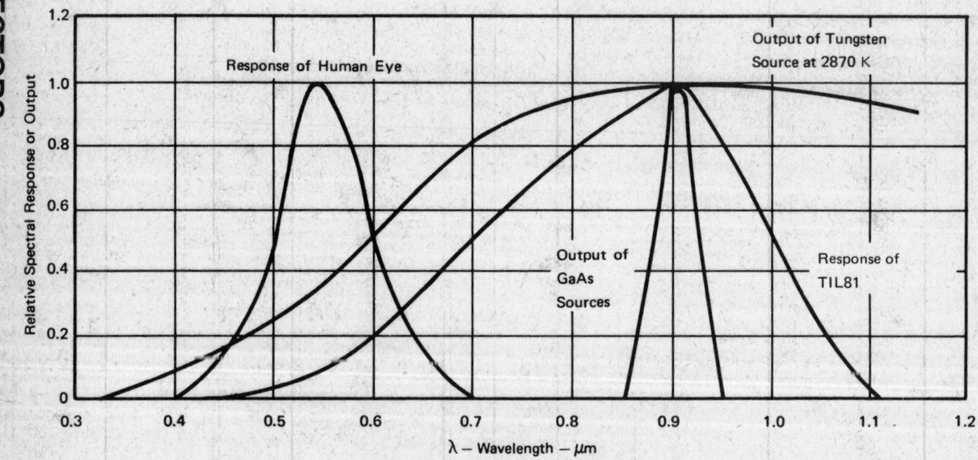

FIGURE 10

TEXAS INSTRUMENTS
INCORPORATED
POST OFFICE BOX 225012 • DALLAS, TEXAS 75265

- **This processing applies only to devices ordered under the part number TIL81HR2**
- **For electrical and mechanical specifications, refer to page 5-9**

This processing and lot acceptance follow the sequence of tests in MIL-S-19500 for JANTX types. This is not to be construed to be a JANTX-qualified part. A detail specification is available upon request through your TI Field Sales Office or Optoelectronics Marketing at the following address:

Texas Instruments Incorporated
Optoelectronics Marketing
P.O. Box 225012, MS 12
Dallas, Texas 75265
Phone: (214) 995-3821

TEST	MIL-STD-750 TEST METHOD
100% Processing	
Storage: T_A = 125°C, t = 24 h	1032
Temperature Cycle: −55°C to 125°C, 10 cycles	1051
Constant Acceleration: 20,000 G, Y_1 axis	2006
High-Temperature Reverse Bias:	
V_{CE} = 20 V, T_A = 125°C, t = 48 h	1039
Power Burn-in:	
P_D = 250 mW, t = 168 h	1039
Hermetic Seal, Fine	1071 Cond. G or H
Hermetic Seal, Gross	1071 Cond. C or D
External Visual	2071
Product Acceptance	
Group A: LTPD = 5	
External Visual	2071
Electrical: T_A = 25°C	
Electrical: T_A = 100°C	
Group B-1: LTPD = 15	
Solderability	2026
Resistance to Solvents	1022
Group B-2: LTPD = 10	
Thermal Shock	1051 Cond. B-1
Hermetic Seal, Fine	1071 Cond. G or H
Hermetic Seal, Gross	1071 Cond. C or D
Group B-3: LTPD = 5	
Steady-State Operating Life: t = 340 h	1027

TEST	MIL-STD-750 TEST METHOD
Group B-4:	
Decap, Internal Visual; Design Verification	
1 Device/0 Failure	2075
Bond Strength LTPD = 20 (C = 0)	2037 Cond. A
Group B-5: Not Applicable	
Group B-6: LTPD = 7	
High-Temperature Life (Nonoperating)	1032
t = 340 h	
(Group C Tests are run on one lot every six months)	
Group C-1: LTPD = 15	
Physical Dimensions	2066
Group C-2: LTPD = 10	
Thermal Shock (Glass Strain)	1056 Cond. A
Terminal Strength	2036 Cond. E
Hermetic Seal, Fine	1071 Cond. G or H
Hermetic Seal, Gross	1071 Cond. C or D
Moisture Resistance	1021
External Visual	2071
Group C-3: LTPD = 10	
Shock: 1500 G	2016
Vibration: 50 G	2056
Acceleration: 20000 G	2006
Group C-4: LTPD = 15	
Salt Atmosphere	1041
Group C-5: Not Applicable	
Group C-6: λ = 10	
Steady-State Operating Life: t = 1000 h	1026

D1960, NOVEMBER 1974–REVISED MARCH 1976

TYPE TIL99 N-P-N PLANAR SILICON PHOTOTRANSISTOR

FOR WIDE-ANGLE VIEWING APPLICATIONS

- Recommended for Application in Character Recognition, Tape and Card Readers, Velocity Indicators, and Encoders
- Spectrally and Mechanically Compatible with TIL31B, TIL903, and TIL904 IR Emitter
- Glass-to-Metal-Seal Header
- Base Connection Externally Available
- Saturation Level Directly Compatible with Most TTL/DTL

mechanical data

The device is in a hermetically sealed package with glass window. The outline of the TIL99 is similar to TO-18 except for the window. All TO-18 registration notes also apply to this outline.

THE COLLECTOR IS IN ELECTRICAL CONTACT WITH THE CASE

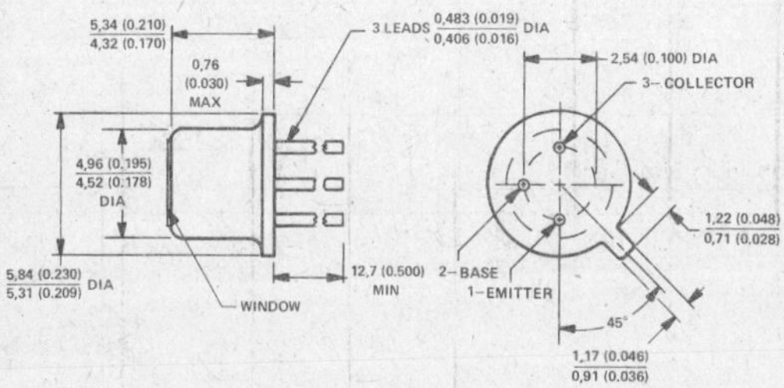

NOTE:
a. The window is Corning 7052 glass or equivalent.
Window dimensions are:
Minimum diameter. . 3,5 (0.140)
Maximum extension from top of case. 0,26 (0.010)

ALL LINEAR DIMENSIONS ARE IN MILLIMETERS AND PARENTHETICALLY IN INCHES.

absolute maximum ratings at 25°C free-air temperature (unless otherwise noted)

Collector-Base Voltage	50 V
Collector-Emitter Voltage	30 V
Emitter-Base Voltage	7 V
Emitter-Collector Voltage	7 V
Continuous Collector Current	50 mA
Continuous Device Dissipation at (or below) 25°C Free-Air Temperature (See Note 1)	250 mW
Operating Free-Air Temperature Range	–65°C to 125°C
Storage Temperature Range	–65°C to 150°C
Lead Temperature 1,6 mm (1/16 Inch) from Case for 10 Seconds	240°C

NOTE 1: Derate linearly to 125°C free-air temperature at the rate of 2.5 mW/°C.

electrical characteristics at 25°C free-air temperature (unless otherwise noted)

PARAMETER			TEST CONDITIONS	MIN	TYP	MAX	UNIT
$V_{(BR)CBO}$	Collector-Base Breakdown Voltage		I_C = 100 μA, I_E = 0, E_e = 0	50			V
$V_{(BR)CEO}$	Collector-Emitter Breakdown Voltage		I_C = 100 μA, I_B = 0, E_e = 0	30			V
$V_{(BR)EBO}$	Emitter-Base Breakdown Voltage		I_E = 100 μA, I_C = 0, E_e = 0	7			V
$V_{(BR)ECO}$	Emitter-Collector Breakdown Voltage		I_E = 100 μA, I_B = 0, E_e = 0	7			V
I_D	Dark Current	Phototransistor Operation	V_{CE} = 10 V, I_B = 0, E_e = 0			0.1	μA
			V_{CE} = 10 V, I_B = 0, E_e = 0, T_A = 100°C		20		
		Photodiode Operation	V_{CE} = 10 V, I_E = 0, E_e = 0			0.01	μA
I_L	Light Current	Phototransistor Operation	V_{CE} = 5 V, I_B = 0, E_e = 20 mW/cm², See Note 2	1	5		mA
		Photodiode Operation	V_{CB} = 0 to 50 V, I_E = 0, E_e = 20 mW/cm², See Note 2		40		μA
h_{FE}	Static Forward Current Transfer Ratio		V_{CE} = 5 V, I_C = 1 mA, E_e = 0		200		
$V_{CE(sat)}$	Collector-Emitter Saturation Voltage		I_C = 0.4 mA, I_B = 0, E_e = 20 mW/cm², See Note 2		0.2		V

NOTE 2: Irradiance (E_e) is the radiant power per unit area incident upon a surface. For these measurements the source is an unfiltered tungsten linear-filament lamp operating at a color temperature at 2870 K.

switching characteristics at 25°C free-air temperature

PARAMETER			TEST CONDITIONS	TYPICAL	UNIT
t_r	Rise Time	Phototransistor Operation	V_{CC} = 5 V, I_L = 800 μA, R_L = 100 Ω, See Test Circuit A of Figure 1	8	μs
t_f	Fall Time			6	
t_r	Rise Time	Photodiode Operation	V_{CC} = 0 to 50 V, I_L = 60 μA, R_L = 100 Ω, See Test Circuit B of Figure 1	350	ns
t_f	Fall Time			500	

PARAMETER MEASUREMENT INFORMATION

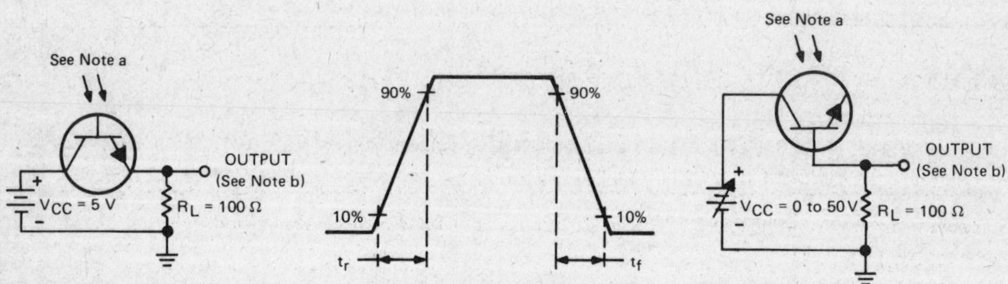

TEST CIRCUIT A
PHOTOTRANSISTOR OPERATION

VOLTAGE WAVEFORM

TEST CIRCUIT B
PHOTODIODE OPERATION

NOTES: a. Input irradiance is supplied by a pulsed gallium arsenide radiant-energy source with rise and fall times less than 50 ns. Incident irradiation is adjusted for specified I_L.

b. The output waveform is monitored on an oscilloscope with the following characteristics: $t_r \leq 25$ ns, $R_{in} \geq 1$ MΩ, $C_{in} \leq 20$ pF.

FIGURE 1

TYPE TIL100
LARGE-AREA SILICON PHOTODIODE

D2478, MAY 1978—REVISED JULY 1978

- High Photosensitivity
- Fast Response
- Low-Cost Plastic Package
- Designed for Infrared Remote-Control Systems
- Compatible with TIL38, TIL39, TIL905, and TIL906 IR Emitters

description

The TIL100 is a high-speed PIN photodiode designed to operate in the reverse-bias mode. It provides low capacitance with high speed and high photosensitivity suitable for near-infrared applications.

mechanical data

The photodiode chip is mounted on a lead frame and molded in a black infrared-transmissive plastic. The active chip area is typically 8,83 millimeters (0.0137 square inches). Its centerline is nominally 4 millimeters (0.157 inch) above the seating plane.

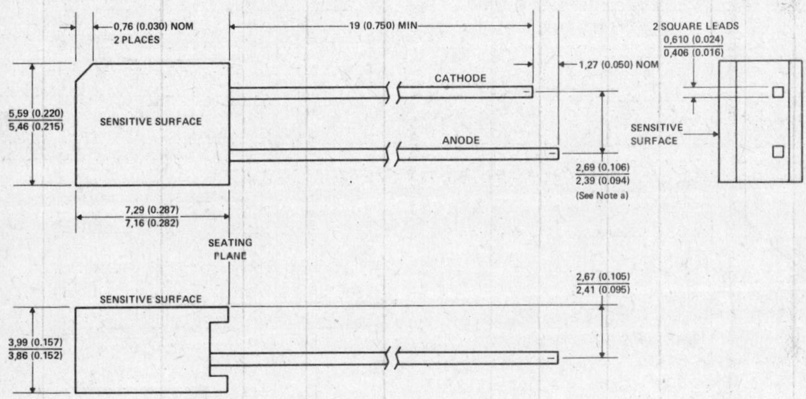

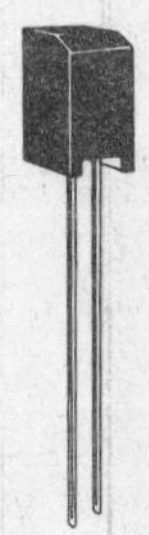

NOTE a: Lead spacing is measured at the seating plane.

ALL DIMENSIONS ARE IN MILLIMETERS AND PARENTHETICALLY IN INCHES

absolute maximum ratings at 25°C free-air temperature (unless otherwise noted)

Reverse Voltage	30 V
Continuous Power Dissipation at (or below) 25°C Free-Air Temperature (See Note 1)	150 mW
Operating Free-Air Temperature Range	−25°C to 80°C
Storage Temperature Range	−25°C to 100°C
Lead Temperature 1.6 mm (1/16 inch) from Case for 3 Seconds	260°C

NOTE 1: Derate linearly to 80°C free-air temperature at the rate of 2.73 mW/°C.

Copyright © 1978 by Texas Instruments Incorporated

TEXAS INSTRUMENTS
INCORPORATED
POST OFFICE BOX 225012 • DALLAS, TEXAS 75265

electrical characteristics at 25°C free-air temperature

	PARAMETER	TEST CONDITIONS		MIN	TYP	MAX	UNIT
$V_{(BR)}$	Breakdown Voltage	I_R = 100 μA,	E_e† = 0	30			V
I_D	Dark Current	V_R = 10 V,	E_e† = 0		5	50	nA
I_L	Light Current	V_R = 10 V,	E_e† = 250 μW/cm² at 940 nm	10	15		μA
C_T	Total Capacitance	V_R = 3 V,	E_e† = 0, f = 1 MHz		35	50	pF
t_r	Rise Time	V_R = 10 V,	R_L = 1 kΩ		100		ns
t_f	Fall Time	V_R = 10 V,	R_L = 1 kΩ		100		ns

†Irradiance (E_e) is the radiant power per unit area incident on a surface.

TYPICAL CHARACTERISTICS

REVERSE CURRENT
vs
IRRADIANCE

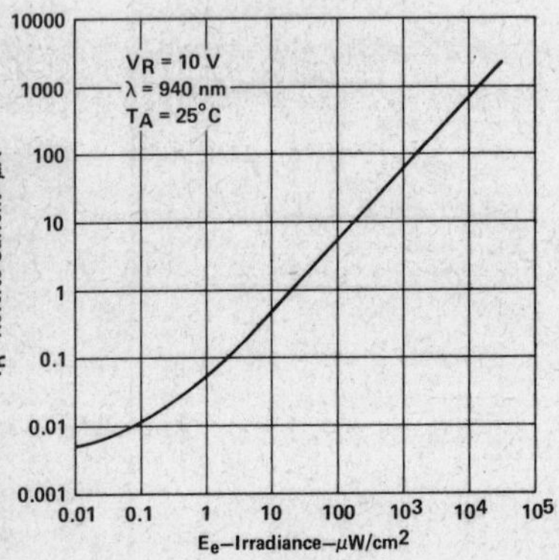

FIGURE 1

TOTAL CAPACITANCE
vs
REVERSE VOLTAGE

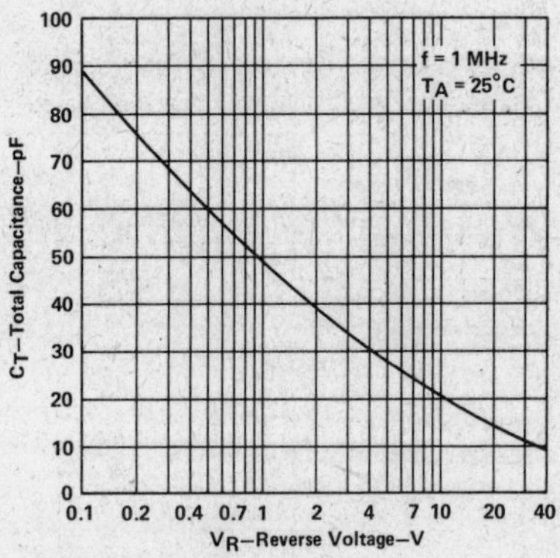

FIGURE 2

5 PHOTODETECTORS

D2559, JULY 1980

- Recommended for Applications Requiring Low-Cost Discrete Phototransistors
- Spectrally and Mechanically Compatible with TIL40 Infrared Emitter
- Designed for use in Housings or Printed Circuit Boards

mechanical data

This device has a clear molded plastic body.

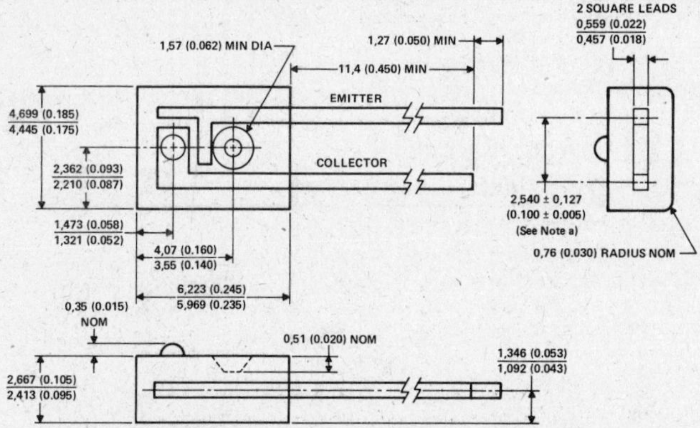

NOTE a: Lead spacing is measured where the leads emerge from the package.

ALL DIMENSIONS ARE IN MILLIMETERS AND PARENTHETICALLY IN INCHES

CL-67

absolute maximum ratings at 25°C free-air temperature (unless otherwise noted)

Collector-Emitter Voltage	30 V
Emitter-Collector Voltage	7 V
Continuous Collector Current	50 mA
Continuous Device Dissipation at (or below) 25°C Free-Air Temperature (see Note 1)	50 mW
Operating Free-Air Temperature Range	−40°C to 80°C
Storage Temperature Range	−40°C to 100°C
Lead Temperature 1,6 mm (1/16 inch) from Case for 5 Seconds	240°C

electrical characteristics at free-air temperature

PARAMETER		TEST CONDITIONS	MIN	TYP	MAX	UNIT
$V_{(BR)CEO}$	Collector-Emitter Breakdown Voltage	$I_C = 100\ \mu A$, $E_e = 0$	30			V
$V_{(BR)ECO}$	Emitter-Collector Breakdown Voltage	$I_E = 100\ \mu A$, $E_e = 0$	7			V
I_D	Dark Current	$V_{CE} = 5\ V$, $E_e = 0$			100	nA
I_L	Light Current	$V_{CE} = 5\ V$, $E_e = 500\ \mu W/cm^2$, See Note 2	100	400		µA
$V_{CE(sat)}$	Collector-Emitter Saturation Voltage	$I_C = 80\ \mu A$, $E_e = 500\ \mu W/cm^2$, See Note 2		0.15		V

switching characteristics at 25°C free-air temperature

PARAMETER		TEST CONDITIONS		TYP	MAX	UNIT
t_r	Rise Time	$V_{CC} = 10\ V$,	$I_L = 100\ \mu A$,	25		µs
t_f	Fall Time	$R_L = 1\ k\Omega$,	See Figure 1	25		µs

NOTES: 1. Derate linearly to 80°C free-air temperature at the rate of 0.91 mW/°C.
2. Irradiance (E_e) is the radiant power per unit area incident upon a surface. For these measurements the source is an infrared-emitting diode, wavelength at peak emission is 930 nm, and spectral bandwidth is 45 nm.

Copyright © 1980 by Texas Instruments Incorporated

TEXAS INSTRUMENTS
INCORPORATED
POST OFFICE BOX 225012 • DALLAS, TEXAS 75265

PARAMETER MEASUREMENT INFORMATION

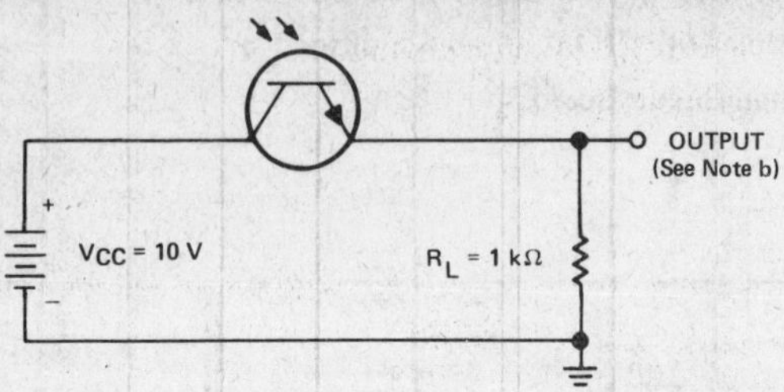

TEST CIRCUIT

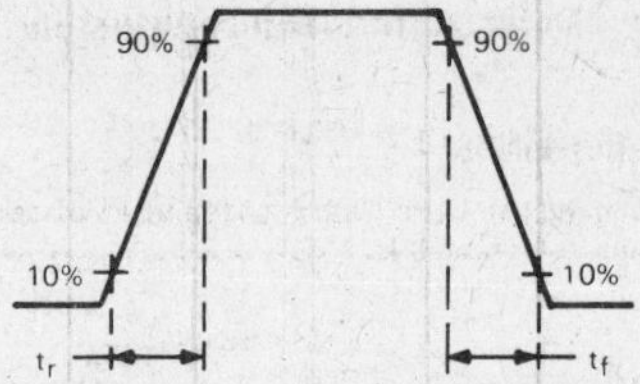

OUTPUT VOLTAGE WAVEFORM

FIGURE 1

NOTES: a. Input irradiance is supplied by a pulsed gallium arsenide infrared emitter with rise and fall times of less than 50 ns. Incident irradiation is adjusted for I_L = 100 μA.

b. Output waveform is monitored on an oscilloscope with the following characteristics: $t_r \leqslant 25$ ns, $r_{in} \geqslant 1$ MΩ, $C_{in} \leqslant 20$ pF.

TYPICAL CHARACTERISTICS

NORMALIZED LIGHT CURRENT vs ANGULAR DISPLACEMENT

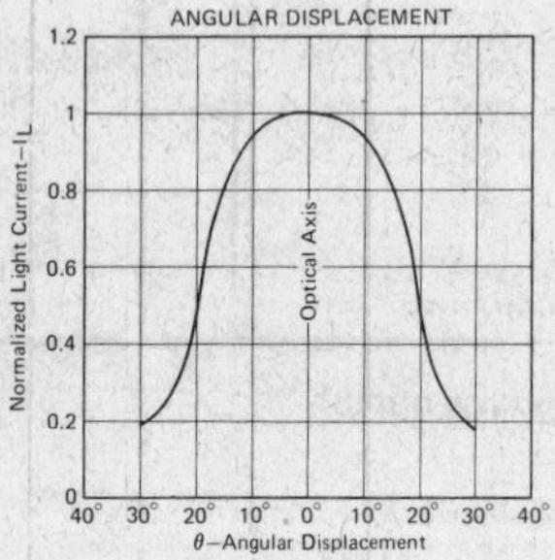

FIGURE 2

DARK CURRENT vs FREE-AIR TEMPERATURE

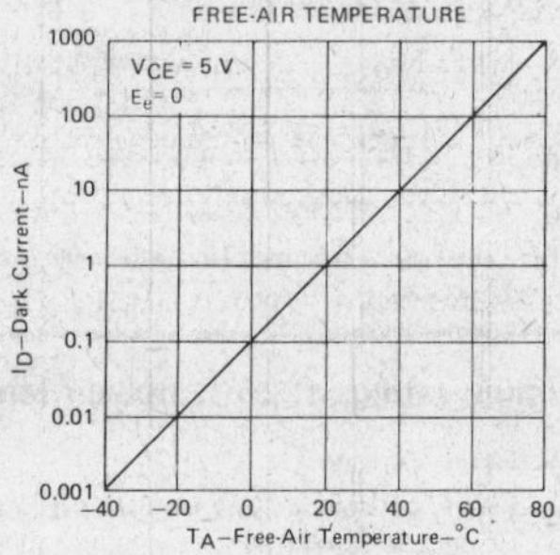

FIGURE 3

COLLECTOR CURRENT vs COLLECTOR-EMITTER VOLTAGE

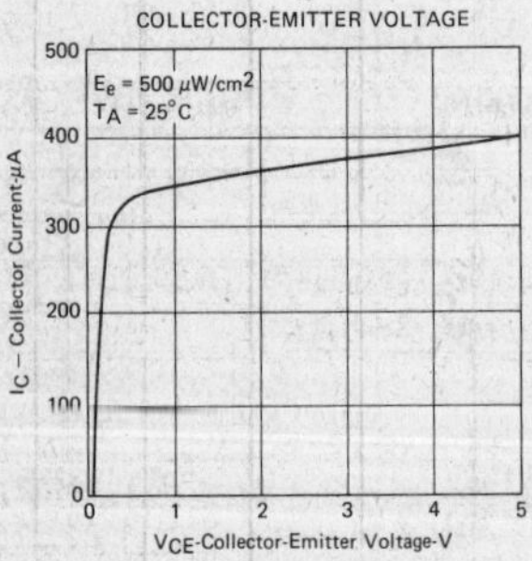

FIGURE 4

COUPLING CHARACTERISTICS OF TIL40 WITH TIL411

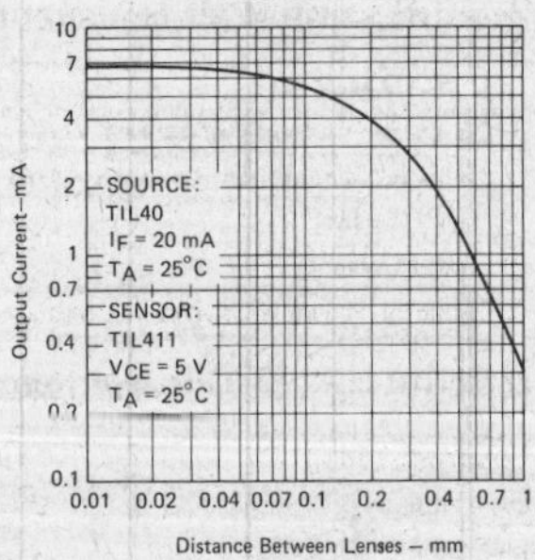

FIGURE 5

NOTE 2: Irradiance (E_e) is the radiant power per unit area incident upon a surface. For these measurements the source is an infrared-emitting diode, wavelength at peak emission is 930 nm, and spectral bandwidth is 45 nm.

780

- Recommended for Applications Requiring Low-Cost Discrete Phototransistors
- Spectrally and Mechanically Compatible with TIL40 Infrared Emitter
- Designed for use in Housings or Printed Circuit Boards

mechanical data

This device has a blue tinted molded plastic body.

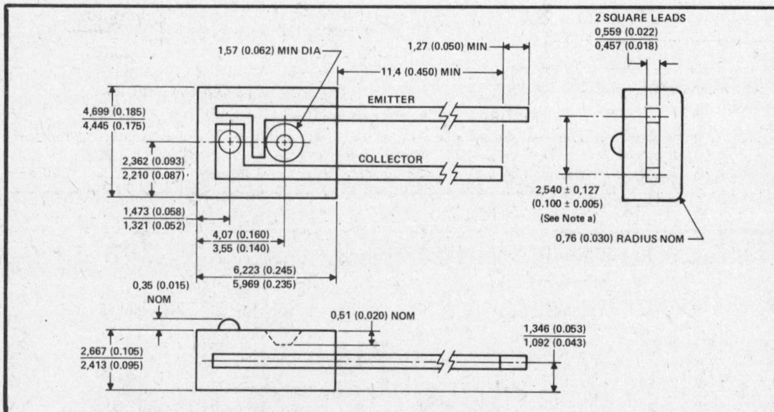

NOTE a: Lead spacing is measured where the leads emerge from the package.

ALL DIMENSIONS ARE IN MILLIMETERS AND PARENTHETICALLY IN INCHES.

5

PHOTODETECTORS

absolute maximum ratings at 25°C free-air temperature (unless otherwise noted)

Collector-Emitter Voltage	30 V
Emitter-Collector Voltage	7 V
Continuous Collector Current	50 mA
Continuous Device Dissipation at (or below) 25°C Free-Air Temperature (see Note 1)	50 mW
Operating Free-Air Temperature Range	−40°C to 80°C
Storage Temperature Range	−40°C to 100°C
Lead Temperature 1,6 mm (1/16 inch) from Case for 5 Seconds	240°C

electrical characteristics at free-air temperature

PARAMETER		TEST CONDITIONS	MIN	TYP	MAX	UNIT
$V_{(BR)CEO}$	Collector-Emitter Breakdown Voltage	$I_C = 100\ \mu A$, $E_e = 0$	30			V
$V_{(BR)ECO}$	Emitter-Collector Breakdown Voltage	$I_E = 100\ \mu A$, $E_e = 0$	7			V
I_D	Dark Current	$V_{CE} = 5$ V, $E_e = 0$			100	nA
I_L	Light Current	$V_{CE} = 1$ V, $E_e = 100\ \mu W/cm^2$, See Note 2	0.5	8		mA
$V_{CE(sat)}$	Collector-Emitter Saturation Voltage	$I_C = 500\ \mu A$, $E_e = 100\ \mu W/cm^2$, See Note 2		0.6		V

switching characteristics at 25°C free-air temperature

PARAMETER		TEST CONDITIONS		TYP	MAX	UNIT
t_r	Rise Time	$V_{CC} = 5$ V,	$I_L = 500\ \mu A$,	1		ms
t_f	Fall Time	$R_L = 1\ k\Omega$,	See Figure	1		ms

NOTES: 1. Derate linearly to 80°C free-air temperature at the rate of 0.91 mW/°C.

2. Irradiance (E_e) is the radiant power per unit area incident upon a surface. For these measurements the source is an infrared-emitting diode, wavelength at peak emission is 930 nm, and spectral bandwidth is 45 nm.

Copyright © 1980 by Texas Instruments Incorporated

PARAMETER MEASUREMENT INFORMATION

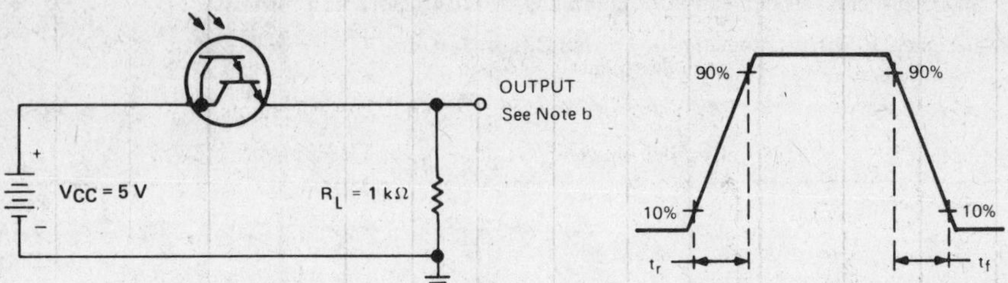

TEST CIRCUIT

OUTPUT VOLTAGE WAVEFORM

FIGURE 1

NOTES: a. Input irradiance is supplied by a pulsed gallium arsenide infrared emitter with rise and fall times of less than 50 ns. Incident irradiation is adjusted for I_L = 500 μA.

b. Output waveform is monitored on an oscilloscope with the following characteristics: $t_r \leq 25$ ns, $R_{in} \geq 1$ MΩ, $C_{in} \leq 20$ pF.

TYPICAL CHARACTERISTICS

RELATIVE PHOTON INTENSITY
vs
ANGULAR DISPLACEMENT

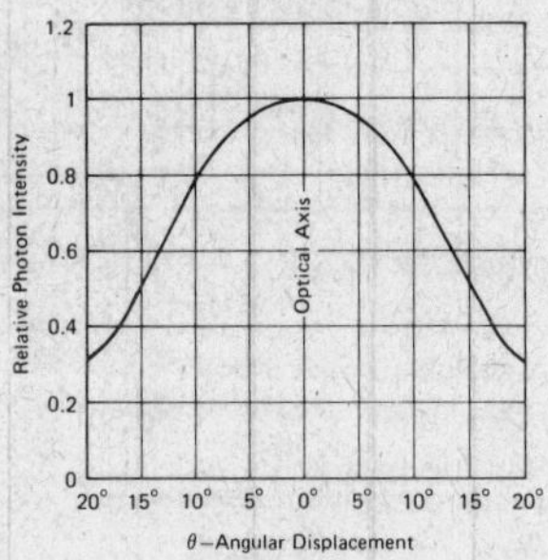

FIGURE 2

DARK CURRENT
vs
FREE-AIR TEMPERATURE

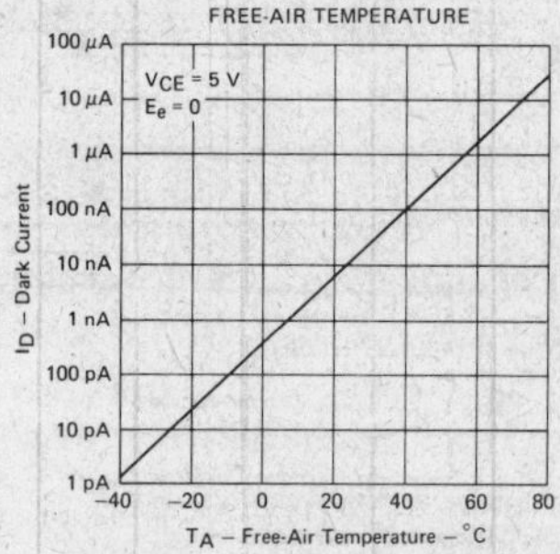

FIGURE 3

COLLECTOR CURRENT
vs
COLLECTOR-EMITTER VOLTAGE

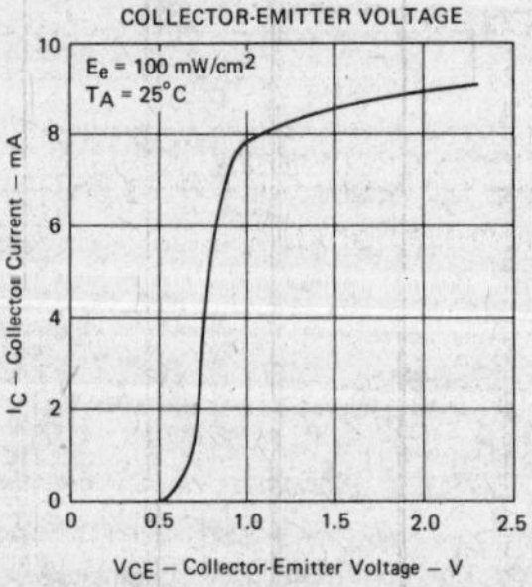

FIGURE 4

COUPLING CHARACTERISTICS
OF
TIL40 WITH TIL412

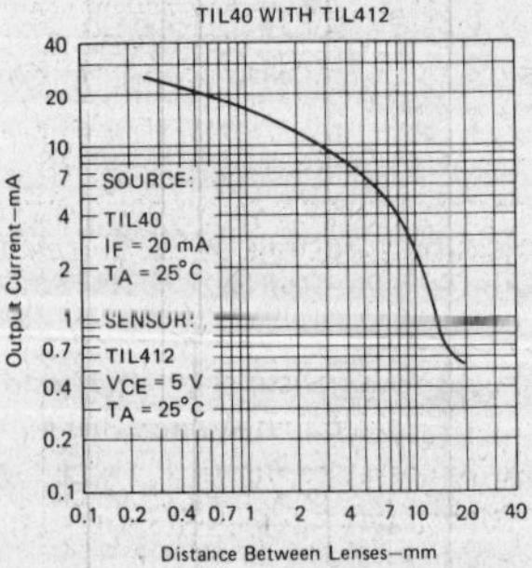

FIGURE 5

TEXAS INSTRUMENTS
INCORPORATED
POST OFFICE BOX 225012 • DALLAS, TEXAS 75265

D2588, JULY 1980—REVISED JANUARY 1983

- **High Photosensitivity**
- **Fast Response**
- **Low-Cost Plastic Package**
- **Designed for Infrared Remote-Control Systems**
- **Compatible with TIL38, TIL39, TIL905, and TIL906 IR Emitters**

description

The TIL413 and TIL413S are high-speed PIN photodiodes designed to operate in the reverse-bias mode. These devices provide low capacitance with high speed and high photosensitivity suitable for near-infrared applications.

mechanical data

The photodiode chip is mounted on a lead frame and molded in black infrared-transmissive plastic. The active chip area is typically 4,4 square millimeters (0.0067 square inch). The centerline is nominally 3,8 millimeters (0.150 inch) above the seating plane.

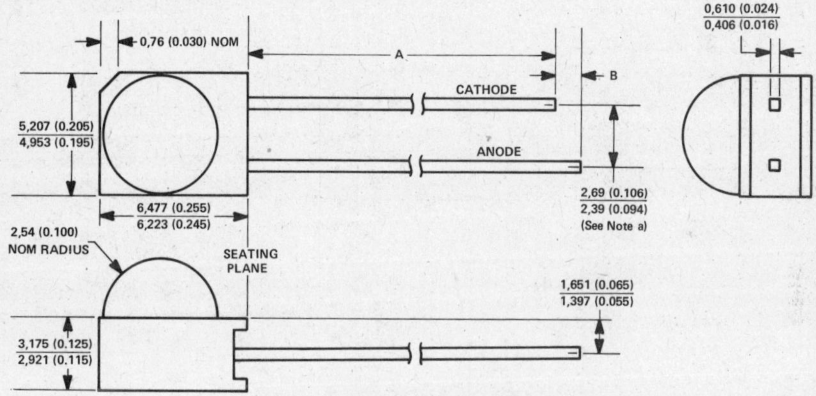

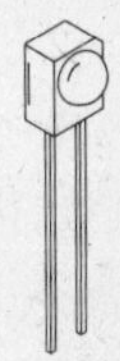

TIL 413

TIL 413S

LEAD LENGTH (A) AND OFFSET (B)

DEVICE	A	B
TIL413	19 (0.750) MIN	1,27 (0.050) MIN
TIL413S	5 (0.200) MIN	0 MIN

NOTE a: Lead spacing is measured at the seating plane.

ALL DIMENSIONS ARE IN MILLIMETERS AND PARENTHETICALLY IN INCHES.

absolute maximum ratings at 25°C free-air temperature (unless otherwise noted)

Reverse Voltage	30 V
Continuous Power Dissipation at (or below) 25°C Free-Air Temperature (See Note 1)	150 mW
Operating Free-Air Temperature Range	−25°C to 80°C
Storage Temperature Range	−25°C to 100°C
Lead Temperature 1,6 mm (1/16 Inch) from Case for 5 Seconds	240°C

NOTE 1: Derate linearly to 80°C free-air temperature at the rate of 2.73 mW/°C.

Copyright © 1983 by Texas Instruments Incorporated

electrical characteristics at 25°C free-air temperature

PARAMETER		TEST CONDITIONS		MIN	TYP	MAX	UNIT
$V_{(BR)}$	Breakdown Voltage	I_R = 100 μA,	E_e† = 0	30			V
I_D	Dark Current	V_R = 10 V,	E_e† = 0		5	50	nA
I_L	Light Current	V_R = 10 V,	E_e† = 250 μW/cm^2, See Note 2	10	15		μA
C_T	Total Capacitance	V_R = 3 V,	E_e† = 0, f = 1 MHz		15	50	pF
t_r	Rise Time	V_R = 10 V,	R_L = 1 kΩ		100		ns
t_f	Fall Time	V_R = 10 V,	R_L = 1 kΩ		100		ns

TYPICAL CHARACTERISTICS

LIGHT CURRENT
vs
IRRADIANCE

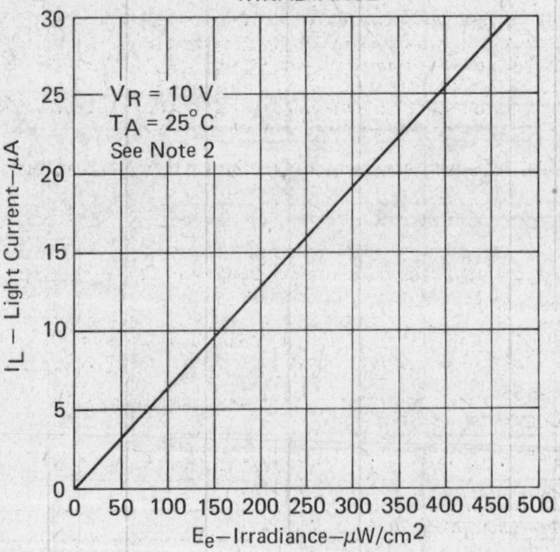

FIGURE 1

NOTE 2: Irradiance (E_e) is the radiant power per unit area incident upon a surface. For these measurements the source is an infrared-emitting diode, wavelength at peak emission is 930 nm, and spectral bandwidth is 45 nm.

780

- Designed for Automatic or Hand Insertion in Sockets or PC Boards
- Recommended for Industrial Applications Requiring Low-Cost Discrete Phototransistors
- Spectrally and Mechanically Matched with TIL38, TIL39, TIL905, and TIL906 IR-Emitting Diodes

mechanical data

This device has a clear molded epoxy body.

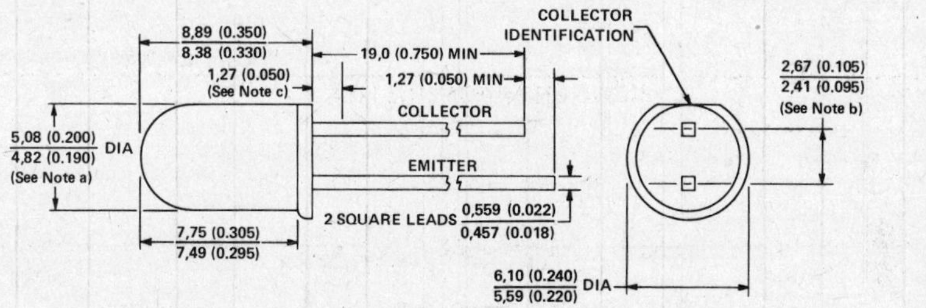

NOTES: a. This diameter is measured 0,127 mm (0.005 inch) from the package shoulder.
b. Lead spacing is measured where the leads emerge from the package.
c. Within this zone the package contour is not controlled.

ALL DIMENSIONS ARE IN MILLIMETERS AND PARENTHETICALLY INCHES

CL-10

absolute maximum ratings at 25°C free-air temperature (unless otherwise noted)

Collector-Emitter Voltage	35 V
Emitter-Collector Voltage	7 V
Continuous Device Dissipation at (or below) 25°C Free-Air Temperature (See Note 1)	50 mW
Operating Free-Air Temperature Range	−40°C to 80°C
Storage Temperature Range	−40°C to 100°C
Lead Temperature 1,6 mm (1/16 Inch) from Case for 5 Seconds	240°C

electrical characteristics at 25°C free-air temperature

PARAMETER		TEST CONDITIONS			MIN	TYP	MAX	UNIT
$V_{(BR)CEO}$	Collector-Emitter Breakdown Voltage	$I_C = 100\ \mu A$,	$E_e = 0$		35			V
$V_{(BR)ECO}$	Emitter-Collector Breakdown Voltage	$I_E = 100\ \mu A$,	$E_e = 0$		7			V
I_D	Dark Current	$V_{CE} = 10$ V,	$E_e = 0$				50	nA
I_L	Light Current	$V_{CE} = 5$ V,	$E_e = 250\ \mu W/cm^2$,	See Note 2	100	700		μA
$V_{CE(sat)}$	Collector-Emitter Saturation Voltage	$I_C = 100\ \mu A$,	$E_e = 250\ \mu W/cm^2$,	See Note 2		0.1		V

switching characteristics at 25°C free-air temperature

PARAMETER		TEST CONDITIONS		TYP	UNIT
t_r	Rise Time	$V_{CC} = 30$ V,	$I_L = 800\ \mu A$,	8	μs
t_f	Fall Time	$R_L = 1\ k\Omega$,	See Figure 1	7	μs

NOTES: 1. Derate linearly to 80°C free-air temperature at the rate of 0.91 mW/°C.
2. Irradiance (E_e) is the radiant power per unit area incident upon a surface. For these measurements the source is an infrared-emitting diode, wavelength at peak emission is 930 nm, and spectral bandwidth is 45 nm.

Copyright © 1980 by Texas Instruments Incorporated

PARAMETER MEASUREMENT INFORMATION

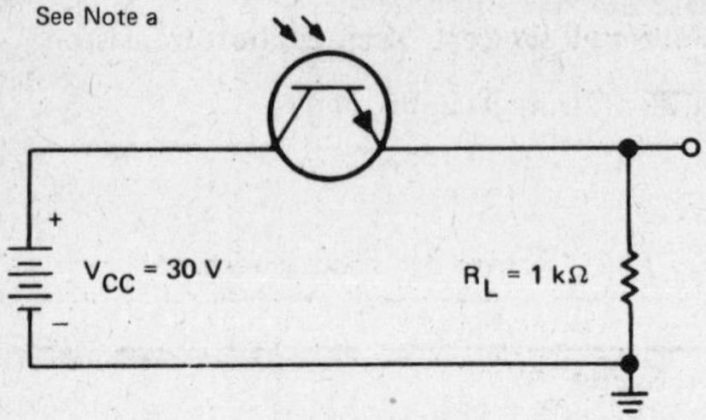

TEST CIRCUIT

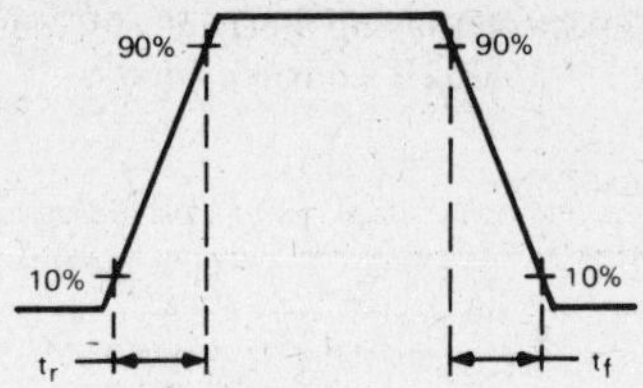

OUTPUT VOLTAGE WAVEFORM

FIGURE 1

NOTES: a. Input irradiance is supplied by a pulsed gallium arsenide infrared emitter with rise and fall times of less than 50 ns. Incident irradiation is adjusted for I_L = 800 μA.

b. Output waveform is monitored on an oscilloscope with the following characteristics: $t_r \leq 25$ ns, $R_{in} \geq 1$ MΩ, $C_{in} \leq 20$ pF.

TYPICAL CHARACTERISTICS

NORMALIZED LIGHT CURRENT
vs
ANGULAR DISPLACEMENT

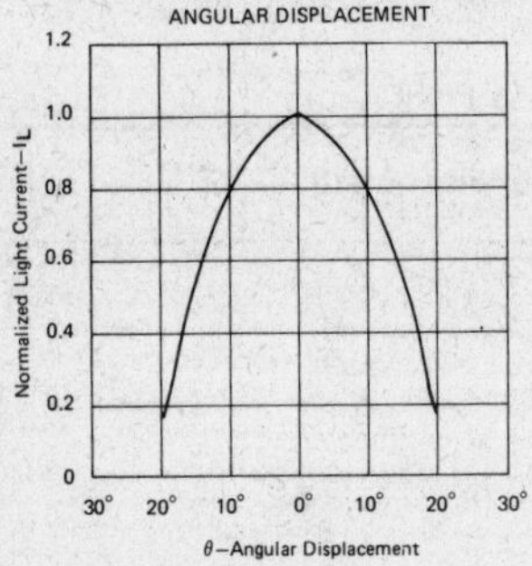

FIGURE 2

DARK CURRENT
vs
FREE-AIR TEMPERATURE

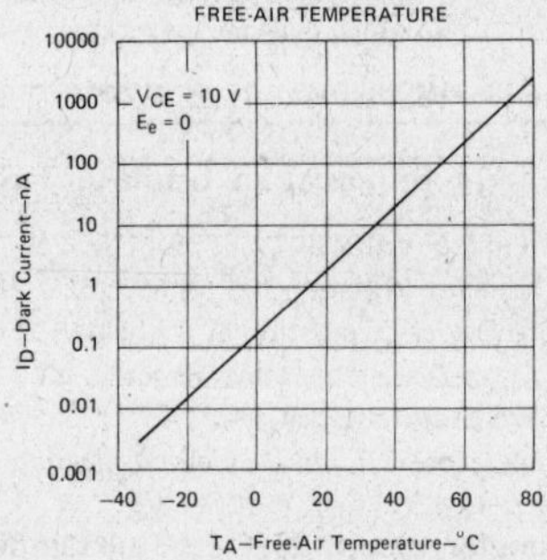

FIGURE 3

COLLECTOR-EMITTER SATURATION VOLTAGE
vs
IRRADIANCE

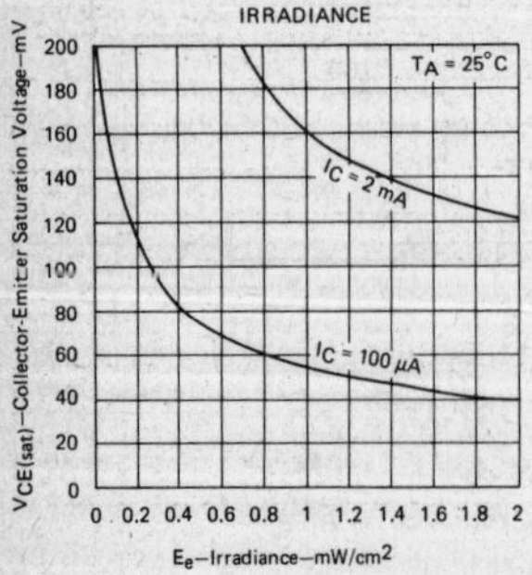

FIGURE 4

COLLECTOR-EMITTER VOLTAGE
vs
IRRADIANCE

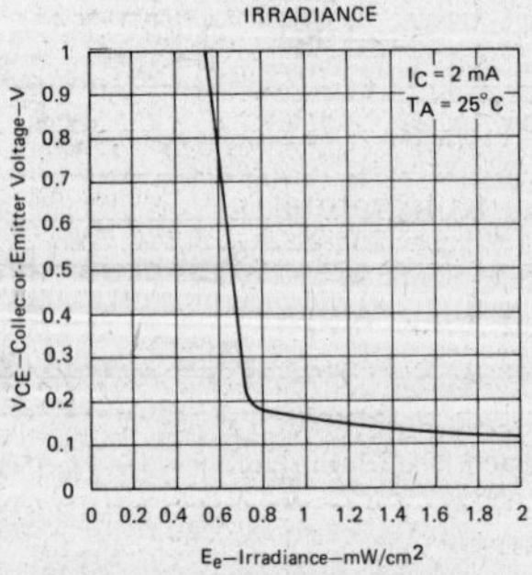

FIGURE 5

COUPLING CHARACTERISTICS
OF
TIL414 WITH TIL39

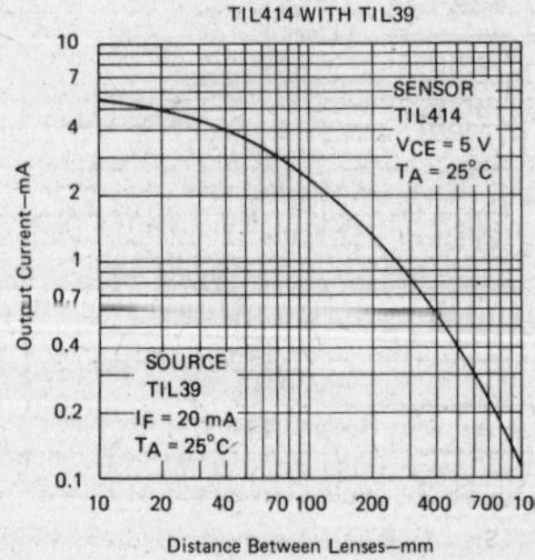

FIGURE 6

D2690, FEBRUARY 1983

- **Recommended for Applications Requiring Low-Cost Discrete Phototransistors**
- **Spectrally and Mechanically Compatible with TIL40 Infrared Emitter**
- **Designed for use in Housings or Printed Circuit Boards**

mechanical data

This device has a clear molded plastic body and is similar to TIL411 except the pinout is reversed.

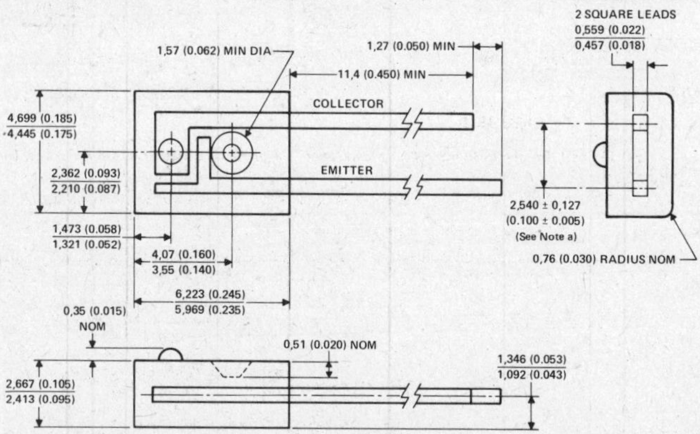

NOTE a: Lead spacing is measured where the leads emerge from the package.

ALL DIMENSIONS ARE IN MILLIMETERS AND PARENTHETICALLY IN INCHES

absolute maximum ratings at 25°C free-air temperature (unless otherwise noted)

Collector-Emitter Voltage	30 V
Emitter-Collector Voltage	7 V
Continuous Collector Current	50 mA
Continuous Device Dissipation at (or below) 25°C Free-Air Temperature (see Note 1)	50 mW
Operating Free-Air Temperature Range	−40°C to 80°C
Storage Temperature Range	−40°C to 100°C
Lead Temperature 1,6 mm (1/16 inch) from Case for 3 Seconds	260°C

electrical characteristics at free-air temperature

PARAMETER		TEST CONDITIONS	MIN	TYP	MAX	UNIT
$V_{(BR)CEO}$	Collector-Emitter Breakdown Voltage	$I_C = 100\ \mu A$, $E_e = 0$	30			V
$V_{(BR)ECO}$	Emitter-Collector Breakdown Voltage	$I_E = 100\ \mu A$, $E_e = 0$	7			V
I_D	Dark Current	$V_{CE} = 5$ V, $E_e = 0$			100	nA
I_L	Light Current	$V_{CE} = 5$ V, $E_e = 500\ \mu W/cm^2$, See Note 2	100	400		μA
$V_{CE(sat)}$	Collector-Emitter Saturation Voltage	$I_C = 80\ \mu A$, $E_e = 500\ \mu W/cm^2$, See Note 2		0.15		V

switching characteristics at 25°C free-air temperature

PARAMETER		TEST CONDITIONS		TYP	MAX	UNIT
t_r	Rise Time	$V_{CC} = 10$ V,	$I_L = 100\ \mu A$,	25		μs
t_f	Fall Time	$R_L = 1\ k\Omega$,	See Figure 1	25		μs

NOTES:
1. Derate linearly to 80°C free-air temperature at the rate of 0.91 mW/°C.
2. Irradiance (E_e) is the radiant power per unit area incident upon a surface. For these measurements the source is an infrared-emitting diode, wavelength at peak emission is 930 nm, and spectral bandwidth is 45 nm.

Copyright © 1983 by Texas Instruments Incorporated.

PARAMETER MEASUREMENT INFORMATION

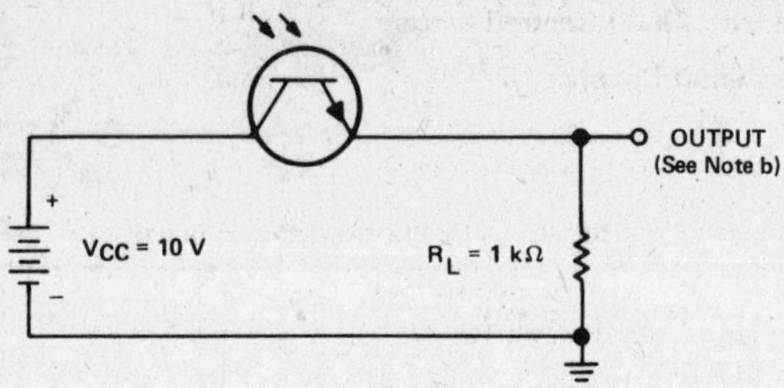

TEST CIRCUIT

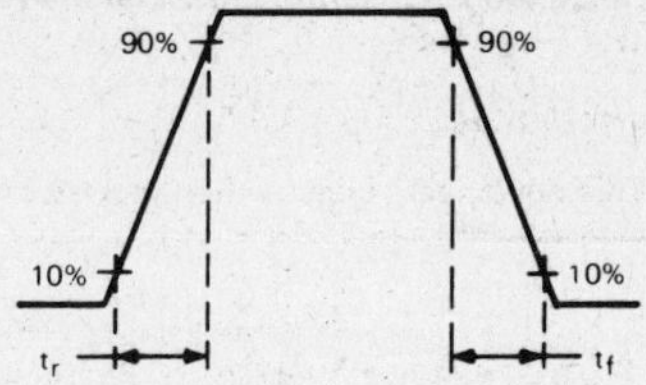

OUTPUT VOLTAGE WAVEFORM

FIGURE 1

NOTES: a. Input irradiance is supplied by a pulsed gallium arsenide infrared emitter with rise and fall times of less than 50 ns. Incident irradiation is adjusted for I_L = 100 µA.

b. Output waveform is monitored on an oscilloscope with the following characteristics: $t_r \leqslant$ 25 ns, $r_{in} \geqslant$ 1 MΩ, $C_{in} \leqslant$ 20 pF.

TYPICAL CHARACTERISTICS

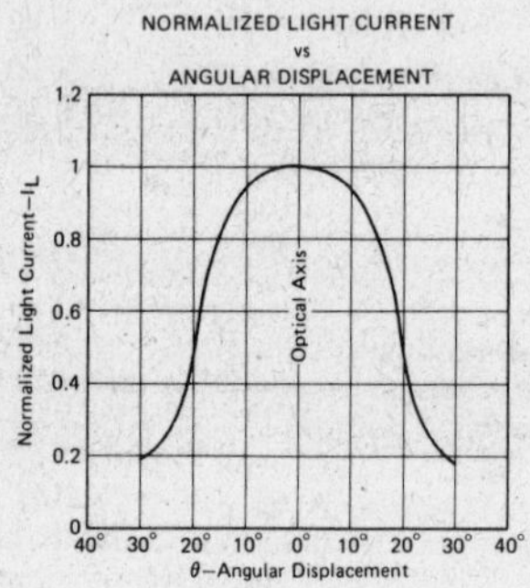

FIGURE 2

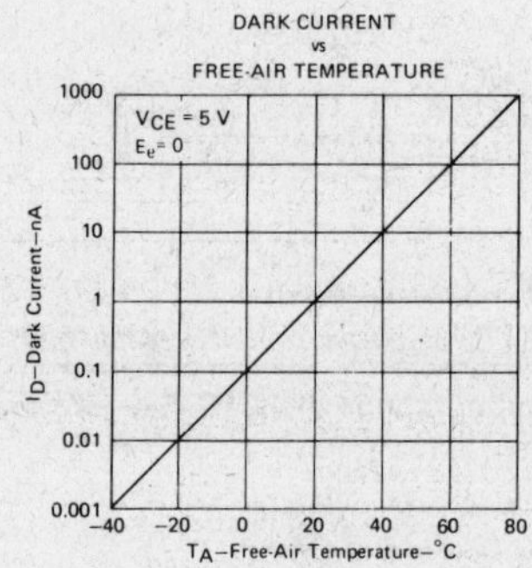

FIGURE 3

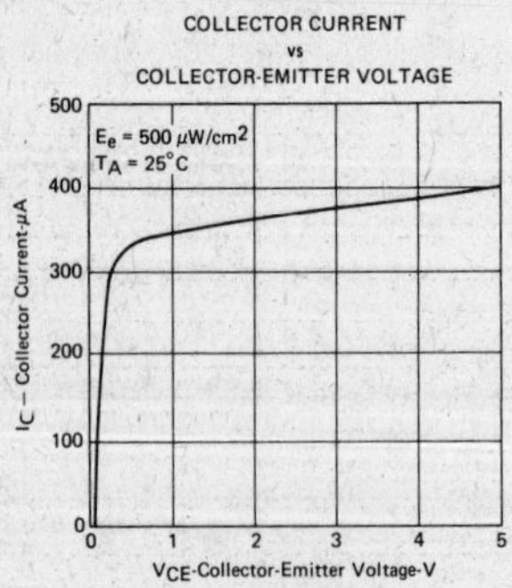

FIGURE 4

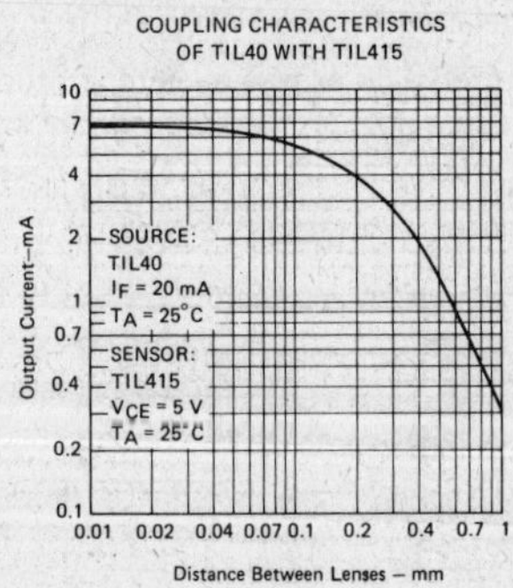

FIGURE 5

- **Recommended for Applications Requiring Low-Cost Discrete Phototransistors**
- **Spectrally and Mechanically Matched with TIL40 Infrared Emitter**
- **Designed for use in Housings or Printed Circuit Boards**

mechanical data

This device has a blue-tinted molded plastic body and is similar to TIL412 except the pinout is reversed.

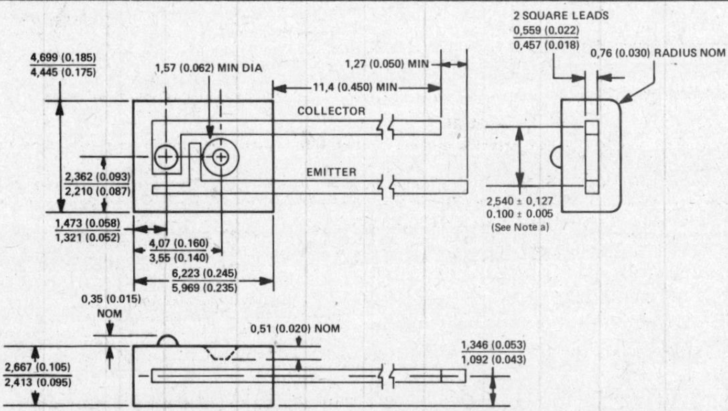

NOTE a: Lead spacing is measured where the leads emerge from the package.

ALL DIMENSIONS ARE IN MILLIMETERS AND PARENTHETICALLY IN INCHES.

absolute maximum ratings at 25 °C free-air temperature (unless otherwise noted)

Collector-Emitter Voltage	30 V
Emitter-Collector Voltage	7 V
Continuous Collector Current	50 mA
Continuous Device Dissipation at (or below) 25 °C Free-Air Temperature (See Note 1)	50 mW
Operating Free-Air Temperature Range	−40 °C to 80 °C
Storage Temperature Range	−40 °C to 100 °C
Lead Temperature 1,6 mm (1/16 inch) from Case for 5 Seconds	260 °C

electrical characteristics at 25 °C free-air temperature

PARAMETER		TEST CONDITIONS	MIN	TYP	MAX	UNIT
$V_{(BR)CEO}$	Collector-Emitter Breakdown Voltage	$I_C = 100\ \mu A$, $E_e = 0$	30			V
$V_{(BR)ECO}$	Emitter-Collector Breakdown Voltage	$I_E = 100\ \mu A$, $E_e = 0$	7			V
I_D	Dark Current	$V_{CE} = 5$ V, $E_e = 0$			100	nA
I_L	Light Current	$V_{CE} = 1$ V, $E_e = 100\ \mu W/cm^2$, See Note 2	0.5	8		mA
$V_{CE(sat)}$	Collector-Emitter Saturation Voltage	$I_C = 500\ \mu A$, $E_e = 100\ \mu W/cm^2$, See Note 2		0.6		V

switching characteristics at 25 °C free-air temperature

PARAMETER		TEST CONDITIONS		TYP	UNIT
t_r	Rise Time	$V_{CC} = 5$ V,	$I_L = 500\ \mu A$,	1	ms
t_f	Fall Time	$R_L = 1\ k\Omega$,	See Figure 1	1	

NOTES:
1. Derate linearly to 80 °C free-air temperature at the rate of 0.91 mW/°C.
2. Irradiance (E_e) is the radiant power per unit area incident upon a surface. For these measurements the source is an infrared-emitting diode, wavelength at peak emission is 930 nm, and spectral bandwidth is 45 nm.

83

Copyright © 1983 by Texas Instruments Incorporated.

PARAMETER MEASUREMENT INFORMATION

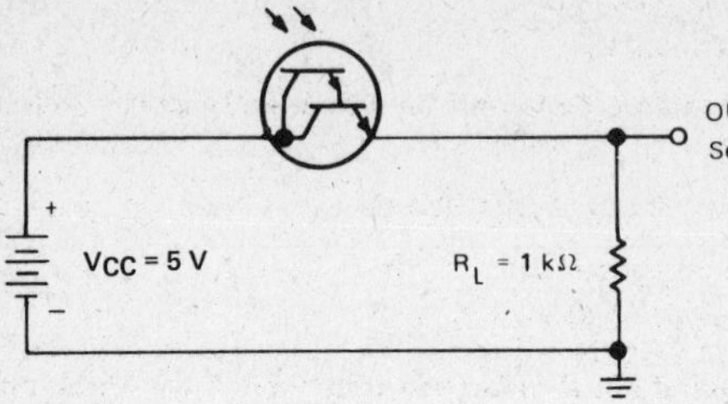

TEST CIRCUIT

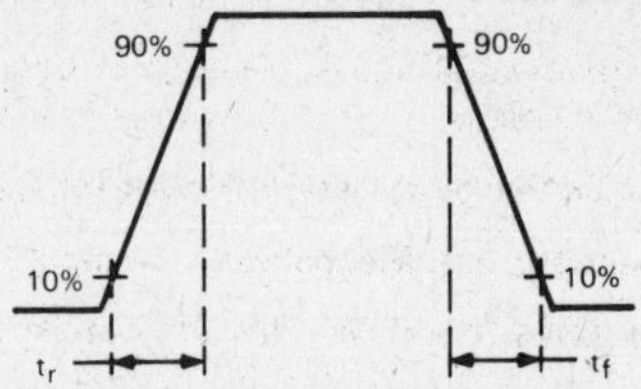

OUTPUT VOLTAGE WAVEFORM

FIGURE 1

NOTES: a. Input irradiance is supplied by a pulsed gallium arsenide infrared emitter with rise and fall times of less than 50 ns. Incident irradiation is adjusted for I_L = 500 μA.

b. Output waveform is monitored on an oscilloscope with the following characteristics: $t_r \leqslant 25$ ns, $R_{in} \geqslant 1$ MΩ, $C_{in} \leqslant 20$ pF.

TYPICAL CHARACTERISTICS

RELATIVE PHOTON INTENSITY
vs
ANGULAR DISPLACEMENT

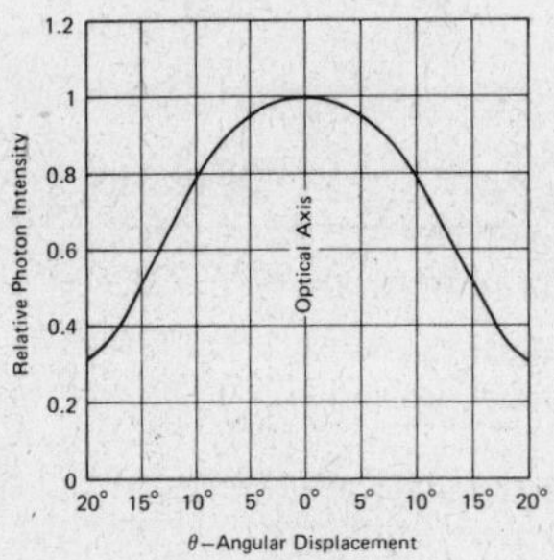

FIGURE 2

DARK CURRENT
vs
FREE-AIR TEMPERATURE

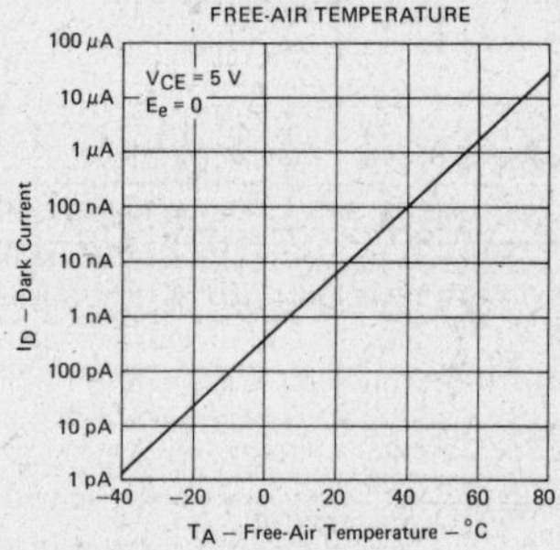

FIGURE 3

COLLECTOR CURRENT
vs
COLLECTOR-EMITTER VOLTAGE

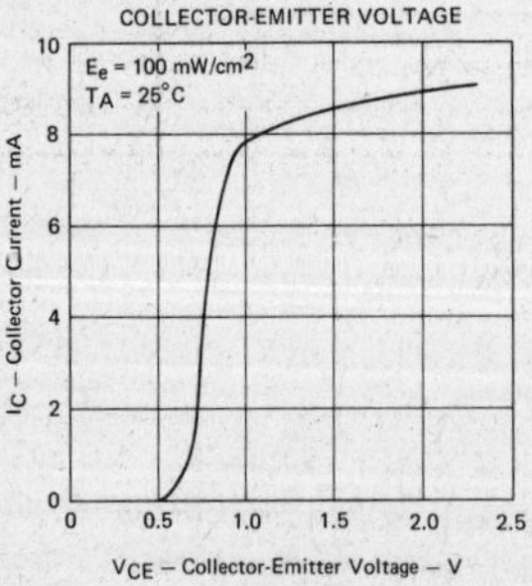

FIGURE 4

COUPLING CHARACTERISTICS
OF
TIL40 WITH TIL416

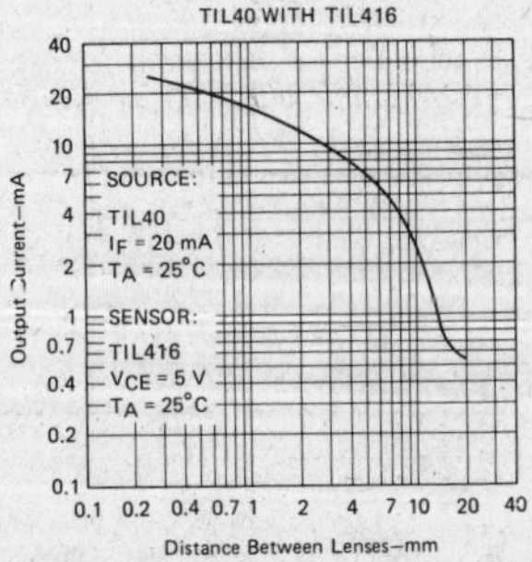

FIGURE 5

TEXAS INSTRUMENTS
INCORPORATED
POST OFFICE BOX 225012 • DALLAS, TEXAS 75265

D1971, NOVEMBER 1974—REVISED FEBRUARY 1983

DESIGNED FOR HIGH-DENSITY READ OUT

- Hermetically Sealed Pill Package
- Recommended for Application in Character Recognition, Tape and Card Readers, Velocity Indicators, and Encoders
- Unique Package Design Allows for Assembly into Printed Circuit Boards
- Spectrally and Mechanically Compatible with TIL23 thru TIL25
- Saturation Level Directly Compatible with most TTL
- TIL604HR2* Includes High-Reliability Processing and Lot Acceptance (See Page 5-39 for Summary of Processing)

mechanical data

TIL601 THRU TIL604, LS600

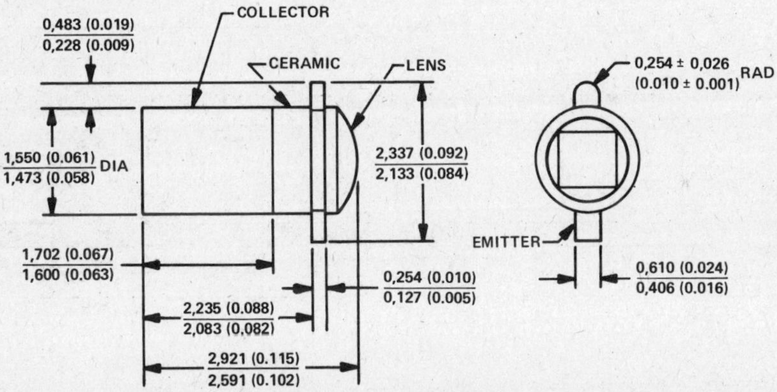

ALL DIMENSIONS ARE IN MILLIMETERS AND PARENTHETICALLY IN INCHES

FOUR TIMES ACTUAL SIZE

*All electrical and mechanical specifications for the TIL604 also apply for TIL604HR2.

absolute maximum ratings at 25 °C case temperature (unless otherwise noted)

Collector-Emitter Voltage	50 V
Emitter-Collector Voltage	7 V
Continuous Device Dissipation at (or below) 25 °C Case Temperature (See Note 1)	50 mW
Operating Case Temperature Range	−65 °C to 125 °C
Storage Temperature Range	−65 °C to 150 °C
Soldering Temperature (10 seconds)	240 °C

electrical characteristics at 25 °C case temperature (unless otherwise noted)

PARAMETER		TEST CONDITIONS	TYPE	MIN	TYP	MAX	UNIT
$V_{(BR)CEO}$	Collector-Emitter Breakdown Voltage	I_C = 100 μA, E_e = 0	ALL	50			V
$V_{(BR)ECO}$	Emitter-Collector Breakdown Voltage	I_E = 100 μA, E_e = 0	ALL	7			V
I_D	Dark Current	V_{CE} = 30 V, E_e = 0	ALL			25	nA
		V_{CE} = 30 V, E_e = 0, T_C = 100 °C	ALL		3		μA
I_L	Light Current	V_{CE} = 5 V, E_e = 20 mW/cm^2 See Note 2	TIL601	0.5		3	mA
			TIL602	2		5	
			TIL603	4		8	
			TIL604	7			
			LS600	0.8			
$V_{CE(sat)}$	Collector-Emitter Saturation Voltage	I_C = 0.4 mA, E_e = 20 mW/cm^2 See Note 2	ALL		0.15		V

NOTES: 1. Derate linearly to 125 °C at the rate of 0.5 mW/°C.
2. Irradiance (E_e) is the radiant power per unit area incident upon a surface. For this measurement the source is an unfiltered tungsten linear-filament lamp operating at a color temperature of 2870 K.

switching characteristics at 25 °C case temperature

PARAMETER		TEST CONDITIONS	TYP	UNIT
t_r	Rise Time	V_{CC} = 30 V, I_L = 800 μA, R_L = 1 kΩ, See Figure 1	8	μs
t_f	Fall Time		6	

PARAMETER MEASUREMENT INFORMATION

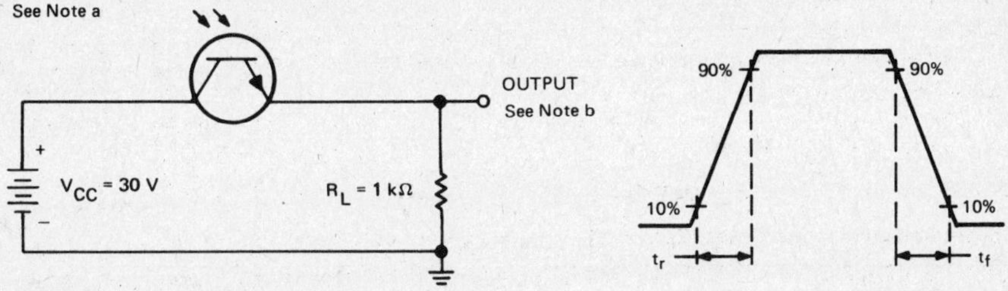

TEST CIRCUIT

OUTPUT VOLTAGE WAVEFORM

FIGURE 1

NOTES: a. Input irradiance is supplied by a pulsed gallium arsenide infrared emitter with rise and fall times of less than 50 ns. Incident irradiation is adjusted for I_L = 800 μA.

b. Output waveform is monitored on an oscilloscope with the following characteristics: $t_r \leqslant 25$ ns, $R_{in} \geqslant 1$ MΩ, $C_{in} \leqslant 20$ pF.

TYPICAL APPLICATION DATA

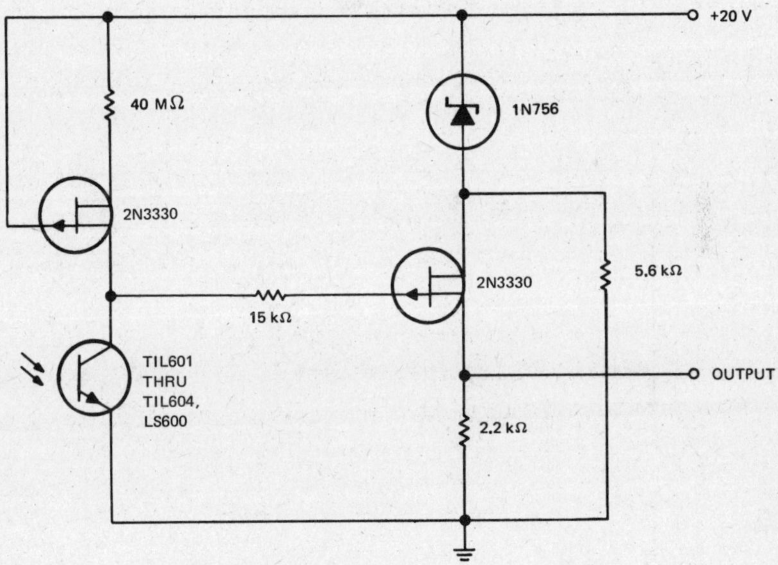

FIGURE 2—LOW-LEVEL DETECTOR AND PREAMPLIFIER

TYPICAL APPLICATION DATA

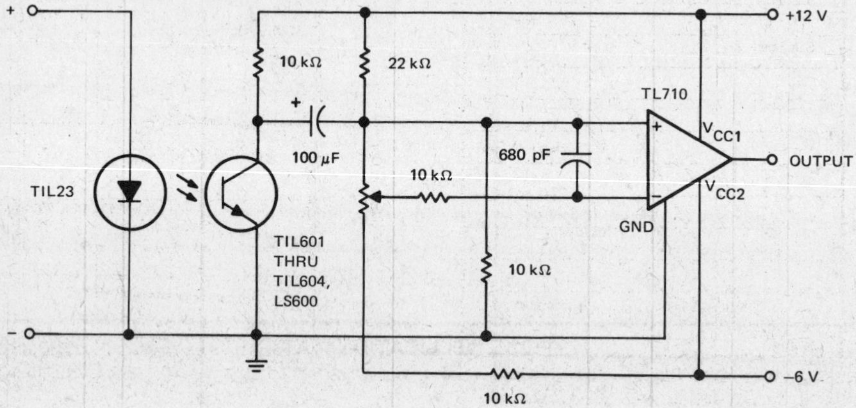

FIGURE 3—OPTICALLY COUPLED AMPLIFIER

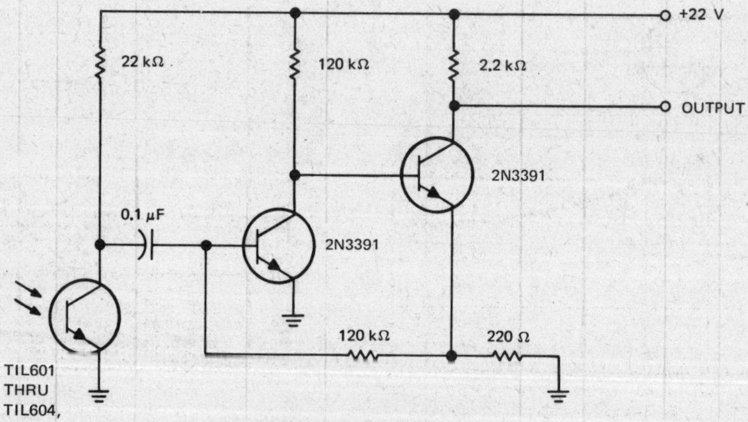

FIGURE 4—LIGHT PULSE DETECTOR

TEXAS INSTRUMENTS
INCORPORATED
POST OFFICE BOX 225012 • DALLAS, TEXAS 75265

TYPICAL CHARACTERISTICS

COLLECTOR CURRENT
vs
COLLECTOR-EMITTER VOLTAGE

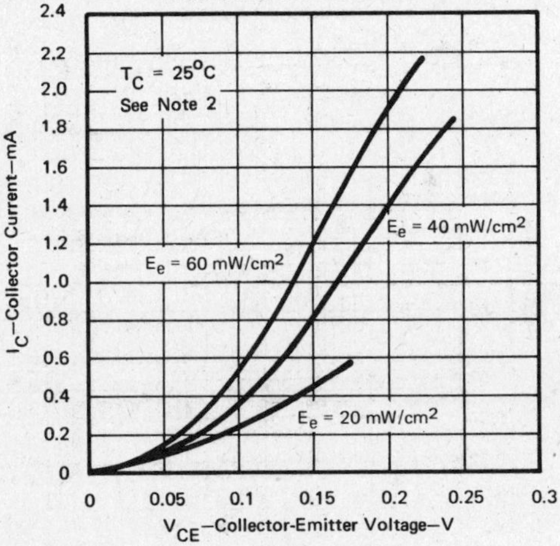

FIGURE 5

COLLECTOR CURRENT
vs
COLLECTOR-EMITTER VOLTAGE

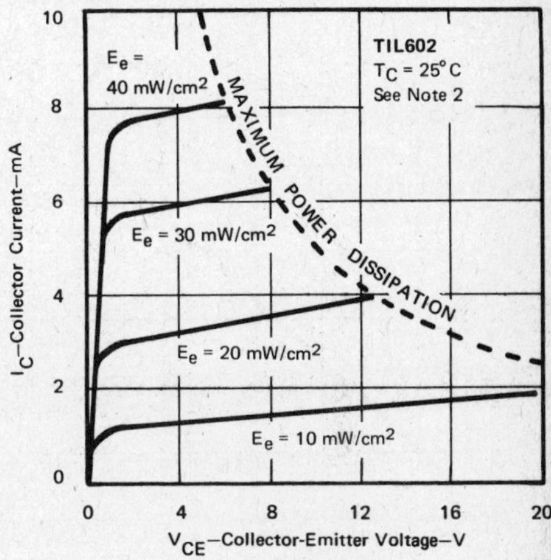

FIGURE 6

COLLECTOR-EMITTER SATURATION VOLTAGE
vs
IRRADIANCE

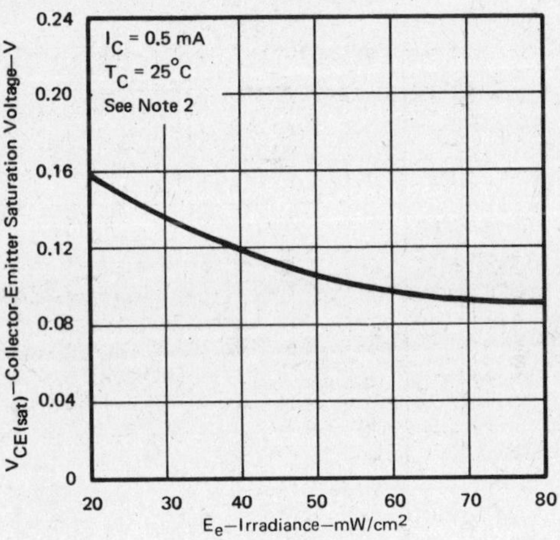

FIGURE 7

COLLECTOR CURRENT
vs
IRRADIANCE

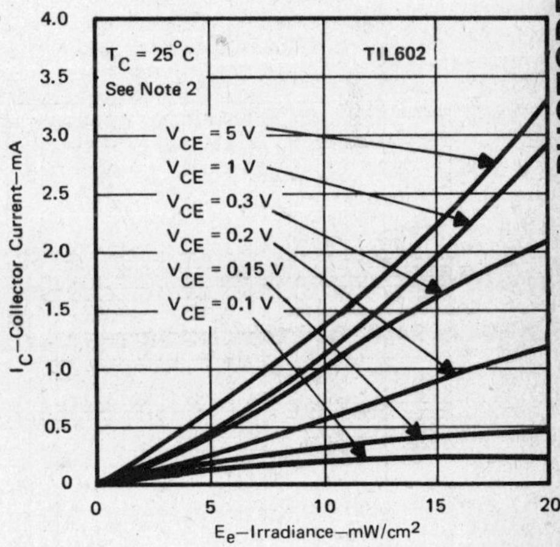

FIGURE 8

NOTE 2: Irradiance (E_e) is the radiant power per unit area incident upon a surface. For this measurement the source is an unfiltered tungsten linear-filament lamp operating at a color temperature of 2870 K.

5
PHOTODETECTORS

TYPICAL CHARACTERISTICS

DARK CURRENT
vs
CASE TEMPERATURE

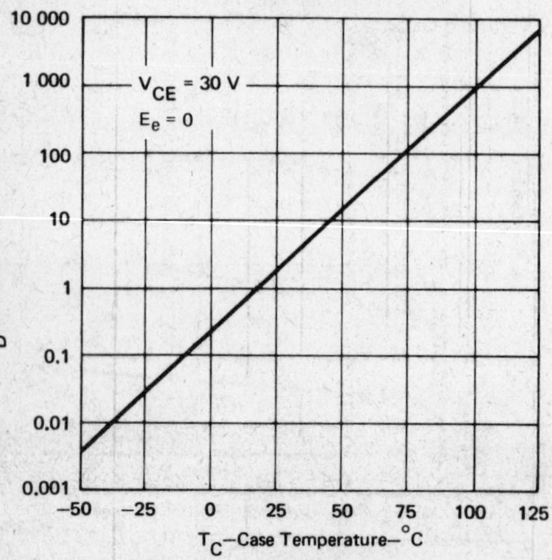

FIGURE 9

RELATIVE OUTPUT
vs
MODULATION FREQUENCY

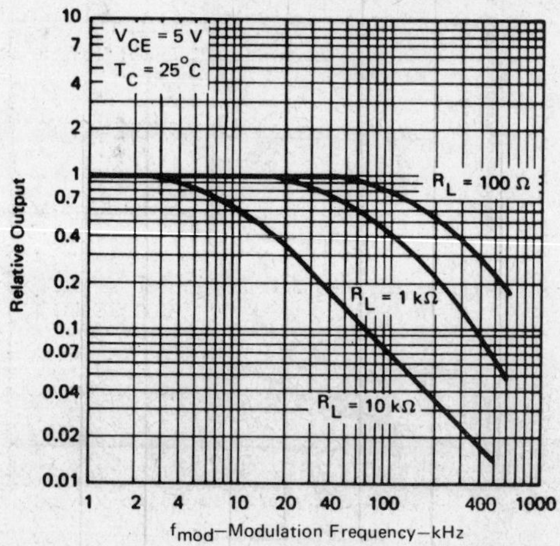

FIGURE 10

COUPLING CHARACTERISTICS
OF TIL23 OR TIL24 WITH TIL602

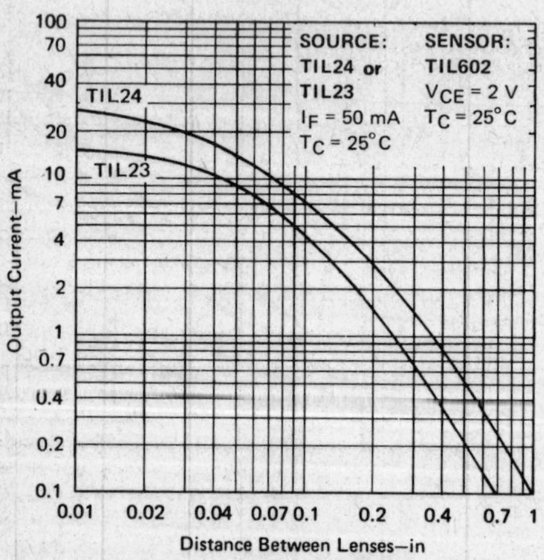

FIGURE 11

RELATIVE OUTPUT
vs
CASE TEMPERATURE OF SOURCE AND SENSOR

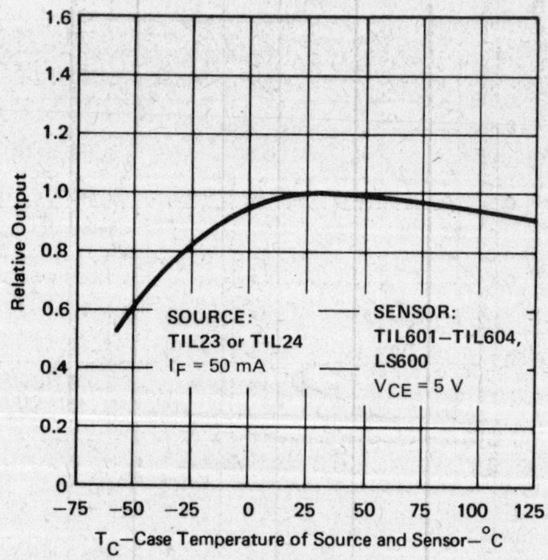

FIGURE 12

5 PHOTODETECTORS

TYPICAL CHARACTERISTICS

NORMALIZED LIGHT CURRENT
vs
ANGULAR DISPLACEMENT

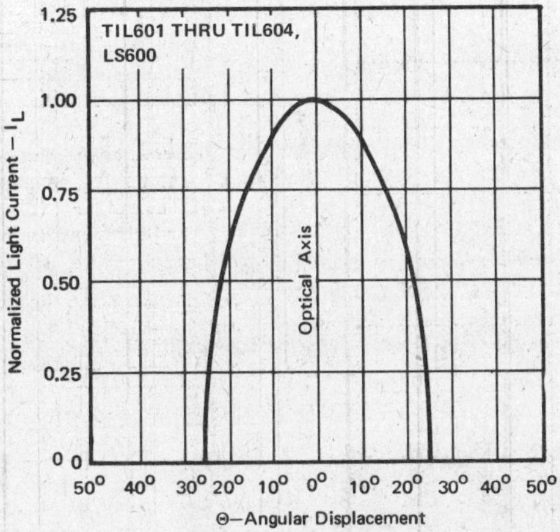

FIGURE 13

RELATIVE SPECTRAL CHARACTERISTICS

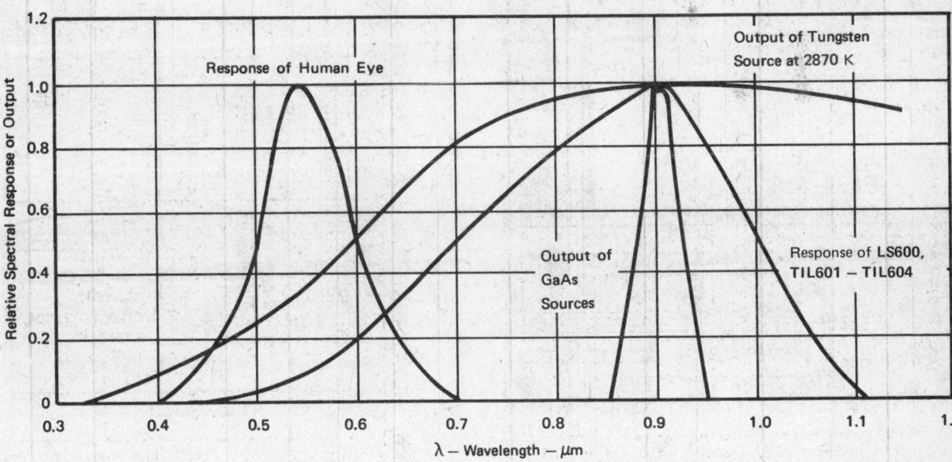

FIGURE 14

TEXAS INSTRUMENTS CUSTOMIZED OPTOELECTRONIC ASSEMBLIES

The TIL601 through TIL604, LS600 series is available mounted in printed circuit boards for custom-designed array or matrix applications. The array is a complete unit, without the problems associated with small, difficult-to-handle components. These arrays can be designed for punched-card or tape readers, position indicators, pattern and character recognition, shaft encoders, and many other special applications.

Texas Instruments custom-array techniques offer many advantages:

- The arrays are preassembled and tested, ready for installation.
- Custom arrays can be manufactured in almost any configuration to allow maximum design flexibility.
- Sensitivity across an entire array will be matched to within 50%.
- Associated components such as ICs and switches can be mounted directly on the printed circuit board.
- Arrays with components firmly soldered into place on both sides of a printed circuit board are more rugged than individually wired sensing devices.

Specifying optoelectronic arrays is easy; all that is required is a print of the array and the desired specifications.

TI sales engineers will assist in developing specifications for special applications.

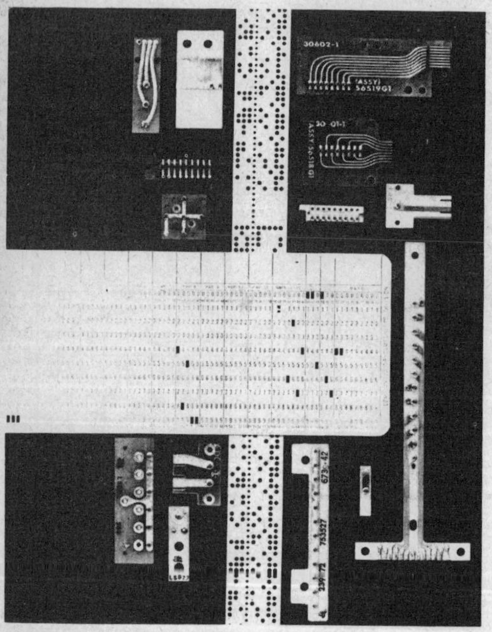

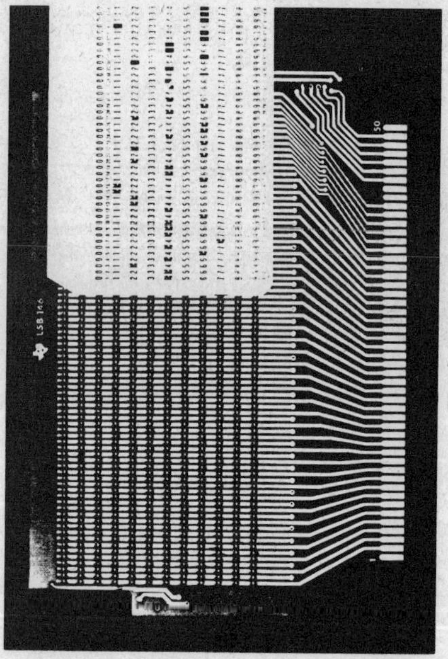

TEXAS INSTRUMENTS
INCORPORATED
POST OFFICE BOX 225012 • DALLAS, TEXAS 75265

- **This processing applies only to devices ordered under the part number TIL604HR2**
- **For electrical and mechanical specifications, refer to page 5-31**

This processing and lot acceptance follow the sequence of tests in MIL-S-19500 for JANTX types. This is not to be construed to be a JANTX-qualified part. A detail specification is available upon request through your TI Field Sales Office or Optoelectronics Marketing at the following address:

Texas Instruments Incorporated
Optoelectronics Marketing
P.O. Box 225012, MS 12
Dallas, Texas 75265
Phone: (214) 995-3821

TEST	MIL-STD-750 TEST METHOD
100% Processing	
Storage: $T_A = 125°C$, t = 24 h	1032
Temperature Cycle: −55°C to 125°C, 10 cycles	1051
Constant Acceleration: 20,000 G, Y_1 axis	2006
High-Temperature Reverse Bias: $V_{CE} = 30$ V, $T_A = 125°C$, t = 48 h	1039
Power Burn-in: $P_D = 50$ mW, t = 168 h	1039
Hermetic Seal, Fine	1071 Cond. G or H
Hermetic Seal, Gross	1071 Cond. C or D
External Visual	2071
Product Acceptance	
Group A: LTPD = 5	
External Visual	2071
Electrical: $T_A = 25°C$	
Electrical: $T_A = 100°C$	
Group B-1: LTPD = 15	
Solderability	2026
Group B-2: LTPD = 10	
Thermal Shock	1051 Cond. B-1
Hermetic Seal, Fine	1071 Cond. G or H
Hermetic Seal, Gross	1071 Cond. C or D
Group B-3: LTPD = 5	
Steady-State Operating Life: t = 340 h	1027

TEST	MIL-STD-750 TEST METHOD
Group B-4:	
Decap, Internal Visual; Design Verification	
1 Device/0 Failure	2075
Bond Strength LTPD = 20 (C = 0)	2037 Cond. A
Group B-5: Not Applicable	
Group B-6: LTPD = 7	
High-Temperature Life (Nonoperating) t = 340 h	1032
(Group C Tests are run on one lot every six months)	
Group C-1: LTPD = 15	
Physical Dimensions	2066
Group C-2: LTPD = 10	
Thermal Shock (Glass Strain)	1056 Cond. A
Hermetic Seal, Fine	1071 Cond. G or H
Hermetic Seal, Gross	1071 Cond. C or D
Moisture Resistance	1021
External Visual	2071
Group C-3: LTPD = 10	
Shock: 1500 G	2016
Vibration: 50 G	2056
Acceleration: 20000 G	2006
Group C-4: LTPD = 15	
Salt Atmosphere	1041
Group C-5: Not Applicable	
Group C 6: λ = 10	
Steady-State Operating Life: t = 1000 h	1026

Avalanche Photodiodes (APDs)

- Quick Reference Guide
- Available with Temperature-Compensation Diode
- Hermetically Sealed Packages

AVALANCHE PHOTODIODES
QUICK REFERENCE GUIDE

DEVICE	ACTIVE AREA	PACKAGE	C_T (TYP)	R_S (TYP)
TIED56	5×10^{-4} cm^2	TO-18	1.2 pF	50 Ω
TIED59	45×10^{-4} cm^2	TO-39	8.5 pF	5 Ω
TIED69	180×10^{-4} cm^2	TO-39	30 pF	5 Ω
TIED87	5×10^{-4} cm^2	TO-12	2.5 pF	50 Ω
TIED88	45×10^{-4} cm^2	TO-12	9 pF	5 Ω
TIED89	180×10^{-4} cm^2	TO-12	30 pF	5 Ω

TEXAS INSTRUMENTS
INCORPORATED
POST OFFICE BOX 225012 • DALLAS, TEXAS 75265

D518, JUNE 1968—REVISED SEPTEMBER 1982

OPTIMIZED FOR HIGH-SPEED DETECTION OF NEAR-INFRARED RADIANT ENERGY

- **Useful from Audio to Microwave Frequencies**
- **Typical Photocurrent Gain of >600**
- **Active Area of 5 X 10^{-4} cm^2 (Diameter = 10 Mils)**
- **Typical System Noise Equivalent Power of 10^{-12} W/$\sqrt{Hz}$ at 1 GHz**

description

The TIED56 is a high-speed, high-resistivity photodiode. It is designed to operate in the reverse-voltage avalanche region just below the breakdown voltage. This results in a photocurrent signal gain of a magnitude dependent on the reverse voltage. The signal gain ahead of the input noise of typical amplifiers provides for enhancement of the signal-to-noise ratio in most optical receiver systems.

mechanical data

The device is in a hermetically sealed package with a glass lens or window. The outline is similar to TO-18.

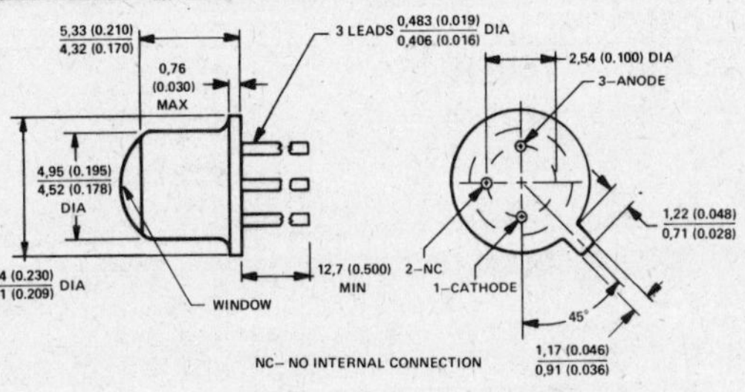

ALL LINEAR DIMENSIONS ARE IN MILLIMETERS AND PARENTHETICALLY IN INCHES.

THE ANODE IS IN ELECTRICAL CONTACT WITH THE CASE.

MAXIMUM WINDOW EXTENSION FROM TOP OF CASE IS 1,02 (0.040) MINIMUM DIAMETER IS 3,81 (0.150).

THE WINDOW IS BORO-SILICATE GLASS.

absolute maximum ratings

Continuous Power Dissipation at (or below) 25°C Case Temperature (See Note 1)	100 mW
Storage Temperature Range	−65°C to 150°C
Lead Temperature 1,6 mm (1/16) Inch from Case for 10 Seconds	230°C

NOTE 1: Derate linearly to 125°C case temperature at the rate of 1 mW/°C.

Copyright © 1982 by Texas Instruments Incorporated

TEXAS INSTRUMENTS
INCORPORATED
POST OFFICE BOX 225012 • DALLAS, TEXAS 75265

TYPE TIED56
SILICON AVALANCHE PHOTODIODE

operating characteristics at 25°C case temperature

PARAMETER		TEST CONDITIONS ‡		MIN	TYP	MAX	UNIT
Breakdown Voltage, $V_{(BR)}$		$I_R = 100\ \mu A$,	$E_e = 0$	155	170	185	V
Dark Current †	Bulk	$M = 100$,	$E_e = 0$		5	30	pA
	Surface				0.8	10	nA
Temperature Coefficient of Breakdown Voltage, $\alpha_{V(BR)}$		$I_R = 100\ \mu A$, See Note 2	$E_e = 0$,		200		mV/°C
Photocurrent Gain at Avalanche Noise Threshold, M_T		$\lambda = 900$ nm,	See Note 3	200	>600		
Total Capacitance, C_T		$V_R = 100$ V,	f = 1 MHz		1.2	3	pF
Series Resistance		f = 0.9 GHz			50		Ω
Radiant Responsivity, R_e		$\lambda = 900$ nm, $f_{mod} = 15$ MHz,	$M = 100$, $\Phi_e \leq 0.1$ mW		20		A/W
		$\lambda = 900$ nm, $f_{mod} = 10$ MHz,	$M = 1$, $\Phi_e \leq 0.1$ mW	0.15			A/W

†Dark current is the sum of surface current and gain M times the bulk current.

‡E_e is the incident radiant power per unit area.

NOTES: 2. Temperature coefficient is determined by the formula:

$$\alpha_{V(BR)} = \frac{V_{(BR)} @ 125^\circ C - V_{(BR)} @ -55^\circ C}{125^\circ C - (-55^\circ C)}$$

3. Gain M_T is measured at the reverse voltage at which the noise deviates from the theoretical linear characteristic. See Figure 1. Radiant flux is as required to give a photocurrent of 0.1 nA rms at $V_R = 40$ V.

TYPICAL CHARACTERISTICS

SIGNAL POWER AND NOISE POWER vs PHOTOCURRENT GAIN

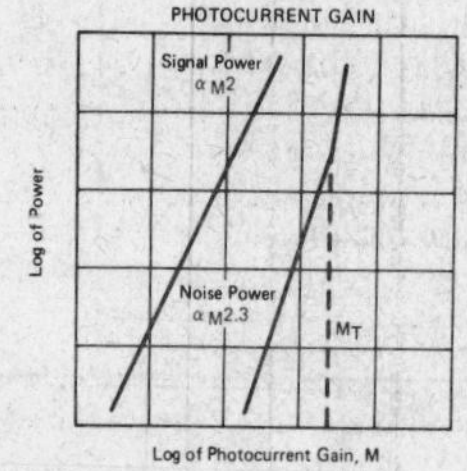

FIGURE 1

PHOTOCURRENT GAIN vs REVERSE VOLTAGE

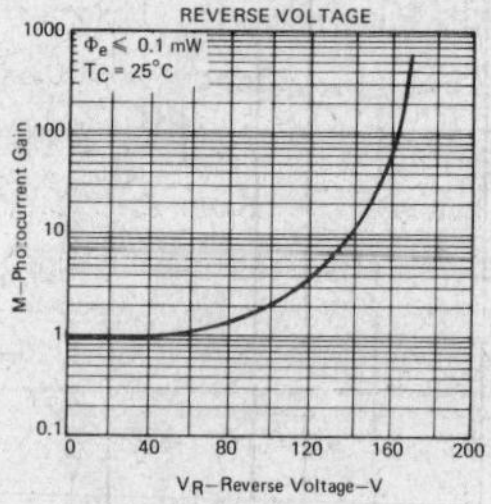

FIGURE 2

RADIANT RESPONSIVITY vs WAVELENGTH

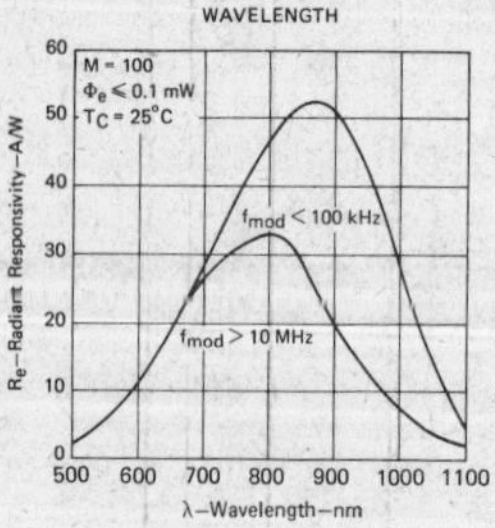

FIGURE 3

TOTAL CAPACITANCE vs REVERSE VOLTAGE

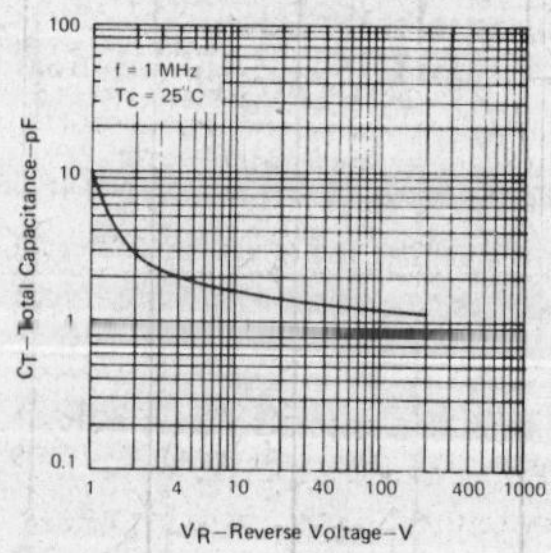

FIGURE 4

BIBLIOGRAPHY: Biard, J.R. and W.N. Shaunfield: A Model of the Avalanche Photodiode, *IEEE Transactions on Electron Devices*, vol. ED-14, no. 5, pp. 233-238, May 1967.

6 APDs

D696, JUNE 1971—REVISED JUNE 1982

OPTIMIZED FOR HIGH-SPEED DETECTION OF NEAR-INFRARED RADIANT ENERGY

- **Isolated Case for Shielding**
- **Useful from Audio to Microwave Frequencies**
- **Typical Photocurrent Gain of >600**
- **Active Area of 4.5 X 10^{-3} cm^2 (Diameter = 30 Mils)**
- **Typical System Noise Equivalent Power of 2 X 10^{-13} W/$\sqrt{}$Hz at 30 MHz Bandwidth**

description

The TIED59 is a high-speed, high-resistivity photodiode. It is designed to operate in the reverse-voltage avalanche region just below the breakdown voltage. This results in photocurrent signal gain of a magnitude dependent on the reverse voltage. The signal gain ahead of the input noise of typical amplifiers provides for enhancement of the signal-to-noise ratio in most optical receiver systems. The TIED59 is similar to TIED56 except that it has a larger active area making it more useful in lens systems with small f-numbers or where focusing is a problem.

mechanical data

The device is in a hermetically sealed welded case similar to, but slightly shorter then, JEDEC TO-39. The window is borosilicate glass. Its nominal dimensions are: diameter, 6,6 mm (0.260 inch); thickness, 1,5 mm (0.060 inch); and distance from front surface of the window to the active area, 1,9 mm (0.075 inch).

THE ACTIVE ELEMENTS ARE ELECTRICALLY INSULATED FROM THE CASE

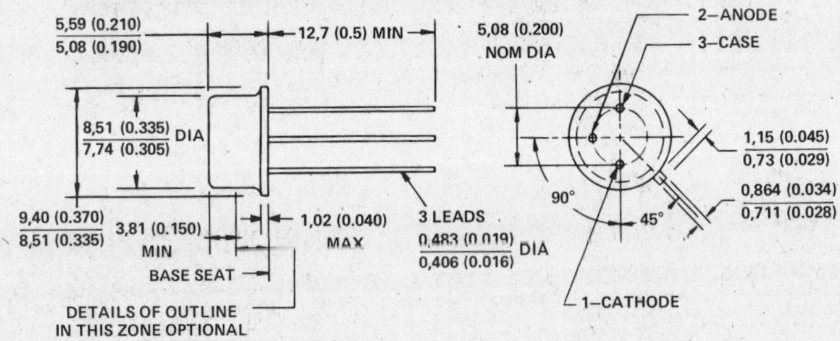

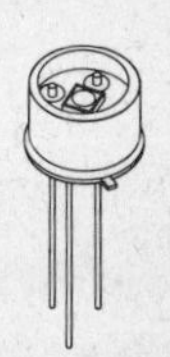

ALL LINEAR DIMENSIONS ARE IN MILLIMETERS AND PARENTHETICALLY IN INCHES.

absolute maximum ratings

Continuous Power Dissipation at (or below) 25°C Case Temperature (See Note 1)	100 mW
Storage Temperature Range	−65°C to 150°C
Lead Temperature 1,6 mm (1/16 inch) from Case for 10 Seconds	230°C

NOTE 1: Derate linearly to 125°C case temperature at the rate of 1 mW/°C.

operating characteristics at 25°C case temperature

PARAMETER		TEST CONDITIONS‡		MIN	TYP	MAX	UNIT
Breakdown Voltage, $V_{(BR)}$		I_R = 100 μA,	E_e = 0	155	170	185	V
Dark Current†	Bulk	M = 100,	E_e = 0		60	150	pA
	Surface				2	20	nA
Temperature Coefficient of Breakdown Voltage, $\alpha_{V(BR)}$		I_R = 100 μA, See Note 2	E_e = 0,		200		mV/°C
Photocurrent Gain at Avalanche Noise Threshold, M_T		λ = 900 nm,	See Note 3	200	>600		
Total Capacitance, C_T		V_R = 100 V,	f = 1 MHz		8.5	12	pF
Series Resistance		f = 0.9 GHz			5		Ω
Radiant Responsivity, R_e		λ = 900 nm, f_{mod} = 15 MHz,	M = 100, $\Phi_e \leqslant$ 0.1 mW		20		A/W
		λ = 900 nm, f_{mod} = 10 MHz,	M = 1, $\Phi_e \leqslant$ 0.1 mW	0.15			

NOTES: 2. Temperature coefficient is determined by the formula:

$$\alpha_{V(BR)} = \frac{V_{(BR)} @ 125°C - V_{(BR)} @ -55°C}{125°C - (-55°C)}$$

3. Gain M_T is measured at the reverse voltage at which the noise deviates from the theoretical linear characteristic. See Figure 1. Radiant flux is as required to give a photocurrrent of 0.1 nA rms at V_R = 40 V.

† Dark current is the sum of surface current and gain M times the bulk current. ‡ E_e is the incident radiant power per unit area.

TYPICAL CHARACTERISTICS

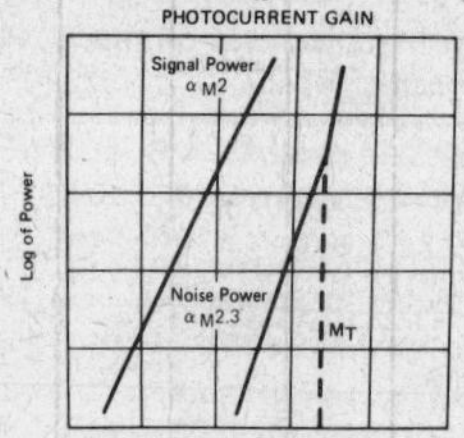

FIGURE 1

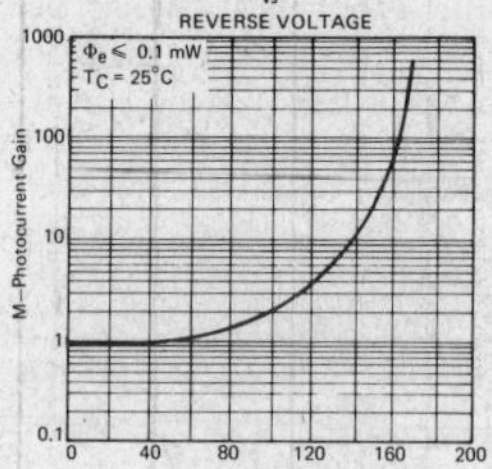

FIGURE 2

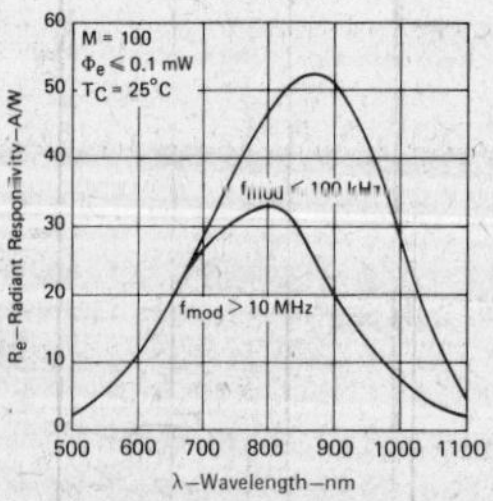

FIGURE 3

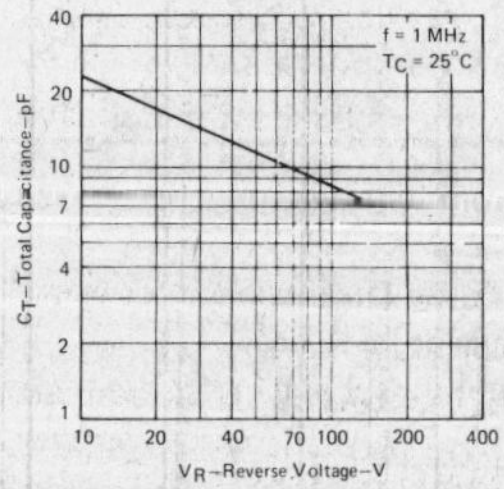

FIGURE 4

TEXAS INSTRUMENTS
INCORPORATED
POST OFFICE BOX 225012 • DALLAS, TEXAS 75265

OPTIMIZED FOR HIGH-SPEED DETECTION OF NEAR-INFRARED RADIANT ENERGY

- **Useful from Audio to Microwave Frequencies**
- **Typical Photocurrent Gain of > 600**
- **Active Area of 1.8 X 10^{-2} cm^2 (Diameter = 60 Mils)**
- **Typical System Noise Equivalent Power of 2 X 10^{-13} W/$\sqrt{Hz}$ at 30-MHz Bandwidth with TIEF151 Amplifier**

description

The TIED69 is a high-speed, high-resistivity photodiode. It is designed to operate in the reverse-voltage avalanche region just below the breakdown voltage. This results in a photocurrent signal gain of a magnitude dependent on the reverse voltage. The signal gain ahead of the input noise of typical amplifiers provides for enhancement of the signal-to-noise ratio in most optical receiver systems. The TIED69 is similar to the TIED56 and TIED59 except that it has a larger active area.

mechanical data

The device is in a hermetically sealed welded case similar to, but slightly shorter than, JEDEC TO-39. The window is borosilicate glass. Nominal lens dimensions are: diameter 6,6 mm (0.260 inch); thickness, 1,5 mm (0.060 inch); and distance from front surface of the window to the active area 1,9 mm (0.075 inch).

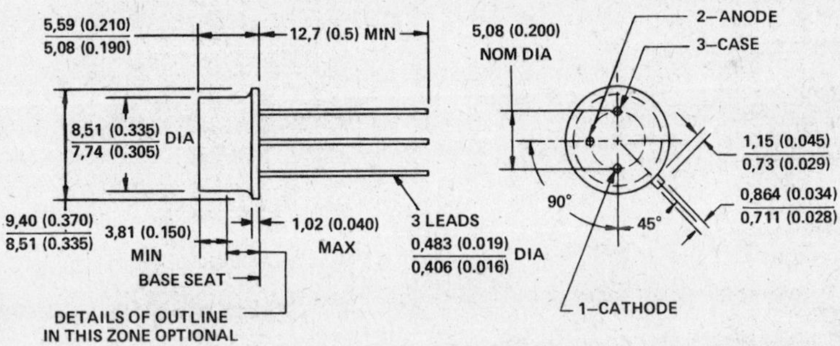

absolute maximum ratings

Continuous Power Dissipation at (or below) 25°C Case Temperature (See Note 1) . . . . 100 mW
Storage Temperature Range . . . . −65°C to 150°C
Lead Temperature 1,6 mm (1/16 Inch) from Case for 10 Seconds . . . . 230°C

NOTE 1: Derate linearly to 125°C case temperature at the rate of 1 mW/°C.

Copyright © 1982 by Texas Instruments Incorporated

operating characteristics at 25°C case temperature

PARAMETER		TEST CONDITIONS‡		MIN	TYP	MAX	UNIT
Breakdown Voltage, $V_{(BR)}$		I_R = 100 μA,	E_e = 0	155	170	185	V
Dark Current†	Bulk	M = 100,	E_e = 0		140	700	pA
	Surface				3.5	40	nA
Temperature Coefficient of Breakdown Voltage, $\alpha_{V(BR)}$		I_R = 100 μA, See Note 2	E_e = 0,		200		mV/°C
Photocurrent Gain at Avalanche Noise Threshold, M_T		λ = 900 nm,	See Note 3	200	>600		
Total Capacitance, C_T		V_R = 100 V,	f = 1 MHz		30	45	pF
Series Resistance		f = 0.9 GHz			5		Ω
Radiant Responsivity, R_e		λ = 900 nm, f_{mod} = 15 MHz,	M = 100, $\Phi_e \leq$ 0.1 mW		20		A/W
		λ = 900 nm, f_{mod} = 10 MHz,	M = 1, $\Phi_e \leq$ 0.1 mW	0.15			

NOTES: 2. Temperature coefficient is determined by the formula:

$$\alpha_{V(BR)} = \frac{V_{(BR)} @ 125°C - V_{(BR)} @ -55°C}{125°C - (-55°)}$$

3. Gain M_T is measured at the reverse voltage at which the noise deviates from the theoretical linear characteristic. See Figure 1. Radiant flux is as required to give a photocurrent of 0.1 nA rms at V_R = 40 V.

†Dark Current is the sum of surface current and gain M times the bulk current.

‡E_e is the incident radiant power per unit area.

6 APDs

TYPICAL CHARACTERISTICS

SIGNAL POWER AND NOISE POWER vs PHOTOCURRENT GAIN

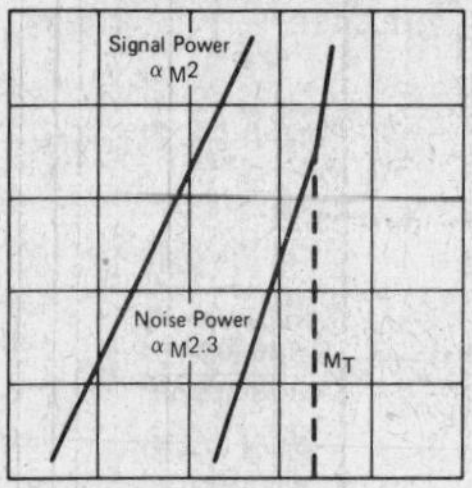

FIGURE 1

PHOTOCURRENT GAIN vs REVERSE VOLTAGE

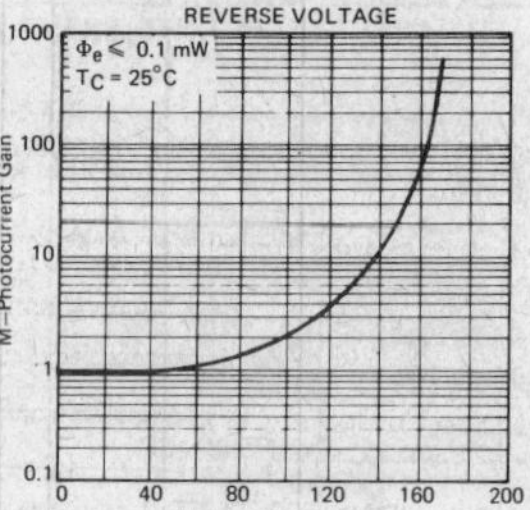

FIGURE 2

RADIANT RESPONSIVITY vs WAVELENGTH

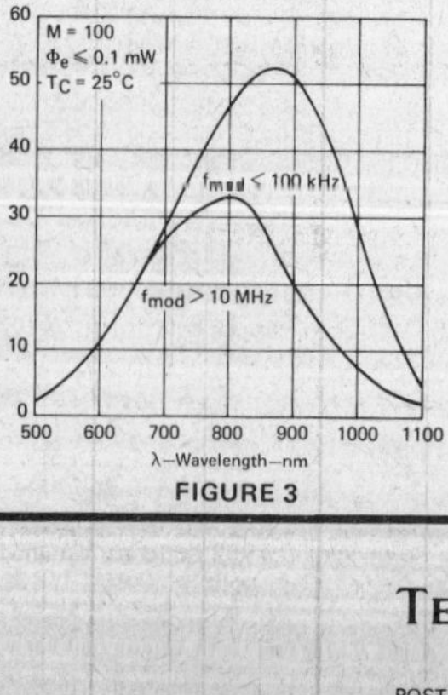

FIGURE 3

TOTAL CAPACITANCE vs REVERSE VOLTAGE

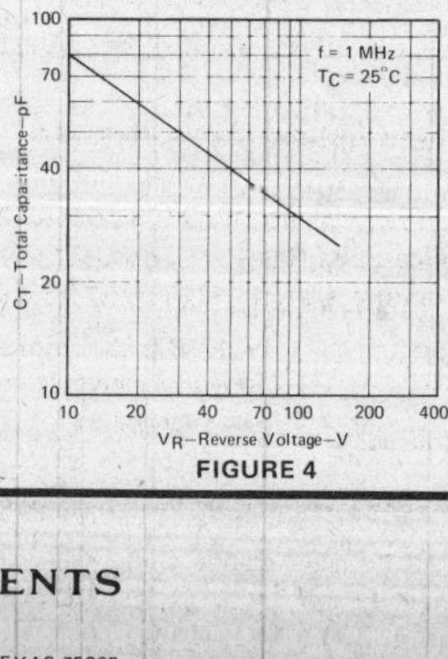

FIGURE 4

TEXAS INSTRUMENTS
INCORPORATED
POST OFFICE BOX 225012 • DALLAS, TEXAS 75265

TYPES TIED87, TIED88, TIED89 SILICON AVALANCHE PHOTODIODE/REFERENCE DIODE PAIRS

OPTIMIZED FOR HIGH-SPEED DETECTION OF NEAR-INFRARED RADIANT ENERGY

- **Photodiode and Reference Diode with Matched Breakdown Characteristics**
- **Both Diodes Mounted on Common Ceramic but Isolated Electrically**
- **Choice of Active Areas of Photodiode:**

  5×10^{-4} cm^2 (Diameter = 10 Mils) for TIED87
  4.5×10^{-3} cm^2 (Diameter = 30 Mils) for TIED88
  1.8×10^{-2} cm^2 (Diameter = 60 Mils) for TIED89

description

These diode pairs consist of an avalanche photodiode (APD) and a small reference diode that have been manufactured together to ensure close matching of both the breakdown voltages and the temperature coefficients of the two diodes. This makes it possible to build a temperature-compensating bias circuit that will hold the avalanche gain constant over wide temperature variations.

mechanical data

Each diode pair is in a hermetically sealed welded case similar to, but slightly shorter than, JEDEC TO-12. The window is borosilicate glass. Its dimensions are: diameter, 6,6 mm (0.260 inch); thickness, 1,5 mm (0.060 inch); and distance from front surface of the window to the active area, 1,9 mm (0.075 inch).

THE REFERENCE DIODE CATHODE IS IN ELECTRICAL CONTACT WITH THE CASE

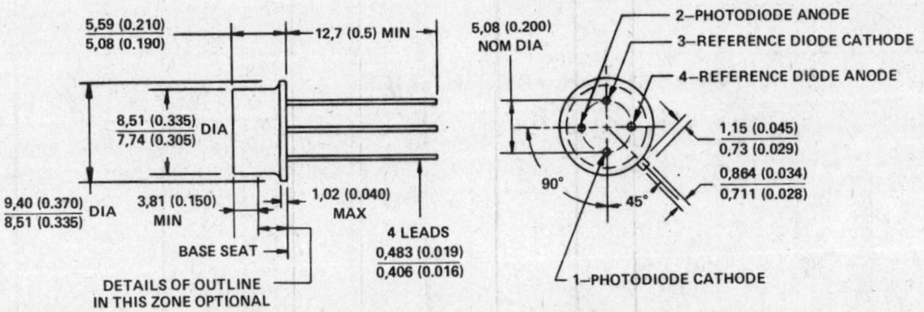

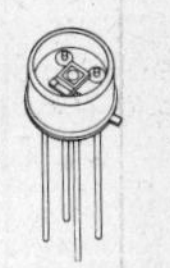

ALL DIMENSIONS ARE IN MILLIMETERS AND PARENTHETICALLY IN INCHES.

absolute maximum ratings

Continuous Power Dissipation at (or below) 25°C Case Temperature, Each Diode (See Note 1)	50 mW
Storage Temperature Range	−65°C to 150°C
Lead Temperature 1,6 mm (1/16 inch) from Case for 10 Seconds	230°C

NOTE 1: Derate linearly to 125°C case temperature at the rate of 0.5 mW/°C.

6 APDs

Copyright © 1982 by Texas Instruments Incorporated

operating characteristics at 25°C case temperature (unless otherwise noted)

PARAMETER		TEST CONDITIONS‡	TIED87 MIN	TIED87 TYP	TIED87 MAX	TIED88 MIN	TIED88 TYP	TIED88 MAX	TIED89 MIN	TIED89 TYP	TIED89 MAX	UNIT
Breakdown Voltage, $V_{(BR)}$	Photodiode	I_R = 100 μA, E_e = 0	155	170	185	155	170	185	155	170	185	V
	Reference Diode		155	170	185	155	170	185	155	170	185	
Temperature Coefficient of Breakdown Voltage	Photodiode	I_R = 100 μA, E_e = 0, See Note 2	170	200	230	170	200	230	170	200	230	mV/°C
	Reference Diode		170	200	230	170	200	230	170	200	230	
Breakdown Voltage Matching, $V_{(BR)APD} - V_{(BR)REF}$		I_R = 100 μA, E_e = 0		0	±10		0	±10		0	±10	V
Temperature Coefficient of Operating Voltage Matching		See Figure 5		+2	+6 / −2		+2	+6 / −2		+2	+6 / −2	mV/°C
Dark Current, I_D†	Bulk	M = 100, E_e = 0		5	30		60	150		140	700	pA
	Surface			0.8	10		2	20		3.5	40	nA
Photocurrent Gain at Avalanche Noise Threshold, M_T		λ = 900 nm, See Note 3	200	>600		200	>600		200	>600		
Total Capacitance, C_T	Photodiode	V_R = 100 V, E_e = 0, f = 1 MHz		2.5	4		9	30		30	45	pF
	Reference Diode			3			3			3		
Series Resistance		f = 0.9 GHz, E_e = 0		50			5			5		Ω
Radiant Responsivity, R_e		λ = 900 nm, M = 100, f_{mod} = 15 MHz, $\Phi_e \leq$ 0.1 mW		20			20			20		A/W
		λ = 900 nm, M = 1, f_{mod} = 10 MHz, $\Phi_e \leq$ 0.1 mW	0.15			0.15			0.15			

NOTES: 2. Temperature coefficient, $\alpha_{V(BR)}$, is determined by the formula:

$$\alpha_{V(BR)} = \frac{V_{(BR)} @ 125°C - V_{(BR)} @ -65°C}{125°C - (-65°C)}$$

3. Gain M_T is measured at the reverse voltage at which the noise deviates from the theoretical linear characteristic. See Figure 1. Radiant flux is as required to give a photocurrent of 0.1 nA rms at V_R = 40 V.

†Dark Current is the sum of surface current and gain M times the bulk current.

‡E_e is the incident radiant-power per unit area.

6 APDs

TYPICAL CHARACTERISTICS

SIGNAL POWER AND NOISE POWER
vs
PHOTOCURRENT GAIN

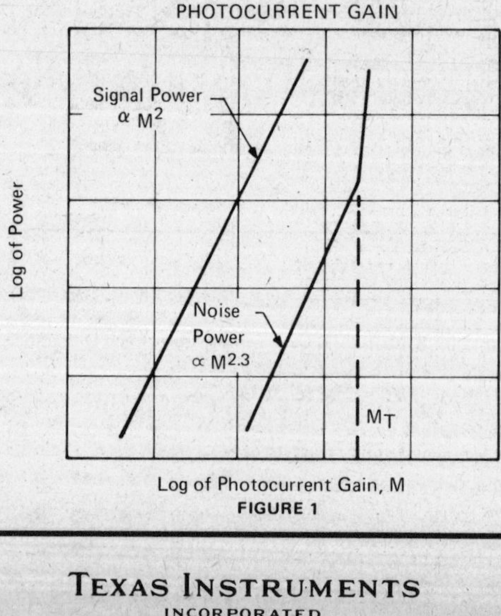

FIGURE 1

TEXAS INSTRUMENTS
INCORPORATED
POST OFFICE BOX 225012 • DALLAS, TEXAS 75265

TYPICAL CHARACTERISTICS

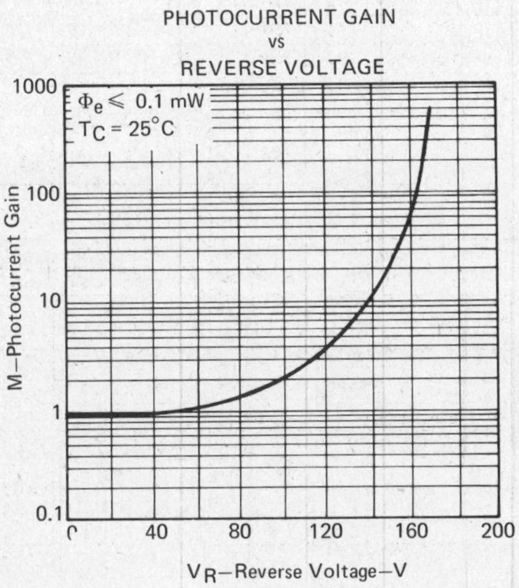

FIGURE 2

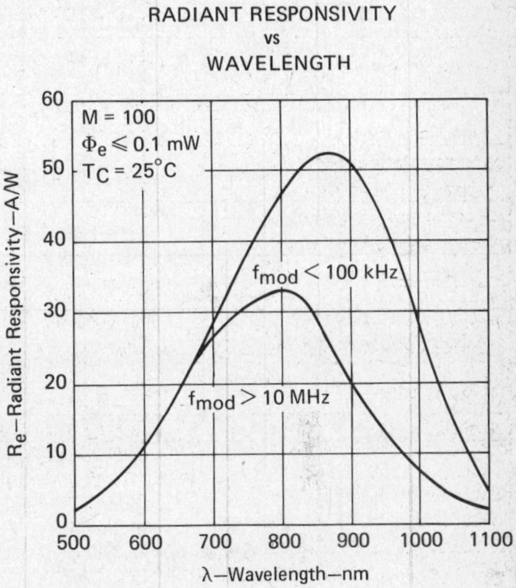

FIGURE 3

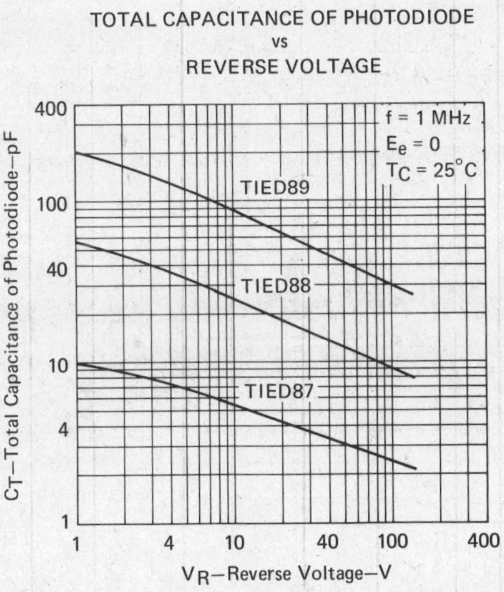

FIGURE 4

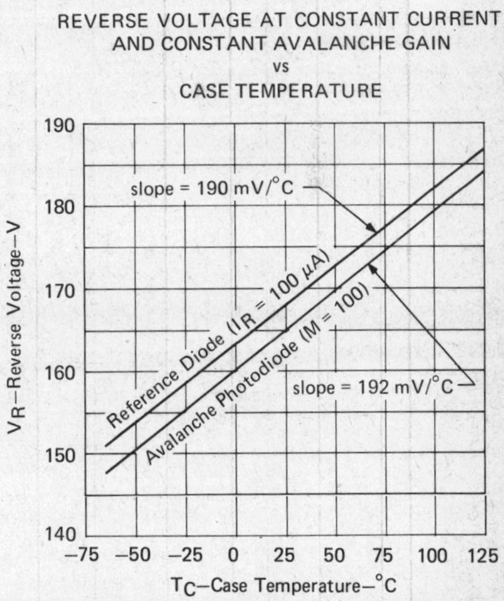

FIGURE 5

6
APDs

TYPICAL APPLICATION DATA

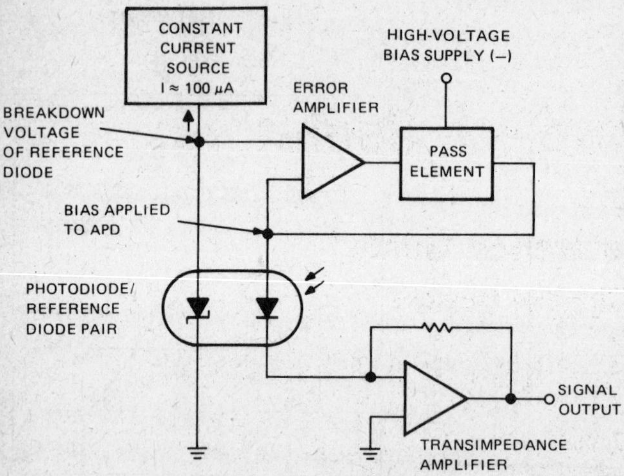

FIGURE 6–BLOCK DIAGRAM OF TEMPERATURE COMPENSATING BIAS CIRCUIT

The bias voltage applied to the APD should closely match the breakdown voltage of the reference diode (within ±10 V). In Figure 7, the bias voltage can be *increased* by *lowering* the value of the gain-control resistor. More importantly, the *change* with temperature in applied bias to the APD should closely match the *change* in breakdown voltage of the reference diode. Typically the temperature coefficient of constant avalanche gain is 2 mV/°C higher than the temperature coefficient of breakdown under conditions of constant current. The temperature coefficient of the bias circuit may be adjusted by insertion of a temperature-sensitive device in the pass element or the error amplifier.

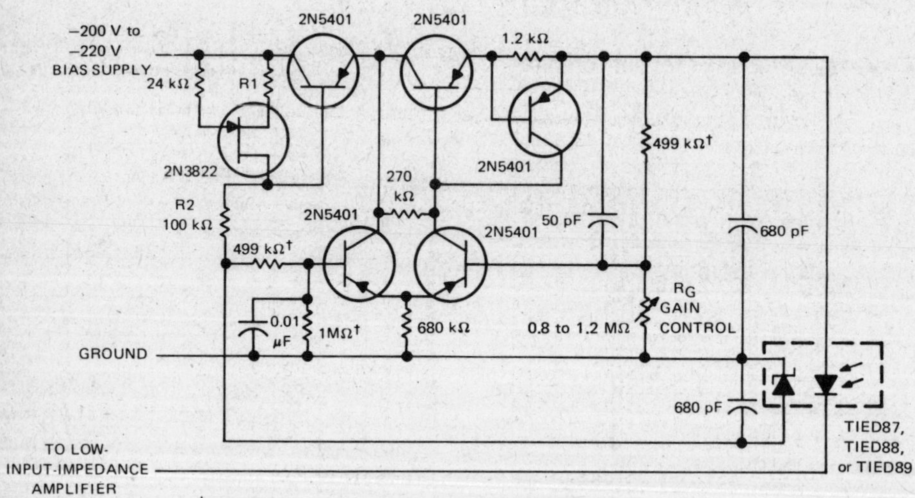

†These resistors are T2 metal film ¼ W 1%. Other resistors are ¼ W carbon.
R1 is selected to give approximately 220 μA current through R2.
Capacitors are ceramic disc, 500 V.

FIGURE 7–SUGGESTED CIRCUIT

TEXAS INSTRUMENTS
INCORPORATED
POST OFFICE BOX 225012 • DALLAS, TEXAS 75265

Optocouplers (Isolators)

- **Quick Reference Guide**
  Single-Channel Devices
- **Low-Cost Plastic (P-DIP) Packages**
- **Metal Cans**
- **JEDEC-Registered Devices**
- **High-Reliability Devices**
  JAN, JANTX, JANTXV Qualified
- **''Super-Couplers''**
- **UL-Approved Devices**

QUICK REFERENCE GUIDE
OPTOCOUPLERS

DEVICE	ISOLATION VOLTAGE (kV)		MINIMUM CTR	FEATURES
	PEAK	RMS	(%)	
3N261	1.0	–	50	JEDEC, Metal can
3N262	1.0	–	100 (500 max)	
3N263	1.0	–	200 (1000 max)	
4N22†	1.0	–	25	JEDEC, Metal can
4N23†	1.0	–	60	
4N24†	1.0	–	100	
4N25§	2.5	–	20	JEDEC, Plastic DIP
4N26	1.5	–	20	
4N27	1.5	–	10	
4N28	0.5	–	10	
4N35§	3.55	2.5	100	JEDEC, Plastic DIP
4N36	2.5	1.75	100	
4N37	1.5	1.05	100	
4N47††	1.0	–	50	JEDEC, Metal can
4N48††	1.0	–	100	
4N49††	1.0	–	200	
MCT2	1.5	–	20	Plastic DIP¶
MCT2E	2.5	–	20	
TIL102	1.0	–	25	Metal can
TIL103	1.0	–	100	
TIL111§	1.5	–	13	Plastic DIP¶
TIL112	1.5	–	2	
TIL113	1.5	–	300	
TIL114	2.5	–	13	
TIL115	2.5	–	2	
TIL116§	2.5	–	20	
TIL117§	2.5	–	50	
TIL118	1.5	–	10	
TIL119§	1.5	–	300	
TIL119A	1.5	–	300	The "A" version has no base connection.
TIL120	1.0	–	25	Metal can¶
TIL121	1.0	–	50	
TIL124	5.0	–	10	High voltage, Plastic DIP¶
TIL125	5.0	–	20	
TIL126	5.0	–	50	
TIL127	5.0	–	300	High voltage, Darlington, Plastic DIP.¶ The "A" version has no base connection.
TIL128	5.0	–	300	
TIL128A	5.0	–	300	
TIL153	3.54	2.5	10	High voltage, Plastic DIP.¶ UL File E-65085
TIL154	3.54	2.5	20	
TIL155§	3.54	2.5	50	
TIL156	3.54	2.5	300	High voltage, Darlington, UL File E-65085, Plastic DIP.¶ The "A" version has no base connection.
TIL157§	3.54	2.5	300	
TIL157A	3.54	2.5	300	

†JAN, JANTX, JANTXV levels to MIL-S-19500/486A USAF are also available.
††JAN, JANTX, JANTXV levels to MIL-S-19500/548
§Available in PEP 3 processing also.
¶Non-silver plated leads are available upon special request.

7 OPTOCOUPLERS

GALLIUM ARSENIDE DIODE INFRARED SOURCE OPTICALLY COUPLED TO A HIGH-GAIN N-P-N SILICON PHOTOTRANSISTOR

- Photon Coupling for Isolator Applications
- Very High Current Transfer Ratio . . . 500%
- High-Voltage Electrical Isolation . . . 1-kV Rating
- Stable Over Wide Temperature Range
- Hermetically Sealed TO-72 Package

description

This optocoupler features an improved current transfer ratio (CTR) at an input of one milliampere making it ideal for coupling with isolation from low-output MOS and CMOS devices to power devices or other systems. Typical applications include motor-speed controls, numeric control systems, meters, and instrumentation.

mechanical data

THE COLLECTOR IS IN ELECTRICAL CONTACT WITH THE CASE

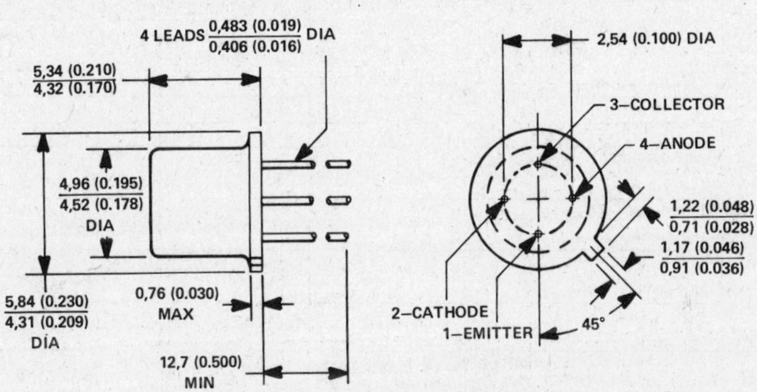

ALL LINEAR DIMENSIONS ARE IN MILLIMETERS AND PARENTHETICALLY IN INCHES.
ALL JEDEC TO-72 DIMENSIONS AND NOTES ARE APPLICABLE

*absolute maximum ratings at 25°C free-air temperature (unless otherwise noted)

Input to-Output Voltage	±1 kV
Collector-Emitter Voltage	40 V
Emitter-Collector Voltage	7 V
Input Diode Reverse Voltage	2 V
Input Diode Continuous Forward Current at (or below) 65°C Free-Air Temperature (See Note 1)	40 mA
Continuous Collector Current	50 mA
Peak Diode Current ($t_W < 1\ \mu s$, PRR < 300 pps)	1 A
Continuous Transistor Power Dissipation at (or below) 25°C Free-Air Temperature (See Note 2)	190 mW
Operating Free-Air Temperature Range	−55°C to 125°C
Storage Temperature Range	−55°C to 125°C
Lead Temperature 1/16 Inch from Case for 10 Seconds	240°C

NOTES: 1. Derate linearly to 125°C free-air temperature at the rate of 0.67 mA/°C.
2. Derate linearly to 125°C free-air temperature at the rate of 1.9 mW/°C.

*JEDEC registered data. This data sheet contains all applicable JEDEC registered data in effect at the time of publication.

TEXAS INSTRUMENTS
INCORPORATED
POST OFFICE BOX 225012 • DALLAS, TEXAS 75265

*electrical characteristics at 25°C free-air temperature (unless otherwise noted)

PARAMETER		TEST CONDITIONS		3N261 MIN	3N261 TYP	3N261 MAX	3N262 MIN	3N262 TYP	3N262 MAX	3N263 MIN	3N263 TYP	3N263 MAX	UNIT
$V_{(BR)CEO}$	Collector-Emitter Breakdown Voltage	I_C = 1 mA, I_F = 0	I_E = 0,	40			40			40			V
$V_{(BR)ECO}$	Emitter-Collector Breakdown Voltage	I_E = 100 µA, I_F = 0	I_C = 0	7			7			7			V
I_R	Input Diode Static Reverse Current	V_R = 2 V				100			100			100	µA
$I_{C(on)}$	On-State Collector Current	V_{CE} = 5 V, I_F = 1 mA		0.5			1		5	2		10	mA
		V_{CE} = 5 V, I_F = 2 mA,	T_A = −55°C	0.7			1.4			2.8			mA
		V_{CE} = 5 V, I_F = 2 mA,	T_A = 100°C	0.5			1			2			mA
		V_{CE} = 5 V, I_F = 10 mA,	See Note 3		50			80			90		mA
$I_{C(off)}$	Off-State Collector Current	V_{CE} = 20 V, I_F = 0			6	100		6	100		6	100	nA
		V_{CE} = 20 V, I_F = 0,	T_A = 100°C		4	100		4	100		4	100	µA
V_F	Input Diode Static Forward Voltage	I_F = 10 mA,	T_A = −55°C	1		1.7	1		1.7	1		1.7	V
		I_F = 10 mA		0.8	1.4	1.5	0.8	1.4	1.5	0.8	1.4	1.5	V
		I_F = 10 mA,	T_A = 100°C	0.7		1.3	0.7		1.3	0.7		1.3	V
$V_{CE(sat)}$	Collector Emitter Saturation Voltage	I_C = 0.5 mA, I_F = 2 mA				0.3							V
		I_C = 1 mA, I_F = 2 mA							0.3				V
		I_C = 2 mA, I_F = 2 mA										0.3	V
r_{IO}	Input-to-Output Internal Resistance	V_{in-out} = ±1 kV,	See Note 4	10^{11}	10^{12}		10^{11}	10^{12}		10^{11}	10^{12}		Ω
C_{io}	Input-to-Output Capacitance	V_{in-out} = 0, See Note 4	f = 1 MHz,		2.5	5		2.5	5		2.5	5	pF

*switching characteristics at 25°C free-air temperature

PARAMETER		TEST CONDITIONS		3N261 MIN	3N261 TYP	3N261 MAX	3N262 MIN	3N262 TYP	3N262 MAX	3N263 MIN	3N263 TYP	3N263 MAX	UNIT
t_r	Rise Time	V_{CC} = 10 V,	$I_{F(on)}$ = 5 mA,		10	20		10	20		15	25	µs
t_f	Fall Time	R_L = 100 Ω,	See Figure 1		10	20		10	20		15	25	µs

NOTES: 3. This parameter must be measured using pulse techniques, t_w = 100 µs, duty cycle ≤ 1%.
4. These parameters are measured between all the input diode leads shorted together and all the phototransistor leads shorted together.

*JEDEC registered data.

PARAMETER MEASUREMENT INFORMATION

Adjust amplitude of input pulse for
$I_{F(on)}$ = 5 mA

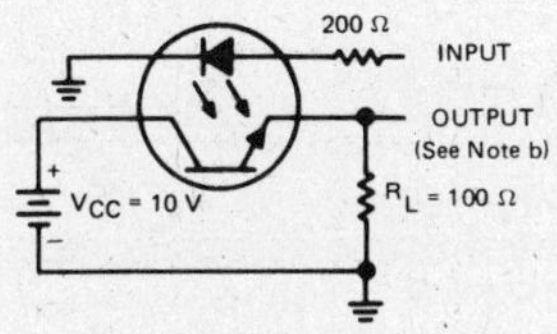

TEST CIRCUIT

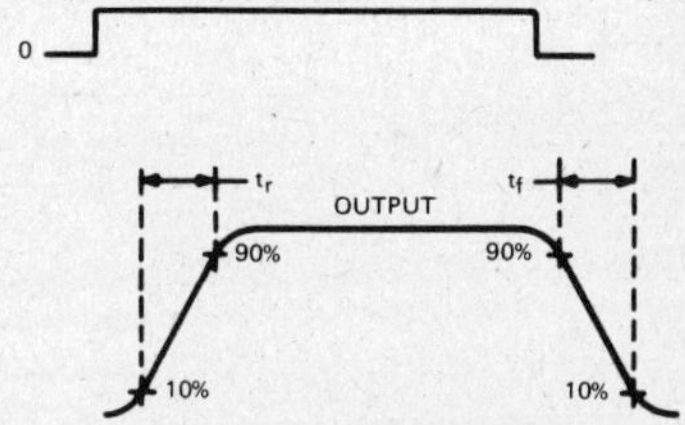

VOLTAGE WAVEFORMS

NOTES: a. The input waveform is supplied by a generator with the following characteristics: z_{out} = 50 Ω, $t_r \leq$ 15 ns, duty cycle ≈ 1%, t_w = 100 μs.

b. Waveforms are monitored on an oscilloscope with the following characteristics: $t_r \leq$ 12 ns, $R_{in} \geq$ 1MΩ, $C_{in} \leq$ 20 pF.

FIGURE 1—SWITCHING TIMES

TYPICAL CHARACTERISTICS

INPUT DIODE FORWARD CONDUCTION CHARACTERISTICS

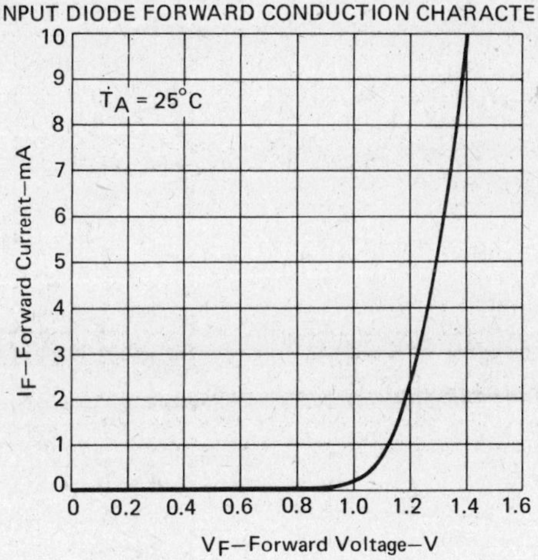

FIGURE 2

TEXAS INSTRUMENTS
INCORPORATED
POST OFFICE BOX 225012 • DALLAS, TEXAS 75265

TYPICAL CHARACTERISTICS

3N262

COLLECTOR CURRENT
vs
COLLECTOR-EMITTER VOLTAGE

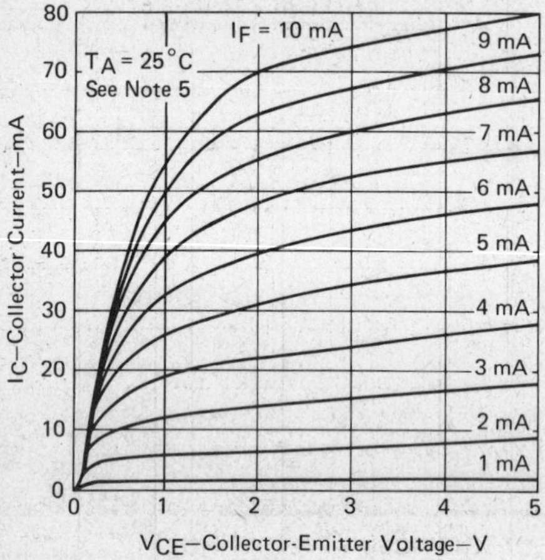

FIGURE 3

3N262

COLLECTOR CURRENT
vs
COLLECTOR-EMITTER VOLTAGE

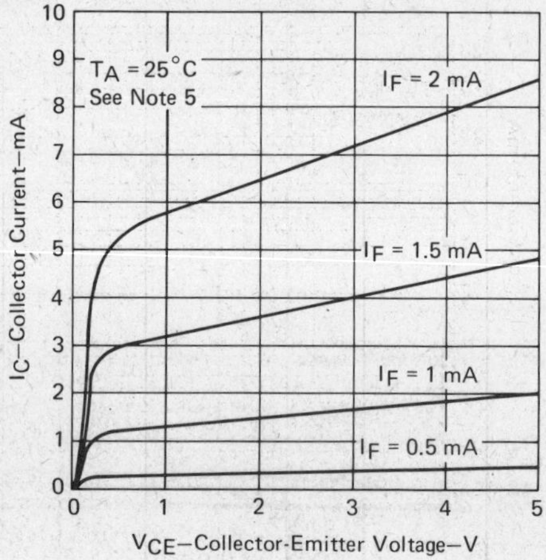

FIGURE 4

3N262

COLLECTOR CURRENT
vs
COLLECTOR-EMITTER VOLTAGE

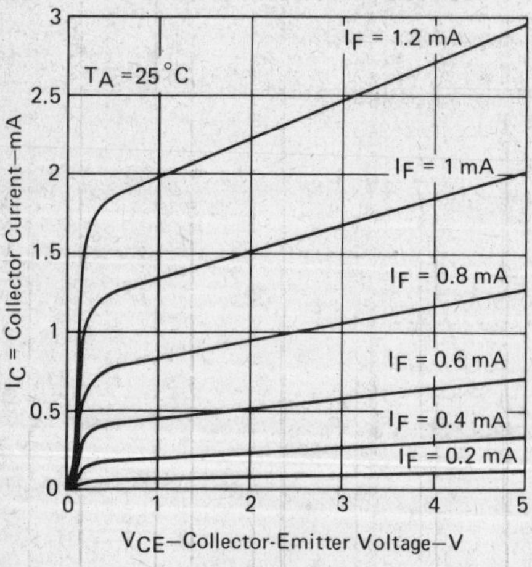

FIGURE 5

3N262

COLLECTOR CURRENT
vs
COLLECTOR-EMITTER VOLTAGE

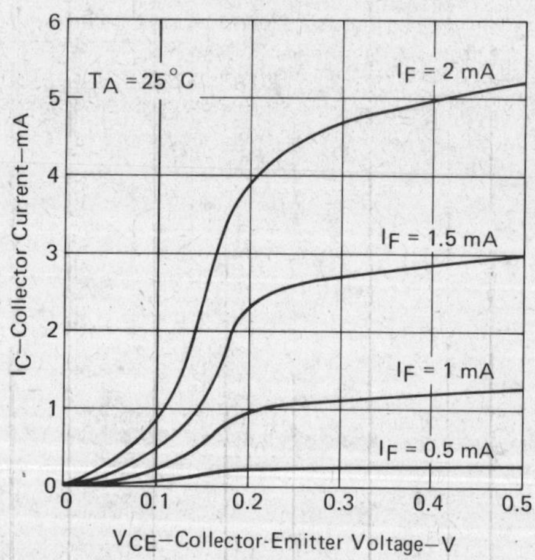

FIGURE 6

NOTE 5: This parameter was measured using pulse techniques. t_w = 100 μs, duty cycle = 1%.

TEXAS INSTRUMENTS
INCORPORATED
POST OFFICE BOX 225012 • DALLAS, TEXAS 75265

TYPICAL CHARACTERISTICS

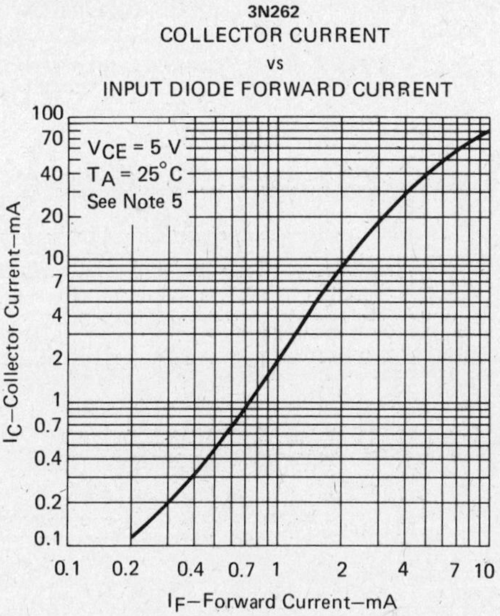

FIGURE 7

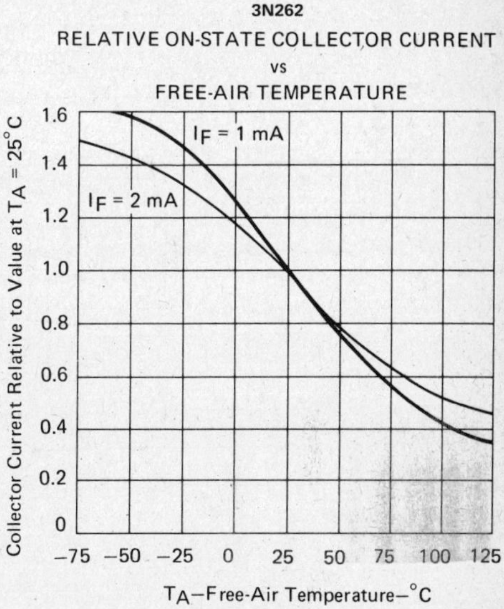

FIGURE 8

NOTE 5: This parameter was measured using pulse techniques. t_w = 100 μs, duty cycle = 1%.

D1424, AUGUST 1973

JEDEC REGISTERED DEVICES
GALLIUM ARSENIDE DIODE INFRARED SOURCE OPTICALLY COUPLED TO A HIGH-GAIN N-P-N SILICON PHOTOTRANSISTOR

- JAN, JAN TX, JAN TXV Versions Available
- Base Lead Provided for Conventional Transistor Biasing
- High Overall Current Gain . . . 1.5 Typ (4N24)
- High-Gain, High-Voltage Transistor . . . h_{FE} = 800 Typ (4N24),
  $V_{(BR)CEO}$ = 35 V Min
- High-Voltage Electrical Isolation . . . 1-kV Rating
- Stable over Wide Temperature Range

*mechanical data

THE COLLECTOR IS IN ELECTRICAL CONTACT WITH THE CASE

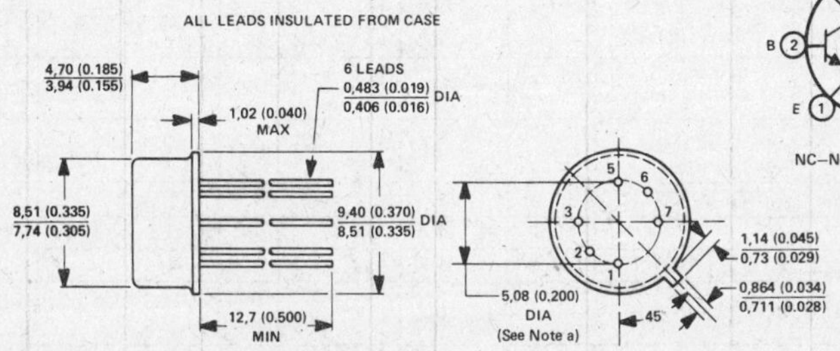

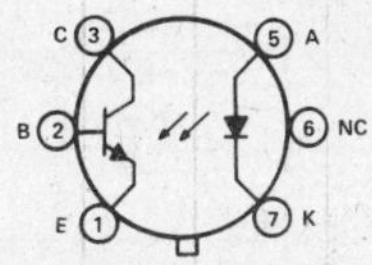

NC—No internal connection

NOTE: a. Leads having maximum diameter shall be within 0,18 mm (0.007 inch) of true position relative to a maximum-width tab when measured in the gaging plane between 1,371 mm (0.054 inch) and 1,397 mm (0.055 inch) below the seating plane.

ALL LINEAR DIMENSIONS ARE IN MILLIMETERS AND PARENTHETICALLY IN INCHES

*absolute maximum ratings at 25°C free-air temperature (unless otherwise noted)

Input-to-Output Voltage	±1 kV
Collector-Base Voltage	35 V
Collector-Emitter Voltage (See Note 1)	35 V
Emitter-Base Voltage	4 V
Input Diode Reverse Voltage	2 V
Input Diode Continuous Forward Current at (or below) 65°C Free-Air Temperature (See Note 2)	40 mA
Continuous Collector Current	50 mA
Peak Diode Current (See Note 3)	1A
Continuous Transistor Power Dissipation at (or below) 25°C Free-Air Temperature (See Note 4)	300 mW
Storage Temperature Range	−55°C to 125°C
Lead Temperature 1,6 mm (1/16 Inch) from Case for 10 Seconds	240°C

NOTES: 1. This value applies with the emitter-base diode open-circuited and the input-diode current equal to zero.
2. Derate linearly to 125°C free-air temperature at the rate of 0.67 mA/°C.
3. This value applies for $t_w \leq 1\ \mu s$, PRR ≤ 300 pps.
4. Derate linearly to 125°C free-air temperature at the rate of 3 mW/°C.

*JEDEC registered data. This data sheet contains all applicable JEDEC registered data in effect at the time of publication.

TEXAS INSTRUMENTS
INCORPORATED
POST OFFICE BOX 225012 • DALLAS, TEXAS 75265

*electrical characteristics at 25°C free-air temperature (unless otherwise noted)

PARAMETER		TEST CONDITIONS	4N22 MIN	4N22 TYP	4N22 MAX	4N23 MIN	4N23 TYP	4N23 MAX	4N24 MIN	4N24 TYP	4N24 MAX	UNIT
$V_{(BR)CBO}$	Collector-Base Breakdown Voltage	$I_C = 100\ \mu A$, $I_E = 0$, $I_F = 0$	35			35			35			V
$V_{(BR)CEO}$	Collector-Emitter Breakdown Voltage	$I_C = 1\ mA$, $I_B = 0$, $I_F = 0$	35			35			35			V
$V_{(BR)EBO}$	Emitter-Base Breakdown Voltage	$I_E = 100\ \mu A$, $I_C = 0$, $I_F = 0$	4			4			4			V
I_R	Input Diode Static Reverse Current	$V_R = 2\ V$			100			100			100	μA
$I_{C(on)}$	On-State Collector Current	$V_{CE} = 5\ V$, $I_B = 0$, $I_F = 2\ mA$	0.15			0.2			0.4			mA
		$V_{CE} = 5\ V$, $I_B = 0$, $I_F = 10\ mA$, $T_A = -55°C$	1			2.5			4			
		$V_{CE} = 5\ V$, $I_B = 0$, $I_F = 10\ mA$	2.5	4		6	8		10	15		
		$V_{CE} = 5\ V$, $I_B = 0$, $I_F = 10\ mA$, $T_A = 100°C$	1			2.5			4			
$I_{C(off)}$	Off-State Collector Current	$V_{CE} = 20\ V$, $I_B = 0$, $I_F = 0$			100			100			100	nA
		$V_{CE} = 20\ V$, $I_B = 0$, $I_F = 0$, $T_A = 100°C$			100			100			100	μA
V_F	Input Diode Static Forward Voltage	$I_F = 10\ mA$, $T_A = -55°C$	1		1.5	1		1.5	1		1.5	V
		$I_F = 10\ mA$	0.8		1.3	0.8		1.3	0.8		1.3	
		$I_F = 10\ mA$, $T_A = 100°C$	0.7		1.2	0.7		1.2	0.7		1.2	
$V_{CE(sat)}$	Collector-Emitter Stauration Voltage	$I_C = 2.5\ mA$, $I_B = 0$, $I_F = 20\ mA$			0.3							V
		$I_C = 5\ mA$, $I_B = 0$, $I_F = 20\ mA$						0.3				
		$I_C = 10\ mA$, $I_B = 0$, $I_F = 20\ mA$									0.3	
r_{IO}	Input-to-Output Internal Resistance	$V_{in\text{-}out} = \pm 1\ kV$, See Note 5	10^{11}			10^{11}			10^{11}			Ω
C_{io}	Input-to-Output Capacitance	$V_{in\text{-}out} = 0$, $f = 1\ MHz$, See Note 5			5			5			5	pF

*switching characteristics at 25°C free-air temperature

PARAMETER		TEST CONDITIONS	4N22 MIN	4N22 TYP	4N22 MAX	4N23 MIN	4N23 TYP	4N23 MAX	4N24 MIN	4N24 TYP	4N24 MAX	UNIT
t_r	Rise Time	$V_{CC} = 10\ V$, $I_{F(on)} = 10\ mA$, $R_L = 100\ \Omega$, See Figure 1			15			15			20	μs
t_f	Fall Time				15			15			20	μs

NOTE 5: These parameters are measured between all the input diode leads shorted together and all the phototransistor leads shorted together.

*JEDEC registered data

*PARAMETER MEASUREMENT INFORMATION

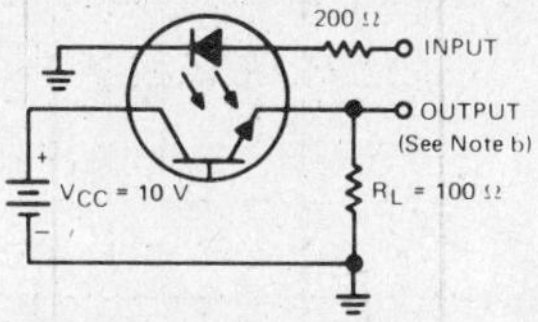

TEST CIRCUIT

Adjust amplitude of input pulse for $I_{F(on)}$ = 10 mA

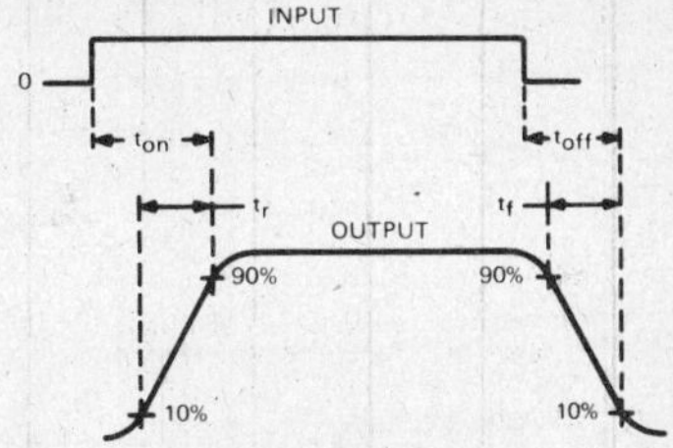

VOLTAGE WAVEFORMS

NOTES: a. The input waveform is supplied by a generator with the following characteristics: Z_{out} = 50 Ω, $t_r \leq$ 15 ns, t_w = 100 μs, duty cycle ≈ 1%.

b. Waveforms are monitored on an oscilloscope with the following characteristics: $t_r \leq$ 12 ns, $R_{in} \geq$ MΩ, $C_{in} \leq$ 20 pF.

*JEDEC registered data

FIGURE 1—SWITCHING TIMES

TYPICAL CHARACTERISTICS

4N22
COLLECTOR CURRENT
vs
COLLECTOR-EMITTER VOLTAGE

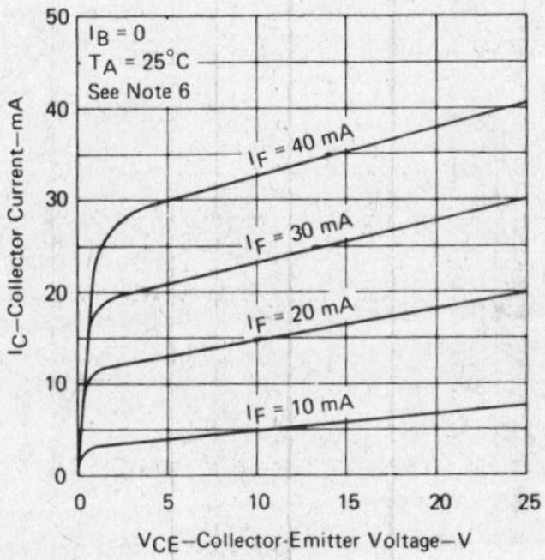

FIGURE 2

4N23
COLLECTOR CURRENT
vs
COLLECTOR-EMITTER VOLTAGE

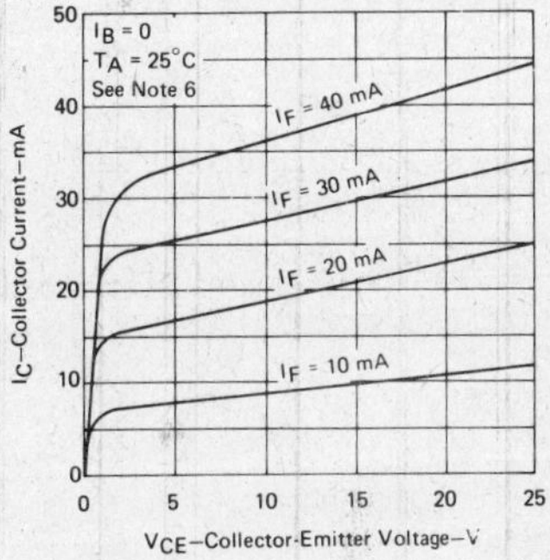

FIGURE 3

4N24
COLLECTOR CURRENT
vs
COLLECTOR-EMITTER VOLTAGE

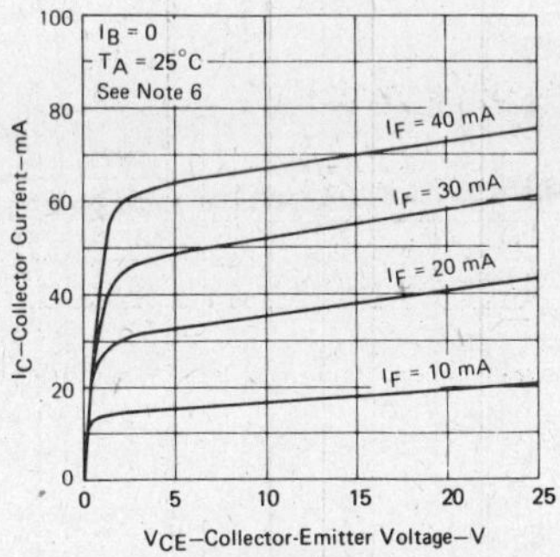

FIGURE 4

NOTE 6: This parameter was measured using pulse techniques. t_w = 100 μs, duty cycle = 1%.

TYPICAL CHARACTERISTICS

INPUT DIODE FORWARD CONDUCTION CHARACTERISTICS

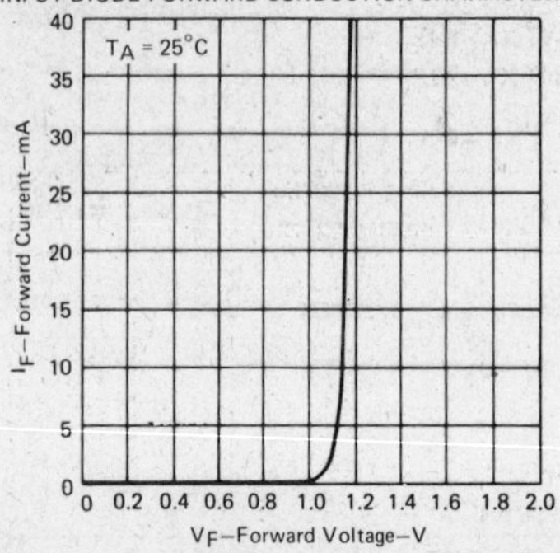

FIGURE 5

NORMALIZED ON-STATE COLLECTOR CURRENT†
vs
FREE-AIR TEMPERATURE

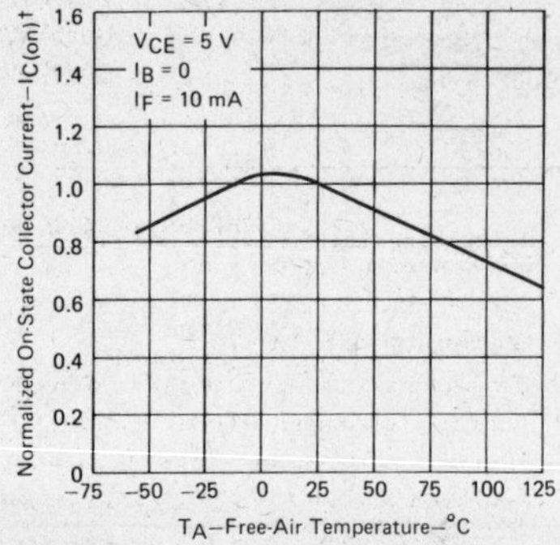

†Normalized to value at T_A = 25° C

FIGURE 6

PHOTOTRANSISTOR COLLECTOR CURRENT
vs
INPUT-DIODE FORWARD CURRENT

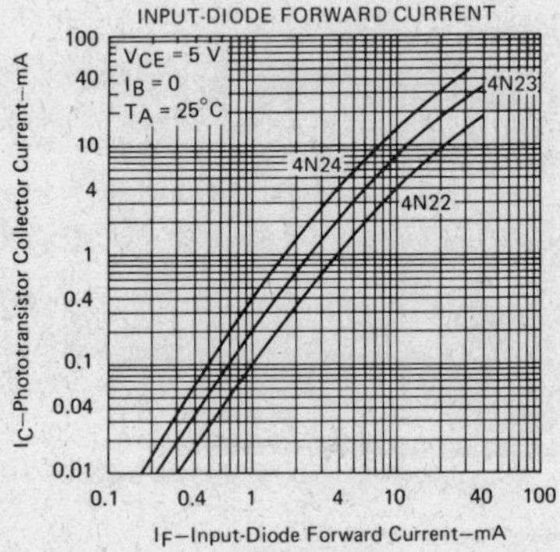

FIGURE 7

OFF-STATE COLLECTOR CURRENT
vs
FREE-AIR TEMPERATURE

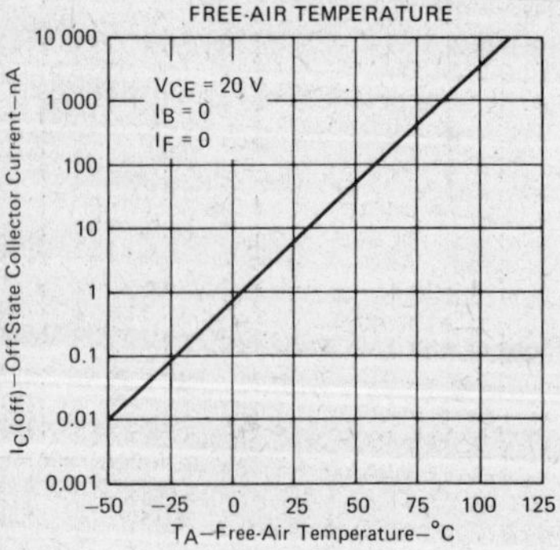

FIGURE 8

4N22, 4N23
AVERAGE SWITCHING TIME
vs
LOAD RESISTANCE

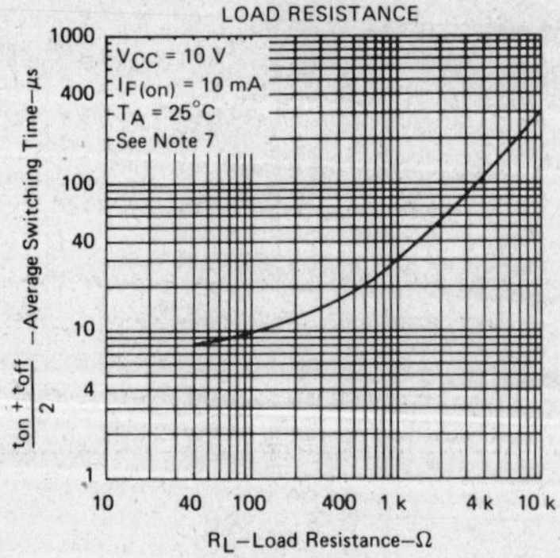

FIGURE 9

NOTE 7: This parameter was measured in the test circuit of Figure 1 with R_L varied between 40 Ω and 10 kΩ.

TEXAS INSTRUMENTS
INCORPORATED
POST OFFICE BOX 225012 • DALLAS, TEXAS 75265

This processing applies only to optocouplers ordered under part numbers shown below:

JAN 4N22, JANTX 4N22, JANTXV 4N22
JAN 4N23, JANTX 4N23, JANTXV 4N23
JAN 4N24, JANTX 4N24, JANTXV 4N24

TEST (PER MIL-S-19500/486A)	MIL-STD-750 TEST METHOD	JAN	JANTX	JANTXV
100% Processing				
Internal visual	2072			X
Storage: $T_A = 125°C$, $t = 72$ h	–		X	X
Temperature cycle: –55°C to 125°C, 10 cycles	1051		X	X
Constant acceleration: 20,000 G, Y_1 axis	2006		X	X
High-temperature reverse bias: $I_F = 0$, $T_A = 125°C$, $V_{CB} = 20$ V, $t = 96$ h	1039		X	X
Power burn-in: $I_F = 40$ mA, $P_D = 275 \pm 25$ mW, $t = 168$ h	1039		X	X
Hermetic seal, fine	1071 Cond. G or H		X	X
Hermetic seal, gross	1071 Cond. C or D		X	X
External visual	2071		X	X
Product Acceptance				
Group A				
External visual: LTPD is 10 for JAN, 7 for JANTX and JANTXV	2071	X	X	X
Electrical: $T_A = 25°C$, LTPD is 7 for JAN, 5 for JANTX and JANTXV	as needed	X	X	X
Electrical: $T_A = 100°C$, LTPD is 10 for JAN, 7 for JANTX and JANTXV	as needed	X	X	X
Electrical: $T_A = -55°C$, LTPD is 10 for JAN, 7 for JANTX and JANTXV	as needed	X	X	X
Group B-1: LTPD = 15				
Solderability	2026	X	X	X
Thermal shock	1051 Cond. B	X	X	X
Thermal shock	1056 Cond. A	X	X	X
Hermetic seal, fine	1071 Cond. G or H	X	X	X
Hermetic seal, gross	1071 Cond. C or D	X	X	X
Moisture resistance	1021	X	X	X
Group B-2: LTPD = 10				
Shock: 1500 G	2016	X	X	X
Vibration: 50 G	2056	X	X	X
Acceleration: 30,000 G	2006	X	X	X
Group B-3: LTPD = 20				
Isolation voltage: $V_{IO} = 150$ V, $T_A = 125°C$, $t = 24$ h	1016	X	X	X
Group B-4: LTPD is 7 for JAN, 5 for JANTX and JANTXV				
High temperature life (nonoperating): $T_A = 125°C$, $t = 340$ h	1032	X	X	X
Group B-5: LTPD is 7 for JAN, 5 for JANTX and JANTXV				
Steady-state operating life: $t = 340$ h	1027	X	X	X

7
OPTOCOUPLERS

TEST (PER MIL-S-19500/486A)	MIL-STD-750 TEST METHOD	JAN	JANTX	JANTXV
(Group C tests are run on one lot every six months)				
Group C-1				
Barometric pressure: LTPD = 10	1001	X	X	X
Group C-2				
Physical dimensions: LTPD = 20	2066	X	X	X
Group C-3 (MIL-STD 202, Method 215)				
Resistance to solvents: LTPD = 10	–	X	X	X
Group C-4				
Terminal strength: LTPD = 10	2036 Cond. E	X	X	X
Group C-5				
Salt atmosphere: LTPD = 10	1041	X	X	X
Group C-6				
High-temperature life (nonoperating): $T_A = 125°C$, t = 1000 h, LTPD is 7 for JAN, 5 for JANTX and JANTXV	1032	X	X	X
Group C-7				
Steady-state operating life: t = 1000 h, LTPD is 7 for JAN, 5 for JANTX and JANTXV	1027	X	X	X

D2493, SEPTEMBER 1978—REVISED MARCH 1983

COMPATIBLE WITH STANDARD TTL INTEGRATED CIRCUITS

- **Gallium Arsenide Diode Infrared Source Optically Coupled to a Silicon N-P-N Phototransistor**
- **High Direct-Current Transfer Ratio**
- **High-Voltage Electrical Isolation . . . 2.5-kV, 1.5-kV, or 0.5-kV Rating**
- **Plastic Dual-In-Line Package**
- **High-Speed Switching . . . t_r = 2 μs, t_f = 2 μs Typical**

mechanical data

The package consists of a gallium arsenide infrared-emitting diode and an n-p-n silicon phototransistor mounted on a 6-lead frame encapsulated within an electrically nonconductive plastic compound. The case will withstand soldering temperature with no deformation and device performance characteristics remain stable when operated in high-humidity conditions. Unit weight is approximately 0.52 grams.

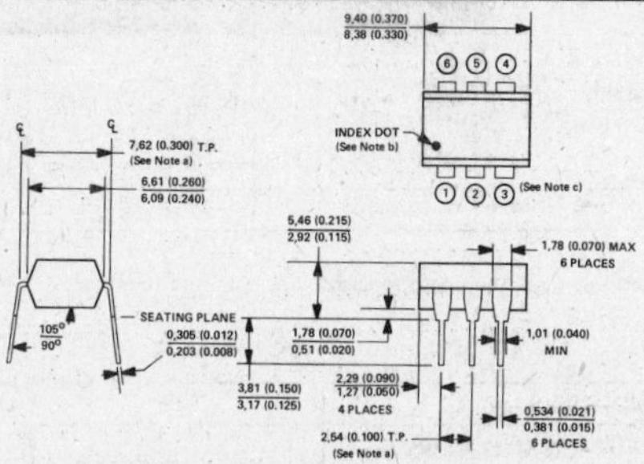

NOTES:

a. Leads are within 0,127 (0.005) radius of true position (T.P.) with maximum material condition and unit installed.

b. Pin 1 identified by index dot.

c. Terminal connections:

1. Anode
2. Cathode
} Infrared-emitting diode
3. No internal connection
4. Emitter
5. Collector
6. Base
} Phototransistor

FALLS WITHIN JEDEC MO-001AM DIMENSIONS
ALL LINEAR DIMENSIONS ARE IN MILLIMETERS AND PARENTHETICALLY IN INCHES

absolute maximum ratings at 25°C free-air temperature (unless otherwise noted)

*Peak Input-to-Output Voltage: 4N25	±2.5 kV
4N26, 4N27	±1.5 kV
4N28	±0.5 kV
*Collector-Base Voltage	70 V
*Collector-Emitter Voltage (See Note 1)	30 V
*Emitter-Collector Voltage	7 V
Emitter-Base Voltage	7 V
*Input-Diode Reverse Voltage	3 V
*Input-Diode Continuous Forward Current at (or below) 25°C Free-Air Temperature (See Note 2)	80 mA
*Input-Diode Peak Forward Current (t_w = 300 μs, duty cycle = 2%)	3 A
*Continuous Power Dissipation at (or below) 25°C Free-Air Temperature:	
Infrared-Emitting Diode (See Note 3)	150 mW
Phototransistor (See Note 3)	150 mW
Total, Infrared-Emitting Diode plus Phototransistor (See Note 4)	250 mW
*Storage Temperature Range	−55°C to 150°C
*Lead Temperature 1,6 mm (1/16 inch) from Case for 10 Seconds	260°C

*JEDEC registered data. This data sheet contains all applicable JEDEC-registered data in effect at the time of publication.

NOTES: 1. This value applies when the base-emitter diode is open-circulated.
2. Derate linearly to 100°C free-air temperature at the rate of 1.33 mA/°C.
3. Derate linearly to 100°C free-air temperature at the rate of 2 mW/°C.
4. Derate linearly to 100°C free-air temperature at the rate of 3.33 mW/°C.

Copyright © 1983 by Texas Instruments Incorporated

electrical characteristics at 25 °C free-air temperature (unless otherwise noted)

PARAMETER		TEST CONDITIONS	4N25, 4N26 MIN	4N25, 4N26 TYP	4N25, 4N26 MAX	4N27,4N28 MIN	4N27,4N28 TYP	4N27,4N28 MAX	UNIT
*$V_{(BR)CBO}$	Collector-Base Breakdown Voltage	I_C = 100 μA, I_E = 0, I_F = 0	70			70			V
*$V_{(BR)CEO}$	Collector-Emitter Breakdown Voltage	I_C = 1 mA, I_B = 0, I_F = 0	30			30			V
*$V_{(BR)ECO}$	Emitter-Collector Breakdown Voltage	I_E = 100 μA, I_B = 0, I_F = 0	7			7			V
*I_R	Input Diode Static Reverse Current	V_R = 3 V			100			100	μA
*$I_{C(on)}$	On-State Collector Current (Phototransistor Operation)	V_{CE} = 10 V, I_B = 0, I_F = 10 mA	2	5		1	3		mA
$I_{C(on)}$	On-State Collector Current (Photodiode Operation)	V_{CB} = 10 V, I_E = 0, I_F = 10 mA		20			20		μA
*$I_{C(off)}$	Off-State Collector Current (Phototransistor Operation)	V_{CE} = 10 V, I_B = 0, I_F = 0		1	50		1	50	nA
*$I_{C(off)}$	Off-State Collector current (Photodiode Operation)	V_{CB} = 10 V, I_E = 0, I_F = 0		0.1	20		0.1	20	nA
*V_F	Input Diode Static Forward Voltage	I_F = 10 mA		1.25	1.5		1.25	1.5	V
*$V_{CE(sat)}$	Collector-Emitter Saturation Voltage	I_C = 2 mA, I_B = 0, I_F = 50 mA		0.25	0.5		0.25	0.5	V
r_{IO}	Input-to-Output Internal resistance	V_{in-out} = ±2.5 kV for 4N25, ±1.5 kV for 4N26, 4N27, ±0.5 kV for 4N28, See Note 5	10^{11}	10^{12}		10^{11}	10^{12}		Ω
C_{io}	Input-to-Output Capacitance	V_{in-out} = 0, f = 1 MHz, See Note 5		1			1		pF

*JEDEC registered data

NOTE 5: These parameters are measured between both input diode leads shorted together and all the phototransistor leads shorted together.

switching characteristics at 25 °C free-air temperature

PARAMETER			TEST CONDITIONS	TYP	UNIT
t_r	Rise Time	Phototransistor Operation	V_{CC} = 10 V, I_B = 0, $I_{C(on)}$ = 2 mA, R_L = 100 Ω, See Test Circuit A of Figure 1	2	μs
t_f	Fall Time			2	
t_r	Rise Time	Photodiode Operation	V_{CC} = 10 V, I_E = 0, $I_{C(on)}$ = 20 μA, R_L = 1 kΩ, See Test Circuit B of Figure 1	1	μs
t_f	Fall Time			1	

PARAMETER MEASUREMENT INFORMATION

Adjust amplitude of input pulse for:
$I_{C(on)}$ = 2 mA (Test Circuit A) or
$I_{C(on)}$ = 20 μA (Test Circuit B)

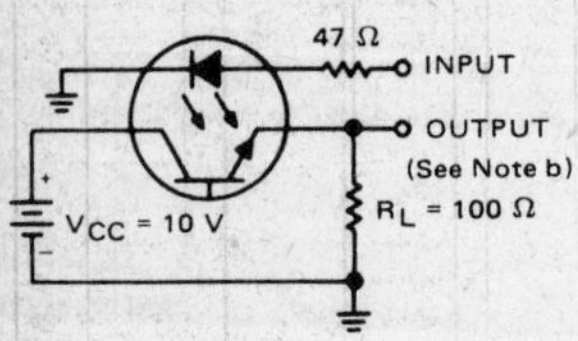

TEST CIRCUIT A
PHOTOTRANSISTOR OPERATION

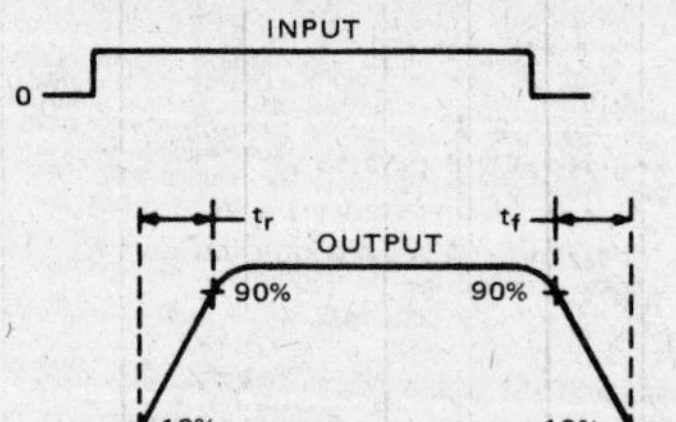

VOLTAGE WAVEFORMS

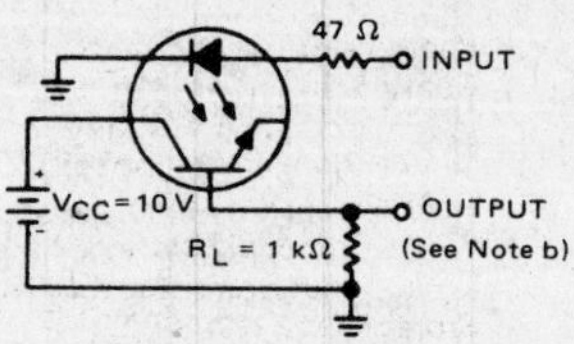

TEST CIRCUIT B
PHOTODIODE OPERATION

NOTES: a. The input waveform is supplied by a generator with the following characteristics: Z_{out} = 50 Ω, t_r ≤ 15 ns, duty cycle ≈ 1%, t_w = 100 μs.
b. The output waveform is monitored on an oscilloscope with the following characteristics: t_r ≤ 12 ns, R_{in} ≥ 1 MΩ, C_{in} ≤ 20 pF.

FIGURE 1 – SWITCHING TIMES

D2657, NOVEMBER 1981—REVISED APRIL 1983

COMPATIBLE WITH STANDARD TTL INTEGRATED CIRCUITS

- Gallium Arsenide Diode Infrared Source Optically Coupled to a Silicon N-P-N Phototransistor
- High Direct-Current Transfer Ratio
- High-Voltage Electrical Isolation . . . 1.5 kV, 2.5 kV, or 3.55 kV Rating
- Plastic Dual-In-Line Package
- High-Speed Switching: $t_r = 7\ \mu s$, $t_f = 7\ \mu s$ Typical
- Typical Applications Include Remote Terminal Isolation, SCR and Triac Triggers, Mechanical Relays, and Pulse Transformers

mechanical

The package consists of a gallium arsenide infrared-emitting diode and an n-p-n silicon phototransistor mounted on a 6-lead frame encapsulated within an electrically nonconductive plastic compound. The case will withstand soldering temperature with no deformation and device performance characteristics remain stable when operated in high-humidity conditions. Unit weight is approximately 0.52 grams.

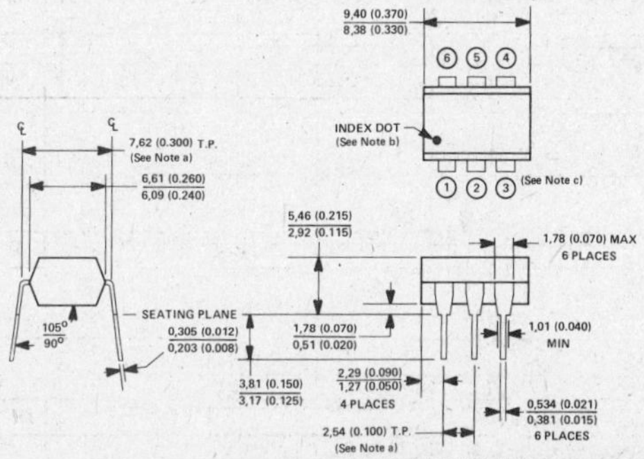

NOTES:
a. Leads are within 0,13 mm (0.005 inch) radius of true position (T.P.) with maximum material condition and unit installed.
b. Pin 1 identified by index dot.
c. Terminal connections:
   1. Anode } Infrared-emitting diode
   2. Cathode
   3. No internal connection
   4. Emitter } Phototransistor
   5. Collector
   6. Base

FALLS WITHIN JEDEC MO-001AM DIMENSIONS
ALL LINEAR DIMENSIONS ARE IN MILLIMETERS AND PARENTHETICALLY IN INCHES

*absolute maximum ratings at 25°C free-air temperature (unless otherwise noted)

	4N35	4N36	4N37
Input-to-Output Peak Voltage (8-ms half sine wave)	3.55 kV	2.5 kV	1.5 kV
Input-to-Output Root-Mean-Square Voltage (8-ms half sine wave)	2.5 kV	1.75 kV	1.05 kV
Collector-Base Voltage	70 V		
Collector-Emitter Voltage (See Note 1)	30 V		
Emitter-Base Voltage	7 V		
Input-Diode Reverse Voltage	6 V		
Input-Diode Forward Current: Continuous	60 mA		
Peak (1 μs, 300 pps)	3 A		
Phototransistor Continuous Collector Current	100 mA		
Continuous Power Dissipation at (or below) 25°C Free-Air Temperature:			
Infrared-Emitting Diode (See Note 2)	100 mW		
Phototransistor (See Note 3)	300 mW		
Continuous Power Dissipation at (or below) 25°C Lead Temperature:			
Infrared-Emitting Diode (See Note 4)	100 mW		
Phototransistor (See Note 5)	500 mW		
Storage Temperature Range	−55°C to 150°C		
Operating Temperature Range	−55°C to 100°C		
Lead Temperature 1,6 mm (1/16 inch) from Case for 10 Seconds	260°C		

NOTES: 1. This value applies when the base-emitter diode is open-circuited.
2. Derate linearly to 100°C free-air temperature at the rate of 1.33 mW/°C.
3. Derate linearly to 100°C free-air temperature at the rate of 4 mW/°C.
4. Derate linearly to 100°C lead temperature at the rate of 1.33 mW/°C. Lead temperature is measured on the collector lead 0.8 mm (1/32 inch) from the case.
5. Derate linearly to 100°C lead temperature at the rate of 6.7 mW/°C.

*JEDEC registered data. This sheet contains all applicable registered data in effect at the time of publication.

Copyright © 1983 by Texas Instruments Incorporated

electrical characteristics at 25°C free-air temperature

PARAMETER		TEST CONDITIONS		MIN	TYP	MAX	UNIT
$V_{(BR)CBO}$	Collector-Base Breakdown Voltage	$I_C = 100\ \mu A$, $I_F = 0$	$I_E = 0$,	70*			V
$V_{(BR)CEO}$	Collector-Emitter Breakdown Voltage	$I_C = 10$ mA, $I_F = 0$	$I_B = 0$,	30*			V
$V_{(BR)EBO}$	Emitter-Base Breakdown Voltage	$I_E = 100\ \mu A$, $I_F = 0$	$I_C = 0$,	7*			V
I_R	Input Diode Static Reverse Current	$V_R = 6$ V				10*	μA
I_{IO}	Input-to-Output Current	V_{IO} = rated peak value,	t = 8 ms			100	μA
$I_{C(on)}$	On-State Collector Current	$V_{CE} = 10$ V, $I_B = 0$	$I_F = 10$ mA,	10*			mA
		$V_{CE} = 10$ V, $I_B = 0$,	$I_F = 10$ mA, $T_A = -55°C$	4*			
		$V_{CE} = 10$ V, $I_B = 0$,	$I_F = 10$ mA $T_A = 100°C$	4*			
$I_{C(off)}$	Off-State Collector Current	$V_{CE} = 10$ V, $I_B = 0$	$I_F = 0$,		1	50	nA
		$V_{CE} = 30$ V, $I_B = 0$,	$I_F = 0$, $T_A = 100°C$			500*	μA
h_{FE}	Transistor Static Forward Current Transfer Ratio	$V_{CE} = 5$ V, $I_F = 0$	$I_C = 10$ mA,		500		
V_F	Input Diode Static Forward Voltage	$I_F = 10$ mA		0.8*		1.5*	V
		$I_F = 10$ mA,	$T_A = -55°C$	0.9*		1.7*	
		$I_F = 10$ mA, $T_A = 100°C$		0.7*		1.4*	
$V_{CE(sat)}$	Collector-Emitter Saturation Voltage	$I_C = 0.5$ mA, $I_B = 0$	$I_F = 10$ mA,			0.3*	V
r_{IO}	Input-to-Output Internal Resistance	$V_{IO} = 500$ V,	See Note 6	10^{11}*			Ω
C_{io}	Input-to-Output Capacitance	$V_{IO} = 0$, See Note 6	f = 1 MHz,		1	2.5*	pF

NOTE 6: These parameters are measured between both input-diode leads shorted together and all the phototransistor leads shorted together.

*switching characteristics at 25°C free-air temperature

PARAMETER		TEST CONDITIONS		MIN	TYP	MAX	UNIT
t_{on}	Turn-on time	$V_{CC} = 10$ V, $R_L = 100\ \Omega$,	$I_{C(on)} = 2$ mA, See Figure 1			10	μs
t_{off}	Turn-off time					10	μs

*JEDEC registered data.

7 OPTOCOUPLERS

TEXAS INSTRUMENTS
INCORPORATED
POST OFFICE BOX 225012 • DALLAS, TEXAS 75265

118

PARAMETER MEASUREMENT INFORMATION

Adjust amplitude of input pulse for $I_{C(on)} = 2$ mA

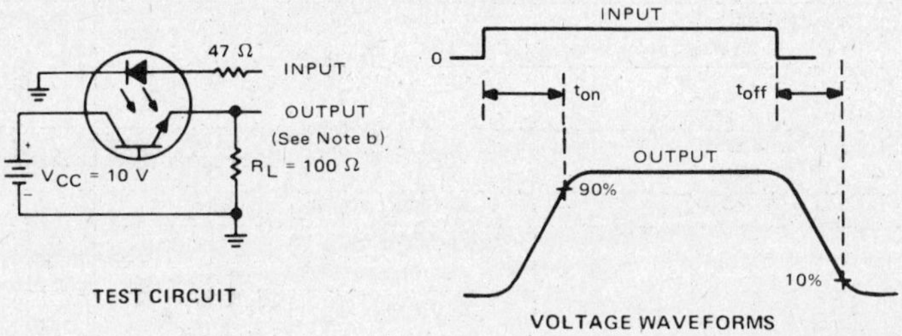

NOTES: a. The input waveform is supplied by a generator with the following characteristics: $Z_{out} = 50\ \Omega$, $t_r \leq 15$ ns, duty cycle ≈ 1%, $t_w = 100\ \mu s$.

b. The output waveform is monitored on an oscilloscope with the following characteristics: $t_r \leq 12$ ns, $R_{in} \geq 1\ M\Omega$, $C_{in} \leq 20$ pF.

FIGURE 1–SWITCHING TIMES

TYPICAL CHARACTERISTICS

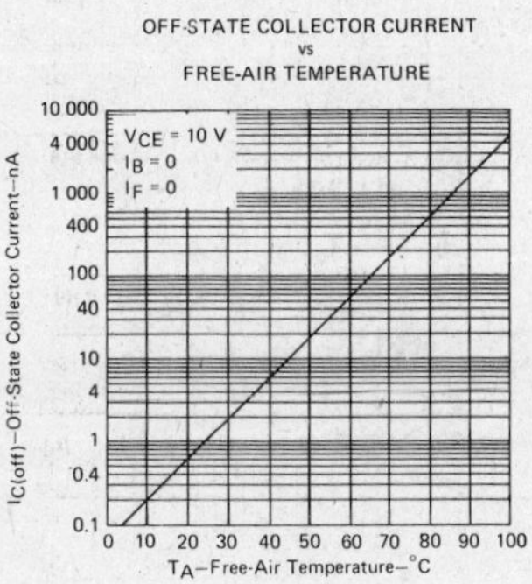

FIGURE 2

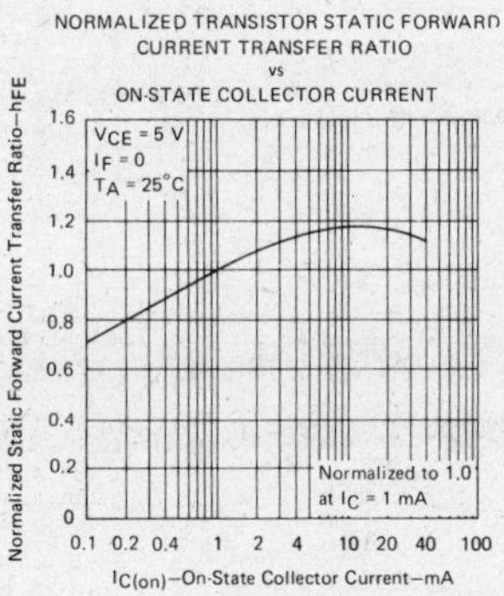

FIGURE 3

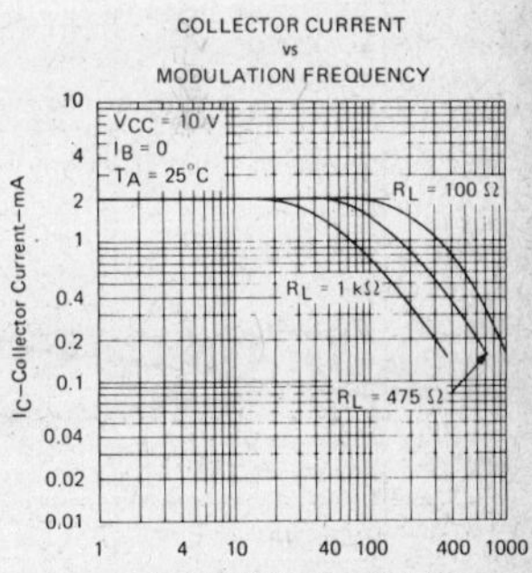

FIGURE 4

TYPICAL CHARACTERISTICS

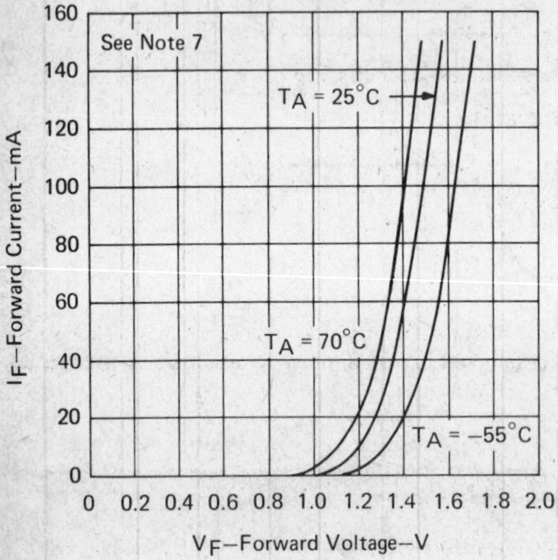

FIGURE 5

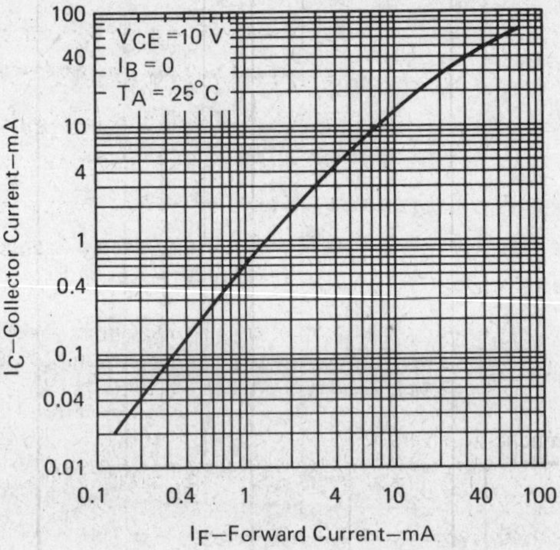

FIGURE 6

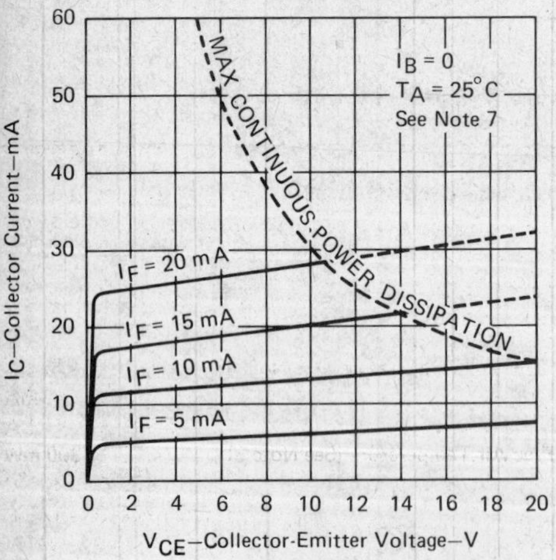

FIGURE 7

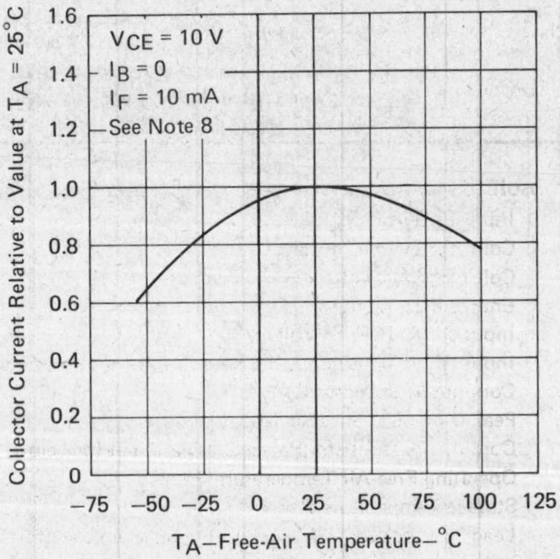

FIGURE 8

NOTES: 7. Pulse operation of input diode is required for operation beyond limits shown by dotted lines.
8. These parameters were measured using pulse techniques. t_w = 1 ms, duty cycle ≤ 2%.

TEXAS INSTRUMENTS
INCORPORATED
POST OFFICE BOX 225012 • DALLAS, TEXAS 75265

D2413, FEBRUARY 1978 – REVISED SEPTEMBER 1981

GALLIUM ARSENIDE DIODE INFRARED SOURCE OPTICALLY COUPLED TO A HIGH-GAIN N-P-N SILICON PHOTOTRANSISTOR

- JAN, JANTX, JANTXV Versions Available
- Very High Current Transfer Ratio . . . 500% Typical (4N49)
- Photon Coupling for Isolator Applications
- Base Lead Provided for Conventional Transistor Biasing
- High-Speed Photodiode-Mode Operation
- High-Voltage Electrical Isolation . . . 1-kV Rating
- Stable over Wide Temperature Range
- Hermetically Sealed Package

description

This opto coupler features an improved current transfer ratio (CTR) at an input of one milliampere making it ideal for coupling with isolation from low-output MOS and CMOS devices to power devices or other systems. Typical applications include motor-speed controls, numeric control systems, meters, and instrumentation.

*mechanical data

THE COLLECTOR IS IN ELECTRICAL CONTACT WITH THE CASE

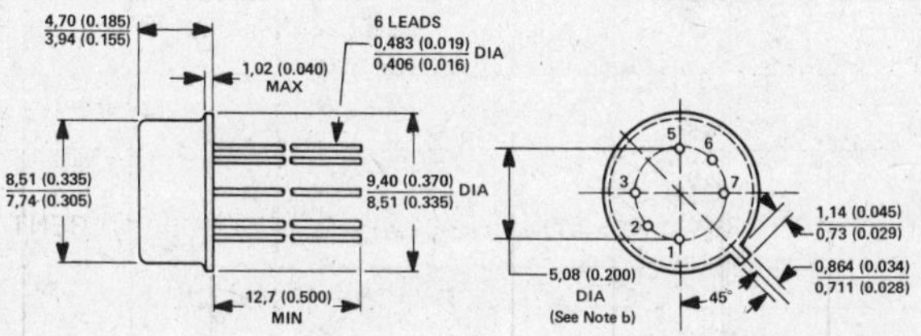

NOTES: a. All linear dimensions are in millimeters and parenthetically in inches.

b. Leads having maximum diameter shall be within 0,18 mm (0.007 inches) of true position relative to a maximum-width tab when measured in the gaging plane between 1,371 mm (0.054 inches) and 1,397 mm (0.055 inches) below the seating plane.

*absolute maximum ratings at 25°C free-air temperature (unless otherwise noted)

Input-to-Output Voltage	±1 kV
Collector-Emitter Voltage	40 V
Collector-Base Voltage	45 V
Emitter-Base Voltage	7 V
Input Diode Reverse Voltage	2 V
Input Diode Continuous Forward Current at (or below) 65°C Free-Air Temperature (See Note 1)	40 mA
Continuous Collector Current	50 mA
Peak Diode Current (See Note 2)	1 A
Continuous Transistor Power Dissipation at (or below) 25°C Free-Air Temperature (See Note 3)	300 mW
Operating Free-Air Temperature Range	–55°C to 125°C
Storage Temperature Range	–55°C to 125°C
Lead Temperature 1/16 Inch (1.6 mm) from Case for 10 Seconds	240°C

NOTES: 1. Derate linearly to 125°C free-air temperature at the rate of 0.67 mA/°C.

2. This values applies for $t_w \leq 1\ \mu s$, PRR ≤ 300 pps.

3. Derate linearly to 125°C free-air temperature at the rate of 3 mW/°C.

*JEDEC registered data. This data sheet contains all applicable JEDEC registered data in effect at the time of publication.

Copyright © 1981 by Texas Instruments Incorporated

***electrical characteristics at 25°C free-air temperature (unless otherwise noted)**

PARAMETER		TEST CONDITIONS		4N47 MIN	4N47 TYP	4N47 MAX	4N48 MIN	4N48 TYP	4N48 MAX	4N49 MIN	4N49 TYP	4N49 MAX	UNIT
$V_{(BR)CBO}$	Collector-Base Breakdown Voltage	I_C = 100 μA, I_F = 0	I_E = 0,	45			45			45			V
$V_{(BR)CEO}$	Collector-Emitter Breakdown Voltage	I_C = 1 mA, I_F = 0	I_B = 0,	40			40			40			V
$V_{(BR)EBO}$	Emitter-Base Breakdown Voltage	I_E = 100 μA, I_F = 0	I_C = 0,	7			7			7			V
I_R	Input Diode Static Reverse Current	V_R = 2 V				100			100			100	μA
$I_{C(on)}$	On-State Collector Current (Phototransistor Mode)	V_{CE} = 5 V, I_F = 1 mA	I_B = 0,	0.5			1		5	2		10	mA
		V_{CE} = 5 V, I_F = 2 mA,	I_B = 0, T_A = −55°C	0.7			1.4			2.8			
		V_{CE} = 5 V, I_F = 2 mA,	I_B = 0, T_A = 100°C	0.5			1			2			
		V_{CE} = 5 V, I_F = 10 mA,	I_B = 0, See Note 4		50			80			90		
$I_{C(on)}$	On-State Collector Current (Photodiode Mode)	V_{CB} = 5 V, I_E = 0	I_F = 10 mA,	30	80		30	80		30	80		μA
$I_{C(off)}$	Off-State Collector Current (Phototransistor Mode)	V_{CE} = 20 V, I_F = 0	I_B = 0,		6	100		6	100		6	100	nA
		V_{CE} = 20 V, I_F = 0,	I_B = 0, T_A = 100°C		4	100		4	100		4	100	μA
$I_{C(off)}$	Off-State Collector Current (Photodiode Mode)	V_{CB} = 20 V, I_F = 0	I_E = 0,		1	10		1	10		1	10	nA
V_F	Input Diode Static Forward Voltage	I_F = 10 mA,	T_A = −55°C	1		1.7	1		1.7	1		1.7	V
		I_F = 10 mA		0.8	1.4	1.5	0.8	1.4	1.5	0.8	1.4	1.5	
		I_F = 10 mA,	T_A = 100°C	0.7		1.3	0.7		1.3	0.7		1.3	
$V_{CE(sat)}$	Collector-Emitter Saturation Voltage	I_C = 0.5 mA, I_F = 2 mA	I_B = 0,			0.3							V
		I_C = 1 mA, I_F = 2 mA	I_B = 0,						0.3				
		I_C = 2 mA, I_F = 2 mA	I_B = 0,									0.3	
r_{io}	Input-to-Output Internal Resistance	V_{in-out} = ±1 kV,	See Note 5	10^{11}	10^{12}		10^{11}	10^{12}		10^{11}	10^{12}		Ω
C_{io}	Input-to-Output Capacitance	V_{in-out} = 0, See Note 5	f = 1 MHz,		2.5	5		2.5	5		2.5	5	pF

switching characteristics at 25°C free-air temperature (See Figure 1)

PARAMETER		TEST CONDITIONS		4N47 MIN	4N47 TYP	4N47 MAX	4N48 MIN	4N48 TYP	4N48 MAX	4N49 MIN	4N49 TYP	4N49 MAX	UNIT
*t_r	Rise Time	V_{CC} = 10 V,	$I_{F(on)}$ = 5 mA,		10	20		10	20		15	25	μs
*t_f	Fall Time	R_L = 100 Ω,	Test Circuit A		10	20		10	20		15	25	μs
t_r	Rise Time	V_{CC} = 10 V,	$I_{F(on)}$ = 5 mA,		1	3		1	3		1	3	μs
t_f	Fall Time	R_L = 100 Ω,	Test Circuit B		1	3		1	3		1	3	μs

NOTES: 4. This parameter must be measured using pulse techniques, t_w = 100 μs, duty cycle ≤ 1%.
5. These parameters are measured between all the input diode leads shorted together and all the phototransistor leads shorted together.

*JEDEC registered data

TEXAS INSTRUMENTS
INCORPORATED
POST OFFICE BOX 225012 • DALLAS, TEXAS 75265

PARAMETER MEASUREMENT INFORMATION

Adjust amplitude of input pulse for
$I_{F(on)}$ = 5 mA

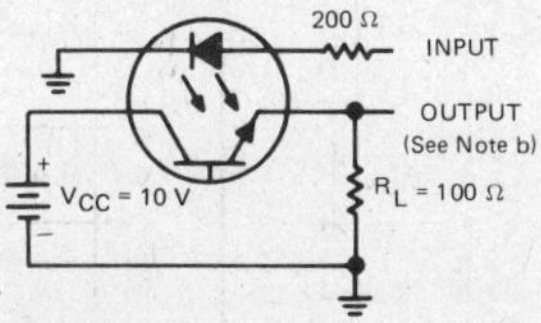

TEST CIRCUIT A
PHOTOTRANSISTOR OPERATION

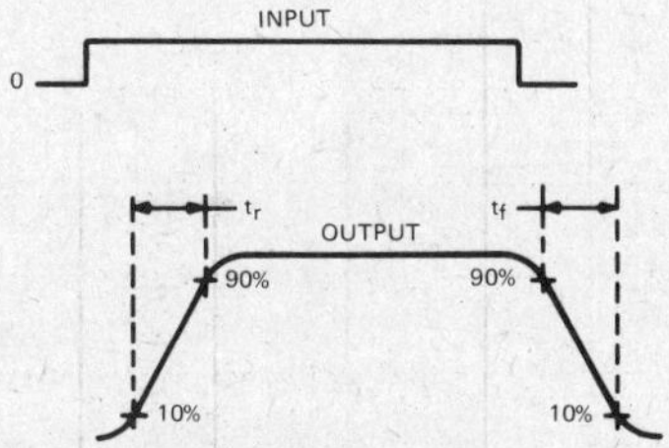

VOLTAGE WAVEFORMS

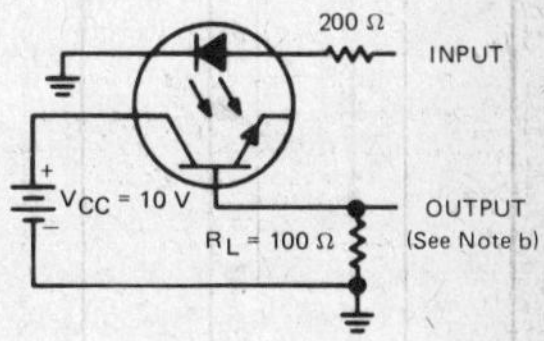

TEST CIRCUIT B
PHOTODIODE OPERATION

NOTES: a. The input waveform is supplied by a generator with the following characteristics: z_{out} = 50 Ω, $t_r \leq 15$ ns, duty cycle ≈ 1%. For Test Circuit A, t_w = 100 μs. For Test Circuit B, t_w = 1 μs.

b. Waveforms are monitored on an oscilloscope with the following characteristics: $t_r \leq 12$ ns, $R_{in} \geq 1$ MΩ, $C_{in} \leq 20$ pF.

FIGURE 1–SWITCHING TIMES

TYPICAL CHARACTERISTICS

INPUT DIODE FORWARD CONDUCTION CHARACTERISTICS

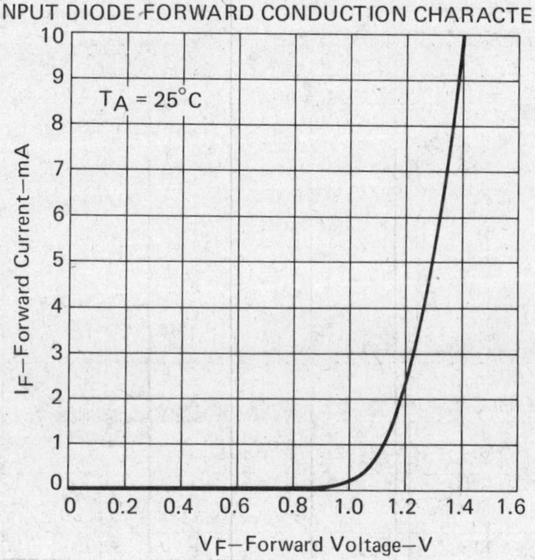

FIGURE 2

TYPICAL CHARACTERISTICS

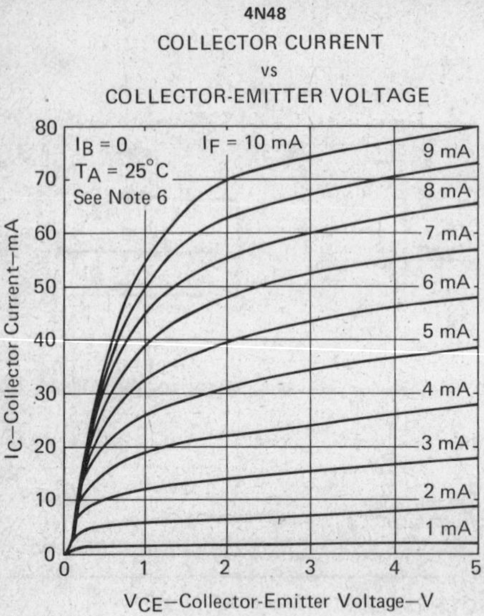

FIGURE 3

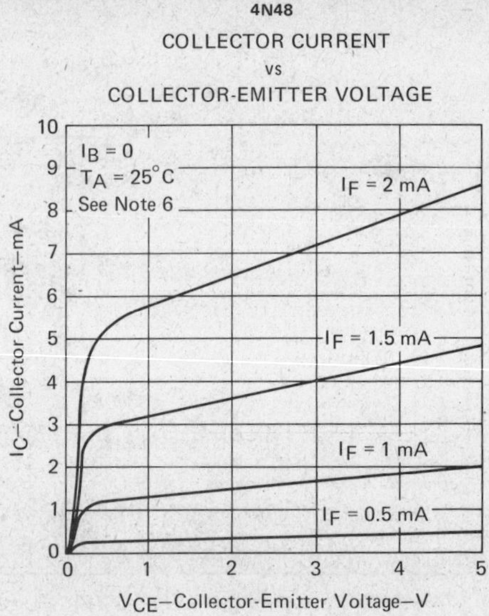

FIGURE 4

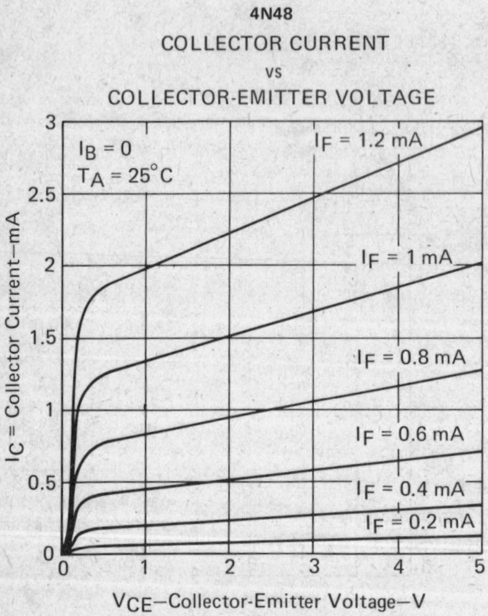

FIGURE 5

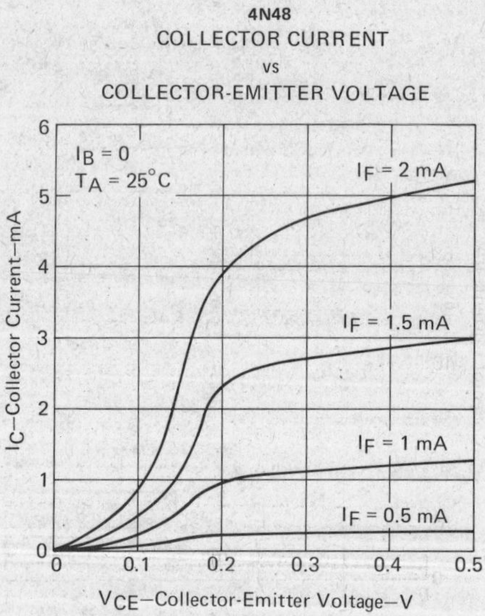

FIGURE 6

NOTE 6: This parameter was measured using pulse techniques. t_w = 100 μs, duty cycle = 1%.

TEXAS INSTRUMENTS
INCORPORATED
POST OFFICE BOX 225012 • DALLAS, TEXAS 75265

TYPICAL CHARACTERISTICS

4N48

COLLECTOR CURRENT
vs
INPUT DIODE FORWARD CURRENT

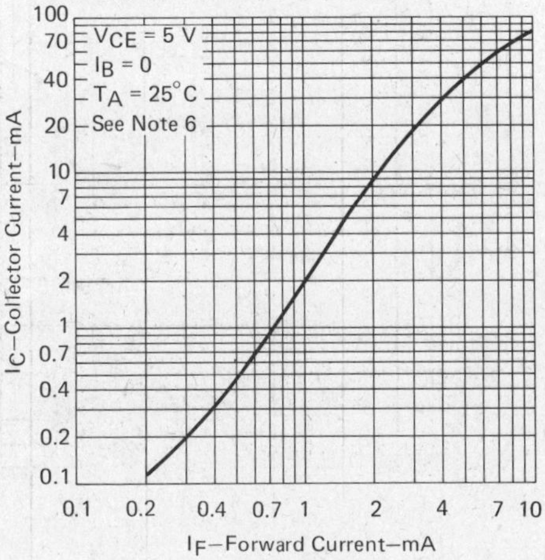

FIGURE 7

4N48

TRANSISTOR COLLECTOR CHARACTERISTICS

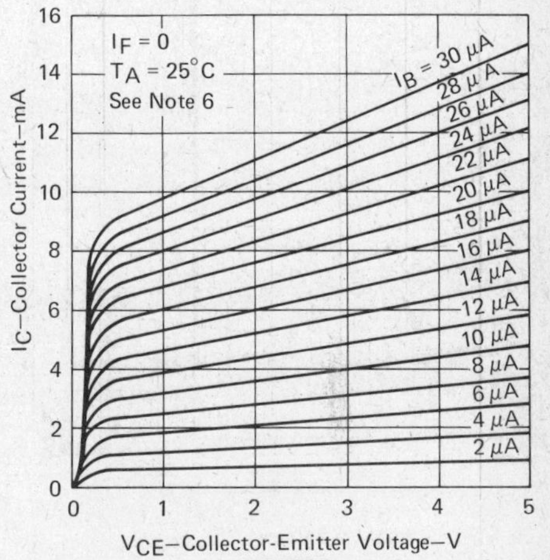

FIGURE 8

COLLECTOR CURRENT
vs
COLLECTOR-BASE VOLTAGE
(PHOTODIODE OPERATION)

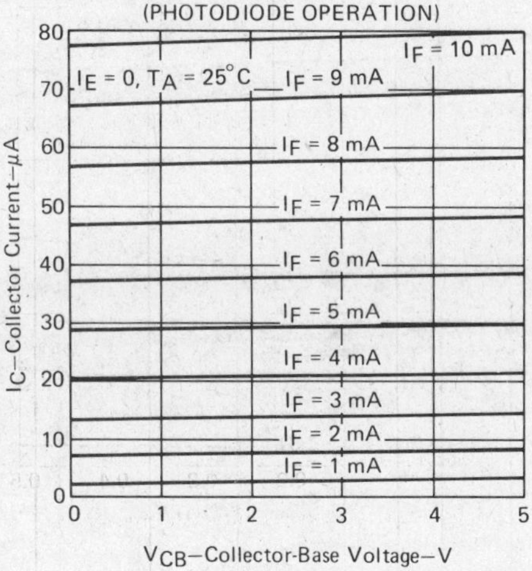

FIGURE 9

RELATIVE ON-STATE COLLECTOR CURRENT
vs
FREE-AIR TEMPERATURE

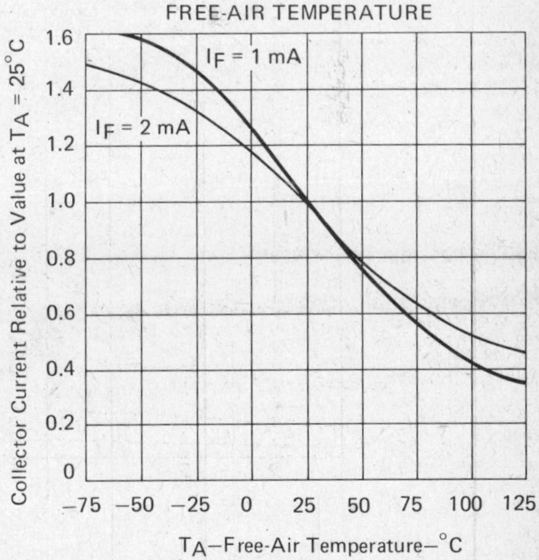

FIGURE 10

NOTE 6: This parameter was measured using pulse techniques. t_w = 100 μs, duty cycle = 1%.

This processing applies only to optocouplers ordered under part numbers shown below:

JAN4N47, JANTX4N47, JANTXV4N47
JAN4N48, JANTX4N48, JANTXV4N48
JAN4N49, JANTX4N49, JANTXV4N49

TEST (PER MIL-S-19500/548)	MIL-STD-750 TEST METHOD	JAN	JANTX	JANTXV
100% Processing				
Internal Visual	2072			X
Storage: T_A = 125 °C, t = 24 hr	1032		X	X
Temperature Cycle: −55 °C to 125 °C, 10 cycles	1051		X	X
Constant Acceleration: 20,000 G, Y_1 axis	2006		X	X
High-Temperature Reverse Bias: I_F = 0, T_A = 125 °C, V_{CB} = 36 V, t = 48 hr	1039		X	X
Power Burn-in: I_F = 40 mA, P_D = 275 ±25 mW, t = 168 hr	1039		X	X
Hermetic Seal, Fine	1071 Cond. G or H		X	X
Hermetic Seal, Gross	1071 Cond. C or D		X	X
Monitored Thermal Shock	Para. 4.2.1.1.*	X	X	X
External Visual	2071		X	X
Product Acceptance				
Group A: LTPD = 5				
External Visual	2071	X	X	X
Electrical: T_A = 25 °C	as needed	X	X	X
Electrical: T_A = 100 °C	as needed	X	X	X
Electrical: T_A = −55 °C	as needed	X	X	X
Group B-1: LTPD = 15				
Solderability	2026	X	X	X
Resistance to Solvents	1022	X	X	X
Group B-2: LTPD = 10				
Thermal Shock	1051 Cond. B-1	X	X	X
Hermetic Seal, Fine	1071 Cond. G or H	X	X	X
Hermetic Seal, Gross	1071 Cond. C or D	X	X	X
Group B-3:				
Isolation Voltage: V_{IO} = 150 V, T_A = 125 °C, t = 24, LTPD = 20	1016	X	X	X
Steady State Operating Life: t = 340 hr, LTPD = 5	1027	X	X	X
Group B-4:				
Decap, Internal Visual; Design Verification		X	X	X
1 Device/0 Failure	2075	X	X	X
Bond Strength LTPD = 20 (C = 0)	2037 Cond. A	X	X	X
Group B-5: Not Applicable				
Group B-6: LTPD = 7				
High-Temperature Life (Nonoperating) t = 340 hr	1032	X	X	X

*MIL-S-19500/548

TEST (PER MIL-S-19500/548)	MIL-STD-750 TEST METHOD	JAN	JANTX	JANTXV
(Group C Tests are run on one lot every six months)				
Group C-1: LTPD = 15				
Physical Dimensions	2066	X	X	X
Group C-2: LTPD = 10				
Thermal Shock (Glass Strain)	1056 Cond. A	X	X	X
Terminal Strength	2036 Cond. E	X	X	X
Hermetic Seal, Fine	1071 Cond. G or H	X	X	X
Hermetic Seal, Gross	1071 Cond. C or D	X	X	X
Moisture Resistance	1021	X	X	X
External Visual	2071	X	X	X
Group C-3: LTPD = 10				
Shock: 1500 G	2016	X	X	X
Vibration: 50 G	2056	X	X	X
Acceleration: 30000 G	2006	X	X	X
Group C-4: LTPD = 15				
Salt Atmosphere	1041	X	X	X
Group C-5: Not Applicable				
Group C-6: λ = 10				
Steady State Operating Life	1026	X	X	X

TEXAS INSTRUMENTS
INCORPORATED
POST OFFICE BOX 225012 • DALLAS, TEXAS 75265

COMPATIBLE WITH STANDARD TTL INTEGRATED CIRCUITS

- **Gallium Arsenide Diode Infrared Source Optically Coupled to a Silicon N-P-N Phototransistor**
- **High Direct-Current Transfer Ratio**
- **Base Lead Provided for Conventional Transistor Biasing**
- **High-Voltage Electrical Isolation . . . 1.5-kV or 3.55-kV Rating**
- **Plastic Dual-In-Line Package**
- **High-Speed Switching: $t_r = 5\ \mu s$, $t_f = 5\ \mu s$ Typical**
- **Designed to be Interchangeable with General Instruments MCT2 and MCT2E**

mechanical data

The package consists of a gallium arsenide infrared-emitting diode and an n-p-n silicon phototransistor mounted on a 6-lead frame encapsulated within an electrically nonconductive plastic compound. The case will withstand soldering temperature with no deformation and device performance characteristics remain stable when operated in high-humidity conditions. Unit weight is approximately 0.52 grams.

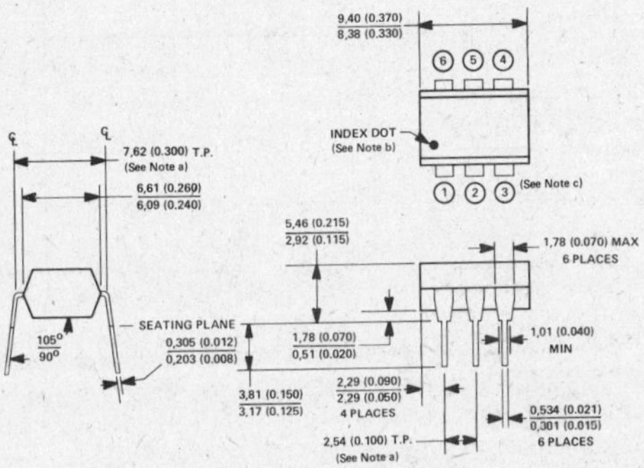

NOTES:

a. Leads are within 0,13 mm (0.005 inch) radius of true position (T.P.) with maximum material condition and unit installed.

b. Pin 1 identified by index dot.

c. Terminal connections:

1. Anode — Infrared-emitting diode
2. Cathode — Infrared-emitting diode
3. No internal connection
4. Emitter — Phototransistor
5. Collector — Phototransistor
6. Base — Phototransistor

FALLS WITHIN JEDEC MO-001AM DIMENSIONS

ALL LINEAR DIMENSIONS ARE IN MILLIMETERS AND PARENTHETICALLY IN INCHES

absolute maximum ratings at 25°C free-air temperature (unless otherwise noted)

Input-to-Output Voltage: MCT2	±1.5 kV
MCT2E	±3.55 kV
Collector-Base Voltage	70 V
Collector-Emitter Voltage (See Note 1)	30 V
Emitter-Collector Voltage	7 V
Emitter-Base Voltage	7 V
Input-Diode Reverse Voltage	3 V
Input-Diode Continuous Forward Current	60 mA
Input-Diode Peak Forward Current ($t_w \leq 1$ ns, PRF $\leq$ 300 Hz)	3 A
Continuous Power Dissipation at (or below) 25°C Free-Air Temperature	
Infrared-Emitting Diode (See Note 2)	200 mW
Phototransistor (See Note 2)	200 mW
Total, Infrared-Emitting Diode plus Phototransistor (See Note 3)	250 mW
Operating Free-Air Temperature Range	−55°C to 100°C
Storage Temperature Range	−55°C to 150°C
Lead Temperature 1,6 mm (1/16 inch) from Case for 10 Seconds	260°C

NOTES: 1. This value applies when the base-emitter diode is open-circuited.
2. Derate linearly to 100°C free-air temperature at the rate of 2.67 mW/°C.
3. Derate linearly to 100°C free-air temperature at the rate of 3.33 mW/°C.

electrical characteristics at 25 °C free-air temperature

PARAMETER			TEST CONDITIONS		MCT2, MCT2E MIN	TYP	MAX	UNIT
$V_{(BR)CBO}$	Collector-Base Breakdown Voltage		I_C = 10 μA, I_E = 0, I_F = 0		70			V
$V_{(BR)CEO}$	Collector-Emitter Breakdown Voltage		I_C = 1 mA, I_B = 0, I_F = 0		30			V
$V_{(BR)ECO}$	Emitter-Collector Breakdown Voltage		I_E = 100 μA, I_B = 0, I_F = 0		7			V
I_R	Input Diode Static Reverse Current		V_R = 3 V				10	μA
$I_{C(on)}$	On-State Collector Current	Phototransistor Operation	V_{CE} = 10 V, I_F = 10 mA, I_B = 0		2	5		mA
		Photodiode Operation	V_{CB} = 10 V, I_F = 10 mA, I_E = 0			20		μA
$I_{C(off)}$	Off-State Collector Current	Phototransistor Operation	V_{CE} = 10 V, I_F = 0 I_B = 0			1	50	nA
		Photodiode Operation	V_{CB} = 10 V, I_F = 0, I_E = 0			0.1	20	nA
h_{FE}	Transistor Static Forward Current Transfer Ratio		V_{CE} = 5 V, I_C = 100 μA, I_F = 0	MCT2		250		
				MCT2E	100	300		
V_F	Input Diode Static Forward Voltage		I_F = 20 mA			1.25	1.5	V
$V_{CE(sat)}$	Collector-Emitter Saturation Voltage		I_C = 2 mA, I_F = 16 mA, I_B = 0			0.25	4	V
r_{IO}	Input-to-Output Internal Resistance		$V_{in\text{-}out}$ = ±1.5 kV for MCT2 ±3.55 kV for MCT2E See Note 4		10^{11}			Ω
C_{io}	Input-to-Output Capacitance		$V_{in\text{-}out}$ = 0, f = 1 MHz, See note 4			1		pF

NOTE 4: These parameters are measured between both input-diode leads shorted together and all the phototransistor leads shorted together.

switching characteristics at 25 °C free-air temperature

PARAMETER			TEST CONDITIONS	MCT2, MCT2E MIN	TYP	MAX	UNIT
t_r	Rise Time	Phototransistor Operation	V_{CC} = 10 V, $I_{C(on)}$ = 2 mA, R_L = 100 Ω, See Test Circuit A of Figure 1		5		μs
t_f	Fall Time						
t_r	Rise Time	Photodiode Operation	V_{CC} = 10 V, $I_{C(on)}$ = 20 μA, R_L = 1 kΩ, See Test Circuit B of Figure 1		1		μs
t_f	Fall Time						

TEXAS INSTRUMENTS
INCORPORATED
POST OFFICE BOX 225012 • DALLAS, TEXAS 75265

PARAMETER MEASUREMENT INFORMATION

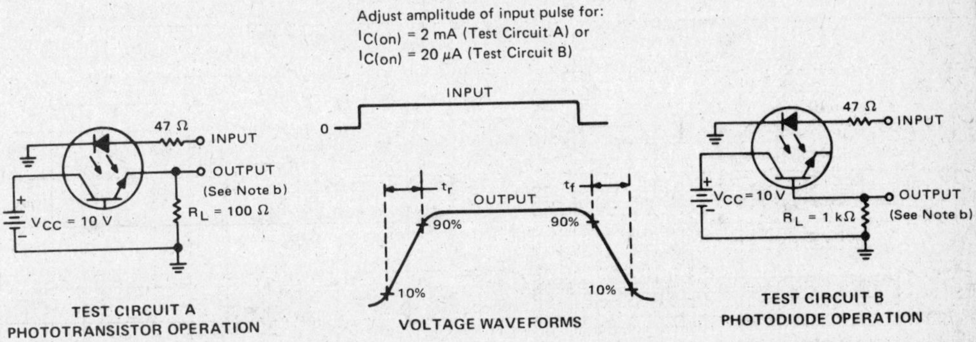

NOTES: a. The input waveform is supplied by a generator with the following characteristics: Z_{out} = 50 Ω, $t_r \leq$ 15 ns, duty cycle ≈ 1%, t_w = 100 μs.

b. The output waveform is monitored on an oscilloscope with the following characteristics: $t_r \leq$ 12 ns, $R_{in} \geq$ 1 MΩ, $C_{in} \leq$ 20 pF.

FIGURE 1—SWITCHING TIMES

TYPICAL CHARACTERISTICS

COLLECTOR CURRENT
vs
INPUT-DIODE FORWARD CURRENT

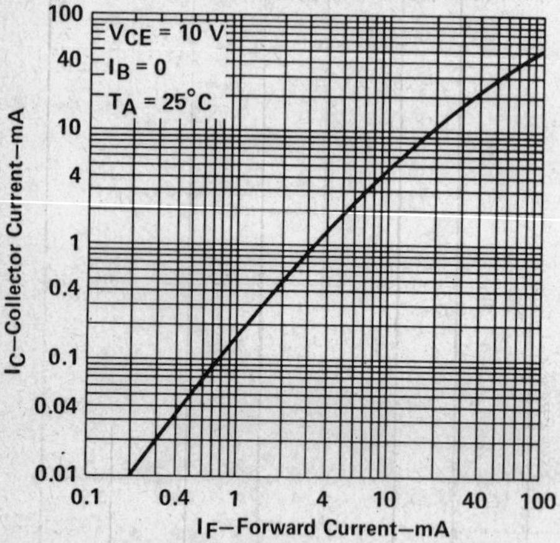

FIGURE 2

COLLECTOR CURRENT
vs
COLLECTOR-EMITTER VOLTAGE

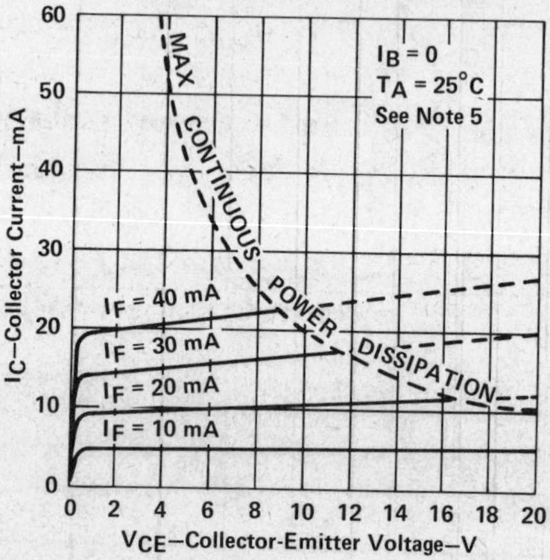

FIGURE 3

RELATIVE ON-STATE COLLECTOR CURRENT
vs
FREE-AIR TEMPERATURE

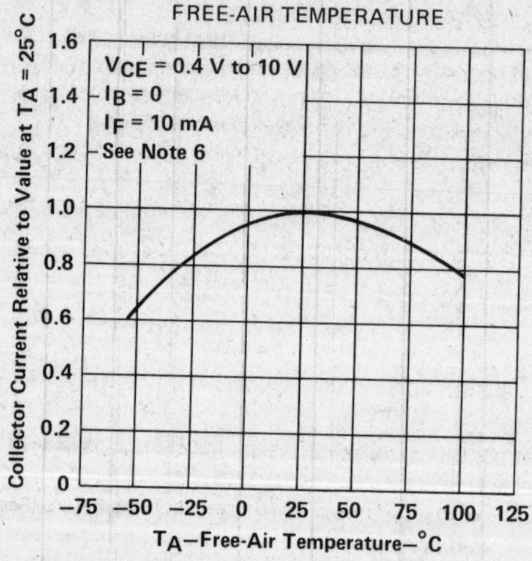

FIGURE 4

NOTES: 5. Pulse operation of input diode is required for operation beyond limits shown by dotted lines.
6. These parameters were measured using pulse techniques. t_w = 1 ms, duty cycle ≤ 2%.

TEXAS INSTRUMENTS
INCORPORATED
POST OFFICE BOX 225012 • DALLAS, TEXAS 75265

D910, SEPTEMBER 1970—REVISED NOVEMBER 1974

GALLIUM ARSENIDE DIODE INFRARED SOURCE OPTICALLY COUPLED TO A HIGH-GAIN N-P-N SILICON PHOTOTRANSISTOR

- Photon Coupling for Isolator Applications
- Base Lead Provided for Conventional Transistor Biasing
- High Overall Current Gain . . . 1.5 Typ (TIL103)
- High-Voltage Transistor . . . $V_{(BR)CEO}$ = 35 V Min
- High-Voltage Electrical Isolation . . . 1-kV Rating
- Stable over Wide Temperature Range

mechanical data

THE COLLECTOR IS IN ELECTRICAL CONTACT WITH THE CASE

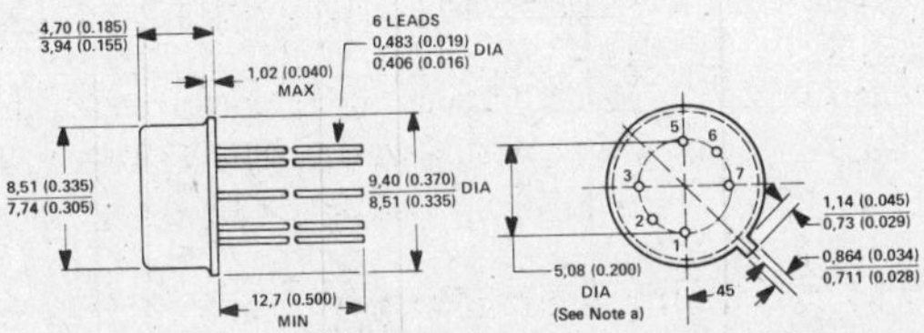

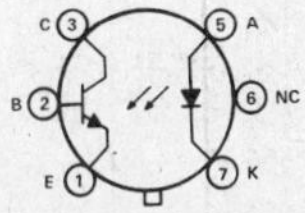

NOTE a: Leads having maximum diameter shall be with 0,18 mm (0.007 inches) of true position relative to a maximum-width tab when measured in the gaging plane between 1,371 mm (0.054 inches) and 1,397 mm (0.055 inches) below the seating plane.

ALL LINEAR DIMENSIONS ARE IN MILLIMETERS AND PARENTHETICALLY IN INCHES.

absolute maximum ratings at 25°C free-air temperature (unless otherwise noted)

Input to Output Voltage	±1 kV
Collector-Emitter Voltage	35 V
Collector-Base Voltage	35 V
Emitter-Base Voltage	4 V
Input Diode Reverse Voltage	2 V
Input Diode Continuous Forward Current at (or below) 65°C Free-Air Temperature (See Note 1)	40 mA
Continuous Collector Current	50 mA
Continuous Transistor Power Dissipation at (or below) 25°C Free-Air Temperature (See Note 2)	300 mW
Storage Temperature Range	−55°C to 125°C
Lead Temperature 1,6 mm (1/16 Inch) from Case for 10 Seconds	240°C

NOTES: 1. Derate linearly to 125°C free-air temperature at the rate of 0.67 mA/°C.
2. Derate linearly to 125°C free-air temperature at the rate of 3 mW/°C.

electrical characteristics at 25°C free-air temperature (unless otherwise noted)

PARAMETER			TEST CONDITIONS	TIL102			TIL103			UNIT
				MIN	TYP	MAX	MIN	TYP	MAX	
$V_{(BR)CBO}$	Collector-Base Breakdown Voltage		$I_C = 100\ \mu A$, $I_E = 0$, $I_F = 0$	35			35			V
$V_{(BR)CEO}$	Collector-Emitter Breakdown Voltage		$I_C = 1$ mA, $I_B = 0$, $I_F = 0$	35			35			V
$V_{(BR)EBO}$	Emitter-Base Breakdown Voltage		$I_E = 100\ \mu A$, $I_C = 0$, $I_F = 0$	4			4			V
I_R	Input Diode Static Reverse Current		$V_R = 2$ V			100			100	μA
$I_{C(on)}$	On-State Collector Current	Phototransistor Operation	$V_{CE} = 5$ V, $I_B = 0$, $I_F = 10$ mA	2.5	6		10	15		mA
		Photodiode Operation	$V_{CB} = 5$ V, $I_E = 0$, $I_F = 10$ mA		40			40		μA
$I_{C(off)}$	Off-State Collector Current	Phototransistor Operation	$V_{CE} = 20$ V, $I_B = 0$, $I_F = 0$		6	100		6	100	nA
			$V_{CE} = 20$ V, $I_B = 0$, $I_F = 0$, $T_A = 100°C$		4			4		μA
		Photodiode Operation	$V_{CB} = 20$ V, $I_E = 0$, $I_F = 0$		0.1			0.1		nA
h_{FE}	Transistor Static Forward Current Transfer Ratio		$V_{CE} = 5$ V, $I_C = 10$ mA, $I_F = 0$		300			500		
V_F	Input Diode Static Forward Voltage		$I_F = 10$ mA			1.3			1.3	V
$V_{CE(sat)}$	Collector-Emitter Saturation Voltage		$I_C = 2.5$ mA, $I_B = 0$, $I_F = 20$ mA			0.3				V
			$I_C = 10$ mA, $I_B = 0$, $I_F = 20$ mA						0.3	
r_{IO}	Input-to-Output Internal Resistance		$V_{in\text{-}out} = \pm 1$ kV, See Note 3	10^{11}	10^{12}		10^{11}	10^{12}		Ω
C_{io}	Input-to-Output Capacitance		$V_{in\text{-}out} = 0$, f = 1 MHz, See Note 3		2.5			2.5		pF

NOTE 3: These parameters are measured between both input diode leads shorted together and all the phototransistor leads shorted together.

switching characteristics at 25°C free-air temperature

PARAMETER			TEST CONDITIONS	TIL102	TIL103	UNIT
				TYP	TYP	
t_r	Rise Time	Phototransistor Operation	$V_{CC} = 20$ V, $I_B = 0$, $I_{C(on)} = 5$ mA, $R_L = 100\ \Omega$, See Test Circuit A of Figure 1	3	6	μs
t_f	Fall Time			3	6	
t_r	Rise Time	Photodiode Operation	$V_{CC} = 20$ V, $I_E = 0$, $I_{C(on)} = 50\ \mu A$, $R_L = 100\ \Omega$, See Test Circuit B of Figure 1	150	150	ns
t_f	Fall Time			150	150	

TEXAS INSTRUMENTS
INCORPORATED
POST OFFICE BOX 225012 • DALLAS, TEXAS 75265

982

PARAMETER MEASUREMENT INFORMATION

Adjust amplitude of input pulse for:
$I_{C(on)}$ = 5 mA (Test Circuit A) or
$I_{C(on)}$ = 50 μA (Test Circuit B)

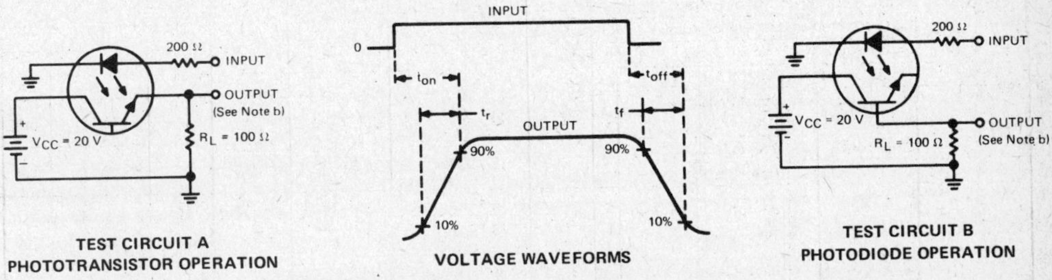

NOTES: a. The input waveform is supplied by a generator with the following characteristics: z_{out} = 50 Ω, $t_r \leq$ 15 ns, duty cycle ≈ 1%. For Test Circuit A, t_w = 100 μs. For Test Circuit B, t_w = 1 μs.

b. Waveforms are monitored on an oscilloscope with the following characteristics: $t_r \leq$ 12 ns, $R_{in} \geq$ 1MΩ, $C_{in} \leq$ 20 pF.

FIGURE 1—SWITCHING TIMES

TYPICAL CHARACTERISTICS

TIL102
COLLECTOR CURRENT
vs
COLLECTOR-EMITTER VOLTAGE

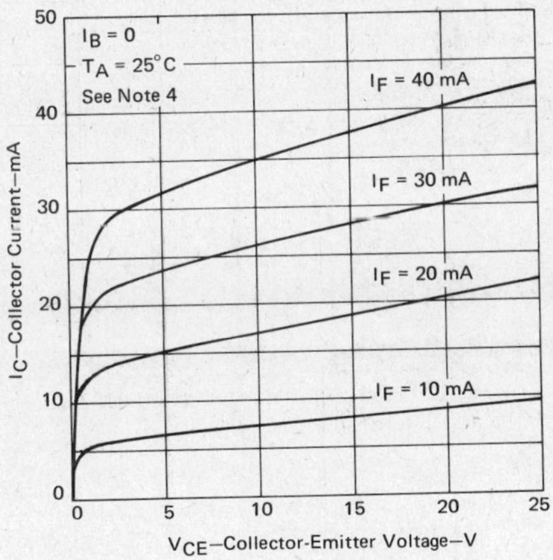

FIGURE 2

TIL103
COLLECTOR CURRENT
vs
COLLECTOR-EMITTER VOLTAGE

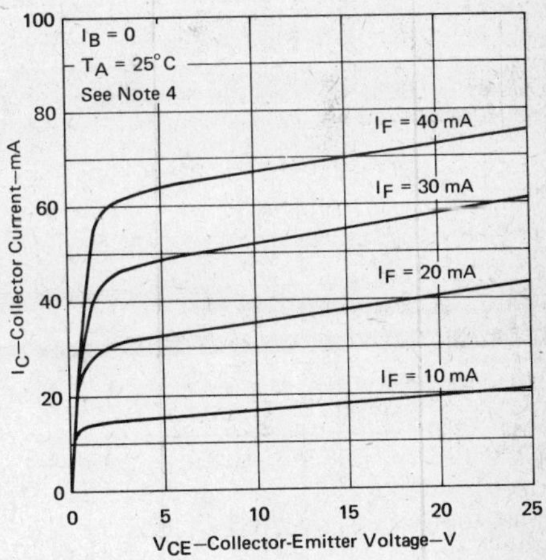

FIGURE 3

NOTE 4: This parameter was measured using pulse techniques. t_w = 100 μs, duty cycle = 1%.

TYPICAL CHARACTERISTICS

INPUT DIODE FORWARD CONDUCTION CHARACTERISTICS

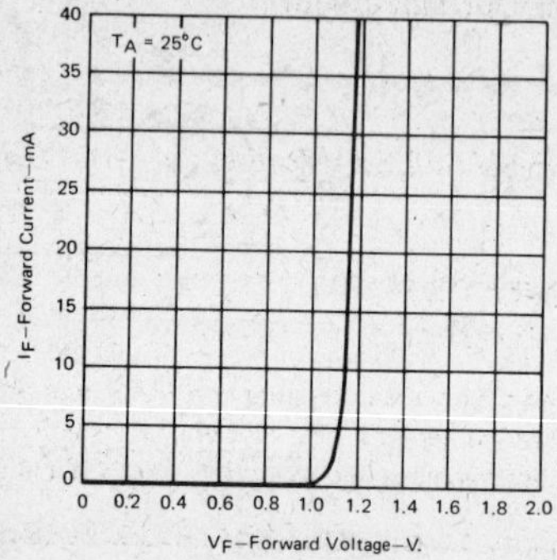

FIGURE 4

NORMALIZED ON-STATE COLLECTOR CURRENT†
vs
FREE-AIR TEMPERATURE

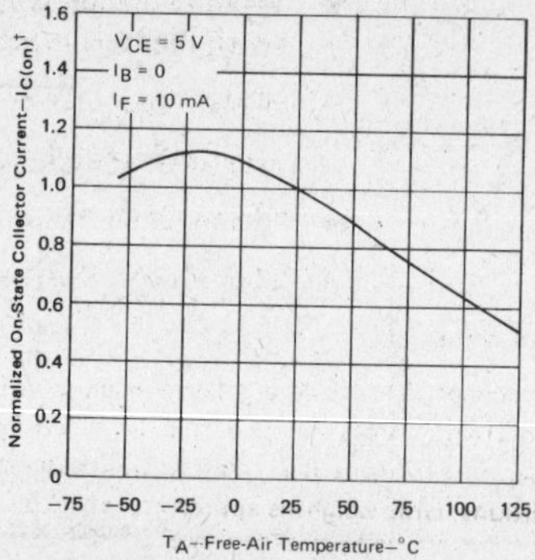

†Normalized to value at T_A = 25°C

FIGURE 5

PHOTOTRANSISTOR COLLECTOR CURRENT
vs
INPUT-DIODE FORWARD CURRENT

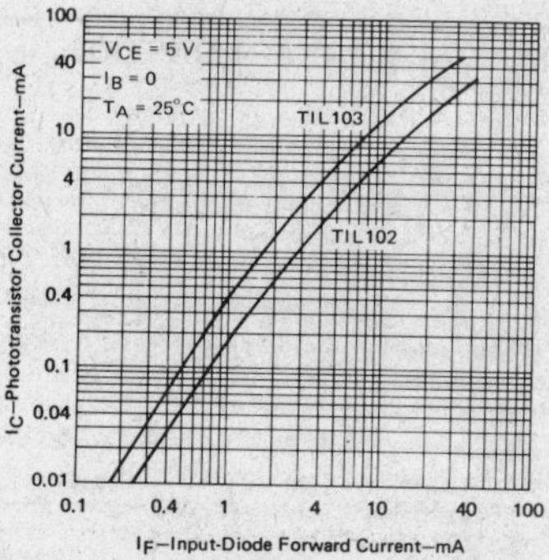

FIGURE 6

OFF-STATE COLLECTOR CURRENT
vs
FREE-AIR TEMPERATURE

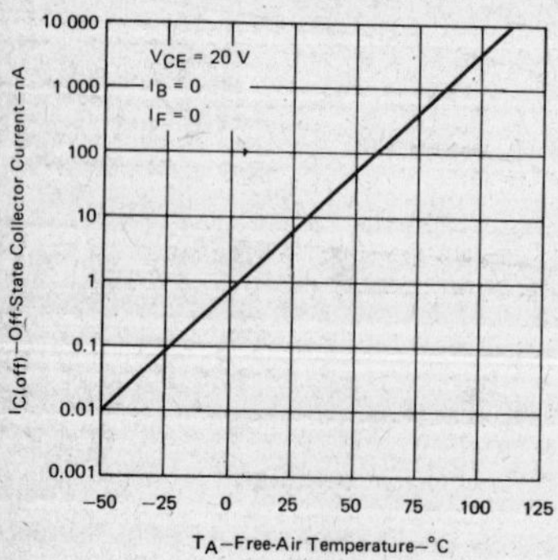

FIGURE 7

TIL102
AVERAGE SWITCHING TIME
vs
LOAD RESISTANCE

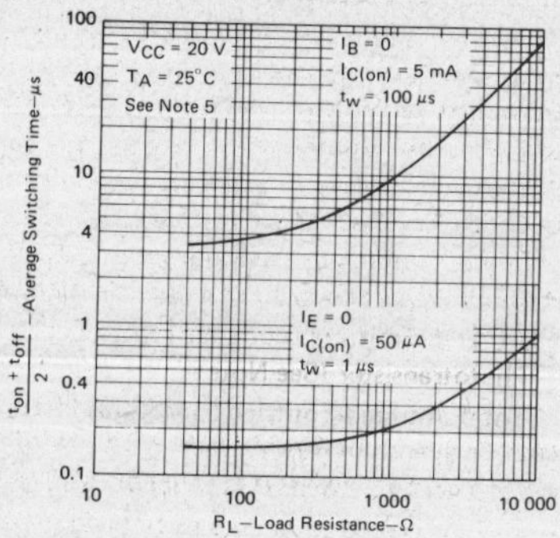

FIGURE 8

NOTE 5: These parameters were measured in Test Circuits A and B of Figure 1 with R_L varied between 40 Ω and 10 kΩ.

TEXAS INSTRUMENTS
INCORPORATED
POST OFFICE BOX 225012 • DALLAS, TEXAS 75265

D1607, NOVEMBER 1973—REVISED FEBRUARY 1983

COMPATIBLE WITH STANDARD TTL INTEGRATED CIRCUITS

- Gallium Arsenide Diode Infrared Source Optically Coupled to a Silicon N-P-N Phototransistor
- High Direct-Current Transfer Ratio
- High-Voltage Electrical Isolation . . . 1.5-kV or 2.5-kV Rating
- Plastic Dual-In-Line Package
- High-Speed Switching: $t_r = 5\ \mu s$, $t_f = 5\ \mu s$ Typical

mechanical data

The package consists of a gallium arsenide infrared-emitting diode and an n-p-n silicon phototransistor mounted on a 6-lead frame encapsulated within an electrically nonconductive plastic compound. The case will withstand soldering temperature with no deformation and device performance characteristics remain stable when operated in high-humidity conditions. Unit weight is approximately 0.52 grams.

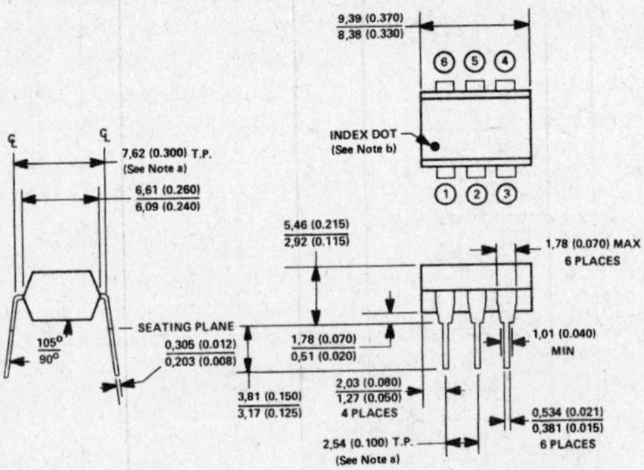

NOTES:

a. Leads are within 0,13 mm (0.005 inch) radius of true position (T.P.) with maximum material condition and unit installed.

b. Pin 1 identified by index dot.

c. Terminal connections:

1. Anode } Infrared-emitting diode
2. Cathode
3. No internal connection
4. Emitter } Phototransistor
5. Collector
6. Base

FALLS WITHIN JEDEC MO-001AM DIMENSIONS

ALL LINEAR DIMENSIONS ARE IN MILLIMETERS AND PARENTHETICALLY IN INCHES

absolute maximum ratings at 25°C free-air temperature (unless otherwise noted)

Input-to-Output Voltage: TIL111	±1.5 kV
TIL114, TIL116, TIL117	±2.5 kV
Collector-Base Voltage	70 V
Collector-Emitter Voltage (See Note 1)	30 V
Emitter-Collector Voltage	7 V
Emitter-Base Voltage	7 V
Input-Diode Reverse Voltage	3 V
Input-Diode Continuous Forward Current at (or below) 25°C Free-Air Temperature (See Note 2)	100 mA
Continuous Power Dissipation at (or below) 25°C Free-Air Temperature:	
Infrared-Emitting Diode (See Note 3)	150 mW
Phototransistor (See Note 4)	150 mW
Total, Infrared-Emitting Diode plus Phototransistor (See Note 5)	250 mW
Storage Temperature Range	−55°C to 150°C
Lead Temperature 1,6 mm (1/16 Inch) from Case for 10 Seconds	260°C

NOTES:
1. This value applies when the base-emitter diode is open-circuited.
2. Derate linearly to 100°C free-air temperature at the rate of 1.33 mA/°C.
3. Derate linearly to 100°C free-air temperature at the rate of 2 mW/°C.
4. Derate linearly to 100°C free-air temperature at the rate of 2 mW/°C.
5. Derate linearly to 100°C free-air temperature at the rate of 3.33 mW/°C.

Copyright © 1983 by Texas Instruments Incorporated

electrical characteristics at 25°C free-air temperature

PARAMETER			TEST CONDITIONS	TIL111 TIL114 MIN	TIL111 TIL114 TYP	TIL111 TIL114 MAX	TIL116 MIN	TIL116 TYP	TIL116 MAX	TIL117 MIN	TIL117 TYP	TIL117 MAX	UNIT
$V_{(BR)CBO}$	Collector-Base Breakdown Voltage		I_C = 10 μA, I_E = 0, I_F = 0	70			70			70			V
$V_{(BR)CEO}$	Collector-Emitter Breakdown Voltage		I_C = 1 mA, I_B = 0, I_F = 0	30			30			30			V
$V_{(BR)EBO}$	Emitter-Base Breakdown Voltage		I_E = 10 μA, I_C = 0, I_F = 0	7			7			7			V
I_R	Input Diode Static Reverse Current		V_R = 3 V			10			10			10	μA
$I_{C(on)}$	On-State Collector Current	Phototransistor Operation	V_{CE} = 0.4 V, I_F = 16 mA, I_B = 0	2	7								mA
			V_{CE} = 10 V, I_F = 10 mA, I_B = 0				2	5		5	9		mA
		Photodiode Operation	V_{CB} = 0.4 V, I_F = 16 mA, I_E = 0	7	20		7	20		7	20		μA
$I_{C(off)}$	Off-State Collector Current	Phototransistor Operation	V_{CE} = 10 V, I_F = 0, I_B = 0		1	50		1	50		1	50	nA
		Photodiode Operation	V_{CB} = 10 V, I_F = 0, I_E = 0		0.1	20		0.1	20		0.1	20	nA
h_{FE}	Transistor Static Forward Current Transfer Ratio		V_{CE} = 5 V, I_C = 10 mA, I_F = 0	100	300					200	550		
			V_{CE} = 5 V, I_C = 100 μA, I_F = 0				100	300					
V_F	Input Diode Static Forward Voltage		I_F = 16 mA		1.2	1.4					1.2	1.4	V
			I_F = 60 mA					1.25	1.5				V
$V_{CE(sat)}$	Collector-Emitter Saturation Voltage		I_C = 2 mA, I_F = 16 mA, I_B = 0		0.25	0.4							V
			I_C = 2.2 mA, I_F = 15 mA, I_B = 0					0.25	0.4				V
			I_C = 0.5 mA, I_F = 10 mA, I_B = 0								0.25	0.4	V
r_{IO}	Input-to-Output Internal Resistance		V_{in-out} = ±1.5 kV for TIL111, ±2.5 kV for all others, See Note 6	10^{11}			10^{11}			10^{11}			Ω
C_{io}	Input-to-Output Capacitance		V_{in-out} = 0, f = 1 MHz, See Note 6		1	1.3		1	1.3		1	1.3	pF

NOTE 6: These parameters are measured between both input-diode leads shorted together and all the phototransistor leads shorted together.

switching characteristics at 25°C free-air temperature

PARAMETER			TEST CONDITIONS	TIL111 TIL114 MIN	TIL111 TIL114 TYP	TIL111 TIL114 MAX	TIL116 MIN	TIL116 TYP	TIL116 MAX	TIL117 MIN	TIL117 TYP	TIL117 MAX	UNIT
t_r	Rise Time	Phototransistor Operation	V_{CC} = 10 V, $I_{C(on)}$ = 2 mA, R_L = 100 Ω, See Test Circuit A of Figure 1		5	10		5	10		5	10	μs
t_f	Fall Time				5	10		5	10		5	10	μs
t_r	Rise Time	Photodiode Operation	V_{CC} = 10 V, $I_{C(on)}$ = 20 μA, R_L = 1 kΩ, See Test Circuit B of Figure 1		1			1			1		μs
t_f	Fall Time				1			1			1		μs

TEXAS INSTRUMENTS
INCORPORATED
POST OFFICE BOX 225012 • DALLAS, TEXAS 75265

PARAMETER MEASUREMENT INFORMATION

Adjust amplitude of input pulse for:
$I_{C(on)}$ = 2 mA (Test Circuit A) or
$I_{C(on)}$ = 20 µA (Test Circuit B)

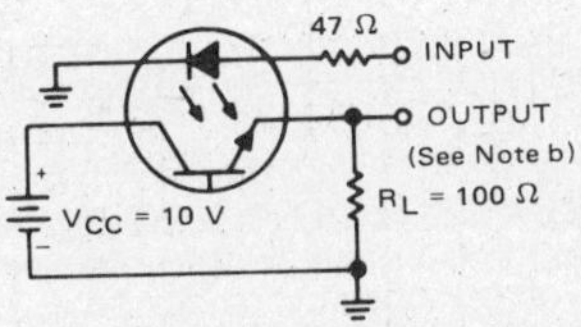

TEST CIRCUIT A
PHOTOTRANSISTOR OPERATION

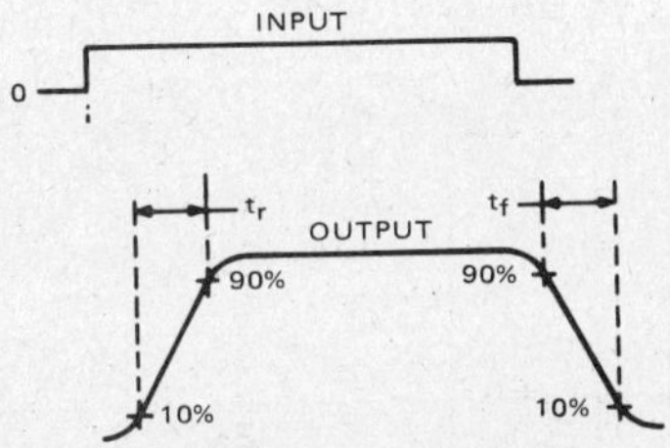

VOLTAGE WAVEFORMS

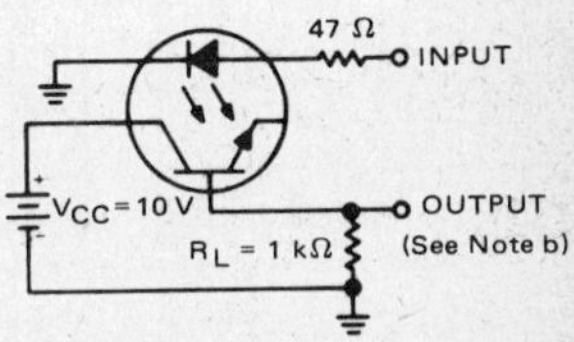

TEST CIRCUIT B
PHOTODIODE OPERATION

NOTES: a. The input waveform is supplied by a generator with the following characteristics: Z_{out} = 50 Ω, $t_r \leqslant$ 15 ns, duty cycle ≈ 1%, t_w = 100 µs.
b. The output waveform is monitored on an oscilloscope with the following characteristics: $t_r \leqslant$ 12 ns, $R_{in} \geqslant$ 1 MΩ, $C_{in} \leqslant$ 20 pF.

FIGURE 1—SWITCHING TIMES

TYPICAL CHARACTERISTICS

TIL111, TIL114

COLLECTOR CURRENT
vs
INPUT-DIODE FORWARD CURRENT

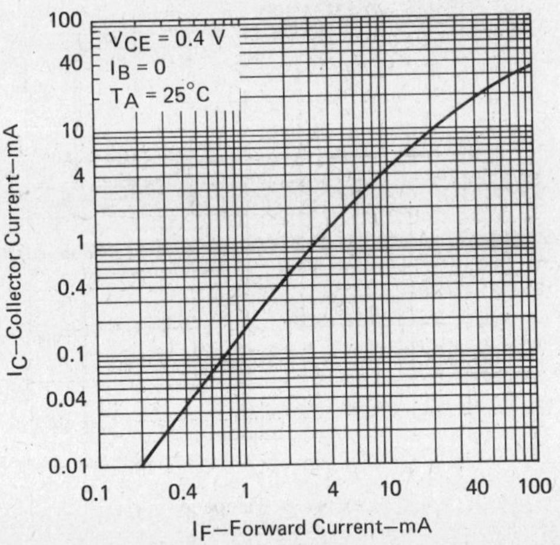

FIGURE 2

TIL116, TIL117

COLLECTOR CURRENT
vs
INPUT-DIODE FORWARD CURRENT

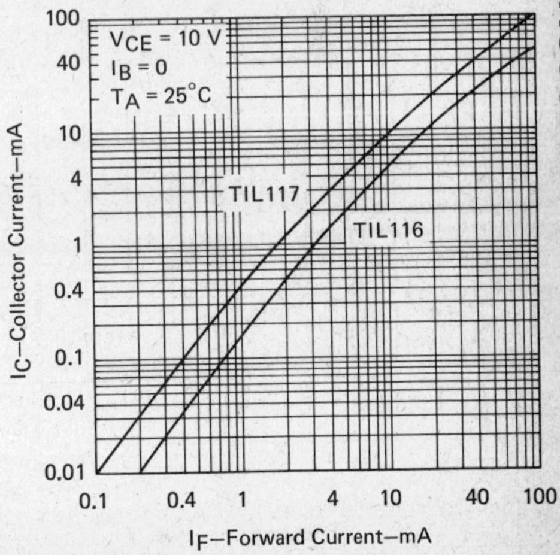

FIGURE 3

TYPICAL CHARACTERISTICS

TIL111, TIL114

COLLECTOR CURRENT
vs
COLLECTOR-EMITTER VOLTAGE

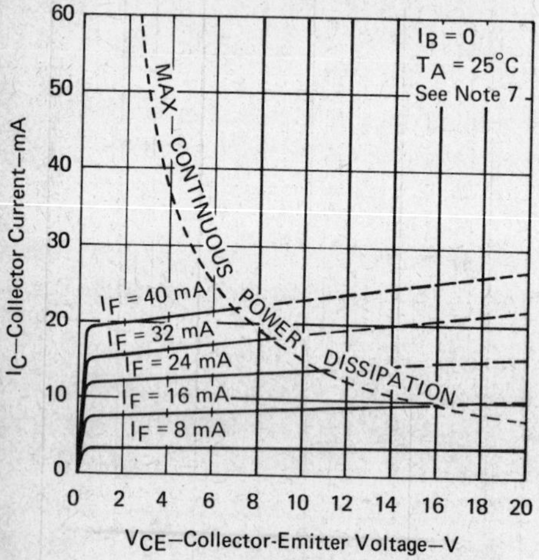

FIGURE 4

TIL116

COLLECTOR CURRENT
vs
COLLECTOR-EMITTER VOLTAGE

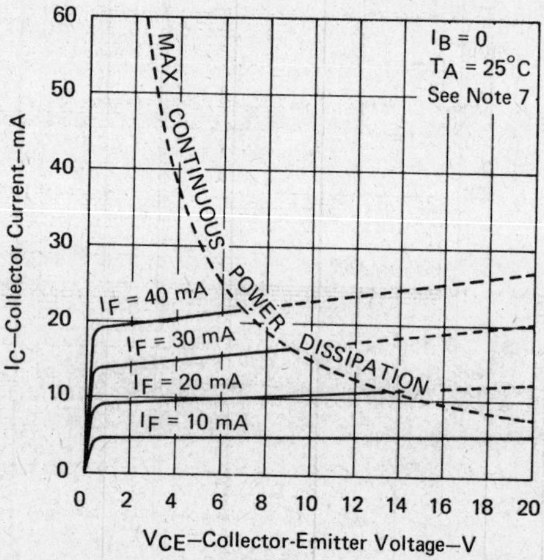

FIGURE 5

TIL117

COLLECTOR CURRENT
vs
COLLECTOR-EMITTER VOLTAGE

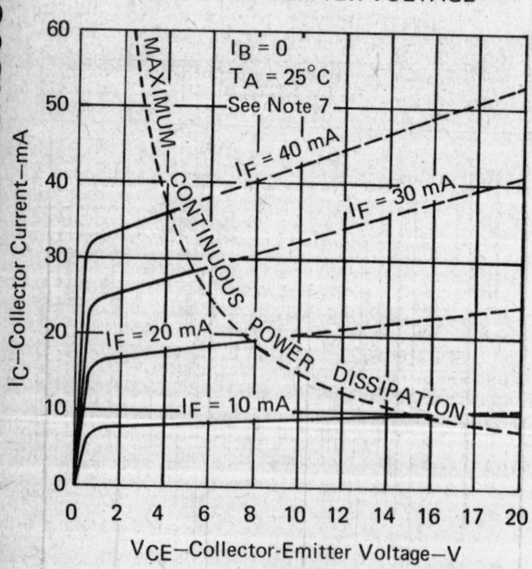

FIGURE 6

RELATIVE ON-STATE COLLECTOR CURRENT
vs
FREE-AIR TEMPERATURE

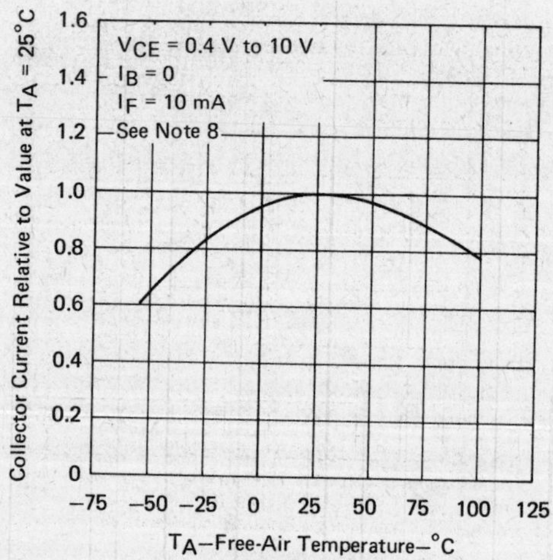

FIGURE 7

NOTES: 7. Pulse operation of input diode is required for operation beyond limits shown by dotted lines.
8. These parameters were measured using pulse techniques. t_w = 1 ms, duty cycle ≤ 2%.

TEXAS INSTRUMENTS
INCORPORATED
POST OFFICE BOX 225012 • DALLAS, TEXAS 75265

TYPICAL CHARACTERISTICS

OFF-STATE COLLECTOR CURRENT
vs
FREE-AIR TEMPERATURE

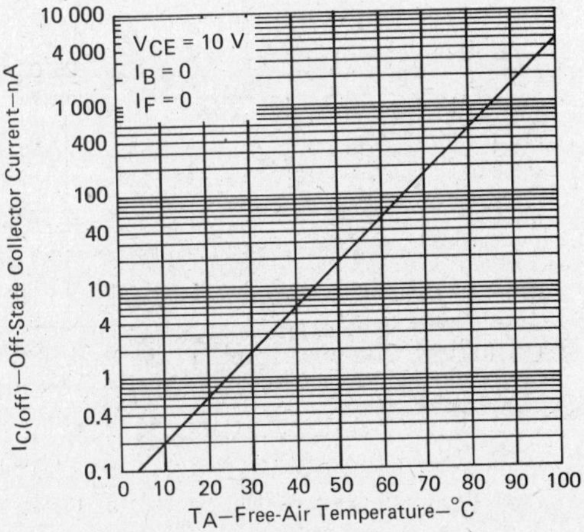

FIGURE 8

NORMALIZED TRANSISTOR STATIC FORWARD CURRENT TRANSFER RATIO
vs
ON-STATE COLLECTOR CURRENT

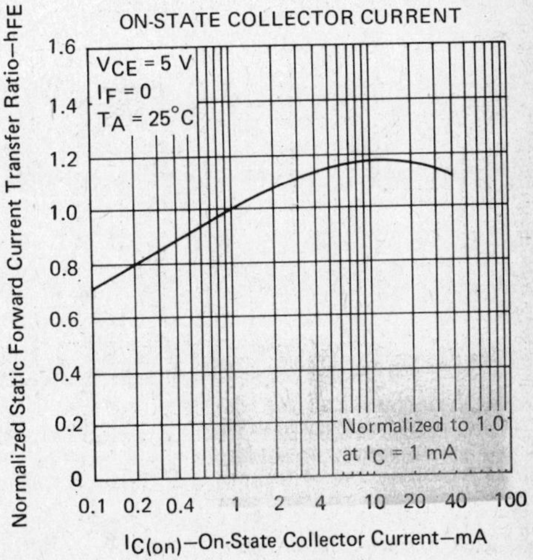

FIGURE 9

INPUT DIODE FORWARD CONDUCTION CHARACTERISTICS

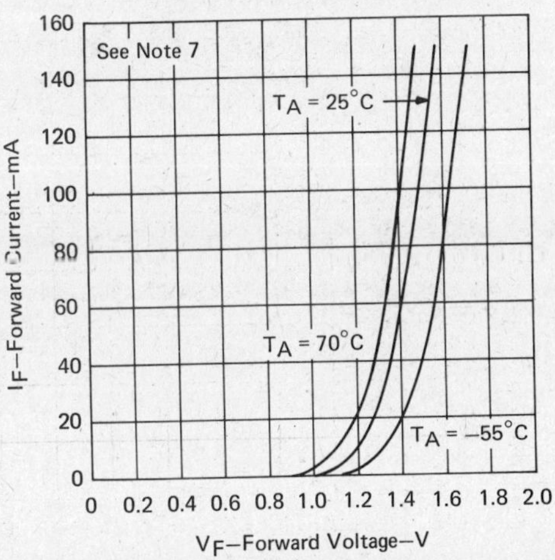

FIGURE 10

COLLECTOR CURRENT
vs
MODULATION FREQUENCY

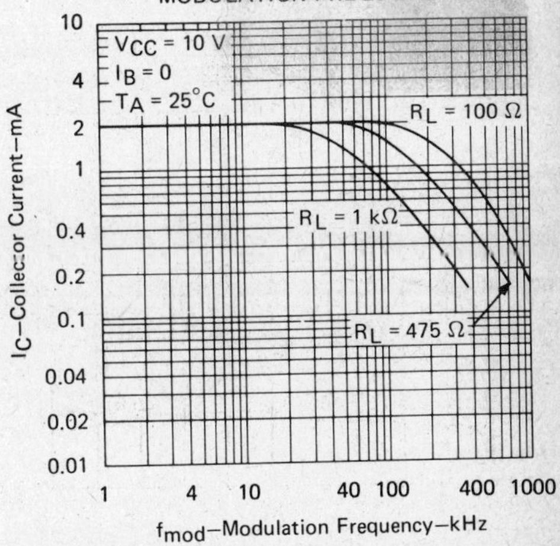

FIGURE 11

NOTE 7: These parameters were measured using pulse techniques. t_w = 1 ms, duty cycle ⩽ 2%

TEXAS INSTRUMENTS
INCORPORATED
POST OFFICE BOX 225012 • DALLAS, TEXAS 75265

7

OPTOCOUPLERS

D1607, NOVEMBER 1973–REVISED FEBRUARY 1983

- **Gallium Arsenide Diode Infrared Source Optically Coupled to a Silicon N-P-N Phototransistor**
- **High Direct-Current Transfer Ratio**
- **Base Lead Provided for Conventional Transistor Biasing (TIL112, TIL115)**
- **High-Voltage Electrical Isolation . . . 1.5-kV or 2.5-kV Rating**
- **Plastic Dual-In-Line Package**
- **High-Speed Switching: $t_r = 2\ \mu s$, $t_f = 2\ \mu s$ Typical**

mechanical data

The package consists of a gallium arsenide infrared-emitting diode and an n-p-n silicon phototransistor mounted on a 6-lead frame encapsulated within an electrically nonconductive plastic compound. The case will withstand soldering temperature with no deformation and device performance characteristics remain stable when operated in high-humidity conditions. Unit weight is approximately 0.52 grams.

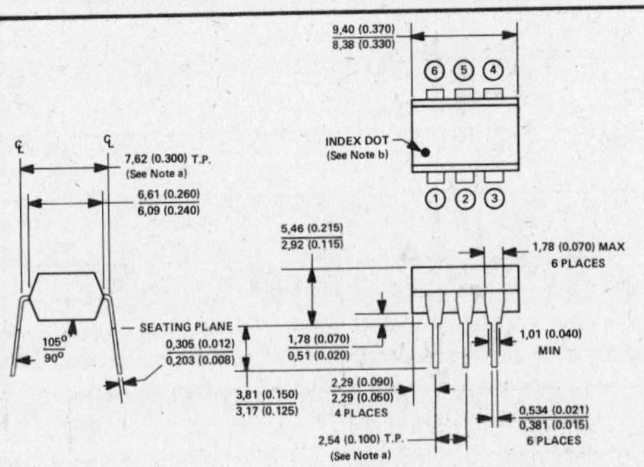

NOTES:
a. Leads are within 0,13 mm (0.005 inch) radius of true position (T.P.) with maximum material condition and unit installed.
b. Pin 1 identified by index dot.

Terminal connections:
1. Anode
2. Cathode } Infrared-emitting diode
3. No internal connection
4. Emitter
5. Collector } Phototransistor
6. Base

FALLS WITHIN JEDEC MO-001AM DIMENSIONS
ALL LINEAR DIMENSIONS ARE IN MILLIMETERS AND PARENTHETICALLY IN INCHES.

absolute maximum ratings at 25°C free-air temperature (unless otherwise noted)

	TIL112	TIL115	TIL118
Input-to-Output Voltage	±1.5 kV	±2.5 kV	±1.5 kV
Collector-Base Voltage	30 V	30 V	
Collector-Emitter Voltage (See Note 1)	20 V	20 V	20 V
Emitter-Collector Voltage	4 V	4 V	4 V
Emitter-Base Voltage	4 V	4 V	
Input-Diode Reverse Voltage	3 V	3 V	3 V
Input-Diode Continuous Forward Current at (or below) 25°C Free-Air Temperature (See Note 2)	← 100 mA →		
Continuous Power Dissipation at (or below) 25°C Free-Air Temperature:			
Infrared-Emitting Diode (See Note 3)	← 150 mW →		
Phototransistor (See Note 4)	← 150 mW →		
Total (Infrared-Emitting Diode plus Phototransistor, See Note 5)	← 250 mW →		
Storage Temperature Range	← −55°C to 150°C →		
Lead Temperature 1,6 mm (1/16 Inch) from Case for 10 Seconds	← 260°C →		

NOTES:
1. This value applies when the base-emitter diode is open-circuited.
2. Derate linearly to 100°C free-air temperature at the rate of 1.33 mA/°C.
3. Derate linearly to 100°C free-air temperature at the rate of 2 mW/°C.
4. Derate linearly to 100°C free-air temperature at the rate of 2 mW/°C.
5. Derate linearly to 100°C free-air temperature at the rate of 3.33 mW/°C.

Copyright © 1983 by Texas Instruments Incorporated

electrical characteristics at 25°C free-air temperature

PARAMETER			TEST CONDITIONS†	TIL112 MIN	TIL112 TYP	TIL112 MAX	TIL115 MIN	TIL115 TYP	TIL115 MAX	TIL118 MIN	TIL118 TYP	TIL118 MAX	UNIT
$V_{(BR)CBO}$	Collector-Base Breakdown Voltage		I_C = 10 µA, I_E = 0, I_F = 0	30			30						V
$V_{(BR)CEO}$	Collector-Emitter Breakdown Voltage		I_C = 1 mA, I_B = 0, I_F = 0	20			20			20			V
$V_{(BR)EBO}$	Emitter-Base Breakdown Voltage		I_E = 10 µA, I_C = 0, I_F = 0	4			4						V
$V_{(BR)ECO}$	Emitter-Collector Breakdown Voltage		I_E = 10 µA, I_F = 0							4			V
$I_{C(on)}$	On-State Collector Current	Phototransistor Operation	V_{CE} = 5 V, I_F = 10 mA, I_B = 0	0.2	2		0.2	2		1	2		mA
		Photodiode Operation	V_{CB} = 5 V, I_F = 10 mA, I_E = 0	2	10		2	10					µA
$I_{C(off)}$	Off-State Collector Current	Phototransistor Operation	V_{CE} = 5 V, I_F = 0, I_B = 0		1	100		1	100		1	100	nA
		Photodiode Operation	V_{CB} = 5 V, I_F = 0, I_E = 0		0.1	50		0.1	50				nA
h_{FE}	Transistor Static Forward Current Transfer Ratio		V_{CE} = 5 V, I_C = 10 mA, I_F = 0	50	200		50	200					
V_F	Input Diode Static Forward Voltage		I_F = 10 mA		1.2	1.5		1.2	1.5		1.2	1.5	V
$V_{CE(sat)}$	Collector-Emitter Saturation Voltage		I_C = 2 mA, I_F = 50 mA, I_B = 0			0.5			0.5			0.5	V
r_{IO}	Input-to-Output Internal Resistance		V_{in-out} = ±1.5 kV, See Note 6	10^{11}						10^{11}			Ω
			V_{in-out} = ±2.5 kV, See Note 6				10^{11}						Ω
C_{io}	Input-to-Output Capacitance		V_{in-out} = 0, f = 1 MHz, See Note 6		1	2		1	2		1	2	pF

NOTE 6: These parameters are measured between both input-diode leads shorted together and all the phototransistor leads shorted together.
†References to the base are not applicable for the TIL118.

switching characteristics at 25°C free-air temperature

PARAMETER			TEST CONDITIONS	TIL112 MIN	TIL112 TYP	TIL112 MAX	TIL115 MIN	TIL115 TYP	TIL115 MAX	TIL118 MIN	TIL118 TYP	TIL118 MAX	UNIT
t_r	Rise Time	Phototransistor Operation	V_{CC} = 10 V, $I_{C(on)}$ = 2 mA, R_L = 100 Ω, See Test Circuit A of Figure 1		2	15		2	15		2	15	µs
t_f	Fall Time				2	15		2	15		2	15	µs
t_r	Rise Time	Photodiode Operation	V_{CC} = 10 V, $I_{C(on)}$ = 20 µA, R_L = 1 kΩ, See Test Circuit B of Figure 1		1			1					µs
t_f	Fall Time				1			1					µs

TEXAS INSTRUMENTS
INCORPORATED
POST OFFICE BOX 225012 • DALLAS, TEXAS 75265

PARAMETER MEASUREMENT INFORMATION

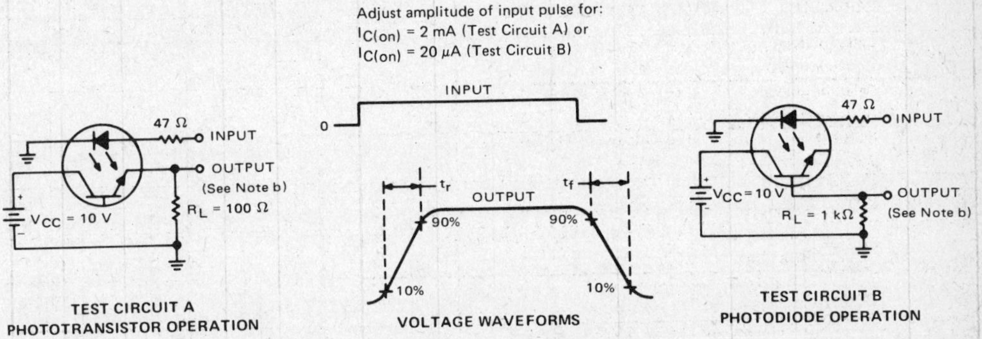

NOTES: a. The input waveform is supplied by a generator with the following characteristics: Z_{out} = 50 Ω, $t_r \leqslant$ 15 ns, duty cycle ≈ 1%, t_w = 100 μs.

b. The output waveform is monitored on an oscilloscope with the following characteristics: $t_r \leqslant$ 12 ns, $R_{in} \geqslant$ 1 MΩ, $C_{in} \leqslant$ 20 pF.

FIGURE 1–SWITCHING TIMES

TYPICAL CHARACTERISTICS

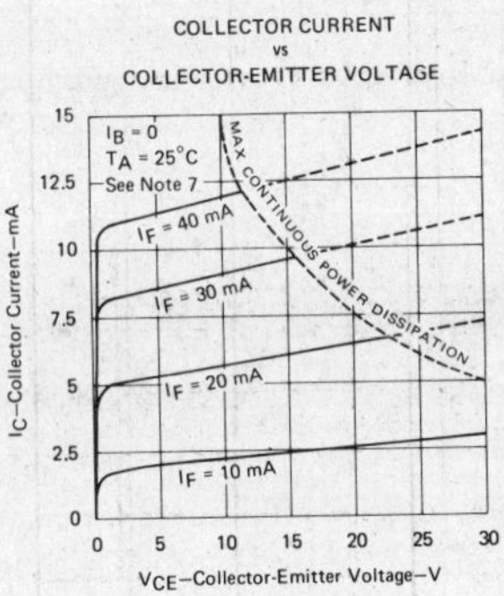

FIGURE 2

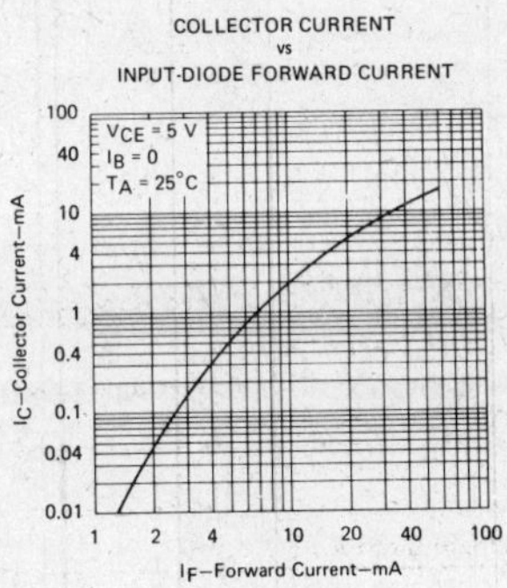

FIGURE 3

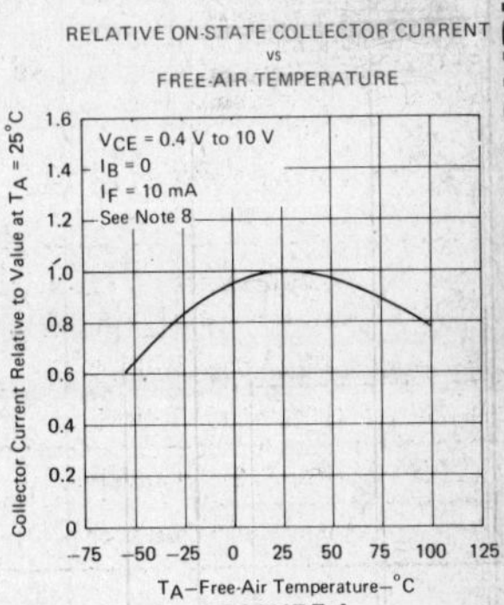

FIGURE 4

NOTES: 7. Pulse operation of input diode is required for operation beyond limits shown by dotted lines.

8. These parameters were measured using pulse techniques t_w = 1 ms, duty cycle $\leqslant$ 2%.

TEXAS INSTRUMENTS
INCORPORATED
POST OFFICE BOX 225012 • DALLAS, TEXAS 75265

TYPICAL CHARACTERISTICS

OFF-STATE COLLECTOR CURRENT
vs
FREE-AIR TEMPERATURE

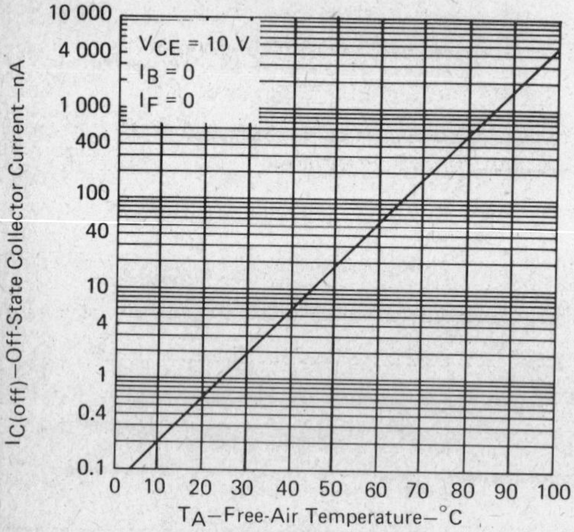

FIGURE 5

NORMALIZED TRANSISTOR STATIC FORWARD
CURRENT TRANSFER RATIO
vs
ON-STATE COLLECTOR CURRENT

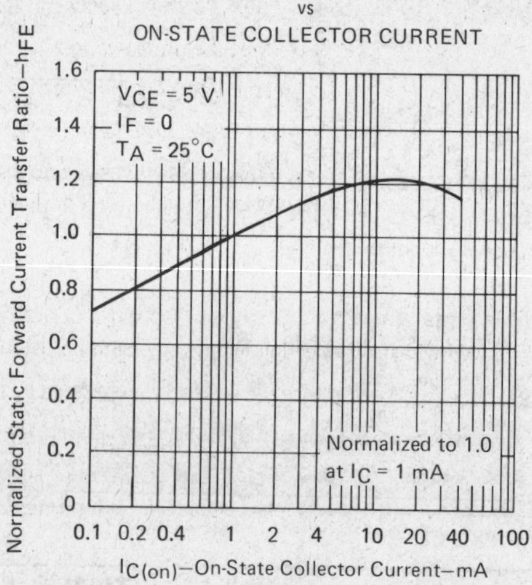

FIGURE 6

INPUT DIODE FORWARD
CONDUCTION CHARACTERISTICS

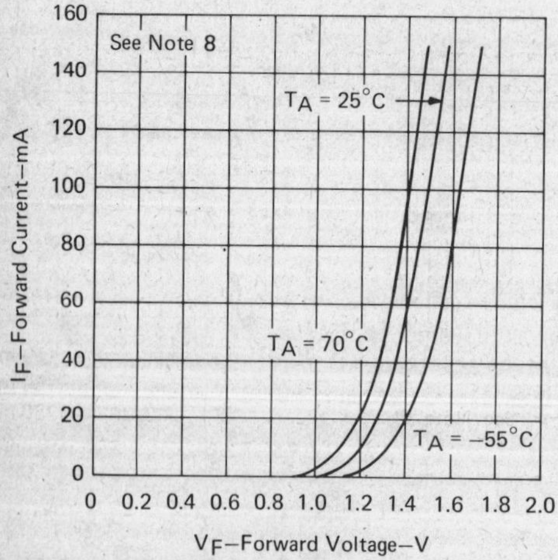

FIGURE 7

COLLECTOR CURRENT
vs
MODULATION FREQUENCY

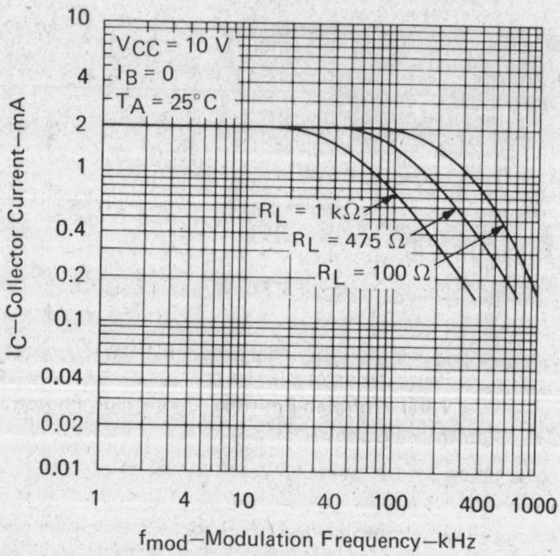

FIGURE 8

NOTE 8: These parameters were measured using pulse techniques. t_w = 1 ms, duty cycle ≤ 2%.

D1499, AUGUST 1981—REVISED FEBRUARY 1983

- Gallium Arsenide Diode Infrared Source Optically Coupled to a Silicon N-P-N Darlington-Connected Phototransistor
- High Direct-Current Transfer Ratio . . . 300% Minimum at 10 mA
- High-Voltage Electrical Isolation . . . 1500-Volt Rating
- Plastic Dual-In-Line Package
- Base Lead Provided on TIL113 for Conventional Transistor Biasing
- No Base Lead Connection on TIL119A for High-EMI Environments
- Typical Applications Include Remote Terminal Isolation, SCR and Triac Triggers, Mechanical Relays, and Pulse Transformers

mechanical data

The package consists of a gallium arsenide infrared-emitting diode and an n-p-n silicon darlington-connected phototransistor mounted on a 6-lead frame encapsulated within an electrically nonconductive plastic compound. The case will withstand soldering temperature with no deformation and device performance characteristics remain stable when operated in high-humidity conditions. Unit weight is approximately 0.52 grams.

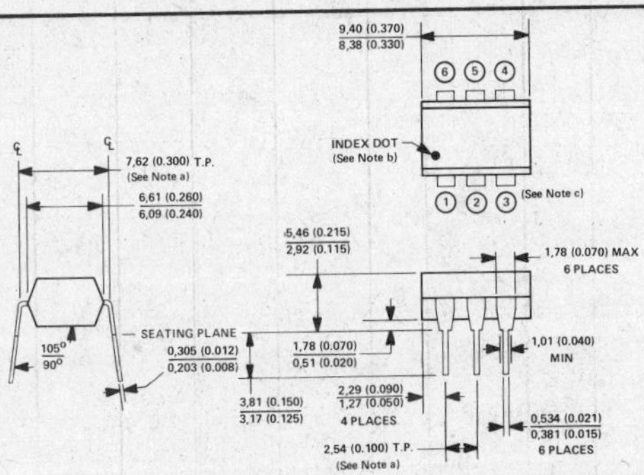

NOTES:
a. Leads are within 0,13 mm (0.005 inch) radius of true position (T.P.) with maximum material condition and unit installed.
b. Pin 1 identified by index dot.

Terminal connections:
1. Anode } Infrared-emitting diode
2. Cathode
3. No internal connection
4. Emitter } Photo-transistor
5. Collector
6. TIL113 base
TIL119 base or no internal connection at the option of TI
TIL119A no internal connection

FALLS WITHIN JEDEC MO-001AM DIMENSIONS
ALL LINEAR DIMENSIONS ARE IN MILLIMETERS AND PARENTHETICALLY IN INCHES

absolute maximum ratings at 25°C free-air temperature (unless otherwise noted)

Input-to-Output Voltage	±1.5 kV
Collector-Base Voltage (TIL113)	30 V
Collector-Emitter Voltage (See Note 1)	30 V
Emitter-Collector Voltage	7 V
Emitter-Base Voltage (TIL113)	7 V
Input-Diode Reverse Voltage	3 V
Input-Diode Continuous Forward Current at (or below) 25°C Free-Air Temperature (See Note 2)	100 mA
Continuous Power Dissipation at (or below) 25°C Free-Air Temperature:	
Infrared-Emitting Diode (See Note 3)	150 mW
Phototransistor (See Note 4)	150 mW
Total (Infrared-Emitting Diode plus Phototransistor, See Note 5)	250 mW
Storage Temperature Range	−55°C to 150°C
Lead Temperature 1,6 mm (1/16 Inch) from Case for 10 Seconds	260°C

NOTES:
1. This value applies when the base-emitter diode is open-circuited.
2. Derate linearly to 100°C free-air temperature at the rate of 1.33 mA/°C.
3. Derate linearly to 100°C free-air temperature at the rate of 2 mW/°C.
4. Derate linearly to 100°C free-air temperature at the rate of 2 mW/°C.
5. Derate linearly to 100°C free-air temperature at the rate of 3.33 mW/°C.

Copyright © 1983 by Texas Instruments Incorporated

TEXAS INSTRUMENTS
INCORPORATED
POST OFFICE BOX 225012 • DALLAS, TEXAS 75265

electrical characteristics at 25°C free-air temperature

PARAMETER		TEST CONDITIONS†	TIL113 MIN	TIL113 TYP	TIL113 MAX	TIL119, TIL119A MIN	TIL119, TIL119A TYP	TIL119, TIL119A MAX	UNIT
$V_{(BR)CBO}$	Collector-Base Breakdown Voltage	$I_C = 10\ \mu A$, $I_E = 0$, $I_F = 0$	30						V
$V_{(BR)CEO}$	Collector-Emitter Breakdown Voltage	$I_C = 1\ mA$, $I_B = 0$, $I_F = 0$	30			30			V
$V_{(BR)EBO}$	Emitter-Base Breakdown Voltage	$I_E = 10\ \mu A$, $I_C = 0$, $I_F = 0$	7						V
$V_{(BR)ECO}$	Emitter-Collector Breakdown Voltage	$I_E = 10\ \mu A$, $I_F = 0$				7			V
$I_{C(on)}$	On-State Collector Current	$V_{CE} = 1\ V$, $I_B = 0$, $I_F = 10\ mA$	30	100					mA
		$V_{CE} = 2\ V$, $I_F = 10\ mA$				30	160		
$I_{C(off)}$	Off-State Collector Current	$V_{CE} = 10\ V$, $I_B = 0$, $I_F = 0$			100			100	nA
h_{FE}	Transistor Static Forward Current Transfer Ratio	$V_{CE} = 1\ V$, $I_C = 10\ mA$, $I_F = 0$		15,000					
V_F	Input Diode Static Forward Voltage	$I_F = 10\ mA$			1.5			1.5	V
$V_{CE(sat)}$	Collector-Emitter Saturation Voltage	$I_C = 125\ mA$, $I_B = 0$, $I_F = 50\ mA$			1.2				V
		$I_C = 10\ mA$, $I_F = 10\ mA$						1	
r_{IO}	Input-to-Output Internal Resistance	$V_{in-out} = \pm 1.5\ kV$, See Note 6	10^{11}			10^{11}			Ω
C_{io}	Input-to-Output Capacitance	$V_{in-out} = 0$, f = 1 MHz, See Note 6		1	1.3		1	1.3	pF

NOTE 6: These parameters are measured between both input-diode leads shorted together and all the phototransistor leads shorted together.

† References to the base are not applicable to TIL119 or TIL119A.

switching characteristics at 25°C free-air temperature

PARAMETER		TEST CONDITIONS	TIL113 MIN	TIL113 TYP	TIL113 MAX	TIL119, TIL119A MIN	TIL119, TIL119A TYP	TIL119, TIL119A MAX	UNIT
t_r	Rise Time	$V_{CC} = 15\ V$, $I_{C(on)} = 125\ mA$,		300					µs
t_f	Fall Time	$R_L = 100\ \Omega$, See Figure 1		300					
t_r	Rise Time	$V_{CC} = 10\ V$, $I_{C(on)} = 2.5\ mA$,					300		µs
t_f	Fall Time	$R_L = 100\ \Omega$, See Figure 1					300		

PARAMETER MEASUREMENT INFORMATION

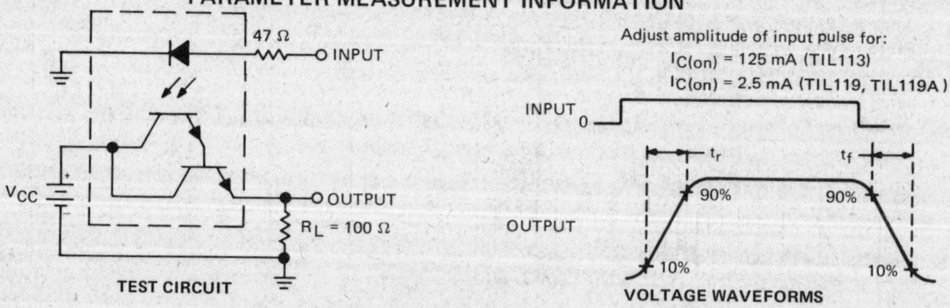

NOTES: a. The input waveform is supplied by a generator with the following characteristics: $Z_{out} = 50\ \Omega$, $t_r \leq 15$ ns, duty cycle ≈ 1%, $t_w = 500\ \mu s$.

b. The output waveform is monitored on an oscilloscope with the following characteristics: $t_r \leq 12$ ns, $R_{in} \geq 1\ M\Omega$, $C_{in} \leq 20$ pF.

FIGURE 1—SWITCHING TIMES

TYPICAL CHARACTERISTICS

COLLECTOR CURRENT
vs
COLLECTOR-EMITTER VOLTAGE

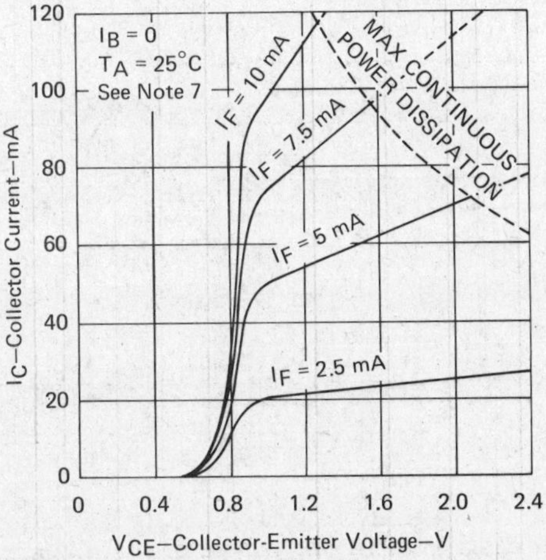

FIGURE 2

COLLECTOR CURRENT
vs
COLLECTOR-EMITTER VOLTAGE

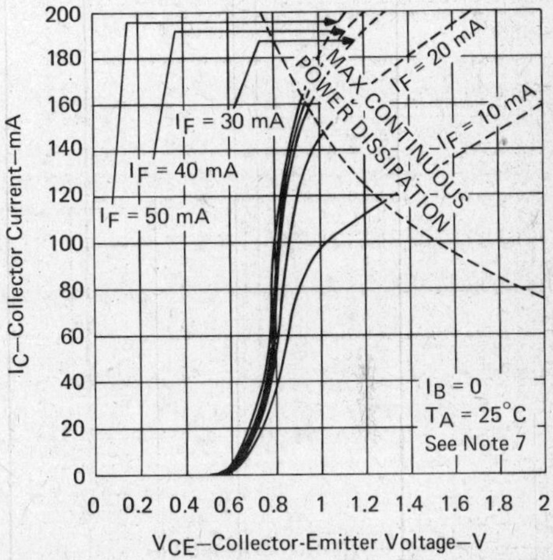

FIGURE 3

COLLECTOR CURRENT
vs
INPUT-DIODE FORWARD CURRENT

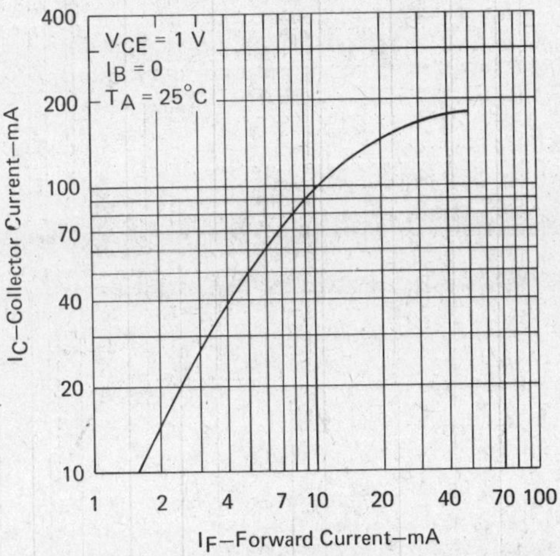

FIGURE 4

OFF-STATE COLLECTOR CURRENT
vs
FREE-AIR TEMPERATURE

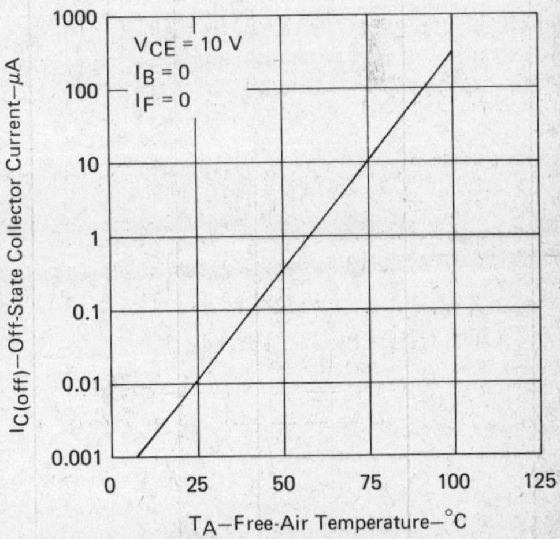

FIGURE 5

NOTE 7: Pulse operation of input diode is required for operation beyond limits shown by dotted line.

TYPICAL CHARACTERISTICS

RELATIVE COLLECTOR-EMITTER SATURATION VOLTAGE vs FREE-AIR TEMPERATURE

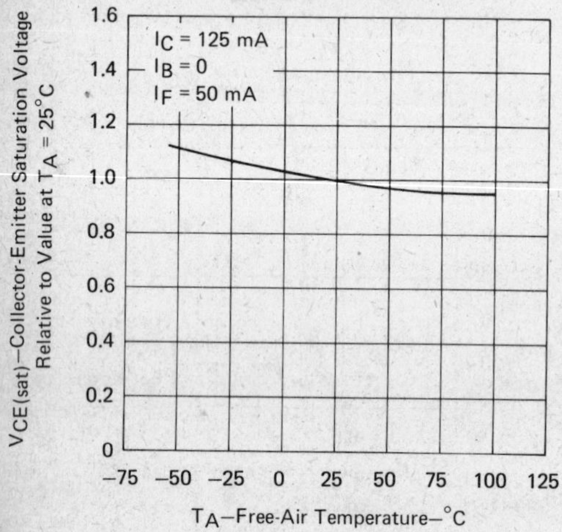

FIGURE 6

TIL113
TRANSISTOR STATIC FORWARD CURRENT TRANSFER RATIO vs COLLECTOR CURRENT

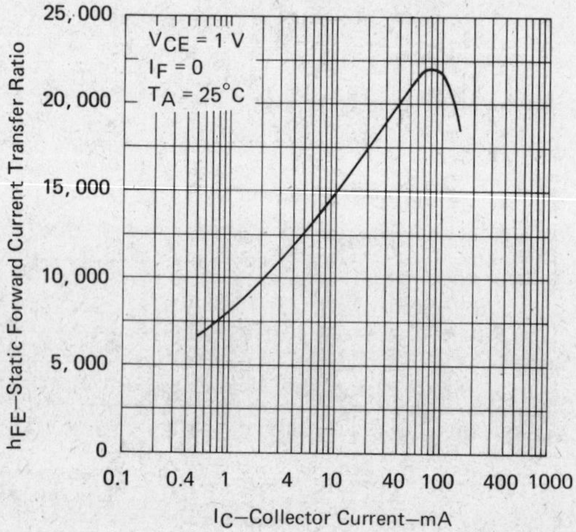

FIGURE 7

INPUT DIODE FORWARD CONDUCTION CHARACTERISTICS

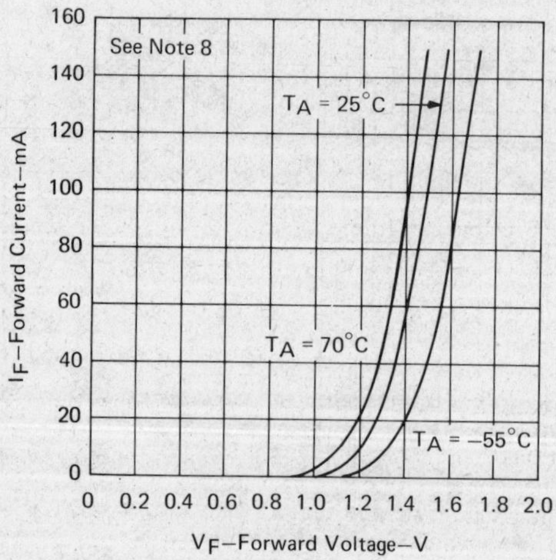

FIGURE 8

NOTE 8: This parameter was measured using pulse techniques. t_w = 1 ms, duty cycle ≤ 2%.

TEXAS INSTRUMENTS
INCORPORATED
POST OFFICE BOX 225012 • DALLAS, TEXAS 75265

GALLIUM ARSENIDE DIODE INFRARED SOURCE OPTICALLY COUPLED TO A HIGH-GAIN N-P-N SILICON PHOTOTRANSISTOR

- Photon Coupling for Isolator Applications
- High Overall Current Gain . . . 1.0 Typ (TIL121)
- High-Gain, High-Voltage Transistor . . . $V_{(BR)CEO}$ = 35 V Min
- High-Voltage Electrical Isolation . . . 1-kV Rating
- Stable Over Wide Temperature Range

mechanical data

THE COLLECTOR IS IN ELECTRICAL CONTACT WITH THE CASE

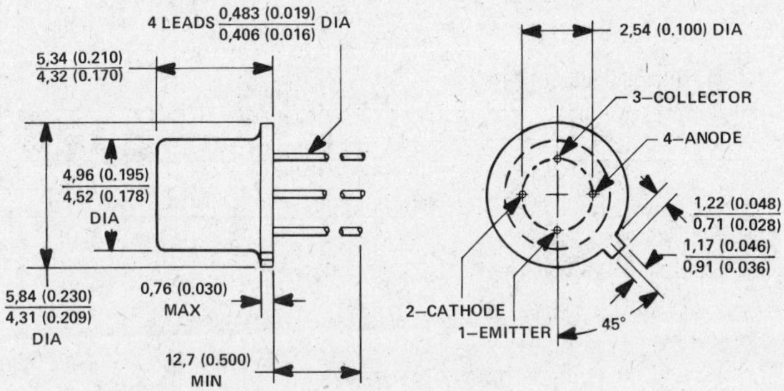

ALL LINEAR DIMENSIONS ARE IN MILLIMETERS AND PARENTHETICALLY IN INCHES.
ALL JEDEC TO-72 DIMENSIONS AND NOTES ARE APPLICABLE

absolute maximum ratings at 25°C free-air temperature (unless otherwise noted)

Input-to-Output Voltage	±1 kV
Collector-Emitter Voltage	35 V
Emitter-Collector Voltage	7 V
Input Diode Reverse Voltage	3 V
Input Diode Continuous Forward Current at (or below) 65°C Free-Air Temperature (See Note 1)	40 mA
Continuous Collector Current	50 mA
Continuous Transistor Power Dissipation at (or below) 25°C Free-Air Temperature (See Note 2)	190 mW
Operating Free-Air Temperature Range	−55°C to 125°C
Storage Temperature Range	−55°C to 150°C
Lead Temperature 1,6 mm (1/16 Inch) from Case for 10 Seconds	240°C

NOTES: 1. Derate linearly to 125°C free-air temperature at the rate of 0.67 mA/°C.
2. Derate linearly to 125°C free-air temperature at the rate of 1.9 mW/°C.

TEXAS INSTRUMENTS
INCORPORATED
POST OFFICE BOX 225012 • DALLAS, TEXAS 75265

electrical characteristics at 25°C free-air temperature (unless otherwise noted)

PARAMETER		TEST CONDITIONS		TIL120			TIL121			UNIT
				MIN	TYP	MAX	MIN	TYP	MAX	
$V_{(BR)CEO}$	Collector-Emitter Breakdown Voltage	$I_C = 1$ mA,	$I_F = 0$	35			35			V
$V_{(BR)ECO}$	Emitter-Collector Breakdown Voltage	$I_E = 100\ \mu A$,	$I_F = 0$	7			7			V
I_R	Input Diode Static Reverse Current	$V_R = 3$ V				100			100	μA
$I_{C(on)}$	On-State Collector Current	$V_{CE} = 5$ V,	$I_F = 10$ mA	2.5	6		5	10		mA
$I_{C(off)}$	Off-State Collector Current	$V_{CE} = 20$ V,	$I_F = 0$		6	100		6	100	nA
		$V_{CE} = 20$ V, $T_A = 100°C$	$I_F = 0$,		4			4		μA
V_F	Input Diode Static Forward Voltage	$I_F = 10$ mA				1.3			1.3	V
$V_{CE(sat)}$	Collector-Emitter Saturation Voltage	$I_C = 2.5$ mA,	$I_F = 20$ mA			0.3				V
		$I_C = 10$ mA,	$I_F = 20$ mA						0.3	
r_{io}	Input-to-Output Internal Resistance	$V_{in\text{-}out} = \pm 1$ kV,	See Note 3	10^{11}	10^{12}		10^{11}	10^{12}		Ω
C_{io}	Input-to-Output Capacitance	$V_{in\text{-}out} = 0$, See Note 3	f = 1 MHz,		2.5			2.5		pF

NOTE 3: These parameters are measured between both input diode leads shorted together and both phototransistor leads shorted together.

switching characteristics at 25°C free-air temperature

PARAMETER		TEST CONDITIONS		TIL120			TIL121			UNIT
				MIN	TYP	MAX	MIN	TYP	MAX	
t_r	Rise Time	$V_{CC} = 20$ V,	$I_{C(on)} = 5$ mA		3	20		6	20	μs
t_f	Fall Time	$R_L = 100\ \Omega$,	See Figure 1		3	20		6	20	μs

PARAMETER MEASUREMENT INFORMATION

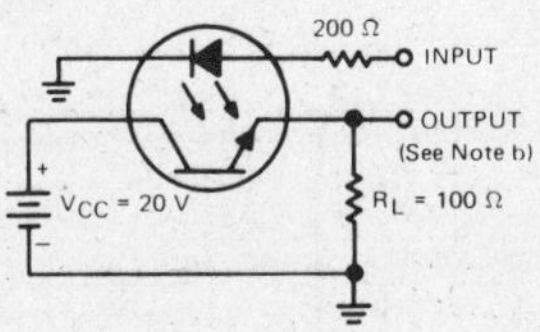

TEST CIRCUIT

Adjust amplitude of input pulse for $I_{C(on)}$ = 5 mA

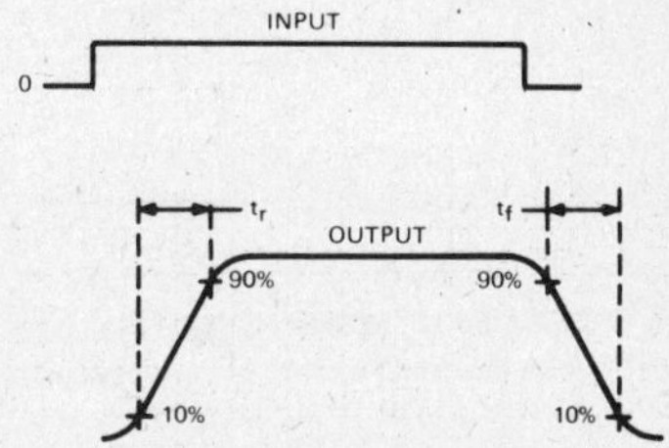

VOLTAGE WAVEFORMS

NOTES: a. The input waveform is supplied by a generator with the following characteristics: z_{out} = 50 Ω, $t_r \leq$ 15 ns, duty cycle ≈ 1%, t_w = 100 μs.

b. Waveforms are monitored on an oscilloscope with the following characteristics: $t_r \leq$ 12 ns, $R_{in} \geq$ 1 MΩ, $C_{in} \leq$ 20 pF.

FIGURE 1—SWITCHING TIMES

TYPICAL CHARACTERISTICS

TIL120
COLLECTOR CURRENT
vs
COLLECTOR-EMITTER VOLTAGE

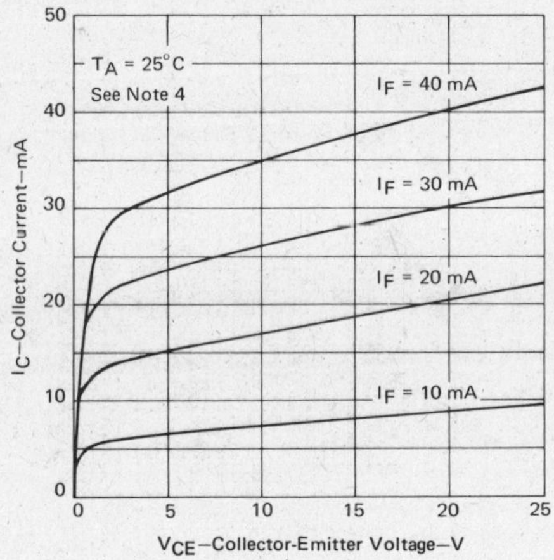

FIGURE 2

TIL121
COLLECTOR CURRENT
vs
COLLECTOR-EMITTER VOLTAGE

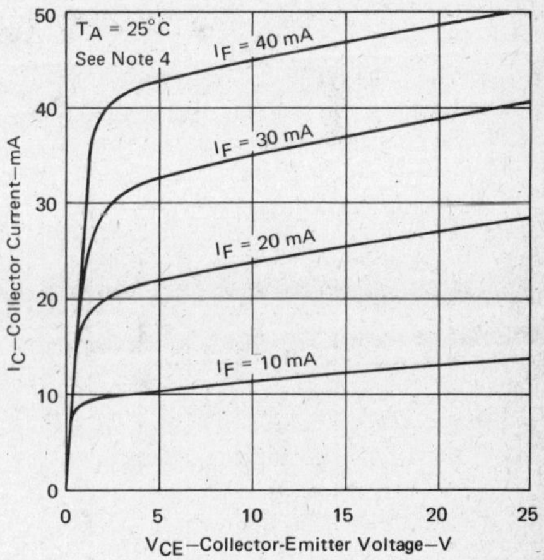

FIGURE 3

NOTE 4: This parameter was measured using pulse techniques. t_w = 100 μs, duty cycle = 1%.

TEXAS INSTRUMENTS
INCORPORATED
POST OFFICE BOX 225012 • DALLAS, TEXAS 75265

TYPICAL CHARACTERISTICS

INPUT DIODE FORWARD CONDUCTION CHARACTERISTICS

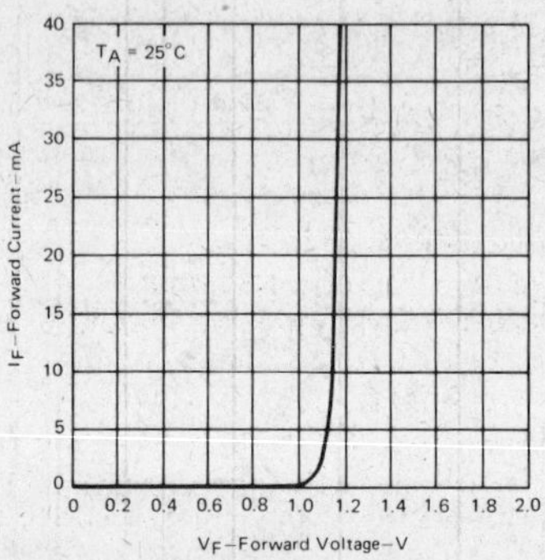

FIGURE 4

NORMALIZED ON-STATE COLLECTOR CURRENT
vs
FREE-AIR TEMPERATURE

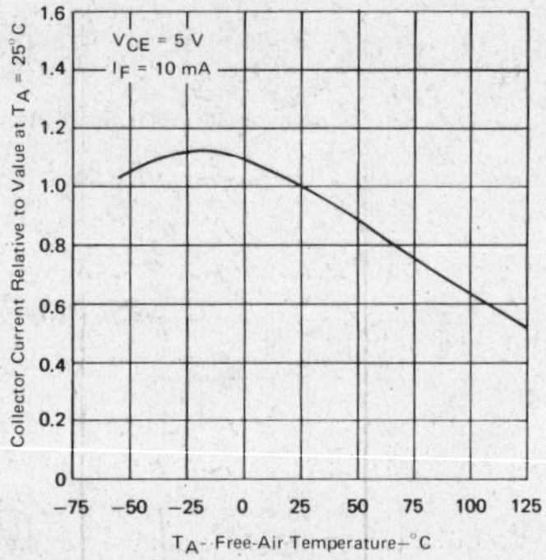

FIGURE 5

PHOTOTRANSISTOR COLLECTOR CURRENT
vs
INPUT-DIODE FORWARD CURRENT

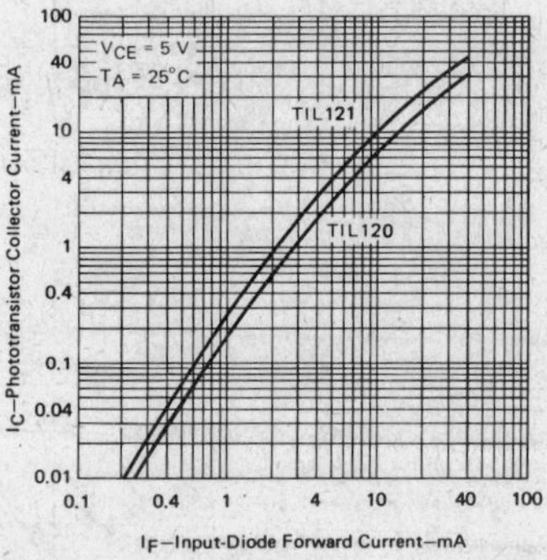

FIGURE 6

OFF-STATE COLLECTOR CURRENT
vs
FREE-AIR TEMPERATURE

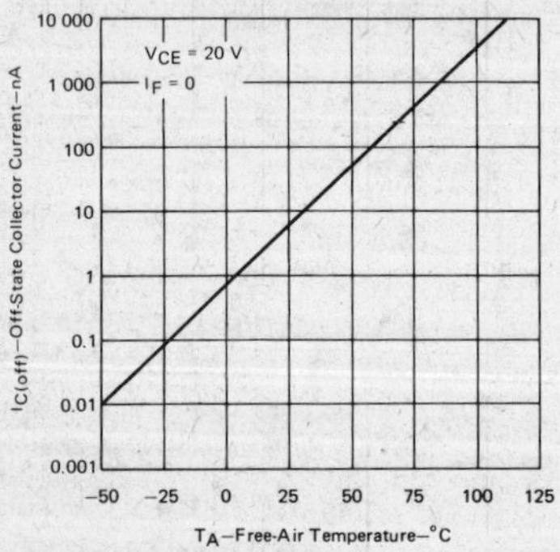

FIGURE 7

TIL120
AVERAGE SWITCHING TIME
vs
LOAD RESISTANCE

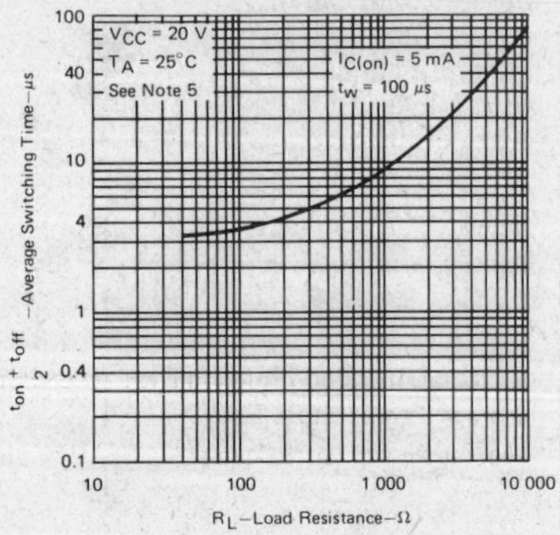

FIGURE 8

NOTE 5: These parameters were measured in the test circuit of Figure 1 with R_L varied between 40 Ω and 10 kΩ.

TEXAS INSTRUMENTS
INCORPORATED
POST OFFICE BOX 225012 • DALLAS, TEXAS 75265

D2227, MAY 1977—REVISED DECEMBER 1982

COMPATIBLE WITH STANDARD TTL INTEGRATED CIRCUITS

- Gallium Arsenide Diode Infrared Source Optically Coupled to a Silicon N-P-N Phototransistor
- High Direct-Current Transfer Ratio
- High-Voltage Electrical Isolation . . . 5000-V Rating
- Plastic Dual-In-Line Package
- High-Speed Switching: $t_r = 2\ \mu s$, $t_f = 2\ \mu s$ Typical
- Typical Applications Include Remote Terminal Isolation, SCR and Triac Triggers, Mechanical Relays, and Pulse Transformers

mechanical data

The package consists of a gallium arsenide infrared-emitting diode and an n-p-n silicon phototransistor mounted on a 6-lead frame encapsulated within an electrically nonconductive plastic compound. The case will withstand soldering temperature with no deformation and device performance characteristics remain stable when operated in high-humidity conditions. Unit weight is approximately 0.52 grams.

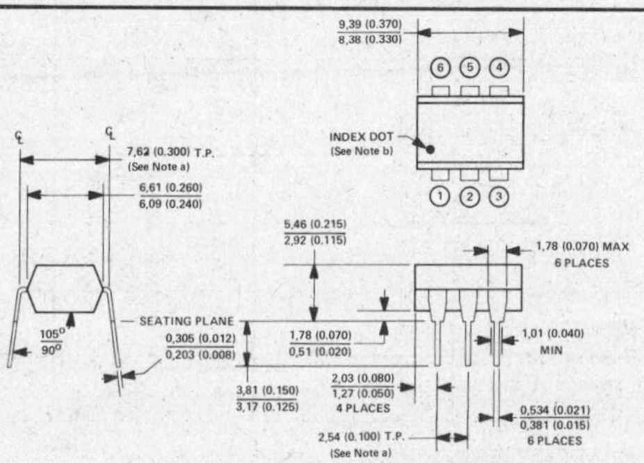

NOTES:
a. Leads are within 0,13 mm (0.005 inch) radius of true position (T.P.) with maximum material condition and unit installed.
b. Pin 1 identified by index dot.

Terminal connections:
1. Anode
2. Cathode } Infrared-emitting diode
3. No internal connection
4. Emitter
5. Collector } Phototransistor
6. Base

FALLS WITHIN JEDEC MO-001 AM DIMENSIONS
ALL LINEAR DIMENSIONS ARE IN MILLIMETERS AND PARENTHETICALLY IN INCHES

absolute maximum ratings at 25°C free-air temperature (unless otherwise noted)

Input-to-Output Voltage	±5 kV
Collector-Base Voltage	70 V
Collector-Emitter Voltage (See Note 1)	30 V
Emitter-Collector Voltage	7 V
Emitter-Base Voltage	7 V
Input-Diode Reverse Voltage	3 V
Input-Diode Continuous Forward Current	100 mA
Continuous Power Dissipation at (or below) 25°C Free-Air Temperature:	
Infrared-Emitting Diode (See Note 2)	150 mW
Phototransistor (See Note 3)	150 mW
Total, Infrared-Emitting Diode plus Phototransistor (See Note 4)	250 mW
Storage Temperature Range	−55°C to 150°C
Lead Temperature 1,6 mm (1/16 inch) from Case for 10 Seconds	260°C

NOTES: 1. This value applies when the base-emitter diode is open-circuited.
2. Derate linearly to 100°C free-air temperature at the rate of 2 mW/°C.
3. Derate linearly to 100°C free-air temperature at the rate of 2 mW/°C.
4. Derate linearly to 100°C free-air temperature at the rate of 3.33 mW/°C.

7

OPTOCOUPLERS

TEXAS INSTRUMENTS
INCORPORATED
POST OFFICE BOX 225012 • DALLAS, TEXAS 75265

electrical characteristics at 25°C free-air temperature

PARAMETER			TEST CONDITIONS	TIL124 MIN	TIL124 TYP	TIL124 MAX	TIL125 MIN	TIL125 TYP	TIL125 MAX	TIL126 MIN	TIL126 TYP	TIL126 MAX	UNIT
$V_{(BR)CBO}$	Collector-Base Breakdown Voltage		$I_C = 10\ \mu A$, $I_E = 0$, $I_F = 0$	70			70			70			V
$V_{(BR)CEO}$	Collector-Emitter Breakdown Voltage		$I_C = 1$ mA, $I_B = 0$, $I_F = 0$	30			30			30			V
$V_{(BR)EBO}$	Emitter-Base Breakdown Voltage		$I_E = 10\ \mu A$, $I_C = 0$, $I_F = 0$	7			7			7			V
I_R	Input Diode Static Reverse Current		$V_R = 3$ V			10			10			10	µA
$I_{C(on)}$	On-State Collector Current	Phototransistor Operation	$V_{CE} = 10$ V, $I_F = 10$ mA, $I_B = 0$	1	3		2	5		5	9		mA
		Photodiode Operation	$V_{CB} = 10$ V, $I_F = 10$ mA, $I_E = 0$	5	20		5	20		5	20		µA
$I_{C(off)}$	Off-State Collector Current	Phototransistor Operation	$V_{CE} = 10$ V, $I_F = 0$ $I_B = 0$		1	50		1	50		1	50	nA
		Photodiode Operation	$V_{CB} = 10$ V, $I_F = 0$, $I_E = 0$		0.1	20		0.1	20		0.1	20	
h_{FE}	Transistor Static Forward Current Transfer Ratio		$V_{CE} = 5$ V, $I_C = 10$ mA, $I_F = 0$	50	100		100	200		100	550		
V_F	Input Diode Static Forward Voltage		$I_F = 10$ mA		1.2	1.4		1.2	1.4		1.2	1.4	V
$V_{CE(sat)}$	Collector-Emitter Saturation Voltage		$I_C = 1$ mA, $I_F = 10$ mA, $I_B = 0$		0.25	0.4		0.25	0.4		0.25	0.4	V
r_{io}	Input-to-Output Internal Resistance		$V_{in\text{-}out} = 500$ V, See Note 5	10^{11}			10^{11}			10^{11}			Ω
C_{io}	Input-to-Output Capacitance		$V_{in\text{-}out} = 0$, f = 1 MHz, See Note 5		1	1.3		1	1.3		1	1.3	pF

NOTE 5: These parameters are measured between both input-diode leads shorted together and all the phototransistor leads shorted together.

switching characteristics at 25°C free-air temperature

PARAMETER			TEST CONDITIONS	MIN	TYP	MAX	UNIT
t_r	Rise Time	Phototransistor Operation	$V_{CC} = 10$ V, $I_{C(on)} = 2$ mA, $R_L = 100\ \Omega$, See Test Circuit A of Figure 1		5	10	µs
t_f	Fall Time				5	10	
t_r	Rise Time	Photodiode Operation	$V_{CC} = 10$ V, $I_{C(on)} = 20\ \mu A$, $R_L = 1\ k\Omega$, See Test Circuit B of Figure 1		1		µs
t_f	Fall Time				1		

PARAMETER MEASUREMENT INFORMATION

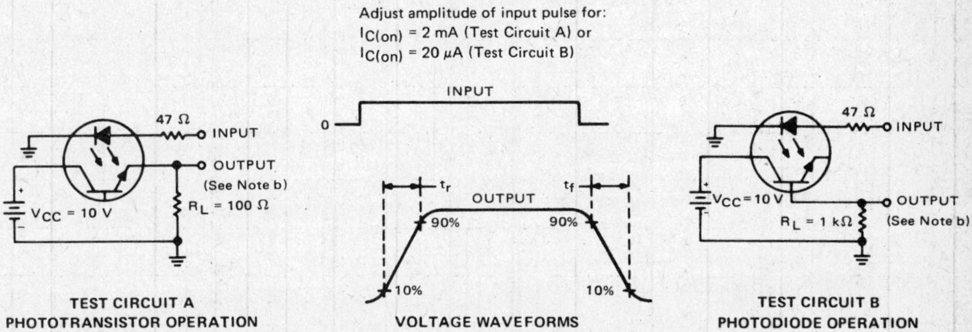

NOTES: a. The input waveform is supplied by a generator with the following characteristics: Z_{out} = 50 Ω, $t_r \leqslant$ 15 ns, duty cycle ≈ 1%, t_w = 100 μs.

b. The output waveform is monitored on an oscilloscope with the following characteristics: $t_r \leqslant$ 12 ns, $R_{in} \geqslant$ 1 MΩ, $C_{in} \leqslant$ 20 pF.

FIGURE 1–SWITCHING TIMES

TYPICAL CHARACTERISTICS

COLLECTOR CURRENT
vs
INPUT-DIODE FORWARD CURRENT

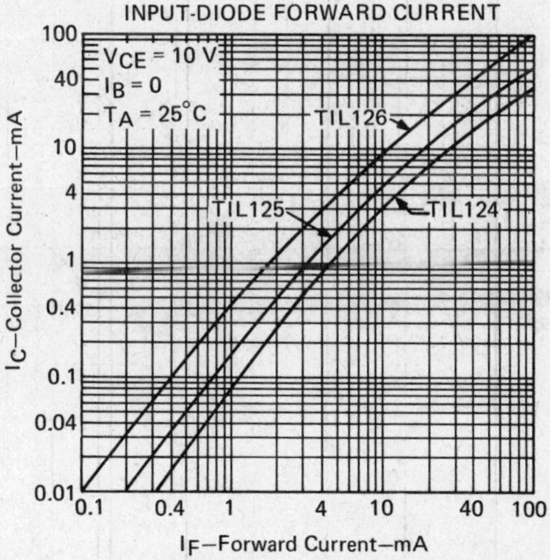

FIGURE 2

TYPICAL CHARACTERISTICS

TIL124

COLLECTOR CURRENT
vs
COLLECTOR-EMITTER VOLTAGE

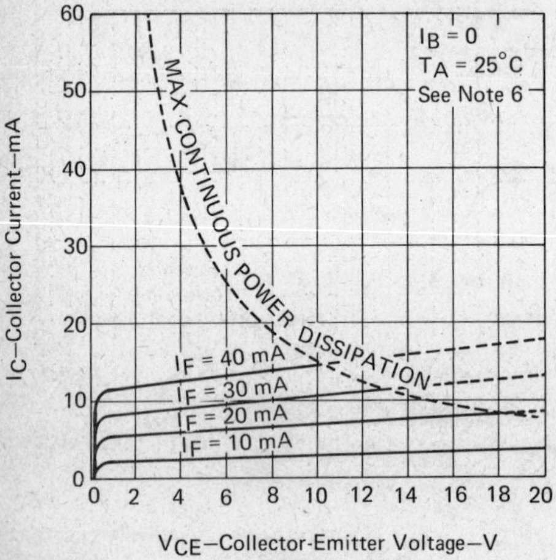

FIGURE 3

TIL125

COLLECTOR CURRENT
vs
COLLECTOR-EMITTER VOLTAGE

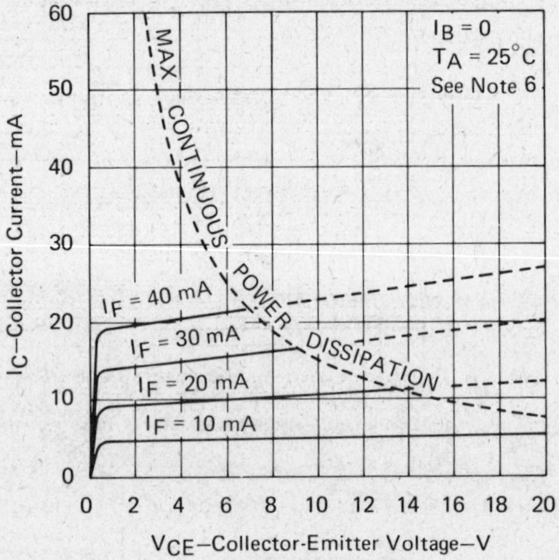

FIGURE 4

TIL126

COLLECTOR CURRENT
vs
COLLECTOR-EMITTER VOLTAGE

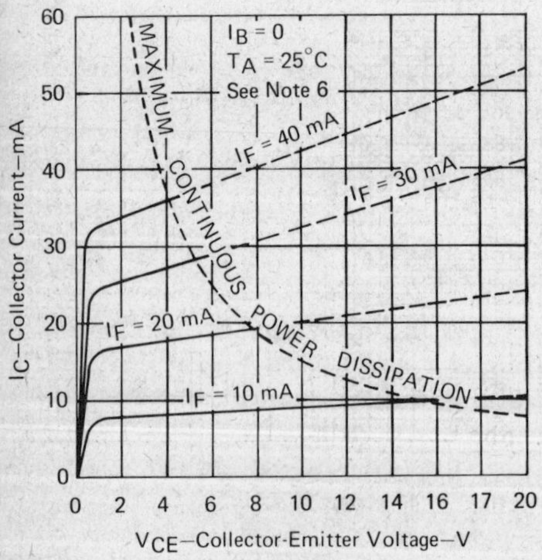

FIGURE 5

RELATIVE ON-STATE COLLECTOR CURRENT
vs
FREE-AIR TEMPERATURE

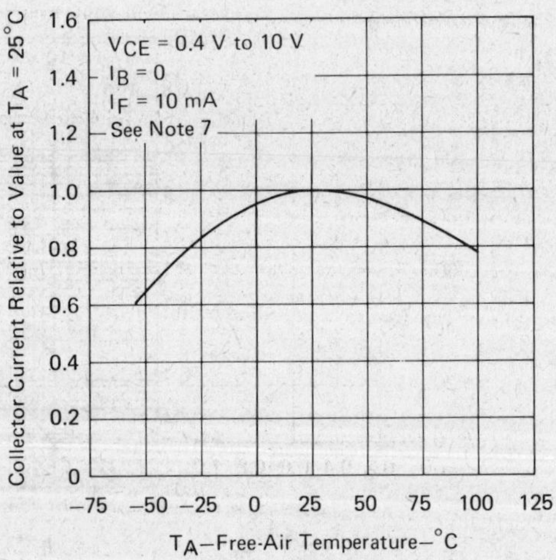

FIGURE 6

NOTES: 6. Pulse operation of input diode is required for operation beyond limits shown by dotted lines.

7. These parameters were measured using pulse techniques. t_w = 1 ms, duty cycle ≤ 2%.

TEXAS INSTRUMENTS
INCORPORATED
POST OFFICE BOX 225012 • DALLAS, TEXAS 75265

TYPICAL CHARACTERISTICS

OFF-STATE COLLECTOR CURRENT
vs
FREE-AIR TEMPERATURE

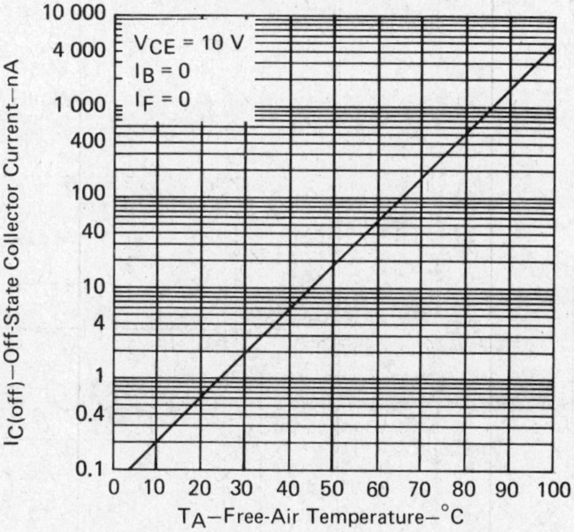

FIGURE 7

NORMALIZED TRANSISTOR STATIC FORWARD
CURRENT TRANSFER RATIO
vs
ON-STATE COLLECTOR CURRENT

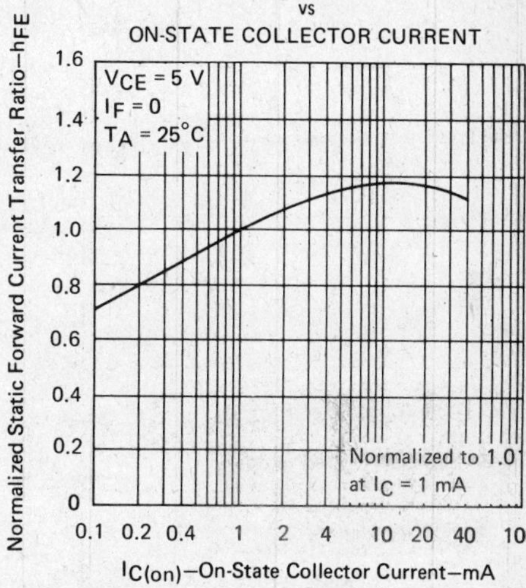

FIGURE 8

INPUT DIODE FORWARD
CONDUCTION CHARACTERISTICS

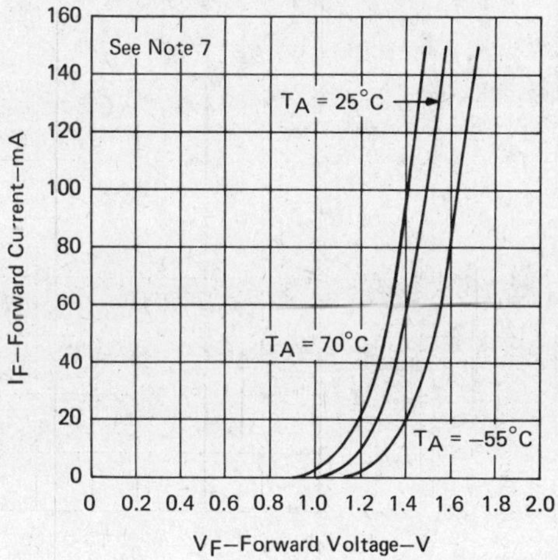

FIGURE 9

COLLECTOR CURRENT
vs
MODULATION FREQUENCY

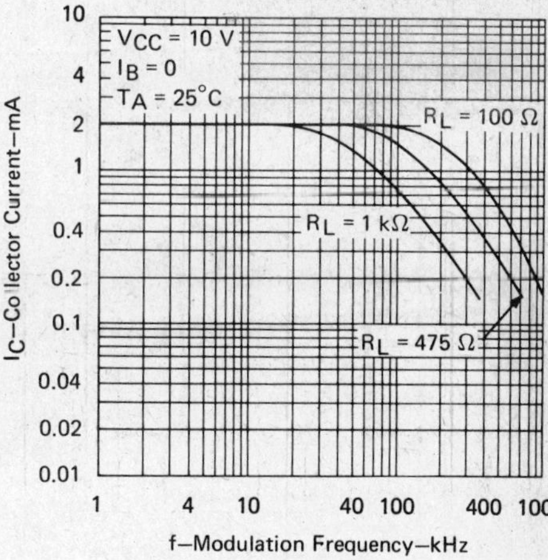

FIGURE 10

NOTE 7: These parameters were measured using pulse techniques. t_w = 1 ms, duty cycle ≤ 2%.

7

OPTOCOUPLERS

D2328, MAY 1977—REVISED DECEMBER 1982

- **Gallium Arsenide Diode Infrared Source Optically Coupled to a Silicon N-P-N Darlington-Connected Phototransistor**
- **High Direct-Current Transfer Ratio . . . 300% Minimum at 10 mA**
- **High-Voltage Electrical Isolation . . . 5000-Volt Rating**
- **Plastic Dual-In-Line Package**
- **Typical Applications Include Remote Terminal Isolation, SCR and Triac Triggers, Mechanical Relays, and Pulse Transformers**
- **No Base Connection on TIL128A for Environments with High Electromagnetic Interference**

mechanical data

The package consists of a gallium arsenide infrared-emitting diode and an n-p-n silicon phototransistor mounted on a 6-lead frame encapsulated within an electrically nonconductive plastic compound. The case will withstand soldering temperature with no deformation,and device performance characteristics remain stable when operated in high-humidity conditions. Unit weight is approximately 0.52 grams.

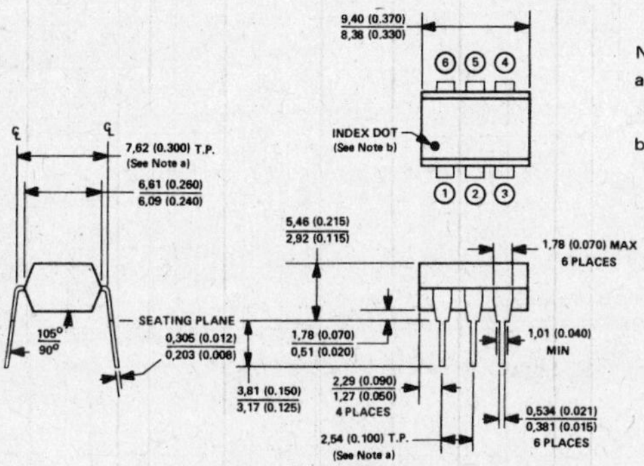

NOTES:
a. Leads are within 0,13 mm (0.005 inch) radius of true position (T.P.) with maximum material condition and unit installed.
b. Pin 1 identified by index dot.

Terminal connections:
1. Anode
2. Cathode } Infrared-emitting diode
3. No internal connection
4. Emitter
5. Collector
6. TIL127: Base
   TIL128: Make no external connection
   TIL128A: No internal connection } Phototransistor

ALL LINEAR DIMENSIONS ARE IN MILLIMETERS AND PARENTHETICALLY IN INCHES.
FALLS WITHIN JEDEC MO-001AM DIMENSIONS

7
OPTOCOUPLERS

absolute maximum ratings at 25°C free-air temperature (unless otherwise noted)

Input-to-Output Voltage	±5 kV
Collector-Base Voltage (TIL127)	30 V
Collector-Emitter Voltage (See Note 1)	30 V
Emitter-Collector Voltage	7 V
Emitter-Base Voltage (TIL127)	7 V
Input-Diode Reverse Voltage	3 V
Input-Diode Continuous Forward Current	100 mA
Continuous Power Dissipation at (or below) 25°C Free-Air Temperature:	
Infrared-Emitting Diode (See Note 2)	150 mW
Phototransistor (See Note 3)	150 mW
Total (Infrared-Emitting Diode plus Phototransistor, See Note 4)	250 mW
Storage Temperature Range	−55°C to 150°C
Lead Temperature 1,6 mm (1/16 Inch) from Case for 10 Seconds	260°C

NOTES: 1. This value applies when the base-emitter diode is open-circuited.
2. Derate linearly to 100°C free-air temperature at the rate of 2 mW/°C.
3. Derate linearly to 100°C free-air temperature at the rate of 2 mW/°C.
4. Derate linearly to 100°C free-air temperature at the rate of 3.33 mW/°C.

Copyright © 1982 by Texas Instruments Incorporated

82

electrical characteristics at 25°C free-air temperature

PARAMETER		TEST CONDITIONS†	TIL127 MIN	TIL127 TYP	TIL127 MAX	TIL128, TIL128A MIN	TIL128, TIL128A TYP	TIL128, TIL128A MAX	UNIT
$V_{(BR)CBO}$	Collector-Base Breakdown Voltage	$I_C = 10\ \mu A$, $I_E = 0$, $I_F = 0$	30						V
$V_{(BR)CEO}$	Collector-Emitter Breakdown Voltage	$I_C = 1$ mA, $I_B = 0$, $I_F = 0$	30			30			V
$V_{(BR)EBO}$	Emitter-Base Breakdown Voltage	$I_E = 10\ \mu A$, $I_C = 0$, $I_F = 0$	7						V
$V_{(BR)ECO}$	Emitter-Collector Breakdown Voltage	$I_E = 10\ \mu A$, $I_F = 0$				7			V
I_R	Input Diode Static Reverse Current	$V_R = 3$ V			10			10	µA
$I_{C(on)}$	On-State Collector Current	$V_{CE} = 1$ V, $I_B = 0$, $I_F = 10$ mA	30	100					mA
		$V_{CE} = 2$ V, $I_F = 10$ mA				30	160		
$I_{C(off)}$	Off-State Collector Current	$V_{CE} = 10$ V, $I_B = 0$, $I_F = 0$			100			100	nA
h_{FE}	Transistor Static Forward Current Transfer Ratio	$V_{CE} = 1$ V, I_C 10 mA, $I_F = 0$		15 000					
V_F	Input Diode Static Forward Voltage	$I_F = 10$ mA			1.5			1.5	V
$V_{CE(sat)}$	Collector-Emitter Saturation Voltage	I_C 125 mA, $I_B = 0$, $I_F = 50$ mA			1.2				V
		$I_C = 10$ mA, $I_F = 10$ mA						1	
r_{IO}	Input-to-Output Internal Resistance	$V_{in-out} = 500$ V, See Note 5	10^{11}			10^{11}			Ω
C_{io}	Input-to-Output Capacitance	$V_{in-out} = 0$, f = 1 MHz, See Note 5		1	1.3		1	1.3	pF

Note 5: These parameters are measured between both input-diode leads shorted together and all the phototransistor leads shorted together.
†References to the base are not applicable to the TIL128 or TIL128A.

switching characteristics at 25°C free-air temperature

PARAMETER		TEST CONDITIONS†	TIL127 MIN	TIL127 TYP	TIL127 MAX	TIL128, TIL128A MIN	TIL128, TIL128A TYP	TIL128, TIL128A MAX	UNIT
t_r	Rise Time	$V_{CC} = 15$ V, $I_{C(on)} = 125$ mA,		300					µs
t_f	Fall Time	$R_L = 100\ \Omega$, See Figure 1		300					
t_r	Rise Time	$V_{CC} = 10$ V, $I_{C(on)} = 2.5$ mA,					300		µs
t_f	Fall Time	$R_L = 100\ \Omega$, See Figure 1					300		

PARAMETER MEASUREMENT INFORMATION

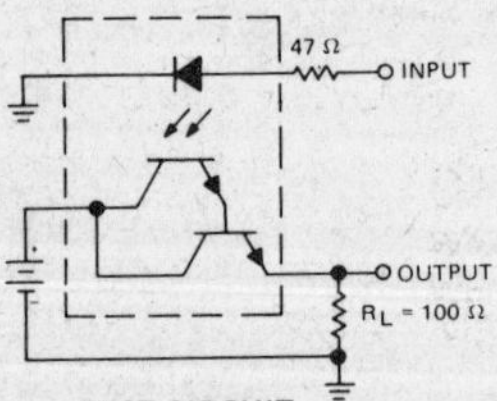

TEST CIRCUIT

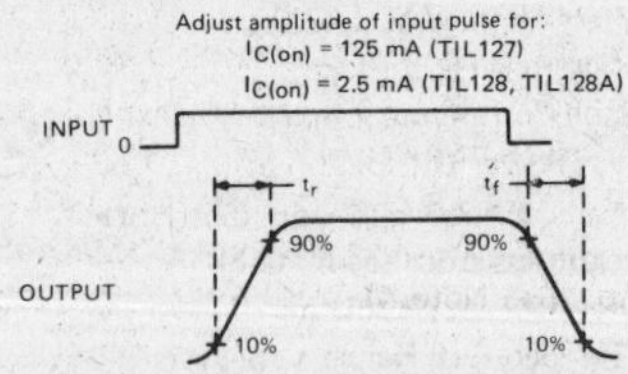

VOLTAGE WAVEFORMS

NOTES: a. The input waveform is supplied by a generator with the following characteristics: $Z_{out} = 50\ \Omega$, $t_r \leq 15$ ns, duty cycle ≈ 1%, $t_w = 500\ \mu s$.
b. The output waveform is monitored on an oscilloscope with the following characteristics: $t_r \leq 12$ ns, $R_{in} \geq 1$ MΩ, $C_{in} \leq 20$ pF.

FIGURE 1–SWITCHING TIMES

TEXAS INSTRUMENTS
INCORPORATED
POST OFFICE BOX 225012 • DALLAS, TEXAS 75265

TYPICAL CHARACTERISTICS

COLLECTOR CURRENT
vs
COLLECTOR-EMITTER VOLTAGE

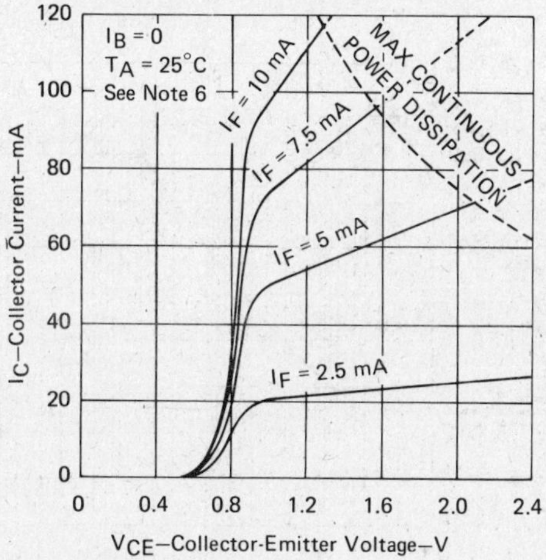

FIGURE 2

COLLECTOR CURRENT
vs
COLLECTOR-EMITTER VOLTAGE

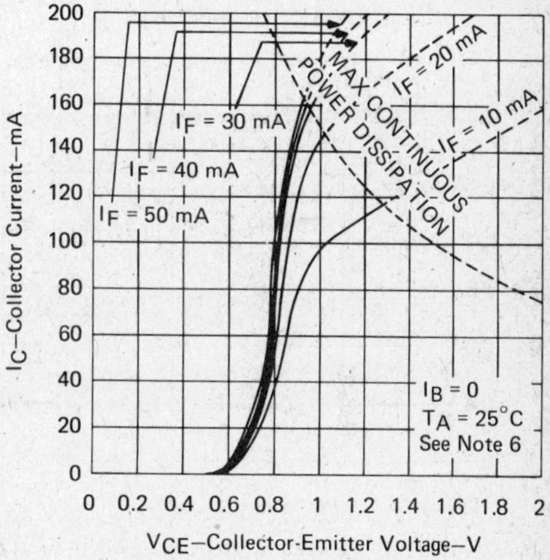

FIGURE 3

COLLECTOR CURRENT
vs
INPUT-DIODE FORWARD CURRENT

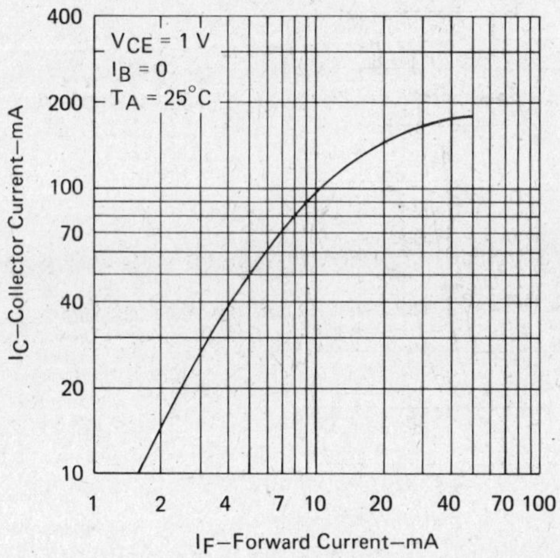

FIGURE 4

OFF-STATE COLLECTOR CURRENT
vs
FREE-AIR TEMPERATURE

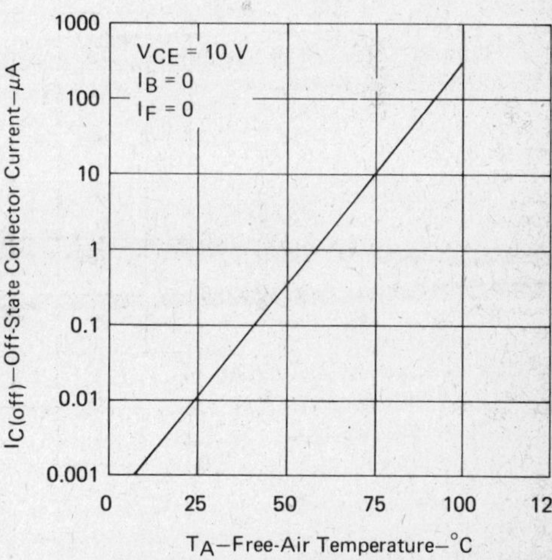

FIGURE 5

NOTE 6: Pulse operation of input diode is required for operation beyond limits shown by dotted line.

7 OPTOCOUPLERS

TYPICAL CHARACTERISTICS

RELATIVE COLLECTOR-EMITTER SATURATION VOLTAGE vs FREE-AIR TEMPERATURE

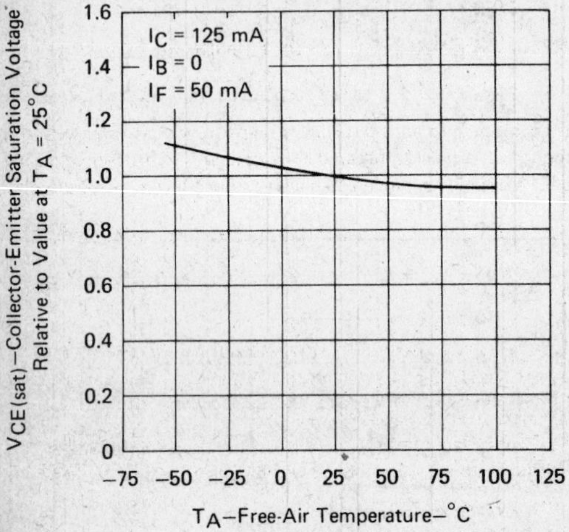

FIGURE 6

TIL127
TRANSISTOR STATIC FORWARD CURRENT TRANSFER RATIO vs COLLECTOR CURRENT

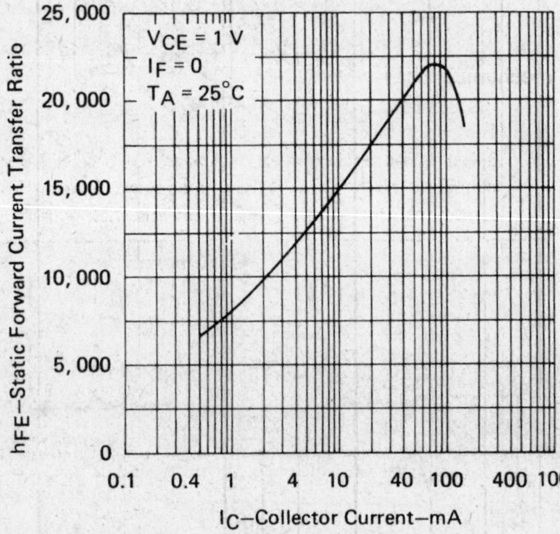

FIGURE 7

INPUT DIODE FORWARD CONDUCTION CHARACTERISTICS

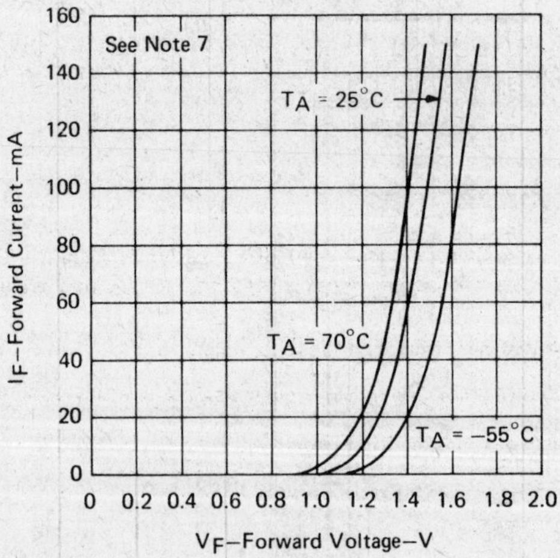

FIGURE 8

NOTE 7: This parameter was measured using pulse techniques. t_w = 1 ms, duty cycle ≤ 2%.

TEXAS INSTRUMENTS
INCORPORATED
POST OFFICE BOX 225012 • DALLAS, TEXAS 75265

D2491, SEPTEMBER–REVISED DECEMBER 1982

UL LISTED – FILE # E65085

- GaAs-Diode Infrared Source Optically Coupled to a Silicon N-P-N Phototransistor
- Direct-Current Transfer Ratio . . . 10% to 50%
- Plug-In Replacements for TIL111 Series
- High-Voltage Electrical Isolation . . . 2500 V RMS (3535 V Peak)

mechanical data

The package consists of a gallium arsenide infrared-emitting diode and an n-p-n silicon phototransistor mounted on a 6-lead frame encapsulated within an electrically nonconductive plastic compound. The case will withstand soldering temperature with no deformation and device performance characteristics remain stable when operated in high-humidity conditions. Unit weight is approximately 0.52 grams.

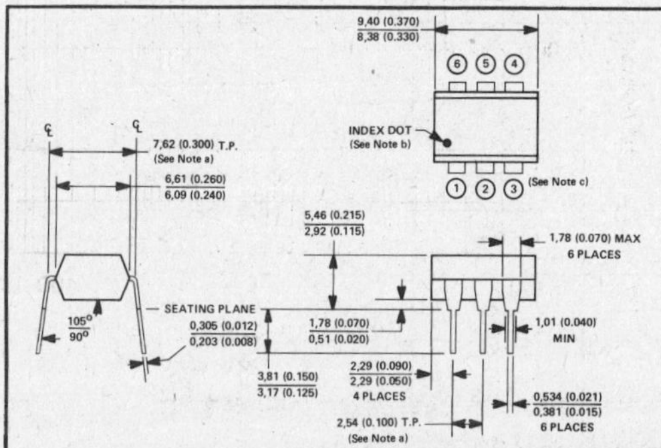

NOTES:
a. Leads are within 0,13 mm (0.005 inch) radius of true position (T.P.) with maximum material condition and unit installed.
b. Pin 1 identified by index dot.

Terminal connections:
1. Anode } Infrared-emitting diode
2. Cathode
3. No internal connection
4. Emitter } Phototransistor
5. Collector
6. Base

FALLS WITHIN JEDEC MO-001AM DIMENSIONS
ALL LINEAR DIMENSIONS ARE IN MILLIMETERS AND PARENTHETICALLY IN INCHES

absolute maximum ratings at 25°C free-air temperature (unless otherwise noted)

Input-to-Output RMS Voltage (See Note 1)	2500 V
Collector-Base Voltage	70 V
Collector-Emitter Voltage (See Note 2)	30 V
Emitter-Collector Voltage	7 V
Emitter-Base Voltage	7 V
Input-Diode Reverse Voltage	3 V
Input-Diode Continuous Forward Current at (or below) 25°C Free-Air Temperature (See Note 3)	100 mA
Continuous Phototransistor Power Dissipation at (or below) 25°C Free-Air Temperature (See Note 4)	150 mW
Storage Temperature Range	−55°C to 150°C
Lead Temperature 1,6 mm (1/16 inch) from Case for 10 Seconds	260°C

NOTES: 1. This rating applies for sine-wave operation at 50 or 60 Hz. Service capability is verified by testing in accordance with UL requirements.
2. This value applies when the base-emitter diode is open-circuited.
3. Derate linearly to 100°C free-air temperature at the rate of 1.33 mA/°C.
4. Derate linearly to 100°C free-air temperature at the rate of 2 mW/°C.

Copyright © 1982 by Texas Instruments Incorporated

282

electrical characteristics at 25°C free-air temperature

PARAMETER			TEST CONDITIONS		TIL153			TIL154			TIL155			UNIT
					MIN	TYP	MAX	MIN	TYP	MAX	MIN	TYP	MAX	
$V_{(BR)CBO}$	Collector-Base Breakdown Voltage		$I_C = 10\ \mu A$, $I_F = 0$	$I_E = 0$,	70			70			70			V
$V_{(BR)CEO}$	Collector-Emitter Breakdown Voltage		$I_C = 1$ mA, $I_F = 0$	$I_B = 0$,	30			30			30			V
$V_{(BR)EBO}$	Emitter-Base Breakdown Voltage		$I_E = 10\ \mu A$, $I_F = 0$	$I_C = 0$,	7			7			7			V
I_R	Input Diode Static Reverse Current		$V_R = 3$ V				10			10			10	μA
$I_{C(on)}$	On-State Collector Current	Phototransistor Operation	$V_{CE} = 10$ V, $I_B = 0$	$I_F = 10$ mA,	1	3		2	5		5	9		mA
		Photodiode Operation	$V_{CB} = 10$ V, $I_E = 0$	$I_F = 10$ mA,		10			10			10		μA
$I_{C(off)}$	Off-State Collector Current	Phototransistor Operation	$V_{CE} = 10$ V, $I_B = 0$	$I_F = 0$,		1	50		1	50		1	50	nA
		Photodiode Operation	$V_{CB} = 10$ V, $I_E = 0$	$I_F = 0$,		0.1	20		0.1	20		0.1	20	nA
h_{FE}	Transistor Static Forward Current Transfer Ratio		$V_{CE} = 5$ V, $I_F = 0$	$I_C = 10$ mA,	50	100		100	200		100	550		
V_F	Input Diode Static Forward Voltage		$I_F = 10$ mA			1.2	1.4		1.2	1.4		1.2	1.4	V
$V_{CE(sat)}$	Collector-Emitter Saturation Voltage		$I_C = 1$ mA, $I_B = 0$	$I_F = 10$ mA,		0.25	0.4		0.25	0.4		0.25	0.4	V
r_{IO}	Input-to-Output Internal Resistance		$V_{in\text{-}out} = 500$ V, See Note 5		10^{11}			10^{11}			10^{11}			Ω
C_{io}	Input-to-Output Capacitance		$V_{in\text{-}out} = 0$, See Note 5	f = 1 MHz,		1	1.3		1	1.3		1	1.3	pF

NOTE 5: These parameters are measured between both input diode leads shorted together and all the phototransistor leads shorted together.

switching characteristics at 25°C free-air temperature

PARAMETER		TEST CONDITIONS			MIN	TYP	MAX	UNIT
t_r Rise Time	Phototransistor Operation	$V_{CC} = 10$ V, See Test Circuit A of Figure 1	$I_{C(on)} = 2$ mA,	$R_L = 100\ \Omega$,		5	10	μs
t_f Fall Time						5	10	μs
t_r Rise Time	Photodiode Operation	$V_{CC} = 10$ V, See Test Circuit B of Figure 1	$I_{C(on)} = 20\ \mu A$,	$R_L = 1\ k\Omega$,		1		μs
t_f Fall Time						1		μs

7 OPTOCOUPLERS

TEXAS INSTRUMENTS
INCORPORATED
POST OFFICE BOX 225012 • DALLAS, TEXAS 75265

12

PARAMETER MEASUREMENT INFORMATION

Adjust amplitude of input pulse for:
$I_{C(on)}$ = 2 mA (Test Circuit A) or
$I_{C(on)}$ = 20 μA (Test Circuit B)

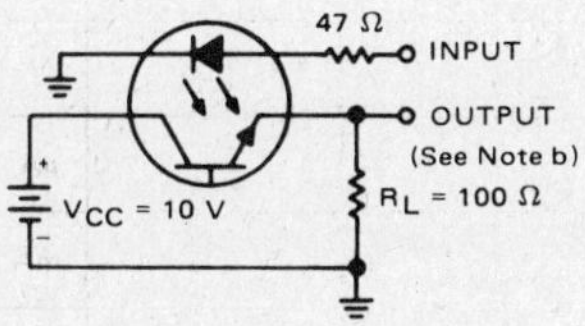

TEST CIRCUIT A
PHOTOTRANSISTOR OPERATION

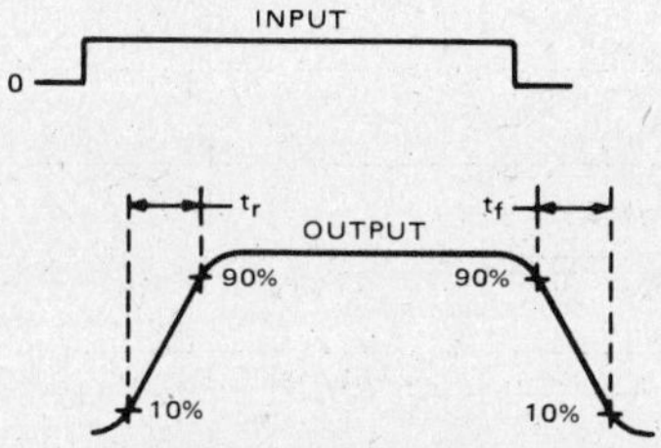

VOLTAGE WAVEFORMS

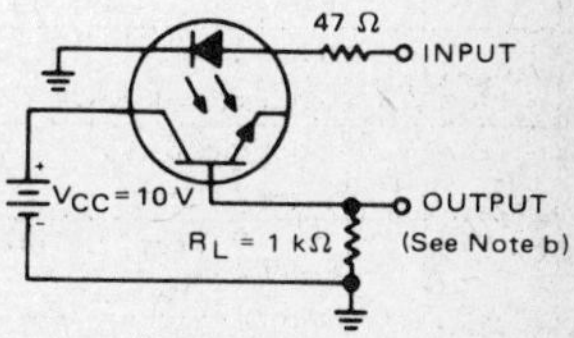

TEST CIRCUIT B
PHOTODIODE OPERATION

NOTES: a. The input waveform is supplied by a generator with the following characteristics: Z_{out} = 50 Ω, $t_r \leq$ 15 ns, duty cycle ≈ 1%, t_w = 100 μs.
b. The output waveform is monitored on an oscilloscope with the following characteristics: $t_r \leq$ 12 ns, $R_{in} \geq$ 1 MΩ, $C_{in} \leq$ 20 pF.

FIGURE 1—SWITCHING TIMES

TYPICAL CHARACTERISTICS

COLLECTOR CURRENT
vs
INPUT-DIODE FORWARD CURRENT

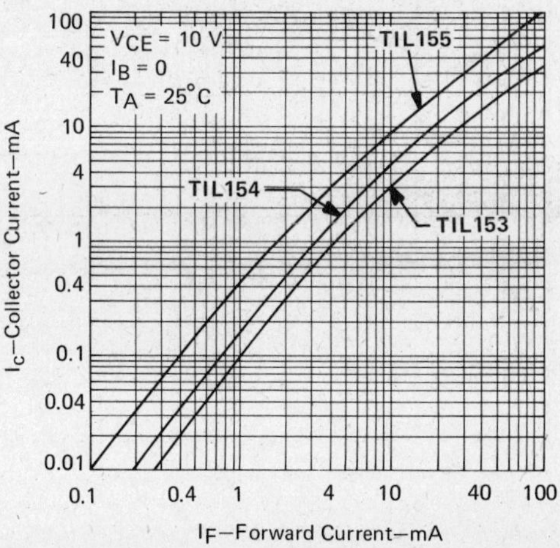

FIGURE 2

7
OPTOCOUPLERS

TYPICAL CHARACTERISTICS

TIL153

COLLECTOR CURRENT
vs
COLLECTOR-EMITTER VOLTAGE

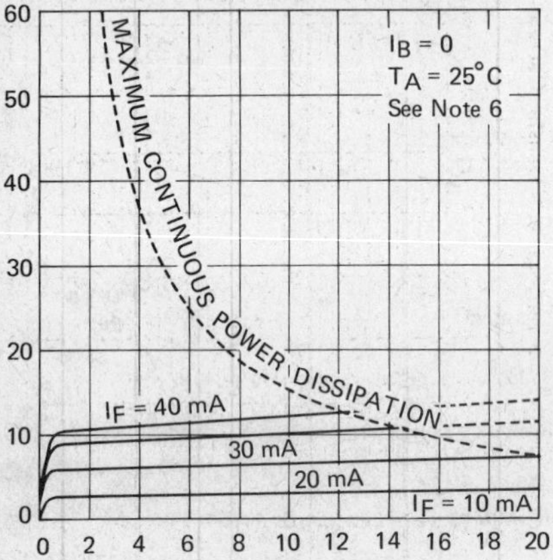

FIGURE 3

TIL154

COLLECTOR CURRENT
vs
COLLECTOR-EMITTER VOLTAGE

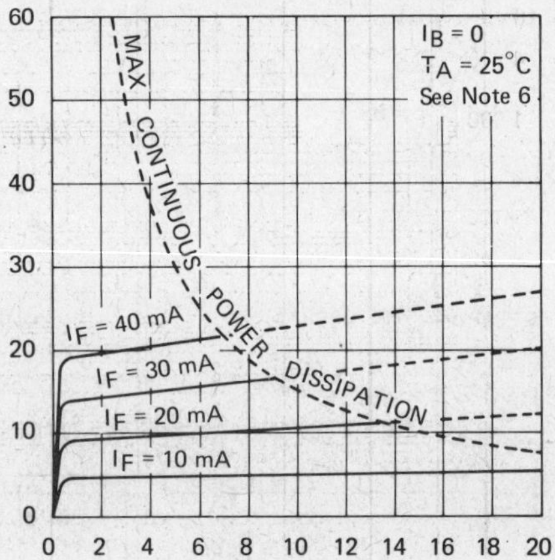

FIGURE 4

TIL155

COLLECTOR CURRENT
vs
COLLECTOR-EMITTER VOLTAGE

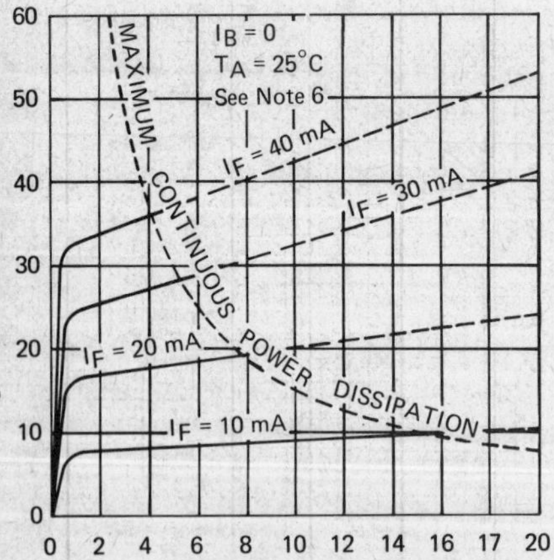

FIGURE 5

RELATIVE ON-STATE COLLECTOR CURRENT
vs
FREE-AIR TEMPERATURE

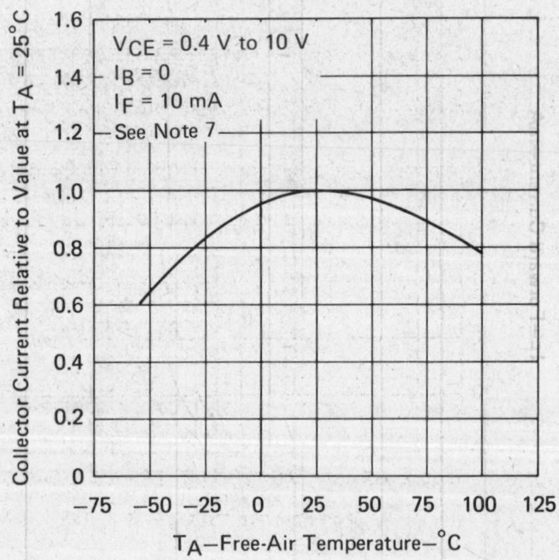

FIGURE 6

NOTES: 6. Pulse operation of input diode is required for operation beyond limits shown by dotted lines.
7. These parameters were measured using pulse techniques. t_w = 1 ms, duty cycle ≤ 2%.

TEXAS INSTRUMENTS
INCORPORATED
POST OFFICE BOX 225012 • DALLAS, TEXAS 75265

TYPICAL CHARACTERISTICS

OFF-STATE COLLECTOR CURRENT
vs
FREE-AIR TEMPERATURE

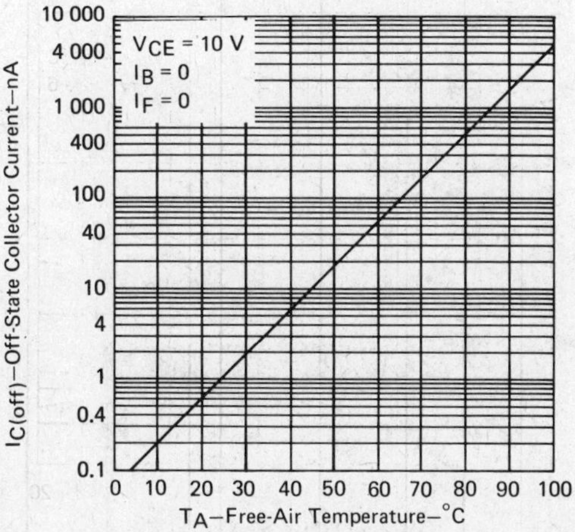

FIGURE 7

NORMALIZED TRANSISTOR STATIC FORWARD
CURRENT TRANSFER RATIO
vs
ON-STATE COLLECTOR CURRENT

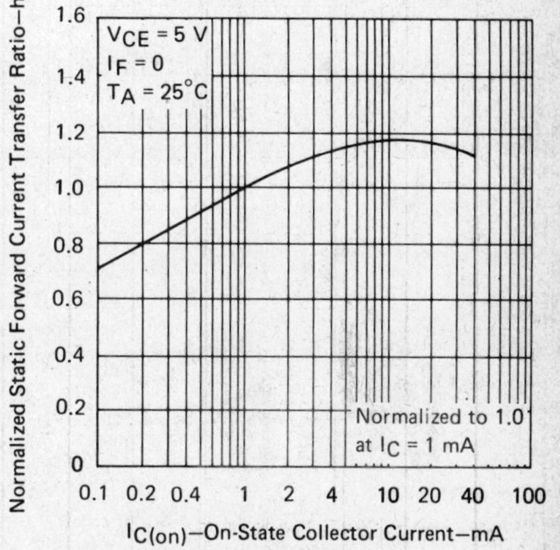

FIGURE 8

INPUT DIODE FORWARD
CONDUCTION CHARACTERISTICS

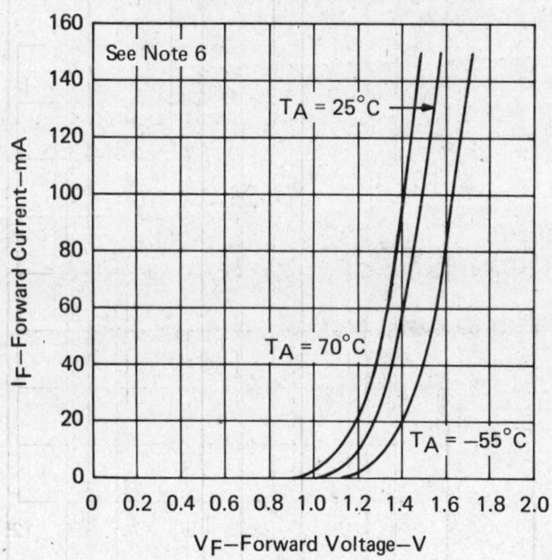

FIGURE 9

COLLECTOR CURRENT
vs
MODULATION FREQUENCY

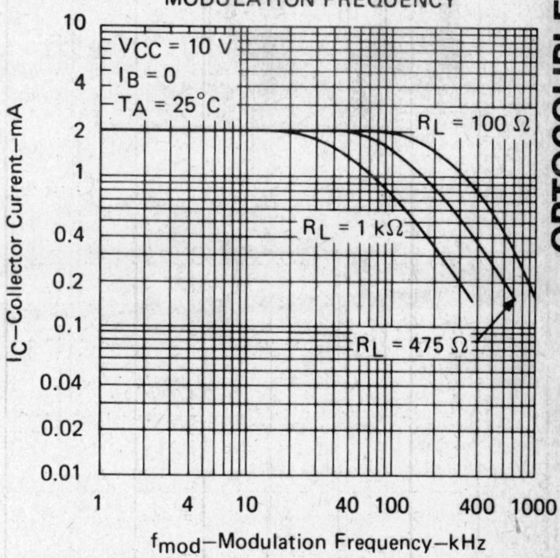

FIGURE 10

NOTE 6: These parameters were measured using pulse techniques. t_w = 1 ms, duty cycle ⩽ 2%

D2492, SEPTEMBER 1978–REVISED FEBRUARY 1983

UL LISTED – FILE #E65085

- GaAs-Diode Light Source Optically Coupled to a Silicon N-P-N Darlington-Connected Phototransistor
- High Direct-Current Transfer Ratio . . . 300% Minimum at 10 mA
- Plug-In Replacement for TIL113, TIL119, and TIL119A
- High-Voltage Electrical Isolation . . . 2500 V RMS (3535 V Peak)
- No Base Connection on TIL157A for Environments with High Electromagnetic Interference

mechanical data

The package consists of a gallium arsenide infrared-emitting diode and an n-p-n silicon darlington-connected phototransistor mounted on a 6-lead frame encapsulated within an electrically nonconductive plastic compound. The case will withstand soldering temperature with no deformation and device performance characteristics remain stable when operated in high humidity conditions. Unit weight is approximately 0.52 grams.

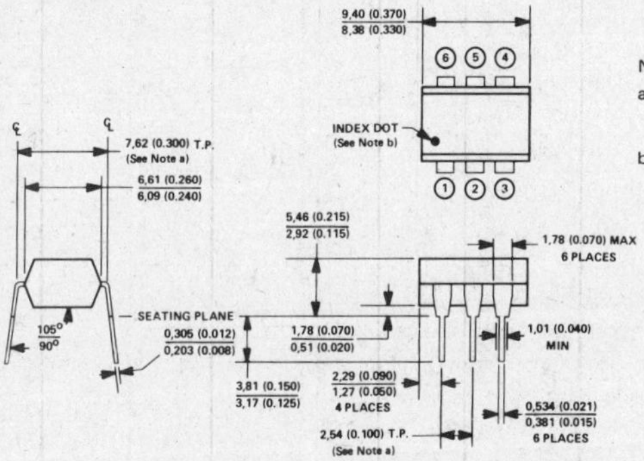

NOTES:
a. Leads are within 0,13 mm (0.005 inch) radius of true position (T.P.) with maximum material condition and unit installed.
b. Pin 1 identified by index dot.

Terminal connections:
1. Anode
2. Cathode } Infrared-emitting diode
3. No internal connection
4. Emitter
5. Collector
6. TIL156 Base
TIL157 Base or no internal connection at the option of TI
TIL157A no internal connection } Phototransistor

FALLS WITHIN JEDEC MO-001AM DIMENSIONS
ALL LINEAR DIMENSIONS ARE IN MILLIMETERS AND PARENTHETICALLY IN INCHES.

absolute maximum ratings at 25°C free-air temperature (unless otherwise noted)

Input-to-Output RMS Voltage (See Note 1)	2500 V
Collector-Base Voltage (TIL156)	30 V
Collector-Emitter Voltage (See Note 2)	30 V
Emitter-Collector Voltage	7 V
Emitter-Base Voltage (TIL156)	7 V
Input-Diode Reverse Voltage	3 V
Input-Diode Continuous Forward Current at (or below) 25°C Free-Air Temperature (See Note 3)	100 mA
Continuous Phototransistor Power Dissipation at (or below) 25°C Free-Air Temperature (See Note 4)	150 mW
Storage Temperature Range	−55°C to 150°C
Lead Temperature 1,6 mm (1/16 inch) from Case for 10 Seconds	260°C

NOTES: 1. This rating applies for sine-wave operation at 50 or 60 Hz. Service capability is verified by testing in accordance with UL requirements.
2. This value applies when the base-emitter diode is open-circuited.
3. Derate linearly to 100°C free-air temperature at the rate of 1.33 mA/°C.
4. Derate linearly to 100°C free-air temperature at the rate of 2 mW/°C.

Copyright © 1983 by Texas Instruments Incorporated

TEXAS INSTRUMENTS
INCORPORATED
POST OFFICE BOX 5012 • DALLAS, TEXAS 75222

electrical characteristics at 25°C free-air temperature

PARAMETER		TEST CONDITIONS†			TIL156 MIN	TIL156 TYP	TIL156 MAX	TIL157, TIL157A MIN	TIL157, TIL157A TYP	TIL157, TIL157A MAX	UNIT
$V_{(BR)CBO}$	Collector-Base Breakdown Voltage	$I_C = 10\ \mu A$,	$I_E = 0$,	$I_F = 0$	30						V
$V_{(BR)CEO}$	Collector-Emitter Breakdown Voltage	$I_C = 1$ mA,	$I_B = 0$,	$I_F = 0$	30			30			V
$V_{(BR)EBO}$	Emitter-Base Breakdown Voltage	$I_E = 10\ \mu A$,	$I_C = 0$,	$I_F = 0$	7						V
$V_{(BR)ECO}$	Emitter-Collector Breakdown Voltage	$I_E = 10\ \mu A$,	$I_F = 0$						7		V
I_R	Input Diode Static Reverse Current	$V_R = 3$ V					10			10	μA
$I_{C(on)}$	On-State Collector Current	$V_{CE} = 1$ V,	$I_B = 0$,	$I_F = 10$ mA	30	100					mA
		$V_{CE} = 2$ V,	$I_F = 10$ mA					30	160		
$I_{C(off)}$	Off-State Collector Current	$V_{CE} = 10$ V,	$I_B = 0$,	$I_F = 0$			100			100	nA
h_{FE}	Transistor Static Forward Current Transfer Ratio	$V_{CE} = 1$ V,	I_C 10 mA,	$I_F = 0$		15 000					
V_F	Input Diode Static Forward Voltage	$I_F = 10$ mA					1.5			1.5	V
$V_{CE(sat)}$	Collector-Emitter Saturation Voltage	I_C 125 mA,	$I_B = 0$,	$I_F = 50$ mA			1.2				V
		$I_C = 10$ mA,	$I_F = 10$ mA							1	
r_{IO}	Input-to-Output Internal Resistance	$V_{in\text{-}out} = 500$ V,	See Note 5		10^{11}			10^{11}			Ω
C_{io}	Input-to-Output Capacitance	$V_{in\text{-}out} = 0$,	f = 1 MHz,	See Note 5		1	1.3		1	1.3	pF

Note 5: These parameters are measured between both input-diode leads shorted together and all the phototransistor leads shorted together.
†References to the base are not applicable to the TIL157 or TIL157A.

switching characteristics at 25°C free-air temperature

PARAMETER		TEST CONDITIONS†		TIL156 MIN	TIL156 TYP	TIL156 MAX	TIL157, TIL157A MIN	TIL157, TIL157A TYP	TIL157, TIL157A MAX	UNIT
t_r	Rise Time	$V_{CC} = 15$ V,	$I_{C(on)} = 125$ mA,		300					μs
t_f	Fall Time	$R_L = 100\ \Omega$,	See Figure 1		300					
t_r	Rise Time	$V_{CC} = 10$ V,	$I_{C(on)} = 2.5$ mA,					300		μs
t_f	Fall Time	$R_L = 100\ \Omega$,	See Figure 1					300		

PARAMETER MEASUREMENT INFORMATION

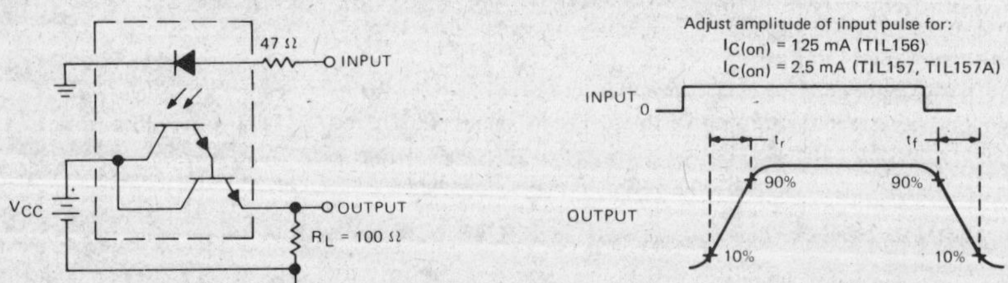

TEST CIRCUIT

VOLTAGE WAVEFORMS

NOTES: a. The input waveform is supplied by a generator with the following characteristics: $Z_{out} = 50\ \Omega$, $t_r \leq 15$ ns, duty cycle ≈ 1%, $t_w = 500\ \mu s$.
b. The output waveform is monitored on an oscilloscope with the following characteristics: $t_r \leq 12$ ns, $R_{in} \geq 1\ M\Omega$, $C_{in} \leq 20$ pF.

FIGURE 1—SWITCHING TIMES

TYPICAL CHARACTERISTICS

COLLECTOR CURRENT
vs
COLLECTOR-EMITTER VOLTAGE

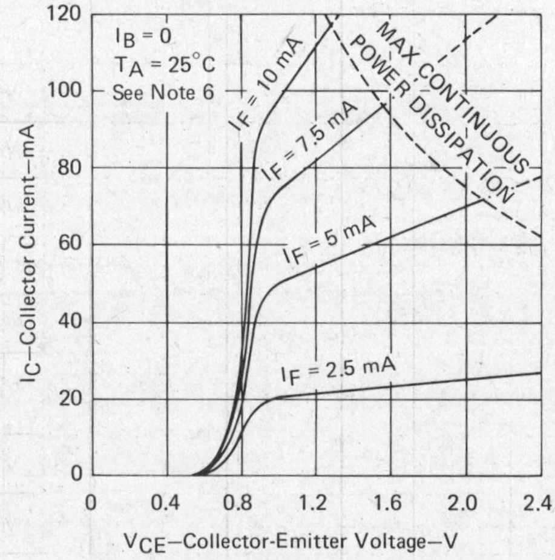

FIGURE 2

COLLECTOR CURRENT
vs
COLLECTOR-EMITTER VOLTAGE

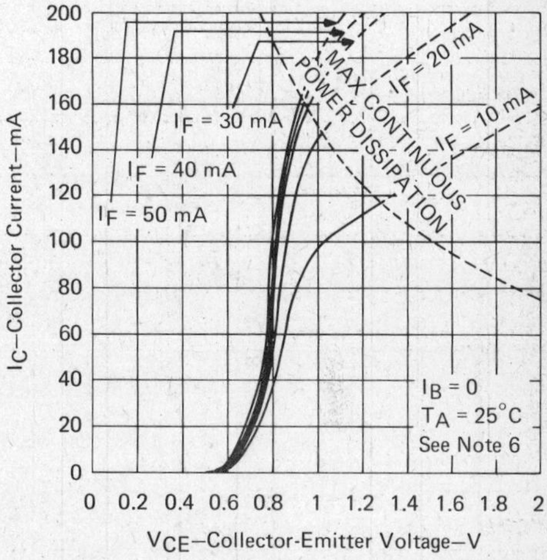

FIGURE 3

COLLECTOR CURRENT
vs
INPUT-DIODE FORWARD CURRENT

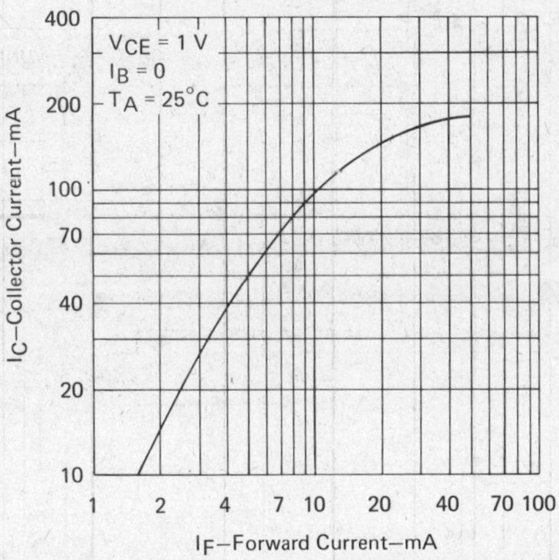

FIGURE 4

OFF-STATE COLLECTOR CURRENT
vs
FREE-AIR TEMPERATURE

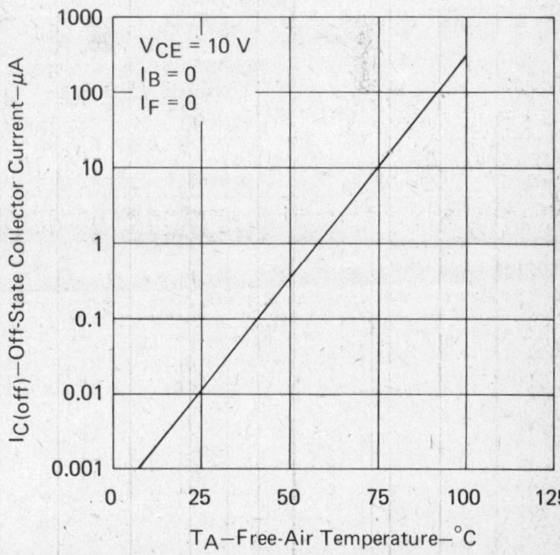

FIGURE 5

NOTE 6: Pulse operation of input diode is required for operation beyond limits shown by dotted line.

TYPICAL CHARACTERISTICS

RELATIVE COLLECTOR-EMITTER SATURATION VOLTAGE vs FREE-AIR TEMPERATURE

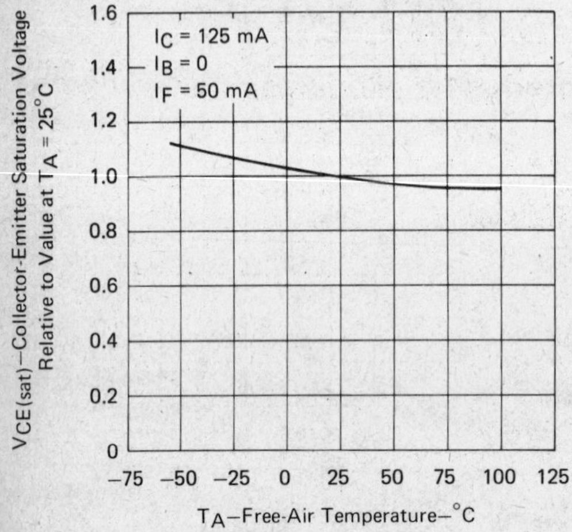

FIGURE 6

TIL156
TRANSISTOR STATIC FORWARD CURRENT TRANSFER RATIO vs COLLECTOR CURRENT

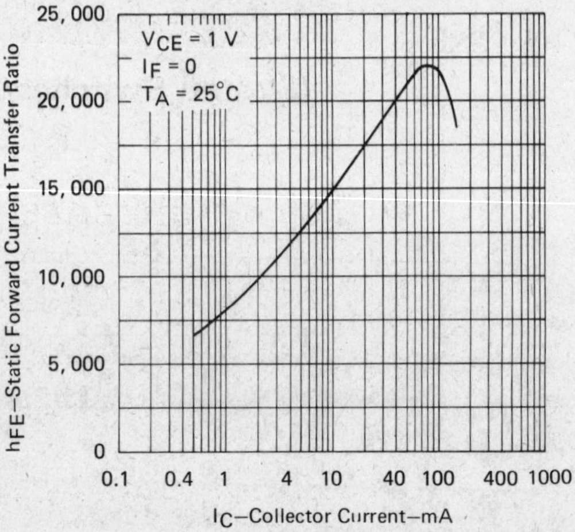

FIGURE 7

INPUT DIODE FORWARD CONDUCTION CHARACTERISTICS

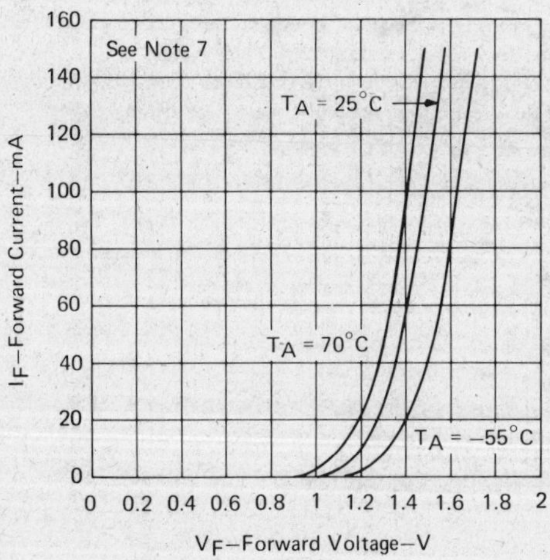

FIGURE 8

NOTE 7: This parameter was measured using pulse techniques. t_w = 1 ms, duty cycle ≤ 2%.

TEXAS INSTRUMENTS
INCORPORATED
POST OFFICE BOX 225012 • DALLAS, TEXAS 75265

Source and Detector Assemblies (SDAs)

(Slotted Switches/Interrupter Modules)

- **Quick Reference Guide**
- **Single-Channel**
  - **Transmissive Designs**
  - **Reflective Designs**
- **Various Packages**
- **Built with Plastic or Hermetic Devices**
- **Bar-Code Read Heads**
- **Custom Designs Available to Meet Specific Needs**

SINGLE-CHANNEL ASSEMBLIES (SWITCHES)
QUICK REFERENCE GUIDE

DEVICE	TYPE	ON-STATE COLLECTOR CURRENT			OFF-STATE COLLECTOR CURRENT		FEATURES§
		MIN $I_{C(on)}$	@ I_F	@ V_{CE}	MAX $I_{C(off)}$	@ V_{CE}	
TIL138	Transmissive Assembly with Mounting Tabs	1.6 mA 0.4 mA	35 mA 15 mA	0.5 V 0.5 V	100 nA	30 V	A TIL32 gallium arsenide IRED and a TIL78 phototransistor
TIL139	Reflective Assembly	10 μA† 100 μA‡	40 mA 40 mA	5 V 5 V	100 nA	30 V	A TIL32 gallium arsenide IRED and a TIL78 phototransistor
TIL143	Transmissive Assembly with Mounting Tabs	600 μA	20 mA	5 V	100 nA	10 V	A TIL40 gallium arsenide IRED and a TIL411 silicon phototransistor
TIL144		200 μA	20 mA	5 V	100 nA	10 V	
TIL145	Transmissive Assembly with Mounting Tabs	2 mA	16 mA	1 V	100 nA	5 V	A TIL40 gallium arsenide IRED and a high-gain TIL412 silicon Darlington phototransistor
TIL146		1.6 mA	50 mA	1 V	100 nA	5 V	
TIL147	Transmissive Assembly	4 mA	20 mA	5 V	100 nA	10 V	Hermetic pill devices mounted in dual-in-line package (TIL23/TIL601 Series)
TIL148		1 mA	20 mA	5 V	100 nA	10 V	
TIL149	Reflective Assembly	25 μA‡	40 mA	5 V	100 nA	15 V	A TIL32 and a TIL78
TIL158	Transmissive Assembly	600 μA	20 mA	5 V	100 nA	10 V	A TIL40 gallium arsenide IRED and a TIL411 silicon phototransistor
TIL159		200 μA	20 mA	5 V	100 nA	10 V	
TIL160	Transmissive Assembly	2 mA	10 mA	2 V	100 nA	5 V	A TIL40 gallium arsenide IRED and a high-gain TIL412 silicon Darlington phototransistor
TIL161		0.5 mA	10 mA	2 V	100 nA	5 V	
TIL167-1	Transmissive Assembly with Mounting Tabs	200 μA	20 mA	5 V	100 nA	10 V	A TIL40 gallium arsenide IRED and a TIL415 silicon phototransistor
TIL167-2		600 μA	20 mA	5 V	100 nA	10 V	
TIL168-1	Transmissive Assembly with Mounting Tabs	0.5 mA	10 mA	2 V	100 nA	5 V	A TIL40 gallium arsenide IRED and a high-gain TIL416 silicon Darlington phototransistor
TIL168-2		2 mA	10 mA	2 V	100 nA	5 V	
TIL169-1	Transmissive Assembly	200 μA	20 mA	5 V	100 nA	10 V	A TIL40 gallium arsenide IRED and a TIL415 silicon phototransistor
TIL169-2		600 μA	20 mA	5 V	100 nA	10 V	
TIL170-1	Transmissive Assembly	0.5 mA	10 mA	2 V	100 nA	5 V	A TIL40 gallium arsenide IRED and a TIL416 silicon Darlington phototransistor
TIL170-2		2 mA	10 mA	2 V	100 nA	5 V	

†Reflective surface is Eastman Kodak (or equivalent) neutral white paper with 90% diffuse reflectance placed 3,81 mm (0.150 inch) from read head.
‡Reflective surface is 0,025-mm (0.001-inch) thick aluminum foil, typical of beginning of tape/end-of-tape strips on magnetic tape surface, placed 3,81 mm (0.150 inch) from read head.
§Selectively matched pairs of the devices listed in this column are currently used in the manufacture of these assemblies. This information is subject to change without notice.

BAR CODE READ HEAD
QUICK REFERENCE GUIDE

DEVICE	TYPE	FEATURES
TIL180	Bar Code Read Head	Capable of reading black and white bar codes: UPC, EAN,CODE 39, HP, MSI, and others.

TEXAS INSTRUMENTS
INCORPORATED
POST OFFICE BOX 225012 • DALLAS, TEXAS 75265

D1089, SEPTEMBER 1971–REVISED MARCH 1983

OPTOELECTRONIC MODULE FOR TRANSMISSIVE SENSING APPLICATIONS

- **Compatible with Standard TTL Integrated Circuits**
- **High-Speed Switching: t_r = 1.5 μs, t_f = 15 μs Typical**
- **Designed for Base or Side Mounting**
- **For Sensing Applications such as Shaft Encoders, Sector Sensors, Level Indicators, and Beginning-of-Tape/End-of-Tape Indicators**

mechanical data

The assembly consists of an infrared-emitting diode and an n-p-n silicon phototransistor mounted in a plastic housing. The assembly will withstand soldering temperature with no deformation and device performance characteristics remain stable when operated in high-humidity conditions. Total assembly weight is approximately 1.5 grams.

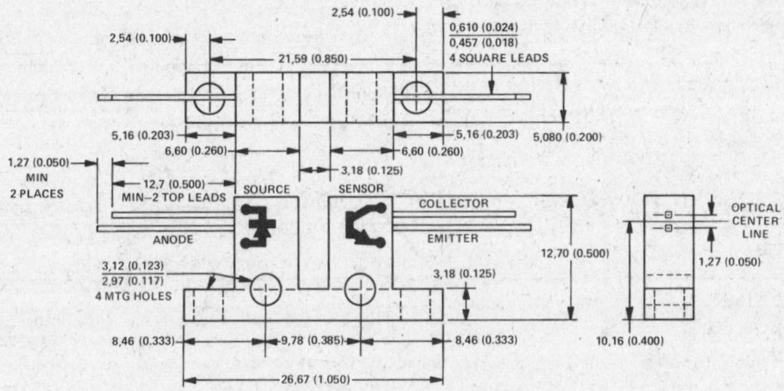

NOTE: Tolerance is ±0,25 mm (0.010 inch) unless otherwise noted.

ALL DIMENSIONS ARE IN MILLIMETERS AND PARENTHETICALLY IN INCHES

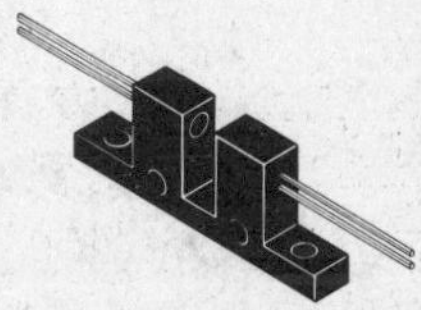

CL-28

absolute maximum ratings at 25°C free-air temperature (unless otherwise noted)

Source Reverse Voltage	2 V
Source Continuous Forward Current (See Note 1)	40 mA
Sensor Collector-Emitter Voltage	50 V
Sensor Emitter-Collector Voltage	7 V
Sensor Continuous Dissipation at (or below) 25°C Free-Air Temperature (See Note 2)	50 mW
Operating Free-Air Temperature	–40°C to 80°C
Storage Temperature Range	–40°C to 85°C
Lead Temperature 1,6 mm (1/16 Inch) from Assembly for 5 Seconds	240°C

NOTES: 1. Derate linearly to 80°C free-air temperature at the rate of 0.73 mA/°C.
2. Derate linearly to 80°C free-air temperature at the rate of 0.91 mW/°C.

Copyright © 1983 by Texas Instruments Incorporated

electrical characteristics at 25°C free-air temperature

PARAMETER		TEST CONDITIONS†	MIN	TYP	MAX	UNIT
$V_{(BR)CEO}$	Collector-Emitter Breakdown Voltage	$I_C = 100\ \mu A$, $I_F = 0$	50			V
$V_{(BR)ECO}$	Emitter-Collector Breakdown Voltage	$I_E = 100\ \mu A$, $I_F = 0$	7			V
$I_{C(off)}$	Off-State Collector Current	$V_{CE} = 30\ V$, $I_F = 0$			100	nA
$I_{C(on)}$	On-State Collector Current	$V_{CE} = 0.5\ V$, $I_F = 15\ mA$	0.4	1		mA
		$V_{CE} = 0.5\ V$, $I_F = 35\ mA$	1.6	4		
V_F	Input-Diode Static Forward Voltage	$I_F = 15\ mA$		1.15	1.5	V
		$I_F = 35\ mA$		1.2		

switching characteristics at 25°C free-air temperature

PARAMETER		TEST CONDITIONS†	MIN	TYP	MAX	UNIT
t_d	Delay Time	$V_{CC} = 30\ V$, $I_{C(on)} = 500\ \mu A$, $R_L = 1\ k\Omega$, See Figure 1		3		μs
t_r	Rise Time			1.5		μs
t_s	Storage Time			0.5		μs
t_f	Fall Time			15		μs

†Stray irradiation outside the range of device sensitivity may be present. A satisfactory condition has been achieved when the parameter being measured approaches a value which cannot be altered by further irradiation shielding.

PARAMETER MEASUREMENT INFORMATION

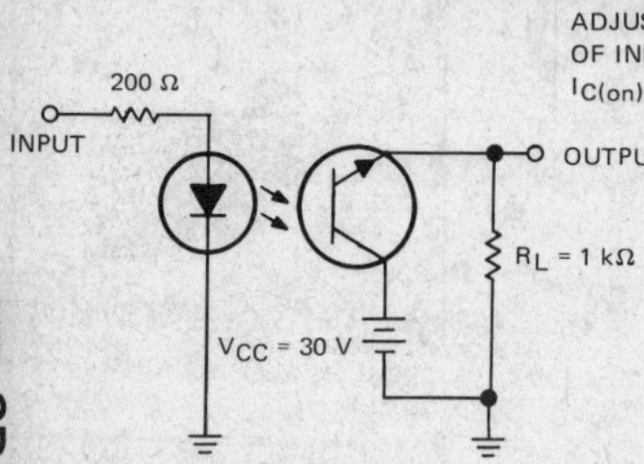

TEST CIRCUIT

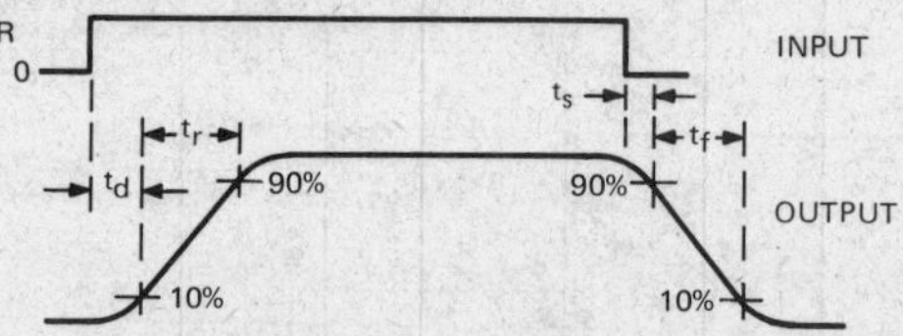

NOTE: The input pulse is supplied by a generator having the following characteristics: $Z_{out} = 50\ \Omega$, $t_r \leq 100$ ns, $t_f \leq 100$ ns, duty cycle ≈ 50%.

VOLTAGE WAVEFORMS

FIGURE 1—SWITCHING TIMES

TYPICAL CHARACTERISTICS

PHOTOTRANSISTOR COLLECTOR CURRENT
vs
INPUT-DIODE FORWARD CURRENT

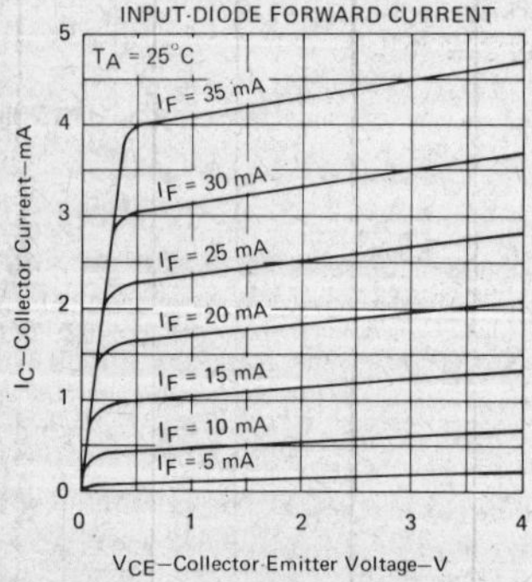

FIGURE 2

PHOTOTRANSISTOR COLLECTOR CURRENT
vs
INPUT-DIODE FORWARD CURRENT

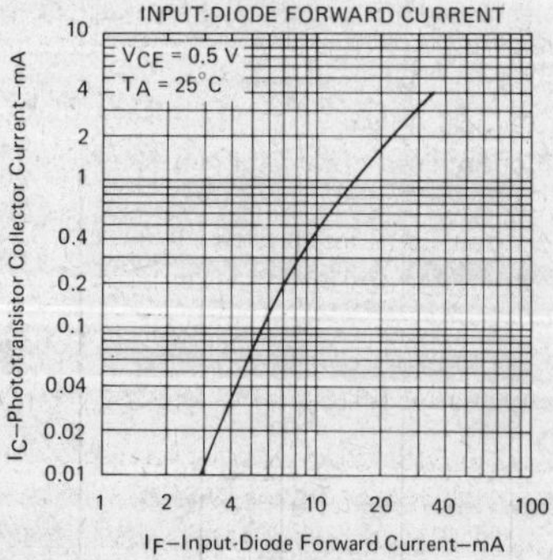

FIGURE 3

TEXAS INSTRUMENTS
INCORPORATED
POST OFFICE BOX 225012 • DALLAS, TEXAS 75265

D1100, SEPTEMBER 1971–REVISED MARCH 1983

OPTOELECTRONIC MODULE FOR REFLECTIVE SENSING APPLICATIONS

- Adaptable for Printed Circuit Board Mounting
- Designed for Sensing Applications such as Line Finders, Batch Counters, Level Indicators, and Beginning-of-Tape/End-of-Tape Indicators

mechanical data

The assembly consists of an infrared emitting diode and an n-p-n silicon phototransistor mounted in a plastic housing. The assembly will withstand soldering temperature with no deformation and device performance characteristics remain stable when operated in high-humidity conditions. Total assembly weight is approximately 1.2 grams.

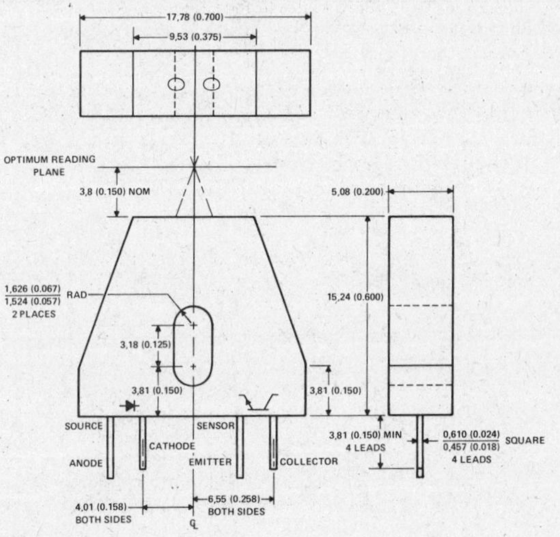

NOTE: Tolerance is ± 0,25 mm (0.010 inch) unless otherwise noted.

ALL DIMENSIONS ARE IN MILLIMETERS AND PARENTHETICALLY IN INCHES

CL-29

absolute maximum ratings at 25°C free-air temperature (unless otherwise noted)

Source Reverse Voltage	2 V
Source Continuous Forward Current (See Note 1)	40 mA
Sensor Collector-Emitter Voltage	50 V
Sensor Emitter-Collector Voltage	7 V
Sensor Continuous Dissipation at (or below) 25°C Free-Air Temperature (See Note 2)	50 mW
Operating Free-Air Temperature Range	−40°C to 80°C
Storage Temperature Range	−40°C to 85°C
Lead Temperature 1,6 mm (1/16 Inch) from Assembly for 5 Seconds	240°C

NOTES: 1. Derate linearly to 80°C free-air temperature at the rate of 0.73 mA/°C.
2. Derate linearly to 80°C free-air temperature at the rate of 0.91 mW/°C.

8 SDAs

Copyright © 1983 by Texas Instruments Incorporated

electrical characteristics at 25°C free-air temperature

PARAMETER		TEST CONDITIONS†			MIN	TYP	MAX	UNIT
$V_{(BR)CEO}$	Collector-Emitter Breakdown Voltage	I_C = 100 μA,	I_F = 0		50			V
$V_{(BR)ECO}$	Emitter-Collector Breakdown Voltage	I_E = 100 μA,	I_F = 0		7			V
$I_{C(off)}$	Off-State Collector Current	V_{CE} = 30 V,	I_F = 0				100	nA
$I_{C(on)}$	On-State Collector Current	V_{CE} = 5 V,	I_F = 40 mA,	See Note 3	10	125		μA
		V_{CE} = 5 V,	I_F = 40 mA,	See Note 4	5	60		
		V_{CE} = 5 V,	I_F = 40 mA,	See Note 5	100	1100		
V_F	Input-Diode Static Forward Voltage	I_F = 40 mA				1.2	1.6	V

†Stray irradiation outside the range of device sensitivity may be present. A satisfactory condition has been achieved when the parameter being measured approaches a value which cannot be altered by further irradiation shielding.

NOTES: 3. Reflective surface is Eastman Kodak (or equivalent) neutral white paper with 90% diffuse reflectance placed 3,81 mm (0.150 inch) from read head.
4. Reflective surface is Mylar‡ (or equivalent) magnetic tape placed 3,81 mm (0.150 inch) from read head.
5. Reflective surface is aluminum foil typical of beginning-of-tape/end-of-tape strips. It is 0,025 mm (0.001 inch) thick and placed 3,81 mm (0.150 inch) from read head.

‡Trademark of E. I. duPont de Nemours, Inc.

TEXAS INSTRUMENTS
INCORPORATED
POST OFFICE BOX 225012 • DALLAS, TEXAS 75265

D1962, NOVEMBER 1974—REVISED MARCH 1983

OPTOELECTRONIC ENCODER ASSEMBLIES FOR TRANSMISSIVE SENSING APPLICATIONS

- **Compatible With Standard TTL Integrated Circuits**
- **High-Speed Switching . . . t_r = 15 μs, t_f = 15 μs Typical**
- **For Counting, Speed Control, Position Sensing, Beginning-of-Tape/End-of-Tape Sensing, and High-Voltage Isolation**
- **Designed for Base Mounting—Standard 7,6 mm (0.300-Inch) Dual-In-Line Pin Spacing**
- **PC Board or Bracket Mounting**
- **Contains Infrared Emitter and Phototransistor**
- **1.02-mm (0.040-inch) Sensor Window Provides High On/Off Resolution**

mechanical data

Each assembly consists of an infrared-emitting diode and an n-p-n silicon phototransistor mounted in a plastic housing. The assembly will withstand soldering temperature with no deformation and device performance characteristics remain stable when operated in high-humidity conditions. Total assembly weight is approximately 1.5 grams.

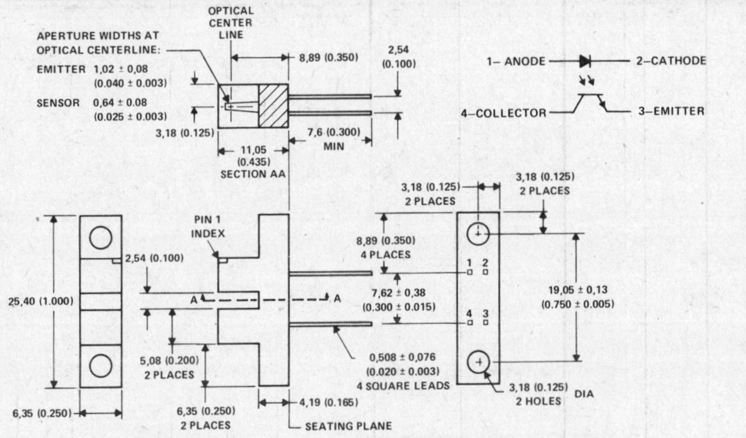

NOTE a: Tolerance is ± 0,26 mm (0.010 inch) unless otherwise noted.

ALL DIMENSIONS ARE IN MILLIMETERS AND PARENTHETICALLY IN INCHES

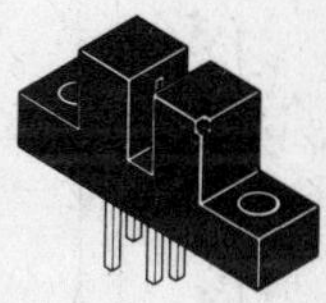

CL-31

absolute maximum ratings at 25°C free-air temperature (unless otherwise noted)

Source Reverse Voltage	3 V
Source Continuous Forward Current (See Note 1)	50 mA
Source Peak Forward Current (See Note 2)	3 A
Source Collector-Emitter Voltage	30 V
Sensor Emitter-Collector Voltage	7 V
Sensor Continuous Dissipation at (or below) 25°C Free-Air Temperature (See Note 3)	50 mW
Source-to-Sensor Voltage	±4 kV
Operating Free-Air Temperature Range	−40°C to 80°C
Storage Temperature Range	−40°C to 100°C
Lead Temperature 1,6 mm (1/16 Inch) from Assembly for 5 Seconds	240°C

NOTES: 1. Derate linearly to 80°C free-air temperature at the rate of 0.91 mA/°C.
2. This value applies for $t_w \leq 1$ μs, PRR ≤ 300 pps.
3. Derate linearly to 80°C free-air temperature at the rate of 0.91 mW/°C.

Copyright © 1983 by Texas Instruments Incorporated

TEXAS INSTRUMENTS
INCORPORATED
POST OFFICE BOX 225012 • DALLAS, TEXAS 75265

electrical characteristics at 25 °C free-air temperature

PARAMETER		TEST CONDITIONS†	TIL143			TIL144			UNIT
			MIN	TYP	MAX	MIN	TYP	MAX	
$V_{(BR)CEO}$	Collector-Emitter Breakdown Voltage	$I_C = 100\ \mu A$, $I_F = 0$	30			30			V
$V_{(BR)ECO}$	Emitter-Collector Breakdown Voltage	$I_E = 100\ \mu A$, $I_F = 0$	7			7			V
$I_{C(off)}$	Off-State Collector Current	$V_{CE} = 10\ V$, $I_F = 0$		5	100		5	100	nA
$I_{C(on)}$	On-State Collector Current	$V_{CE} = 5\ V$, $I_F = 20\ mA$	0.6	1		0.2	0.5		mA
V_F	Input-Diode Static Forward Voltage	$I_F = 50\ mA$		1.35	1.7		1.35	1.7	V

switching characteristics at 25 °C free-air temperature

PARAMETER		TEST CONDITIONS†	MIN	TYP	MAX	UNIT
t_r	Rise Time	$V_{CC} = 5\ V$, $I_{C(on)} = 1\ mA$,		15		μs
t_f	Fall Time	$R_L = 1\ k\Omega$, See Figure 1		15		μs

†Stray irradiation outside the range of device sensitivity may be present. A satisfactory condition has been achieved when the parameter being measured approaches a value which cannot be altered by further irradiation shielding.

PARAMETER MEASUREMENT INFORMATION

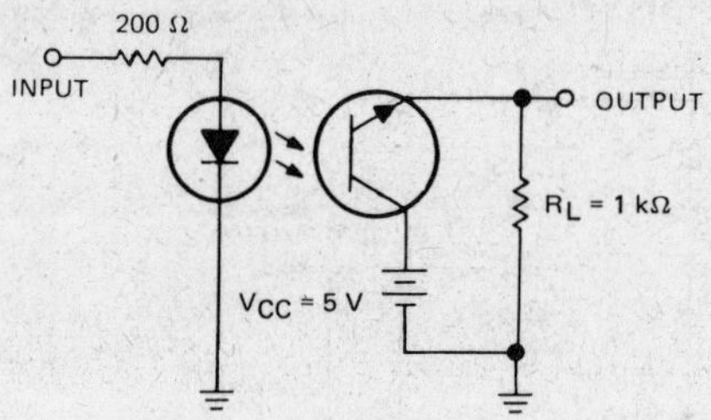

TEST CIRCUIT

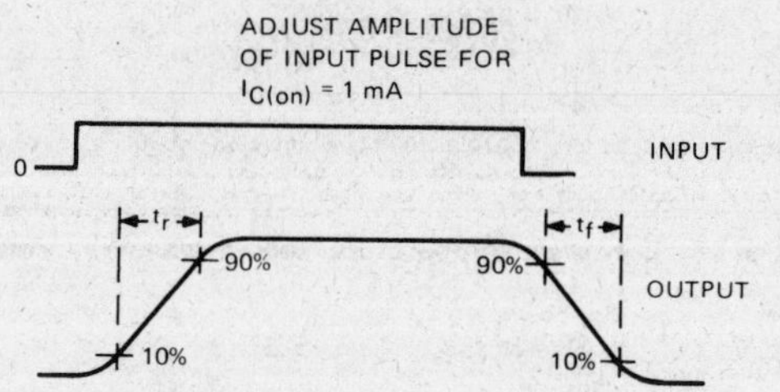

NOTE: The input pulse is supplied by a generator having the following characteristics: $Z_{out} = 50\ \Omega$, $t_r \leq 100$ ns, $t_f \leq 100$ ns, $t_w = 100\ \mu s$, duty cycle ≈2%.

VOLTAGE WAVEFORMS

FIGURE 1—SWITCHING TIMES

D1963, NOVEMBER 1974—REVISED MARCH 1983

HIGH-GAIN OPTOELECTRONIC ENCODER ASSEMBLIES FOR TRANSMISSIVE SENSING APPLICATIONS

- Compatible With Standard TTL Integrated Circuits
- High Current Transfer Ratio . . . 12.5% Min (TIL145)
- For Counting, Speed Control, Position Sensing, Beginning-of-Tape/End-of-Tape Sensing, and High-Voltage Isolation
- Designed for Base Mounting—Standard 7,6 mm (0.300-Inch) Dual-In-Line Pin Spacing
- PC Board or Bracket Mounting
- Contains Infrared Emitter and Silicon Darlington Phototransistor
- 1.02-mm (0.040-inch) Sensor Window Provides High On/Off Resolution

mechanical data

Each assembly consists of an infrared-emitting diode and an n-p-n silicon Darlington phototransistor mounted in a plastic housing. The assembly will withstand soldering temperature with no deformation and device performance characteristics remain stable when operated in high-humidity conditions. Total assembly weight is approximately 1.5 grams.

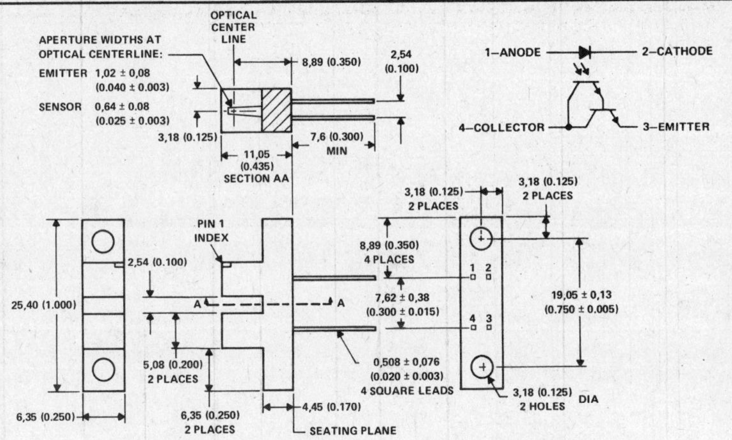

NOTE a. Tolerance is ± 0,26 mm (0.010 inch) unless otherwise noted.

ALL DIMENSIONS ARE IN MILLIMETERS AND PARENTHETICALLY IN INCHES

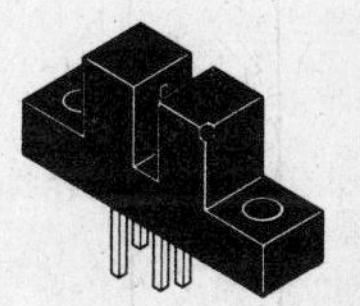

CL-31

absolute maximum ratings at 25 °C free air temperature (unless otherwise noted)

Source Reverse Voltage	3 V
Source Continuous Forward Current (See Note 1)	50 mA
Source Peak Forward Current (See Note 2)	3 A
Sensor Collector-Emitter Voltage	30 V
Sensor Emitter-Collector Voltage	5 V
Sensor Continuous Dissipation at (or below) 25 °C Free-Air Temperature (See Note 3)	100 mW
Source-to-Sensor Voltage	±4 kV
Operating Free-Air Temperature Range	−40° to 80 °C
Storage Temperature Range	−40 °C to 100 °C
Lead Temperature 1,6 mm (1/16 Inch) from Assembly for 5 Seconds	240 °C

NOTES: 1. Derate linearly to 80 °C free-air temperature at the rate of 0.91 mA/°C.
2. This value applies for $t_w \leq 1$ μs, PRR ≤ 300 pps.
3. Derate linearly to 80 °C free-air temperature at the rate of 1.81 mW/°C.

Copyright © 1983 by Texas Instruments Incorporated

electrical characteristics at 25°C free-air temperature

PARAMETER		TEST CONDITIONS†	TIL145			TIL146			UNIT
			MIN	TYP	MAX	MIN	TYP	MAX	
$V_{(BR)CEO}$	Collector-Emitter Breakdown Voltage	$I_C = 100\ \mu A$, $I_F = 0$	30			30			V
$V_{(BR)ECO}$	Emitter-Collector Breakdown Voltage	$I_E = 100\ \mu A$, $I_F = 0$	5			5			V
$I_{C(off)}$	Off-State Collector Current	$V_{CE} = 5\ V$, $I_F = 0$		5	100		5	100	nA
$I_{C(on)}$	On-State Collector Current	$V_{CE} = 1\ V$, $I_F = 16\ mA$	2	5					mA
		$V_{CE} = 1\ V$, $I_F = 50\ mA$				1.6	4		
V_F	Input-Diode Static Forward Voltage	$I_F = 50\ mA$		1.35	1.7		1.35	1.7	V

switching characteristics at 25°C free-air temperature

PARAMETER		TEST CONDITIONS†	MIN	TYP	MAX	UNIT
t_r	Rise Time	$V_{CC} = 5\ V$, $I_{C(on)} = 500\ \mu A$,			3	ms
t_f	Fall Time	$R_L = 1\ k\Omega$, See Figure 1			2.5	ms

†Stray irradiation outside the range of device sensitivity may be present. A satisfactory condition has been achieved when the parameter being measured approaches a value that cannot be altered by further irradiation shielding.

PARAMETER MEASUREMENT INFORMATION

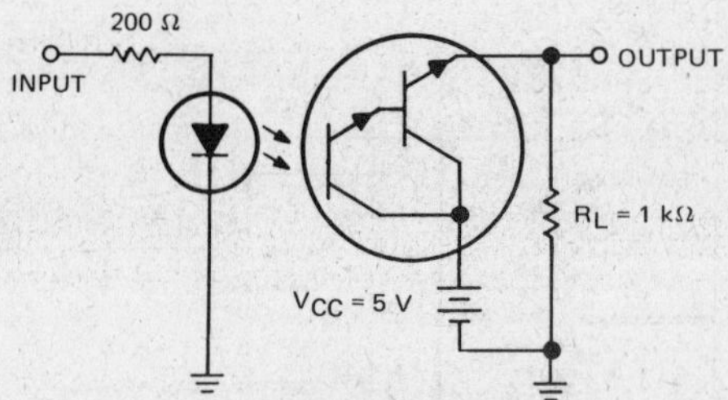

TEST CIRCUIT

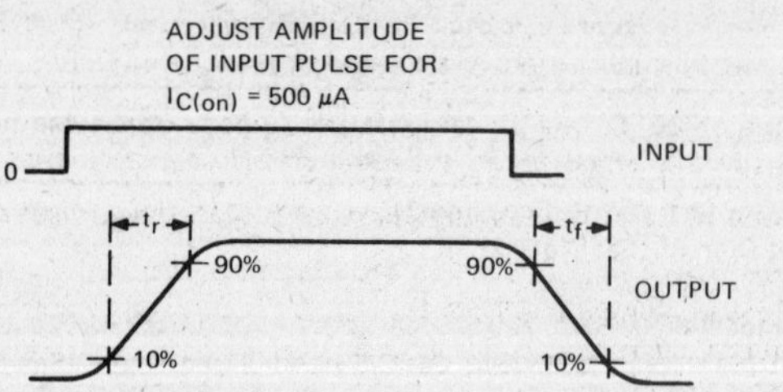

NOTE: The input pulse is supplied by a generator having the following characteristics: $Z_{out} = 50\ \Omega$, $t_r \leq 10\ \mu s$, $t_f \leq 10\ \mu s$, $t_w = 10$ ms, duty cycle ≈ 50%.

VOLTAGE WAVEFORMS

FIGURE 1– SWITCHING TIMES

OPTOELECTRONIC ASSEMBLIES FOR TRANSMISSIVE SENSING APPLICATIONS

- Compatible with Standard TTL Integrated Circuits
- High-Speed Switching: $t_r = 5\ \mu s$, $t_f = 5\ \mu s$ Typical
- Designed for Base Mounting . . . Fits Standard Dual-In-Line-Package Socket
- For Sensing Applications such as Shaft Encoders, Sector Sensors, Level Indicators, and Beginning-of-Tape/End-of-Tape Indicators
- Hermetically Sealed Phototransistor and Infrared-Emitting Diode
- 0,63-mm (0.025-inch) Sensor Aperture Slit Provides High On/Off Resolution
- High Current Transfer Ratio . . . 20% Min (TIL147A)

mechanical data

Each assembly consists of an infrared-emitting diode and an n-p-n silicon phototransistor mounted in a plastic housing. The assembly will withstand soldering temperature with no deformation and device performance characteristics remain stable when operated in high-humidity conditions. Total assembly weight is approximately 1.5 grams. The TIL147A and TIL148A have 0.020-inch-square leads; the TIL147 and TIL148 had 0.020-inch-diameter round leads.

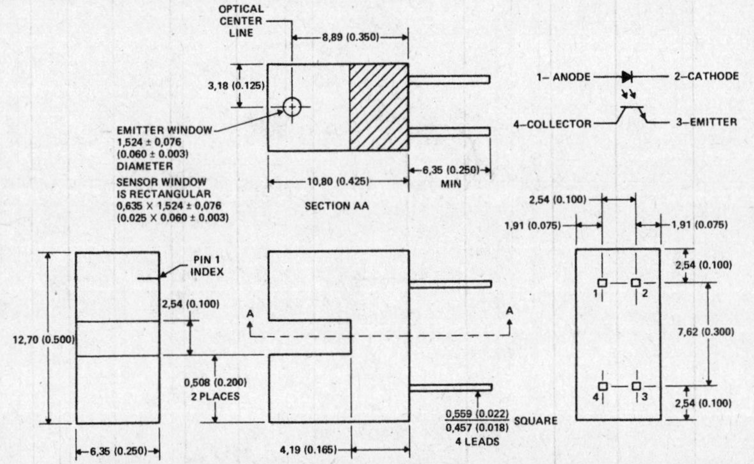

NOTE a: Dimension tolerance is ± 0,26 mm (0.010 inch) unless otherwise noted.
ALL DIMENSIONS ARE IN MILLIMETERS AND PARENTHETICALLY IN INCHES

CL-32

absolute maximum ratings at 25°C free-air temperature (unless otherwise noted)

Source Reverse Voltage	3 V
Source Continuous Forward Current (See Note 1)	100 mA
Sensor Collector-Emitter Voltage	30 V
Sensor Emitter-Collector Voltage	5 V
Sensor Continuous Dissipation at (or below) 25°C Free-Air Temperature (See Note 2)	50 mW
Source-to-Sensor Voltage	±2 kV
Operating Free-Air Temperature Range	–40°C to 80°C
Storage Temperature Range	–40°C to 100°C
Lead Temperature 1,6 mm (1/16 inch) from Assembly for 5 Seconds	240°C

NOTES: 1. Derate linearly to 80°C free-air temperature at the rate of 1.82 mA/°C.
2. Derate linearly to 80°C free-air temperature at the rate of 0.91 mW/°C.

Copyright © 1983 by Texas Instruments Incorporated

TEXAS INSTRUMENTS
INCORPORATED
POST OFFICE BOX 225012 • DALLAS, TEXAS 75265

electrical characteristics at 25°C free-air temperature

PARAMETER		TEST CONDITIONS†	TIL147A		TIL148A		UNIT
			MIN	MAX	MIN	MAX	
$V_{(BR)CEO}$	Collector-Emitter Breakdown Voltage	$I_C = 100\ \mu A$, $I_F = 0$	30		30		V
$V_{(BR)ECO}$	Emitter-Collector Breakdown Voltage	$I_E = 100\ \mu A$, $I_F = 0$	5		5		V
$I_{C(off)}$	Off-State Collector Current	$V_{CE} = 10\ V$, $I_F = 0$		100		100	nA
$I_{C(on)}$	On-State Collector Current	$V_{CE} = 5\ V$, $I_F = 20\ mA$	4		1		mA
V_F	Input-Diode Static Forward Voltage	$I_F = 20\ mA$		1.3		1.3	V
		$I_F = 50\ mA$		1.7		1.7	

switching characteristics at 25°C free-air temperature

PARAMETER		TEST CONDITIONS†	MIN	TYP	MAX	UNIT
t_r	Rise Time	$V_{CC} = 10\ V$, $I_{C(on)} = 1\ mA$, $R_L = 100\ \Omega$, See Figure 1		5		μs
t_f	Fall Time			5		μs

†Stray irradiation outside the range of device sensitivity may be present. A satisfactory condition has been achieved when the parameter being measured approaches a value which cannot be altered by further irradiation shielding.

PARAMETER MEASUREMENT INFORMATION

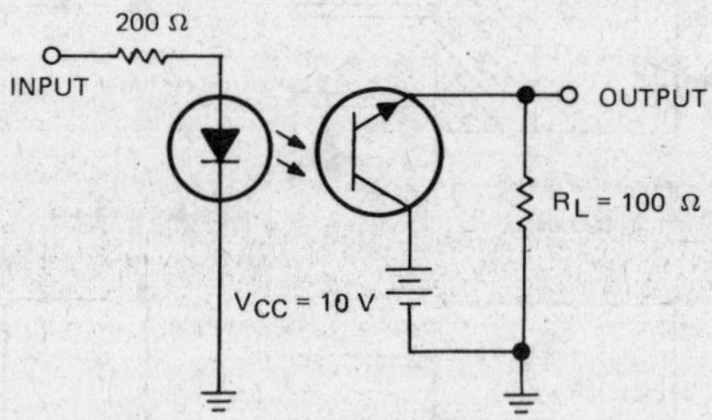

TEST CIRCUIT

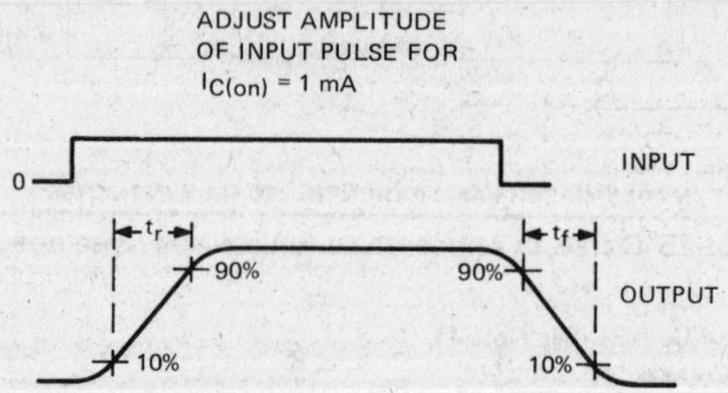

NOTE: The input pulse is supplied by a generator having the following characteristics: $Z_{out} = 50\ \Omega$, $t_r \leqslant 100$ ns, $t_f \leqslant 100$ ns, $t_w = 10\ \mu s$, duty cycle ≈ 2%.

VOLTAGE WAVEFORMS

FIGURE 1–SWITCHING TIMES

D2163, MARCH 1976–REVISED MARCH 1983

OPTOELECTRONIC MODULE FOR REFLECTIVE SENSING APPLICATIONS

- **Adaptable for Printed Circuit Board Mounting**
- **Designed for Sensing Applications such as Line Finders, Batch Counters, Level Indicators, and Beginning-of-Tape/End-of-Tape Indicators**

mechanical data

The assembly consists of an infrared-emitting diode and an n-p-n silicon phototransistor in a plastic housing. The assembly will withstand soldering temperature with no deformation and device performance characteristics remain stable when operated in high-humidity conditions. Total assembly weight is approximately 0.9 grams.

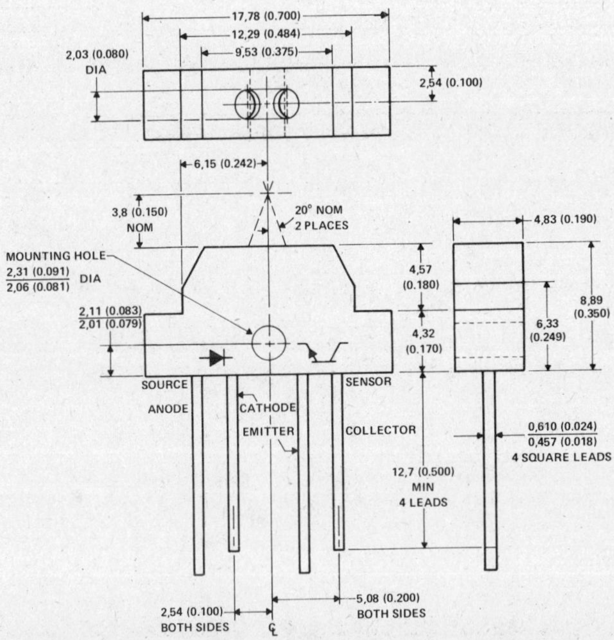

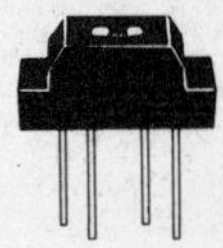

NOTE a: Dimension tolerance is ± 0,025 mm (0.010 inch) unless otherwise noted.

ALL DIMENSIONS ARE IN MILLIMETERS AND PARENTHETICALLY IN INCHES

CL 33

absolute maximum ratings at 25°C free-air temperature (unless otherwise noted)

Source Reverse Voltage	2 V
Source Continuous Forward Current (See Note 1)	40 mA
Sensor Collector-Emitter Voltage	30 V
Sensor Emitter-Collector Voltage	7 V
Sensor Continuous Dissipation at (or below) 25°C Free-Air Temperature (See Note 2)	50 mW
Operating Temperature Range	−40°C to 80°C
Storage Temperature Range	−40°C to 85°C
Lead Temperature 1,6 mm (1/16 inch) from Assembly for 5 Seconds	240°C

NOTES: 1. Derate linearly to 80°C free-air temperature at the rate of 0.73 mA/°C.
2. Derate linearly to 80°C free-air temperature at the rate of 0.91 mW/°C.

Copyright © 1983 by Texas Instruments Incorporated

TEXAS INSTRUMENTS
INCORPORATED
POST OFFICE BOX 225012 • DALLAS, TEXAS 75265

electrical characteristics at 25°C free-air temperature

PARAMETER		TEST CONDITIONS†	MIN	TYP	MAX	UNIT
$V_{(BR)CEO}$	Collector-Emitter Breakdown Voltage	$I_C = 100\ \mu A$, $I_F = 0$	30			V
$V_{(BR)ECO}$	Emitter-Collector Breakdown Voltage	$I_E = 100\ \mu A$, $I_F = 0$	7			V
$I_{C(off)}$	Off-State Collector Current	$V_{CE} = 15\ V$, $I_F = 0$			100	nA
$I_{C(on)}$	On-State Collector Current	$V_{CE} = 5\ V$, $I_F = 40\ mA$, See Note 3	25	275		μA
V_F	Input-Diode Static Forward Voltage	$I_F = 40\ mA$		1.2	1.6	V

†Stray irradiation outside the range of device sensitivity may be present. A satisfactory condition has been achieved when the parameter being measured approaches a value that cannot be altered by further irradiation shielding.

NOTE 3: Reflective surface is aluminum foil typical of beginning-of-tape/end-of-tape strips. It is 0,026 mm (0.001 inch) thick and placed 3,81 mm (0.150 inch) from the read head.

TEXAS INSTRUMENTS
INCORPORATED
POST OFFICE BOX 225012 • DALLAS, TEXAS 75265

D2692, APRIL 1983

OPTOELECTRONIC ENCODER ASSEMBLIES FOR TRANSMISSIVE SENSING APPLICATIONS

- **Compatible With Standard TTL Integrated Circuits**
- **High-Speed Switching . . . t_r = 15 μs, t_f = 15 μs Typical**
- **For Counting, Speed Control, Position Sensing, Beginning-of-Tape/End-of-Tape Sensing, and High-Voltage Isolation**
- **Designed for Base Mounting—Standard 7,6 mm (0.300-Inch) Dual-In-Line Pin Spacing**
- **PC Board or Bracket Mounting**
- **Contains Infrared Emitter and Phototransistor**
- **1.02-mm (0.040-inch) Sensor Window Provides High On/Off Resolution**

mechanical data

Each assembly consists of an infrared-emitting diode and an n-p-n silicon phototransistor mounted in a plastic housing. The assembly will withstand soldering temperature with no deformation and device performance characteristics remain stable when operated in high-humidity conditions. Total assembly weight is approximately 1.5 grams.

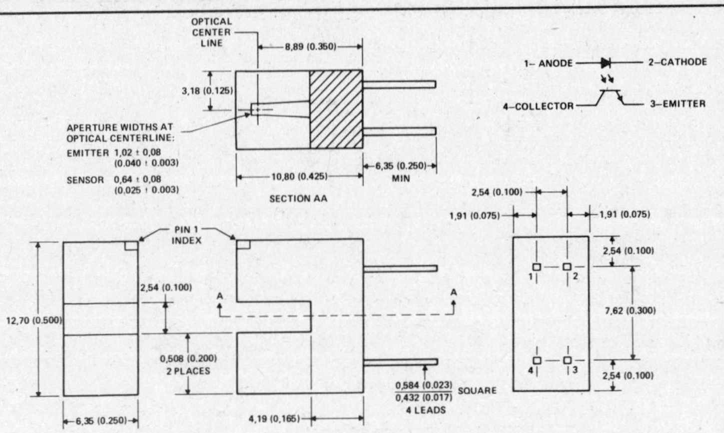

NOTE a: Tolerance is ± 0,26 mm (0.010 inch) unless otherwise noted.
ALL DIMENSIONS ARE IN MILLIMETERS AND PARENTHETICALLY IN INCHES

CL-81

absolute maximum ratings at 25°C free-air temperature (unless otherwise noted)

Source Reverse Voltage	3 V
Source Continuous Forward Current (See Note 1)	40 mA
Source Peak Forward Current (See Note 2)	3 A
Source Collector-Emitter Voltage	30 V
Sensor Emitter-Collector Voltage	7 V
Sensor Continuous Dissipation at (or below) 25°C Free-Air Temperature (See Note 3)	50 mW
Source-to-Sensor Voltage	±4 kV
Operating Free-Air Temperature Range	−40°C to 80°C
Storage Temperature Range	−40°C to 100°C
Lead Temperature 1,6 mm (1/16 Inch) from Assembly for 5 Seconds	240°C

NOTES: 1. Derate linearly to 80°C free-air temperature at the rate of 0.73 mA/°C.
2. This value applies for $t_w \leq 1$ μs, PRR ≤ 300 pps.
3. Derate linearly to 80°C free-air temperature at the rate of 0.91 mW/°C.

Copyright © 1983 by Texas Instruments Incorporated

electrical characteristics at 25°C free-air temperature

PARAMETER		TEST CONDITIONS†	TIL158			TIL159			UNIT
			MIN	TYP	MAX	MIN	TYP	MAX	
$V_{(BR)CEO}$	Collector-Emitter Breakdown Voltage	$I_C = 100\ \mu A$, $I_F = 0$	30			30			V
$V_{(BR)ECO}$	Emitter-Collector Breakdown Voltage	$I_E = 100\ \mu A$, $I_F = 0$	7			7			V
$I_{C(off)}$	Off-State Collector Current	$V_{CE} = 10$ V, $I_F = 0$		5	100		5	100	nA
$I_{C(on)}$	On-State Collector Current	$V_{CE} = 5$ V, $I_F = 20$ mA	0.6	1		0.2	0.5		mA
V_F	Input-Diode Static Forward Voltage	$I_F = 20$ mA		1.2	1.6		1.2	1.6	V

switching characteristics at 25°C free-air temperature

PARAMETER		TEST CONDITIONS†	MIN	TYP	MAX	UNIT
t_r	Rise Time	$V_{CC} = 5$ V, $I_{C(on)} = 1$ mA,		15		μs
t_f	Fall Time	$R_L = 1\ k\Omega$, See Figure 1		15		μs

†Stray irradiation outside the range of device sensitivity may be present. A satisfactory condition has been achieved when the parameter being measured approaches a value which cannot be altered by further irradiation shielding.

PARAMETER MEASUREMENT INFORMATION

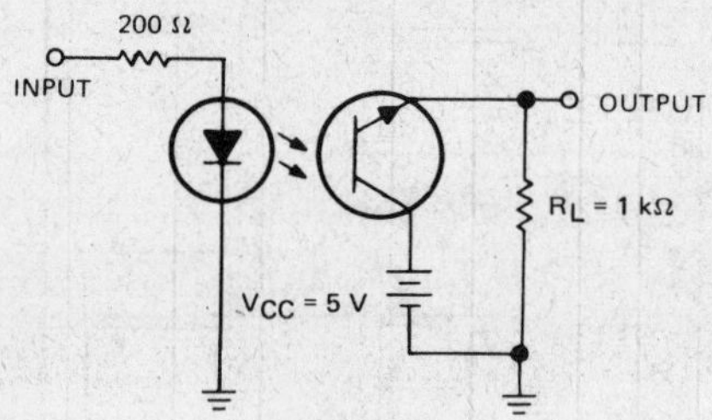

TEST CIRCUIT

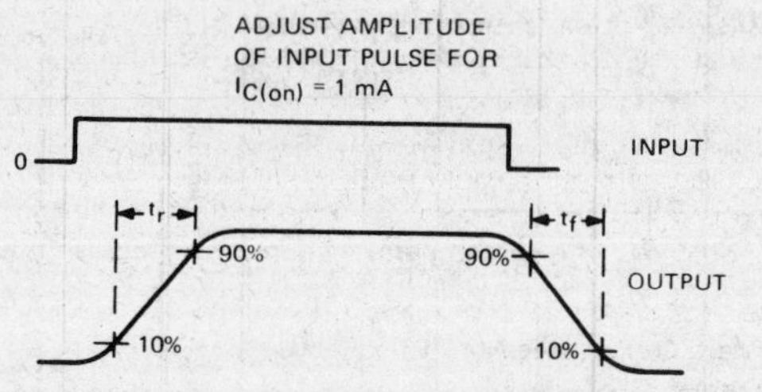

NOTE: The input pulse is supplied by a generator having the following characteristics: $Z_{out} = 50\ \Omega$, $t_r \leq 100$ ns, $t_f \leq 100$ ns, $t_w = 100\ \mu s$, duty cycle ≈2%.

VOLTAGE WAVEFORMS

FIGURE 1—SWITCHING TIMES

HIGH-GAIN OPTOELECTRONIC ENCODER ASSEMBLIES FOR TRANSMISSIVE SENSING APPLICATIONS

- **Compatible With Standard TTL Integrated Circuits**
- **High Current Transfer Ratio . . . 12.5% Min (TIL145)**
- **For Counting, Speed Control, Position Sensing, Beginning-of-Tape/End-of-Tape Sensing, and High-Voltage Isolation**
- **Designed for Base Mounting—Standard 7,6 mm (0.300-Inch) Dual-In-Line Pin Spacing**
- **PC Board or Bracket Mounting**
- **Contains Infrared Emitter and Silicon Darlington Phototransistor**
- **1.02-mm (0.040-inch) Sensor Window Provides High On/Off Resolution**

mechanical data

Each assembly consists of an infrared-emitting diode and an n-p-n silicon Darlington phototransistor mounted in a plastic housing. The assembly will withstand soldering temperature with no deformation and device performance characteristics remain stable when operated in high-humidity conditions. Total assembly weight is approximately 1.5 grams.

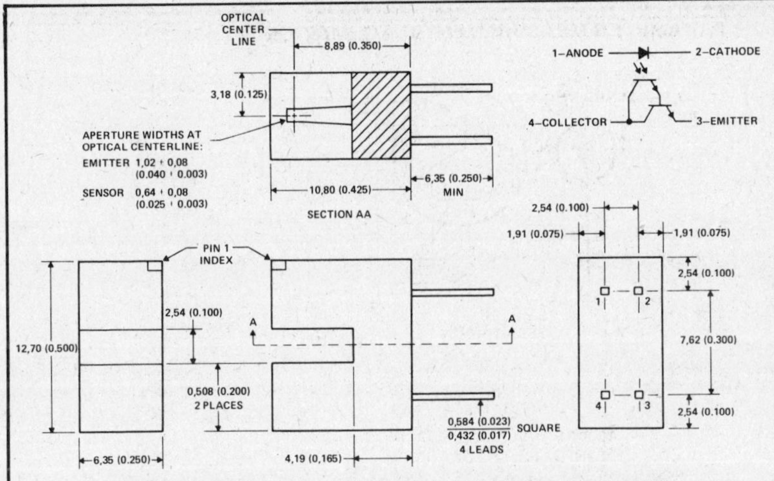

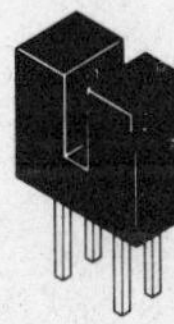

NOTE a: Dimension tolerance is ± 0,26 mm (0.010 inch) unless otherwise noted.
ALL DIMENSIONS ARE IN MILLIMETERS AND PARENTHETICALLY IN INCHES

CL 81

absolute maximum ratings at 25 °C free-air temperature (unless otherwise noted)

Source Reverse Voltage	3 V
Source Continuous Forward Current (See Note 1)	40 mA
Source Peak Forward Current (See Note 2)	3 A
Sensor Collector-Emitter Voltage	30 V
Sensor Emitter-Collector Voltage	7 V
Sensor Continuous Dissipation at (or below) 25 °C Free-Air Temperature (See Note 3)	100 mW
Source-to-Sensor Voltage	±4 kV
Operating Free-Air Temperature Range	−40° to 80°C
Storage Temperature Range	−40°C to 100°C
Lead Temperature 1,6 mm (1/16 Inch) from Assembly for 5 Seconds	240°C

NOTES: 1. Derate linearly to 80 °C free-air temperature at the rate of 0.73 mA/°C.
2. This value applies for $t_w \leq 1$ μs, PRR ≤ 300 pps.
3. Derate linearly to 80 °C free-air temperature at the rate of 1.81 mW/°C.

Copyright © 1983 by Texas Instruments Incorporated

electrical characteristics at 25 °C free-air temperature

PARAMETER		TEST CONDITIONS†	TIL160			TIL161			UNIT
			MIN	TYP	MAX	MIN	TYP	MAX	
$V_{(BR)CEO}$	Collector-Emitter Breakdown Voltage	I_C = 100 μA, I_F = 0	30			30			V
$V_{(BR)ECO}$	Emitter-Collector Breakdown Voltage	I_E = 100 μA, I_F = 0	7			7			V
$I_{C(off)}$	Off-State Collector Current	V_{CE} = 5 V, I_F = 0		5	100		5	100	nA
$I_{C(on)}$	On-State Collector Current	V_{CE} = 2 V, I_F = 10 mA	2	3.5		0.5	1		mA
V_F	Input-Diode Static Forward Voltage	I_F = 20 mA		1.2	1.6		1.2	1.6	V

switching characteristics at 25 °C free-air temperature

PARAMETER		TEST CONDITIONS†	MIN	TYP	MAX	UNIT
t_r	Rise Time	V_{CC} = 5 V, $I_{C(on)}$ = 500 μA, R_L = 1 kΩ, See Figure 1		1		ms
t_f	Fall Time			1		ms

†Stray irradiation outside the range of device sensitivity may be present. A satisfactory condition has been achieved when the parameter being measured approaches a value which cannot be altered by further irradiation shielding.

PARAMETER MEASUREMENT INFORMATION

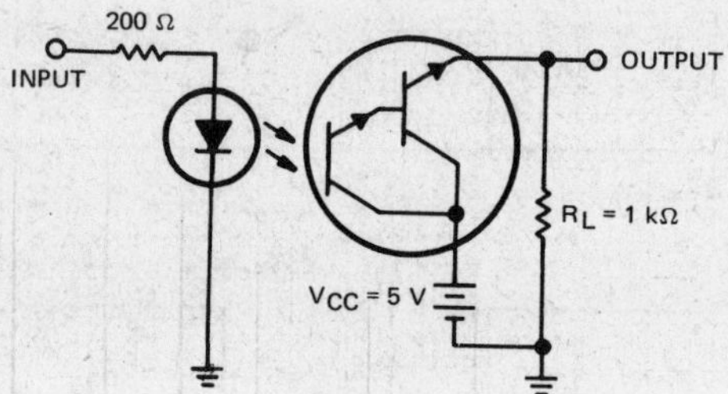

TEST CIRCUIT

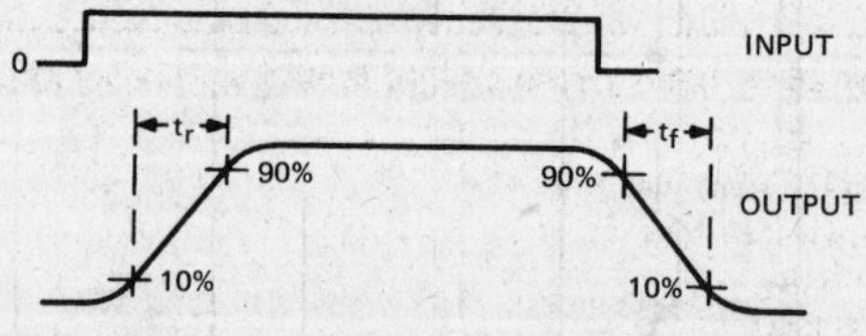

NOTE: The input pulse is supplied by a generator having the following characteristics: Z_{out} = 50 Ω, $t_r \leq 10$ μs, $t_f \leq 10$ μs, t_w = 10 ms, duty cycle ≈ 50%.

VOLTAGE WAVEFORMS

FIGURE 1 – SWITCHING TIMES

8

SDAs

TEXAS INSTRUMENTS
INCORPORATED
POST OFFICE BOX 225012 • DALLAS, TEXAS 75265

OPTOELECTRONIC ENCODER ASSEMBLIES FOR TRANSMISSIVE SENSING APPLICATIONS

- **Compatible With Standard TTL Integrated Circuits**
- **High-Speed Switching . . . t_r = 15 μs, t_f = 15 μs Typical**
- **For Counting, Speed Control, Position Sensing, Beginning-of-Tape/End-of-Tape Sensing, and High-Voltage Isolation**
- **Designed for Base Mounting—Standard 7,6 mm (0.300-Inch) Dual-In-Line Pin Spacing**
- **PC Board or Bracket Mounting**
- **Contains Infrared Emitter and Phototransistor**
- **1.02-mm (0.040-inch) Sensor Window Provides High On/Off Resolution**

mechanical data

Each assembly consists of an infrared-emitting diode and an n-p-n silicon phototransistor mounted in a plastic housing. The assembly will withstand soldering temperature with no deformation and device performance characteristics remain stable when operated in high-humidity conditions. Total assembly weight is approximately 1.5 grams.

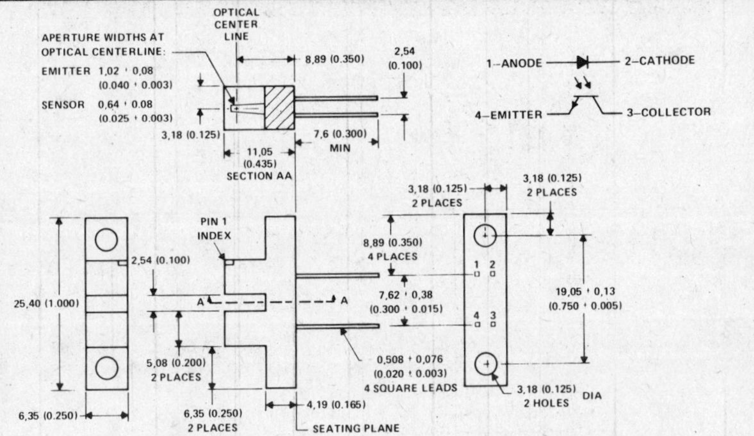

NOTE a: Tolerance is ± 0,26 mm (0.010 inch) unless otherwise noted.
ALL DIMENSIONS ARE IN MILLIMETERS AND PARENTHETICALLY IN INCHES

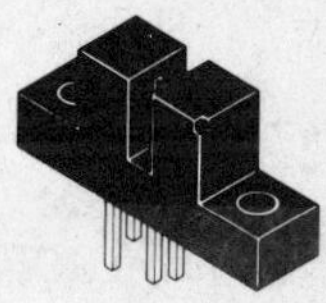

CL 31

absolute maximum ratings at 25°C free-air temperature (unless otherwise noted)

Source Reverse Voltage	3 V
Source Continuous Forward Current (See Note 1)	40 mA
Source Peak Forward Current (See Note 2)	3 A
Source Collector-Emitter Voltage	30 V
Sensor Emitter-Collector Voltage	7 V
Sensor Continuous Dissipation at (or below) 25°C Free-Air Temperature (See Note 3)	50 mW
Source-to-Sensor Voltage	±4 kV
Operating Free-Air Temperature Range	−40°C to 80°C
Storage Temperature Range	−40°C to 100°C
Lead Temperature 1,6 mm (1/16 Inch) from Assembly for 5 Seconds	240°C

NOTES: 1. Derate linearly to 80°C free-air temperature at the rate of 0.73 mA/°C.
2. This value applies for $t_w \leq 1$ μs, PRR ≤ 300 pps.
3. Derate linearly to 80°C free-air temperature at the rate of 0.91 mW/°C.

Copyright © 1983 by Texas Instruments Incorporated

electrical characteristics at 25°C free-air temperature

PARAMETER		TEST CONDITIONS†	TIL167-1			TIL167-2			UNIT
			MIN	TYP	MAX	MIN	TYP	MAX	
$V_{(BR)CEO}$	Collector-Emitter Breakdown Voltage	$I_C = 100\ \mu A$, $I_F = 0$	30			30			V
$V_{(BR)ECO}$	Emitter-Collector Breakdown Voltage	$I_E = 100\ \mu A$, $I_F = 0$	7			7			V
$I_{C(off)}$	Off-State Collector Current	$V_{CE} = 10$ V, $I_F = 0$		5	100		5	100	nA
$I_{C(on)}$	On-State Collector Current	$V_{CE} = 5$ V, $I_F = 20$ mA	0.2	0.5		0.6	1		mA
V_F	Input-Diode Static Forward Voltage	$I_F = 20$ mA		1.2	1.6		1.2	1.6	V

switching characteristics at 25°C free-air temperature

PARAMETER		TEST CONDITIONS†	MIN	TYP	MAX	UNIT
t_r	Rise Time	$V_{CC} = 5$ V, $I_{C(on)} = 1$ mA,		15		μs
t_f	Fall Time	$R_L = 1\ k\Omega$, See Figure 1		15		μs

†Stray irradiation outside the range of device sensitivity may be present. A satisfactory condition has been achieved when the parameter being measured approaches a value which cannot be altered by further irradiation shielding.

PARAMETER MEASUREMENT INFORMATION

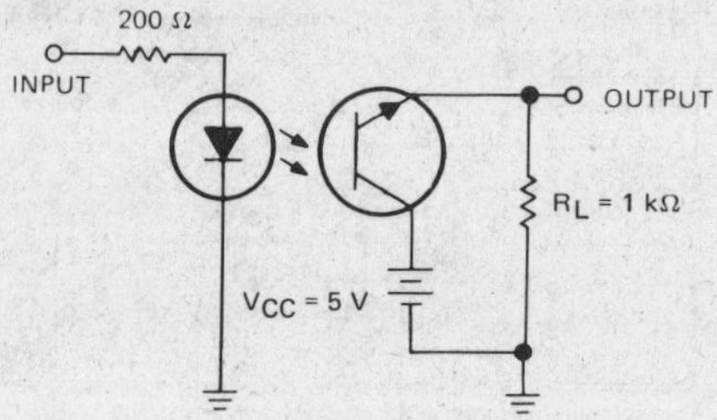

TEST CIRCUIT

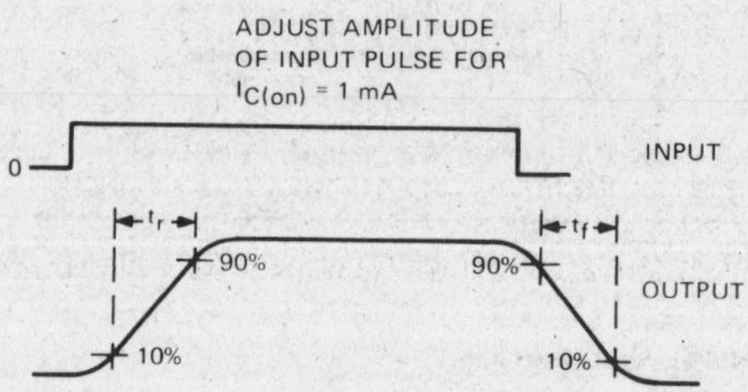

NOTE: The input pulse is supplied by a generator having the following characteristics: $Z_{out} = 50\ \Omega$, $t_r \leq 100$ ns, $t_f \leq 100$ ns, $t_w = 100\ \mu s$, duty cycle ≈2%.

VOLTAGE WAVEFORMS

FIGURE 1–SWITCHING TIMES

TEXAS INSTRUMENTS
INCORPORATED
POST OFFICE BOX 225012 • DALLAS, TEXAS 75265

TYPES 168-1, TIL168-2 SOURCE AND DARLINGTON DETECTOR ASSEMBLIES

HIGH-GAIN OPTOELECTRONIC ENCODER ASSEMBLIES FOR TRANSMISSIVE SENSING APPLICATIONS

- **Compatible With Standard TTL Integrated Circuits**
- **High Current Transfer Ratio . . . 12.5% Min (TIL145)**
- **For Counting, Speed Control, Position Sensing, Beginning-of-Tape/End-of-Tape Sensing, and High-Voltage Isolation**
- **Designed for Base Mounting—Standard 7,6 mm (0.300-Inch) Dual-In-Line Pin Spacing**
- **PC Board or Bracket Mounting**
- **Contains Infrared Emitter and Silicon Darlington Phototransistor**
- **1.02-mm (0.040-inch) Sensor Window Provides High On/Off Resolution**

mechanical data

Each assembly consists of an infrared-emitting diode and an n-p-n silicon Darlington phototransistor mounted in a plastic housing. The assembly will withstand soldering temperature with no deformation and device performance characteristics remain stable when operated in high-humidity conditions. Total assembly weight is approximately 1.5 grams.

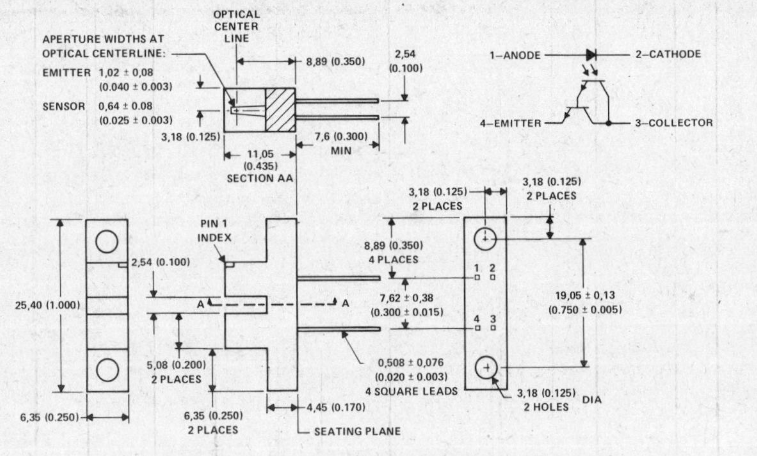

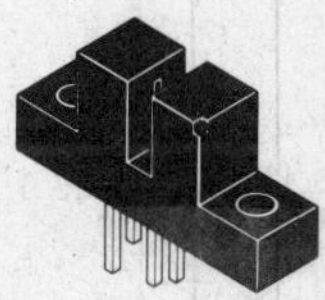

NOTE a: Dimension tolerance is ± 0,26 mm (0.010 inch) unless otherwise noted.
ALL DIMENSIONS ARE IN MILLIMETERS AND PARENTHETICALLY IN INCHES.

CL-31

absolute maximum ratings at 25 °C free-air temperature (unless otherwise noted)

Source Reverse Voltage	3 V
Source Continuous Forward Current (See Note 1)	40 mA
Source Peak Forward Current (See Note 2)	3 A
Sensor Collector-Emitter Voltage	30 V
Sensor Emitter-Collector Voltage	7 V
Sensor Continuous Dissipation at (or below) 25 °C Free-Air Temperature (See Note 3)	100 mW
Source-to-Sensor Voltage	±4 kV
Operating Free-Air Temperature Range	−40° to 80 °C
Storage Temperature Range	−40 °C to 100 °C
Lead Temperature 1,6 mm (1/16 Inch) from Assembly for 5 Seconds	240 °C

NOTES: 1. Derate linearly to 80 °C free-air temperature at the rate of 0.73 mA/°C.
2. This value applies for $t_w \leq 1\ \mu s$, PRR ≤ 300 pps.
3. Derate linearly to 80 °C free-air temperature at the rate of 1.81 mW/°C.

Copyright © 1983 by Texas Instruments Incorporated

3

electrical characteristics at 25 °C free-air temperature

PARAMETER		TEST CONDITIONS†	TIL168-1			TIL168-2			UNIT
			MIN	TYP	MAX	MIN	TYP	MAX	
$V_{(BR)CEO}$	Collector-Emitter Breakdown Voltage	I_C = 100 μA, I_F = 0	30			30			V
$V_{(BR)ECO}$	Emitter-Collector Breakdown Voltage	I_E = 100 μA, I_F = 0	7			7			V
$I_{C(off)}$	Off-State Collector Current	V_{CE} = 10 V, I_F = 0		5	100		5	100	nA
$I_{C(on)}$	On-State Collector Current	V_{CE} = 2 V, I_F = 10 mA	0.5	1		2	3.5		mA
V_F	Input-Diode Static Forward Voltage	I_F = 20 mA		1.2	1.6		1.2	1.6	V

switching characteristics at 25 °C free-air temperature

PARAMETER		TEST CONDITIONS†	MIN	TYP	MAX	UNIT
t_r	Rise Time	V_{CC} = 5 V, $I_{C(on)}$ = 500 μA,		1		ms
t_f	Fall Time	R_L = 1 kΩ, See Figure 1		1		ms

† Stray irradiation outside the range of device sensitivity may be present. A satisfactory condition has been achieved when the parameter being measured approaches a value which cannot be altered by further irradiation shielding.

PARAMETER MEASUREMENT INFORMATION

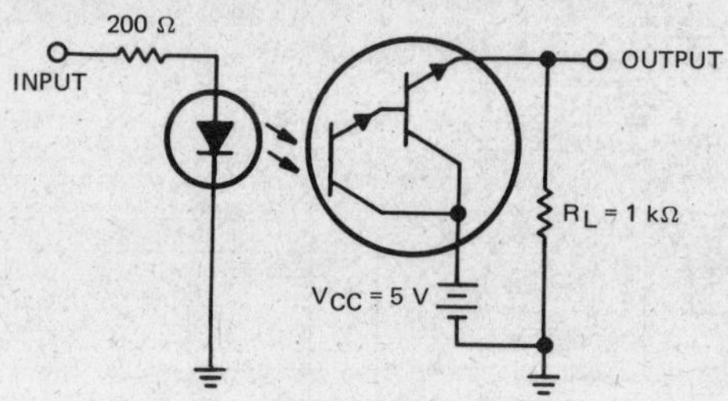

TEST CIRCUIT

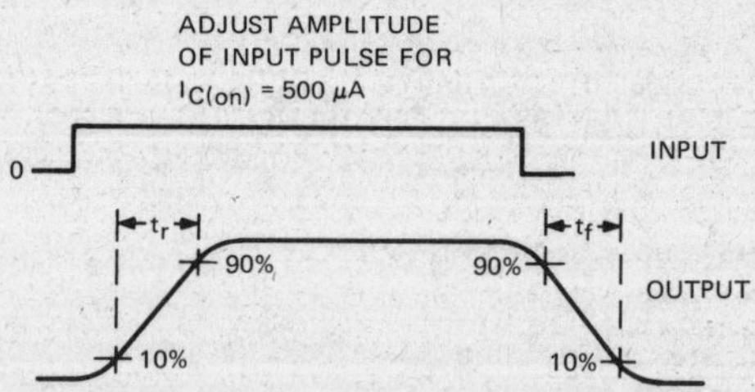

NOTE: The input pulse is supplied by a generator having the following characteristics: Z_{out} = 50 Ω, $t_r \leqslant 10$ μs, $t_f \leqslant 10$ μs, t_w = 10 ms, duty cycle ≈ 50%.

VOLTAGE WAVEFORMS

FIGURE 1 – SWITCHING TIMES

D2696, APRIL 1983

OPTOELECTRONIC ENCODER ASSEMBLIES FOR TRANSMISSIVE SENSING APPLICATIONS

- **Compatible With Standard TTL Integrated Circuits**
- **High-Speed Switching . . . t_r = 15 μs, t_f = 15 μs Typical**
- **For Counting, Speed Control, Position Sensing, Beginning-of-Tape/End-of-Tape Sensing, and High-Voltage Isolation**
- **Designed for Base Mounting—Standard 7,6 mm (0.300-Inch) Dual-In-Line Pin Spacing**
- **PC Board or Bracket Mounting**
- **Contains Infrared Emitter and Phototransistor**
- **1.02-mm (0.040-inch) Sensor Window Provides High On/Off Resolution**

mechanical data

Each assembly consists of an infrared-emitting diode and an n-p-n silicon phototransistor mounted in a plastic housing. The assembly will withstand soldering temperature with no deformation and device performance characteristics remain stable when operated in high-humidity conditions. Total assembly weight is approximately 1.5 grams.

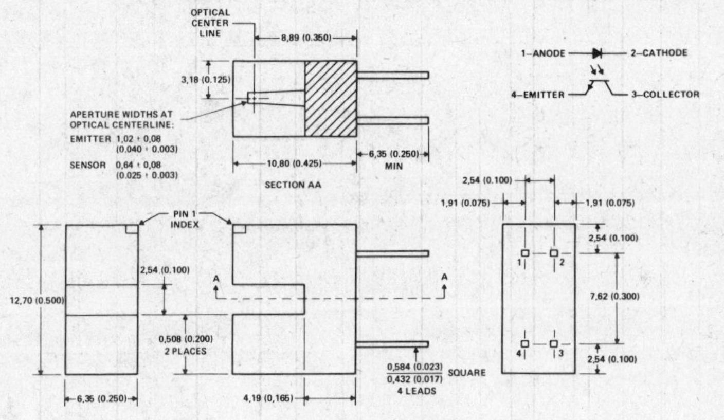

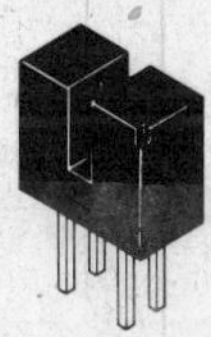

NOTE a: Tolerance is ± 0,26 mm (0.010 inch) unless otherwise noted.
ALL DIMENSIONS ARE IN MILLIMETERS AND PARENTHETICALLY IN INCHES

CL-81

absolute maximum ratings at 25°C free-air temperature (unless otherwise noted)

Source Reverse Voltage	3 V
Source Continuous Forward Current (See Note 1)	40 mA
Source Peak Forward Current (See Note 2)	3 A
Source Collector-Emitter Voltage	30 V
Sensor Emitter-Collector Voltage	7 V
Sensor Continuous Dissipation at (or below) 25°C Free-Air Temperature (See Note 3)	50 mW
Source-to-Sensor Voltage	±4 kV
Operating Free-Air Temperature Range	−40°C to 80°C
Storage Temperature Range	−40°C to 100°C
Lead Temperature 1,6 mm (1/16 Inch) from Assembly for 5 Seconds	240°C

NOTES: 1. Derate linearly to 80°C free-air temperature at the rate of 0.73 mA/°C.
2. This value applies for $t_w \leq 1$ μs, PRR ≤ 300 pps.
3. Derate linearly to 80°C free-air temperature at the rate of 0.91 mW/°C.

Copyright © 1983 by Texas Instruments Incorporated

electrical characteristics at 25 °C free-air temperature

PARAMETER		TEST CONDITIONS†	TIL169-1			TIL169-2			UNIT
			MIN	TYP	MAX	MIN	TYP	MAX	
$V_{(BR)CEO}$	Collector-Emitter Breakdown Voltage	I_C = 100 μA, I_F = 0	30			30			V
$V_{(BR)ECO}$	Emitter-Collector Breakdown Voltage	I_E = 100 μA, I_F = 0	7			7			V
$I_{C(off)}$	Off-State Collector Current	V_{CE} = 10 V, I_F = 0		5	100		5	100	nA
$I_{C(on)}$	On-State Collector Current	V_{CE} = 5 V, I_F = 20 mA	0.2	0.5		0.6	1		mA
V_F	Input-Diode Static Forward Voltage	I_F = 20 mA		1.2	1.6		1.2	1.6	V

switching characteristics at 25 °C free-air temperature

PARAMETER		TEST CONDITIONS†	MIN	TYP	MAX	UNIT
t_r	Rise Time	V_{CC} = 5 V, $I_{C(on)}$ = 1 mA,		15		μs
t_f	Fall Time	R_L = 1 kΩ, See Figure 1		15		μs

†Stray irradiation outside the range of device sensitivity may be present. A satisfactory condition has been achieved when the parameter being measured approaches a value which cannot be altered by further irradiation shielding.

PARAMETER MEASUREMENT INFORMATION

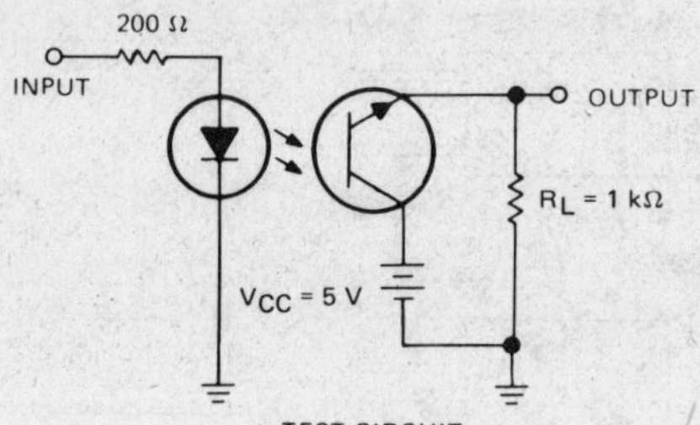

TEST CIRCUIT

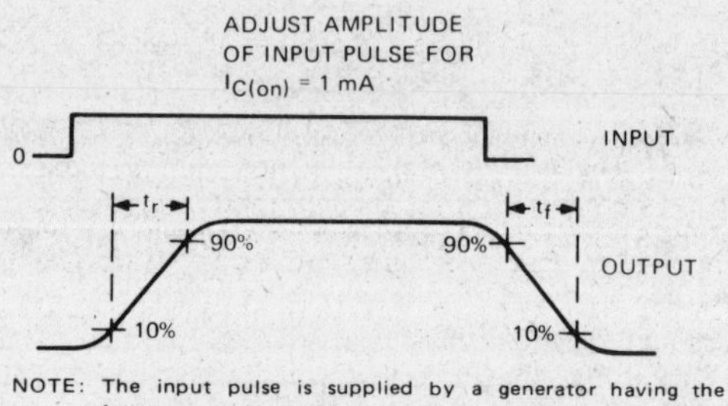

NOTE: The input pulse is supplied by a generator having the following characteristics: Z_{out} = 50 Ω, $t_r \leq 100$ ns, $t_f \leq 100$ ns, t_w = 100 μs, duty cycle ≈2%.

VOLTAGE WAVEFORMS

FIGURE 1–SWITCHING TIMES

HIGH-GAIN OPTOELECTRONIC ENCODER ASSEMBLIES FOR TRANSMISSIVE SENSING APPLICATIONS

- **Compatible With Standard TTL Integrated Circuits**
- **High Current Transfer Ratio . . . 12.5% Min (TIL145)**
- **For Counting, Speed Control, Position Sensing, Beginning-of-Tape/End-of-Tape Sensing, and High-Voltage Isolation**
- **Designed for Base Mounting—Standard 7,6 mm (0.300-Inch) Dual-In-Line Pin Spacing**
- **PC Board or Bracket Mounting**
- **Contains Infrared Emitter and Silicon Darlington Phototransistor**
- **1.02-mm (0.040-inch) Sensor Window Provides High On/Off Resolution**

mechanical data

Each assembly consists of an infrared-emitting diode and an n-p-n silicon Darlington phototransistor mounted in a plastic housing. The assembly will withstand soldering temperature with no deformation and device performance characteristics remain stable when operated in high-humidity conditions. Total assembly weight is approximately 1.5 grams.

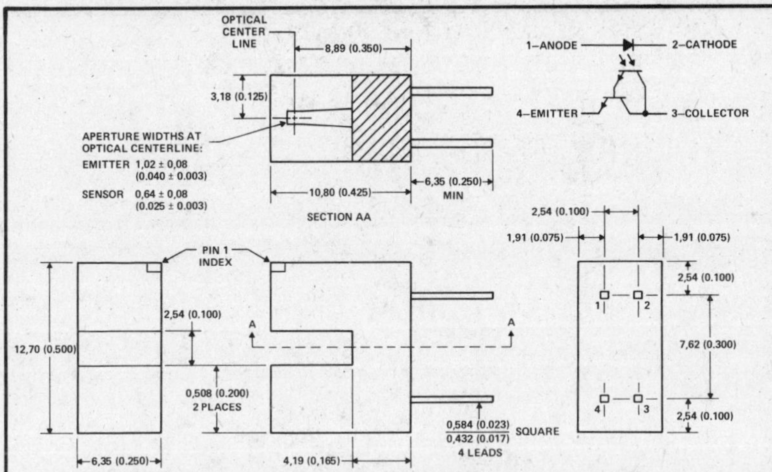

NOTE a: Dimension tolerance is ± 0,26 mm (0.010 inch) unless otherwise noted.
ALL DIMENSIONS ARE IN MILLIMETERS AND PARENTHETICALLY IN INCHES

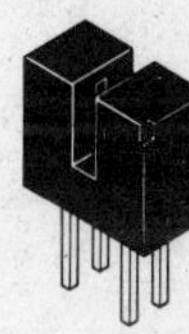

CL-81

absolute maximum ratings at 25°C free-air temperature (unless otherwise noted)

Source Reverse Voltage	3 V
Source Continuous Forward Current (See Note 1)	40 mA
Source Peak Forward Current (See Note 2)	3 A
Sensor Collector-Emitter Voltage	30 V
Sensor Emitter-Collector Voltage	7 V
Sensor Continuous Dissipation at (or below) 25°C Free-Air Temperature (See Note 3)	100 mW
Source-to-Sensor Voltage	±4 kV
Operating Free-Air Temperature Range	−40° to 80°C
Storage Temperature Range	−40°C to 100°C
Lead Temperature 1,6 mm (1/16 Inch) from Assembly for 5 Seconds	240°C

NOTES: 1. Derate linearly to 80°C free-air temperature at the rate of 0.73 mA/°C.
2. This value applies for $t_w \leq 1$ μs, PRR ≤ 300 pps.
3. Derate linearly to 80°C free-air temperature at the rate of 1.81 mW/°C.

Copyright © 1983 by Texas Instruments Incorporated

TEXAS INSTRUMENTS
INCORPORATED
POST OFFICE BOX 225012 • DALLAS, TEXAS 75265

electrical characteristics at 25 °C free-air temperature

PARAMETER		TEST CONDITIONS†	TIL170-1			TIL170-2			UNIT
			MIN	TYP	MAX	MIN	TYP	MAX	
$V_{(BR)CEO}$	Collector-Emitter Breakdown Voltage	I_C = 100 μA, I_F = 0	30			30			V
$V_{(BR)ECO}$	Emitter-Collector Breakdown Voltage	I_E = 100 μA, I_F = 0	7			7			V
$I_{C(off)}$	Off-State Collector Current	V_{CE} = 5 V, I_F = 0		5	100		5	100	nA
$I_{C(on)}$	On-State Collector Current	V_{CE} = 2 V, I_F = 10 mA	0.5	1		2	3.5		mA
V_F	Input-Diode Static Forward Voltage	I_F = 20 mA		1.2	1.6		1.2	1.6	V

switching characteristics at 25 °C free-air temperature

PARAMETER		TEST CONDITIONS†	MIN	TYP	MAX	UNIT
t_r	Rise Time	V_{CC} = 5 V, $I_{C(on)}$ = 500 nA, R_L = 1 kΩ,		1		ms
t_f	Fall Time	See Figure 1		1		ms

† Stray irradiation outside the range of device sensitivity may be present. A satisfactory condition has been achieved when the parameter being measured approaches a value which cannot be altered by further irradiation shielding.

PARAMETER MEASUREMENT INFORMATION

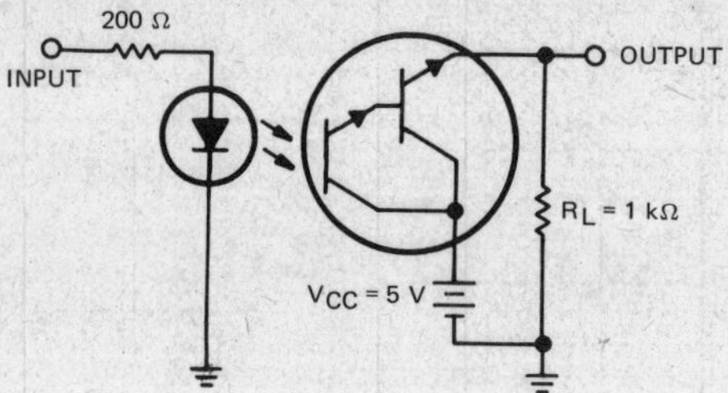

TEST CIRCUIT

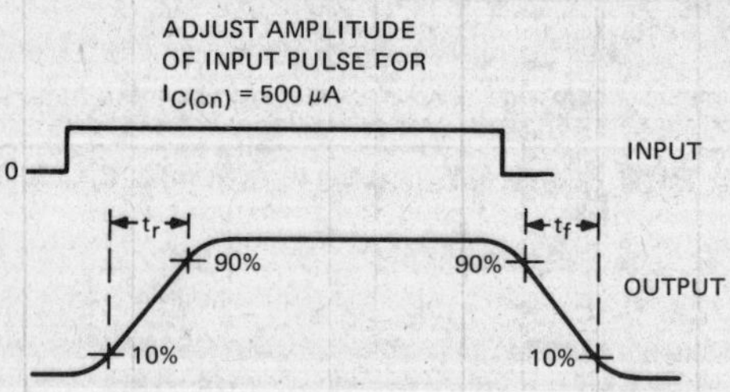

NOTE: The input pulse is supplied by a generator having the following characteristics: Z_{out} = 50 Ω, $t_r \leqslant 10$ μs, $t_f \leqslant 10$ μs, t_w = 10 ms, duty cycle ≈ 50%.

VOLTAGE WAVEFORMS

FIGURE 1 – SWITCHING TIMES

TEXAS INSTRUMENTS
INCORPORATED
POST OFFICE BOX 225012 • DALLAS, TEXAS 75265

INFRARED SENSOR AND EMITTER FOR BAR-CODE READING APPLICATIONS

- **Capable of Reading Black and White Bar Codes, i.e., UPC, EAN, Code 39, HP, and MSI**
- **Designed PCB for Mounting**
- **Contains a Gallium Arsenide Infrared LED and Phototransistor**
- **Reads Offset Press, Dot Matrix, and Printed Codes**
- **Codes Must Be Printed with Inks with a High Carbon Content**

mechanical data

Each assembly contains a Gallium Arsenide Diode that emits light in the 940-nm region and a silicon phototransistor detector. The case is made of high-impact polycarbonate plastic. The assembly weight is approximately 5 grams.

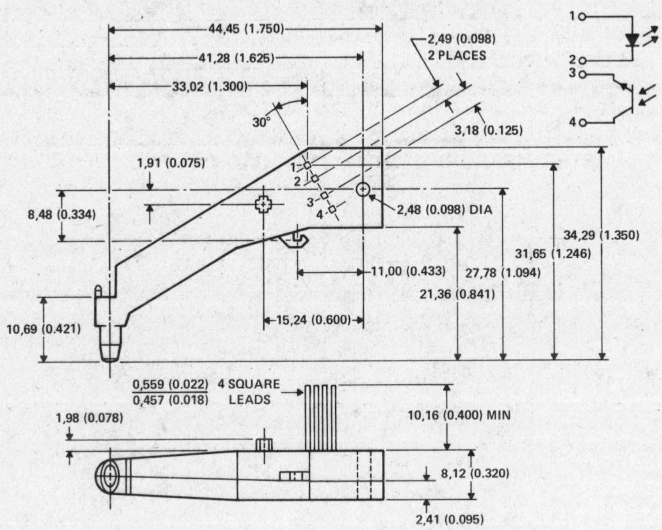

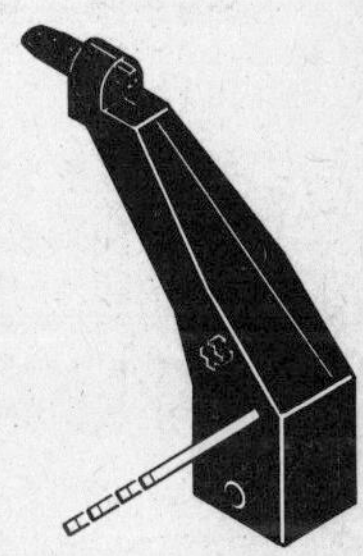

ALL LINEAR DIMENSIONS ARE IN MILLIMETERS AND PARENTHETICALLY IN INCHES

CL-85

absolute maximum ratings at 25°C free-air temperature (unless otherwise noted)

Source Reverse Voltage	2 V
Source Continuous Forward Current	100 mA
Source Peak Forward Current (see Note 1)	1 A
Sensor Collector-Emitter Voltage	25 V
Sensor Emitter-Collector Voltage	5 V
Sensor Continuous Dissipation at (or below) 25°C Free-Air Temperature (see Note 2)	90 mW
Operating Free-Air Temperature Range	−40°C to 70°C
Storage Temperature Range	−40°C to 70°C
Lead Temperature 1,6 mm (1/16 inch) from Assembly for 5 Seconds	240°C

NOTES: 1. This value applies for $t_w \leq 1\ \mu s$, PRR ≤300 pps.
2. Derate linearly to 70°C free-air temperature at the rate of 2 mW/°C.

Copyright © 1983 by Texas Instruments Incorporated

electrical characteristics at 25 °C free-air temperature

PARAMETER		TEST CONDITIONS		MIN	TYP	MAX	UNIT
$V_{(BR)CEO}$	Collector-Emitter Breakdown Voltage	I_C = 100 μA,	I_F = 0	25			V
$V_{(BR)ECO}$	Emitter-Collector Breakdown Voltage	I_E = 100 μA,	I_F = 0	5			V
$I_{C(off)}$	Off-State Collector Current	V_{CE} = 10 V,	I_F = 0			200	nA
$I_{C(on)}$	On-State Collector Current (White Paper)	V_{CE} = 5 V, See Note 3	I_F = 40 mA,	5	30		μA
V_F	Input Diode Static Forward Voltage	I_F = 50 mA			1.2	1.45	V
η	Reading Efficiency (see Note 4)	V_{CE} = 5 V,	I_F = 40 mA	65%		100%	

NOTES: 3. The reflective surface is 9-point chromate paper coated on both sides with low-gloss varnish less than 0,00076 mm (0.0003 inch) thick.

4. This is ratio of (1) the peak-to-peak change in collector current when the red head is scanning a test bar-code pattern to (2) the difference in $I_{C(on)}$ with the read head over white paperand over inked paper. The scanning rate is 767 mm/s (30 in/s), the bar code pattern is comprised of 0,254-mm (0.010-in) bars and spaces, and the ink is Pantone 419C or other high-carbon black ink.

switching characteristics at 25 °C free-air temperature

PARAMETER		TEST CONDITIONS†	MIN	TYP	MAX	UNIT
t_r	Rise Time	V_{CC} = 10 V, $I_{C(on)}$ = 1 mA, R_L = 1 kΩ, See Figure 1		125		μs
t_f	Fall Time			125		μs

†Stray irradiation outside the range of device sensitivity may be present. A satisfactory condition has been achieved when the parameter being measured approaches a value that cannot be altered by further irradiation shielding.

PARAMETER MEASUREMENT INFORMATION

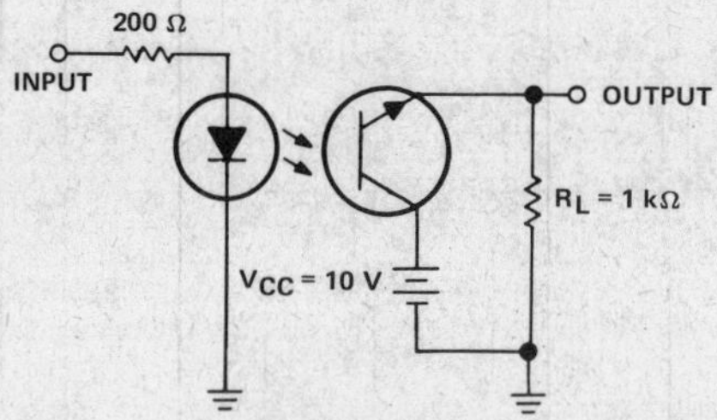

TEST CIRCUIT

ADJUST AMPLITUDE
OF INPUT PULSE FOR
$I_{C(on)}$ = 1 mA

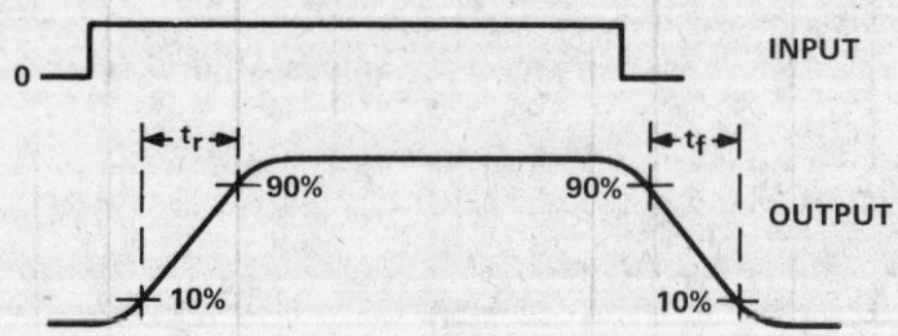

NOTE: The input pulse is supplied by a generator having the following characteristics: Z_{out} = 50 Ω, $t_r \leqslant$ 100 ns, $t_f \leqslant$ 100 ns, t_w = 10 μs, duty cycle ≈ 2%.

VOLTAGE WAVEFORMS

FIGURE 1—SWITCHING TIMES

TEXAS INSTRUMENTS
INCORPORATED
POST OFFICE BOX 225012 • DALLAS, TEXAS 75265

Light-Emitting Diodes

(LEDs/Solid-State Lamps)

- **Quick Reference Guide**
- **Various Plastic Packages**

  T-1

  T-1¾
- **Various Colors Available**

  Red, High-Efficiency Red, Yellow, and Green
- **Panel-Mounting Hardware**

LIGHT-EMITTING DIODES
QUICK REFERENCE GUIDE

DEVICE	COLOR	LENS	BRIGHTNESS MIN (mcd)	@ I_F (mA)	PACKAGE† (LAMP SIZE)	FEATURES
TIL209A	Red	Diffused	0.5	20	CL-9 (T-1)	
TIL212-1	Yellow	Diffused	0.8	20		
TIL212-2	Yellow	Diffused	2.1	20		
TIL216-1	Red	Diffused	2.1	20		
TIL216-2	Red	Diffused	6	20		
TIL220	Red	Diffused	0.8	20	CL-10 (T-1¾)	
TIL221	Red	Clear	1	20		
TIL224-1	Yellow	Diffused	2.1	20	CL-10 (T-1¾)	High intensity
TIL224-2	Yellow	Diffused	6	20		
TIL228-1	Red	Diffused	2.1	20		
TIL228-2	Red	Diffused	6	20		
TIL232-1	Green	Diffused	0.5	20	CL-9 (T-1)	
TIL232-2	Green	Diffused	1.3	20		
TIL234-1	Green	Diffused	0.8	20	CL-10 (T-1¾)	High intensity
TIL234-2	Green	Diffused	2.1	20		
5082-4550	Yellow	Diffused	1	10	CL-10 (T-1¾)	Direct replacements for Hewlett-Packard parts
5082-4555	Yellow	Diffused	2.2	10		
5082-4650	Red	Diffused	1	10		
5082-4655	Red	Diffused	3	10		
5082-4950	Green	Diffused	1	20		
5082-4955	Green	Diffused	2.2	20		

†The following accessories are available: panel mounting bushings TILM1 for CL-9 (T1) and TILM4 for CL-10 (T-1¾).

9 LEDs

TEXAS INSTRUMENTS
INCORPORATED
POST OFFICE BOX 225012 • DALLAS, TEXAS 75265

YELLOW, RED, OR GREEN LIGHT SOURCES

- 90-Degree Viewing Angle
- Rugged Construction
- Solid-State Reliability
- Compatible with TTL Circuitry
- Versatile Mounting in PCB, Panel, or Socket
- Replacements for Popular Hewlett-Packard Devices

DEVICE TYPE	SOURCE	LENS MATERIAL
5082-4550	Yellow	Diffused yellow plastic
5082-4555	Bright yellow	
5082-4650	Red	Diffused red plastic
5082-4655	Bright red	
5082-4950	Green	Diffused-green plastic
5082-4955	Bright green	

mechanical data

These devices are similar in size to lamp style T-1¾ and may be panel mounted using mounting clip TILM4.

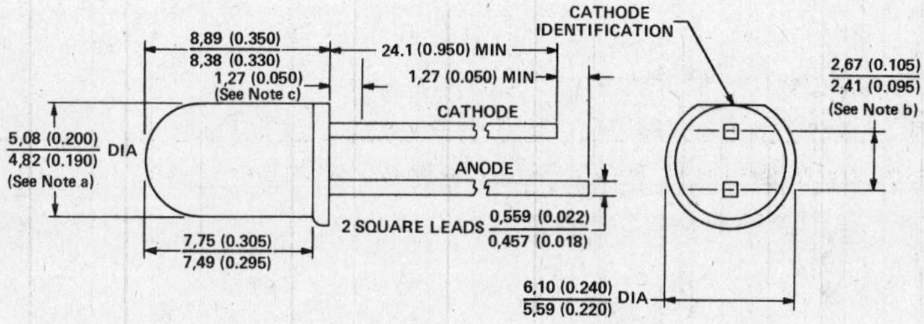

NOTES: a. This diameter is measured 0,13 (0.005) above the shoulder.
b. Lead spacing is measured where the leads emerge from the package.
c. Within this zone the package contour is not controlled.

ALL DIMENSIONS ARE IN MILLIMETERS AND PARENTHETICALLY IN INCHES

CL-10

9 LEDs

absolute maximum ratings at 25°C free-air temperature (unless otherwise noted)

Reverse Voltage at 25°C Free-Air Temperature	5 V
Continuous Forward Current at (or below) 25°C Free-Air Temperature (See Note 1)	50 mA
Operating Free-Air Temperature Range	−40°C to 80°C
Storage Temperature Range	−40°C to 100°C
Lead Temperature 1,6 mm (1/16 Inch) from Case for 3 Seconds	260°C

NOTE 1: Derate linearly to 10 mA at 80°C free-air temperature at the rate of 0.73 mA/°C.

Copyright © 1982 by Texas Instruments Incorporated

'82

electrical characteristics at 25°C free-air temperature

PARAMETER		TEST CONDITIONS	YELLOW 5082-4550 MIN	TYP	MAX	YELLOW 5082-4555 MIN	TYP	MAX	RED 5082-4650 MIN	TYP	MAX	RED 5082-4655 MIN	TYP	MAX	GREEN 5082-4950 MIN	TYP	MAX	GREEN 5082-4955 MIN	TYP	MAX	UNIT
I_V	Luminous Intensity	I_F = 10 mA	1						1						1						mcd
		I_F = 20 mA				2.2						3						2.2			
λ_p	Wavelength at Peak Emission	I_F = 10 mA		583						635						565					nm
		I_F = 20 mA					583						635						565		
θ_{HI}	Half-Intensity Beam Angle			90°			90°			90°			90°			90°			90°		
V_F	Forward Voltage	I_F = 10 mA			3						3						3				V
		I_F = 20 mA						3						3						3	
I_R	Static Reverse Voltage	V_R = 5 V			100			100			100			100			100			100	µA

9 LEDs

TEXAS INSTRUMENTS
INCORPORATED
POST OFFICE BOX 225012 • DALLAS, TEXAS 75265

GALLIUM ARSENIDE PHOSPHIDE LIGHT-EMITTING DIODE

D1637, JUNE 1973—REVISED FEBRUARY 1983

DESIGNED TO EMIT VISIBLE RED LIGHT WHEN FORWARD BIASED

- **Recommended for Application in Visual Indicators, Alpha-Numeric Displays, and Built-In Diagnostics**
- **High Brightness with Solid-State Reliability**
- **Compatible with Most TTL and DTL Circuits**
- **Ideal as Fault or Trouble Indicator**
- **Filled-Epoxy Lens Provides Diffused Source**
- **Ideal for Socket, Printed Circuit Board, and 1.6-mm (1/16-Inch) Panel Mounting Techniques**

mechanical data

This device has a red molded filled-epoxy body. It is similar in size to lamp style T-1 and may be panel-mounted using mounting clip TILM1 (formally TIL209MC).

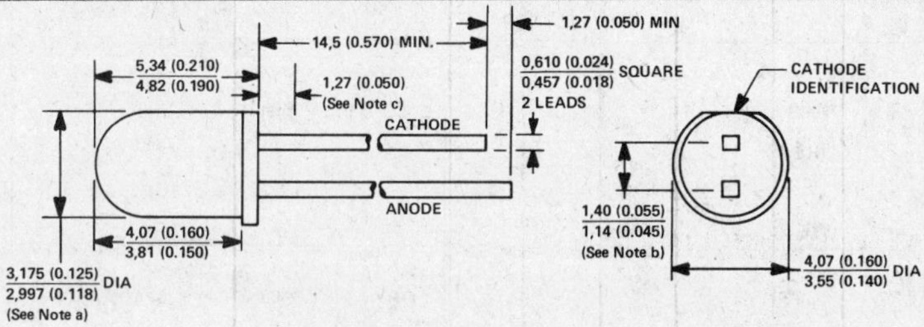

NOTES: a. This diameter is measured 0,127 mm (0.005 inches) from the shoulder.
b. Lead spacing is measured where the leads emerge from the package.
c. Within this zone the package contour is not controlled.

ALL DIMENSIONS ARE IN MILLIMETERS AND PARENTHETICALLY IN INCHES

CL-9

absolute maximum ratings

Reverse Voltage at 25°C Free-Air Temperature . . . . 3 V
Continuous Forward Current at (or below) 25°C Free-Air Temperature (See Note 1) . . . . 40 mA
Operating Free-Air Temperature Range . . . . −40°C to 80°C
Storage Temperature Range . . . . −40°C to 100°C
Lead Temperature 1,6 mm (1/16 Inch) from Case for 3 Seconds . . . . 260°C

operating characteristics at 25°C free-air temperature

	PARAMETER	TEST CONDITIONS	MIN	TYP	MAX	UNIT
I_V	Luminous Intensity (See Note 2)	I_F = 20 mA	0.5			mcd
λ_p	Wavelength at Peak Emission	I_F = 20 mA	630	650	670	nm
V_F	Static Forward Voltage	I_F = 20 mA		1.6	2	V
I_R	Static Reverse Current	V_R = 3 V			100	μA

NOTES: 1. Derate linearly to 80°C free-air temperature at the rate of 0.73 mA/°C.
2. Luminous intensity is measured with a light sensor and filter combination that approximates the CIE (International Commission on Illumination) eye-response curve.

Copyright © 1983 by Texas Instruments Incorporated

TEXAS INSTRUMENTS
INCORPORATED
POST OFFICE BOX 225012 • DALLAS, TEXAS 75265

TYPICAL CHARACTERISTICS

RELATIVE SPECTRAL CHARACTERISTICS

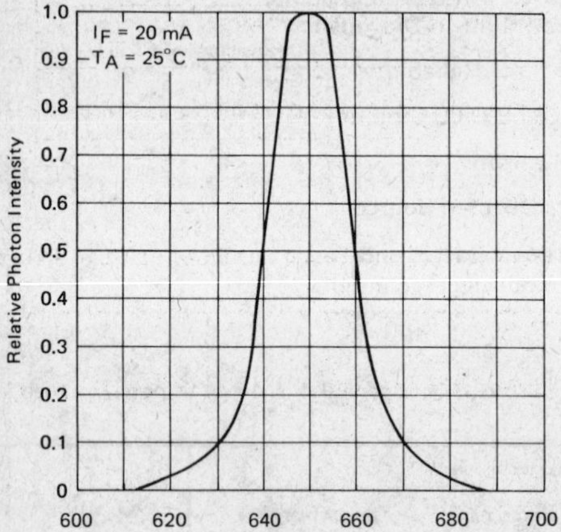

FIGURE 1

RELATIVE LUMINOUS INTENSITY
vs
FREE-AIR TEMPERATURE

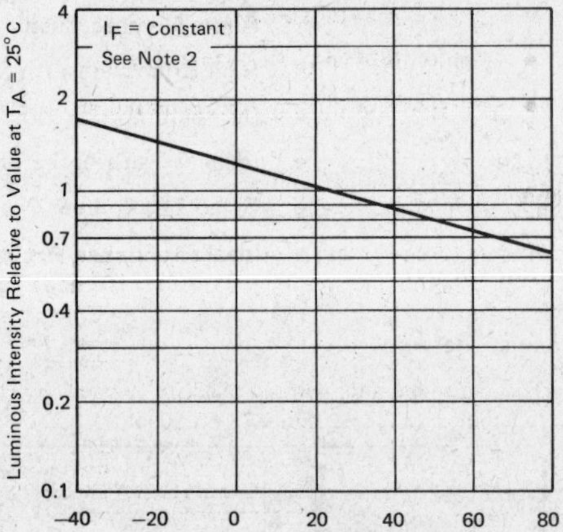

FIGURE 2

RELATIVE LUMINOUS INTENSITY
vs
FORWARD CURRENT

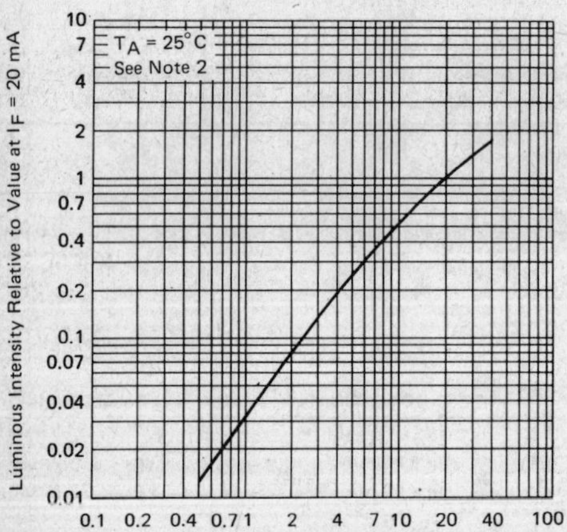

FIGURE 3

FORWARD CONDUCTION CHARACTERISTICS

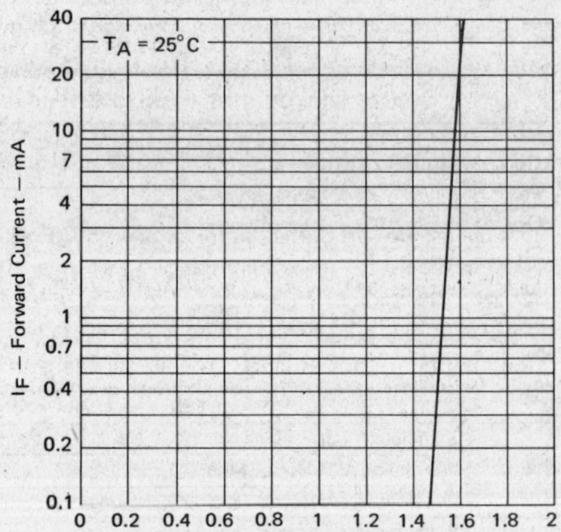

FIGURE 4

NOTE 2: Luminous intensity is measured with a light sensor and filter combination that approximates the CIE (International Commission on Illumination) eye-response curve.

D2500, OCTOBER 1978—REVISED FEBRUARY 1983

YELLOW, RED, OR GREEN HIGH-INTENSITY LIGHT SOURCES

- Good Viewing Angles
- Rugged Construction
- Solid-State Reliability
- Compatible with TTL Circuitry
- Versatile Mounting in PCB, Panel, or Socket

DEVICE	DESCRIPTION
TIL212	Yellow source Diffused yellow plastic body
TIL216	Red source Diffused red plastic body
TIL232	Green source Diffused green plastic body

mechanical data

These devices are similar in size to lamp style T-1 and may be panel-mounted using mounting clip TILM1 (formerely TIL209MC).

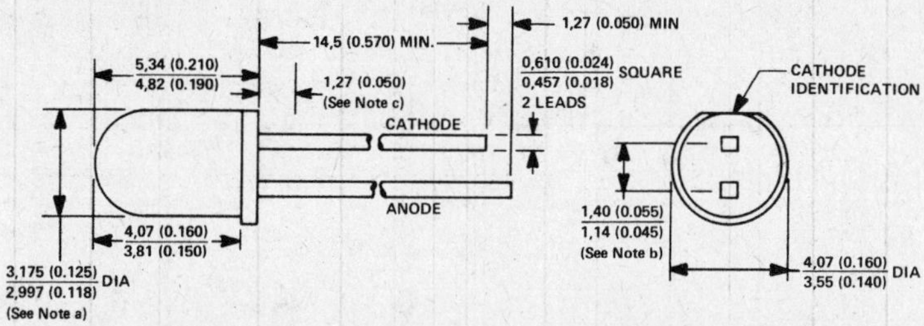

NOTES: a. This diameter is measured 0,127 mm (0.005 inches) from the shoulder.
b. Lead spacing is measured where the leads emerge from the package.
c. Within this zone the package contour is not controlled.

ALL DIMENSIONS ARE IN MILLIMETERS AND PARENTHETICALLY IN INCHES

CL-9

absolute maximum ratings at 25°C free-air temperature (unless otherwise noted)

Reverse Voltage	5 V
Continuous Forward Current (See Note 1)	50 mA
Peak Forward Current (See Note 2)	1 A
Operating Free-Air Temperature Range	−40°C to 80°C
Storage Temperature Range	−40°C to 100°C
Lead Temperature 1,6 mm (1/16 Inch) From Case for 3 Seconds	260°C

NOTES: 1. Derate linearly to 80°C free-air temperature at the rate of 0.91 mA/°C.
2. This value applies for t_w = 1 μs, PRR = 300 Hz.

Copyright © 1983 by Texas Instruments Incorporated

3

operating characteristics at 25°C free-air temperature

PARAMETER		TEST CONDITIONS		MIN	TYP	MAX	UNIT
I_v	Luminous Intensity (See Note 3)	I_F = 20 mA	TIL212-1	0.8			mcd
			TIL212-2	2.1			
			TIL216-1	2.1			
			TIL216-2	6			
			TIL232-1	0.5			
			TIL232-2	1.3			
λ_p	Wavelength at Peak Emission	I_F = 20 mA	TIL212		580		nm
			TIL216		620		
			TIL232		560		
θ_{HI}	Half-Intensity Beam Angle	I_F = 20 mA			60°		
V_F	Static Forward Voltage	I_F = 20 mA				3.2	V
I_R	Static Reverse Current	V_R = 5 V				100	μA

NOTE 3: Luminous intensity is measured with a light sensor and filter combination that approximates the CIE (International Commission on Illumination) eye-response curve.

TYPICAL CHARACTERISTICS

FORWARD CURRENT
vs
FORWARD VOLTAGE

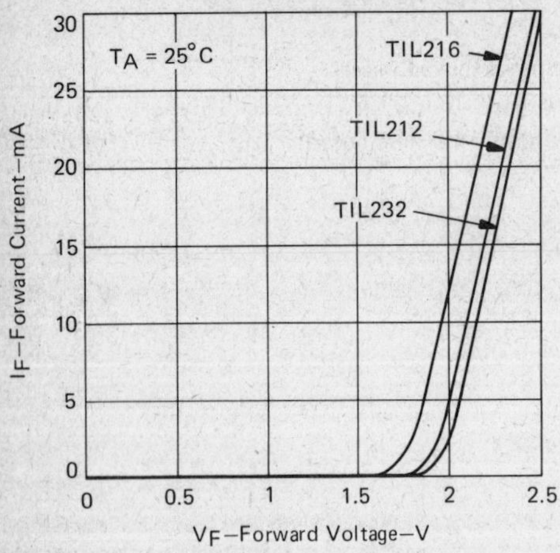

FIGURE 1

RELATIVE LUMINOUS INTENSITY
vs
ANGULAR DISPLACEMENT

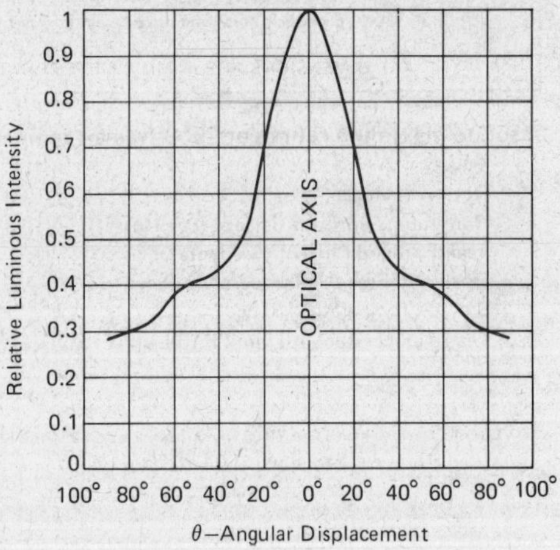

FIGURE 2

GALLIUM ARSENIDE PHOSPHIDE LIGHT-EMITTING DIODES

D1638, JULY 1973–REVISED FEBRUARY 1983

DESIGNED TO EMIT VISIBLE RED LIGHT WHEN FORWARD BIASED

- Recommended for Panel Mounted Visual Indicators
- High Brightness with Solid-State Reliability
- Compatible with Most TTL Circuits
- Leads of TIL220 and TIL221 are Designed to be Wire-Wrapped
- Leads of TIL220S are Designed for PCB Insertion
- Filled-Epoxy Lens of TIL220 and TIL220S Provides Diffused Source
- Clear-Epoxy Lens of TIL221 Provides Pin-Point Source

mechanical data

TIL220 and TIL220S both have red molded filled-epoxy bodies. TIL221 has a colorless clear molded epoxy body. The devices are similar in size to lamp style T1¾ and may be panel mounted using mounting clip TILM4.

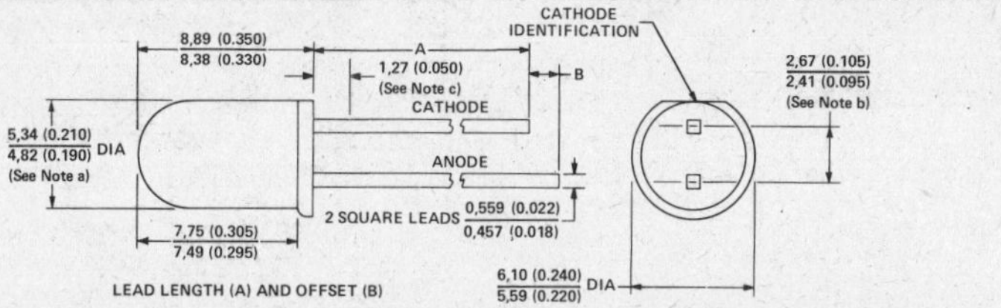

DEVICE	A	B
TIL220, TIL221	19 (0.750) MIN	1,27 (0.050) MIN
TIL220S	3,3 (0.130) MIN	0,25 (0.010) MIN

NOTES: a. This diameter is measured 0,13 (0.005) above the shoulder.
b. Lead spacing is measured where the leads emerge from the package.
c. Within this zone the package contour is not controlled.

ALL DIMENSIONS ARE IN MILLIMETERS AND PARENTHETICALLY IN INCHES

absolute maximum ratings

Reverse Voltage at 25°C Free-Air Temperature	3 V
Continuous Forward Current at (or below) 25°C Free-Air Temperature (See Note 1)	50 mA
Power Dissipation	See Note 2
Operating Free-Air Temperature Range	−40°C to 80°C
Storage Temperature Range	−40°C to 100°C
Lead Temperature 1,6 mm (1/16 Inch) from Case for 3 Seconds	260°C

operating characteristics at 25°C free-air temperature

PARAMETER		TEST CONDITIONS	TIL220, TIL220S			TIL221			UNIT
			MIN	TYP	MAX	MIN	TYP	MAX	
I_V	Luminous Intensity (See Note 3)	I_F = 20 mA	0.8			1			mcd
λ_p	Wavelength at Peak Emission	I_F = 20 mA		650			650		nm
V_F	Static Forward Voltage	I_F = 20 mA		1.6	2		1.6	2	V
I_R	Static Reverse Current	V_R = 3 V			100			100	μA

NOTES: 1. Derate linearly to 80°C free-air temperature at the rate of 0.91 mA/°C.
2. The package is capable of dissipating whatever power ($V_F \times I_F$) is developed at any level of forward current at or below the rated amount. Typical junction-to-free-air thermal resistance, $R_{\theta JA}$, is 230°C/W.
3. Luminous intensity is measured with a light sensor and filter combination that approximates the CIE (International Commission on Illumination) eye-response curve.

Copyright © 1983 by Texas Instruments Incorporated

TYPICAL CHARACTERISTICS

RELATIVE SPECTRAL CHARACTERISTICS

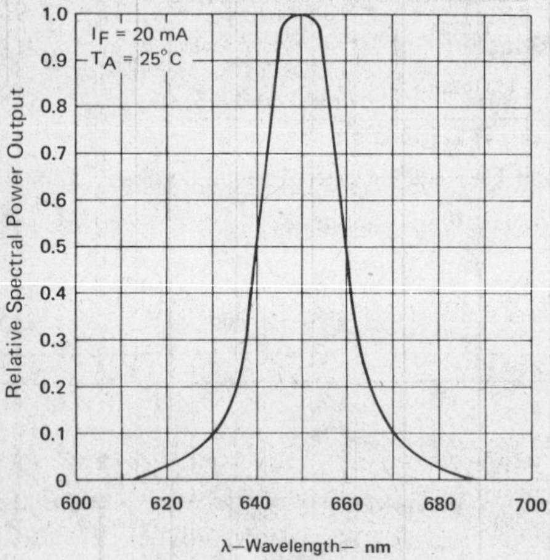

FIGURE 1

RELATIVE LUMINOUS INTENSITY vs FREE-AIR TEMPERATURE

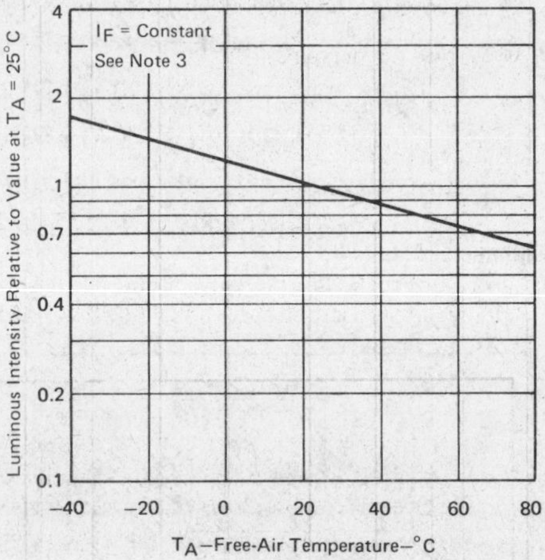

FIGURE 2

RELATIVE LUMINOUS INTENSITY vs FORWARD CURRENT

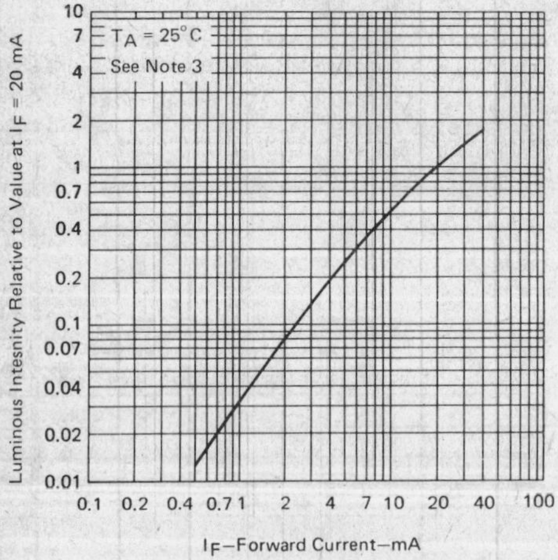

FIGURE 3

FORWARD CONDUCTION CHARACTERISTICS

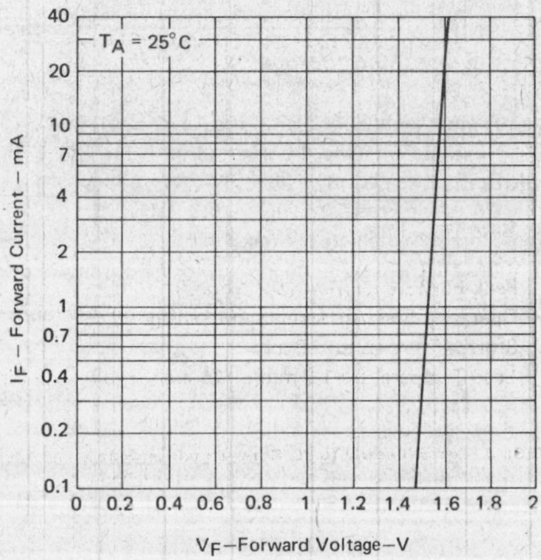

FIGURE 4

NOTE 3: Luminous intensity is measured with a light sensor and filter combination that approximates the CIE (International Commission on Illumination) eye-response curve.

TEXAS INSTRUMENTS
INCORPORATED
POST OFFICE BOX 225012 • DALLAS, TEXAS 75265

YELLOW, RED, OR GREEN HIGH-INTENSITY LIGHT SOURCES

- Good Viewing Angles
- Rugged Construction
- Solid-State Reliability
- Compatible with TTL Circuitry
- Versatile Mounting in PCB, Panel, or Socket

DEVICE	DESCRIPTION
TIL224	Yellow source Diffused yellow plastic body
TIL228	Red source Diffused red plastic body
TIL234	Green source Diffused green plastic body

mechanical data

These devices are similar in size to lamp style T1¾ and may be panel mounted using mounting clip TILM4 (formerly TILM2).

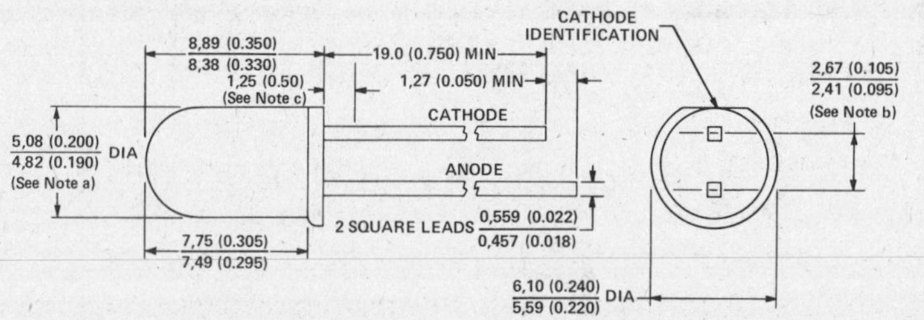

NOTES: a. This diameter is measured 0,127 mm (0.005 inches) from the shoulder.
b. Lead spacing is measured where the leads emerge from the package.
c. Within this zone the package contour is not controlled.

ALL DIMENSIONS ARE IN MILLIMETERS AND PARENTHETICALLY IN INCHES

CL-10

absolute maximum ratings at 25°C free-air temperature (unless otherwise noted)

Reverse Voltage	5 V
Continuous Forward Current (See Note 1)	50 mA
Peak Forward Current (See Note 2)	1 A
Operating Free-Air Temperature Range	−40°C to 80°C
Storage Temperature Range	−40°C to 100°C
Lead Temperature 1,6 mm (1/16 inch) From Case for 3 Seconds	260°C

NOTES: 1. Derate linearly to 80°C free-air temperature at the rate of 0.91 mA/°C.
2. This value applies for t_w = 1 μs, PRR = 300 Hz.

Copyright © 1983 by Texas Instruments Incorporated

operating characteristics at 25°C free-air temperature

	PARAMETER	TEST CONDITIONS		MIN	TYP	MAX	UNIT
I_v	Luminous Intensity (See Note 3)	I_F = 20 mA	TIL224-1	2.1			mcd
			TIL224-2	6			
			TIL228-1	2.1			
			TIL228-2	6			
			TIL234-1	0.8			
			TIL234-2	2.1			
λ_p	Wavelength at Peak Emission	I_F = 20 mA	TIL224		580		nm
			TIL228		620		
			TIL234		560		
θ_{HI}	Half-Intensity Beam Angle	I_F = 20 mA			60°		
V_F	Static Forward Voltage	I_F = 20 mA				3.2	V
I_R	Static Reverse Current	V_R = 5 V				100	μA

NOTE 3: Luminous intensity is measured with a light sensor and filter combination that approximates the CIE (International Commission on Illumination) eye-response curve.

TYPICAL CHARACTERISTICS

FORWARD CURRENT
vs
FORWARD VOLTAGE

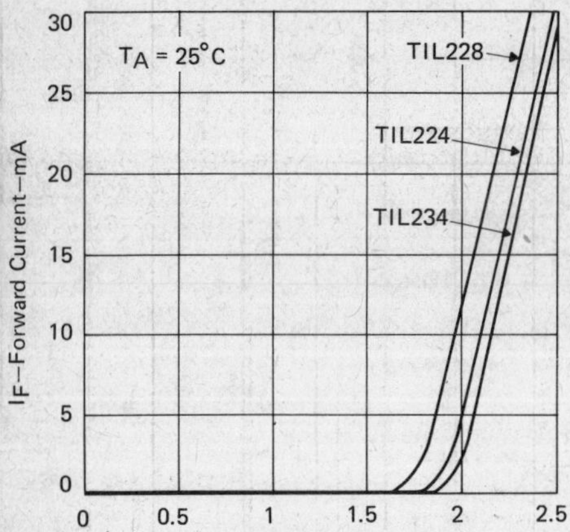

FIGURE 1

RELATIVE LUMINOUS INTENSITY
vs
ANGULAR DISPLACEMENT

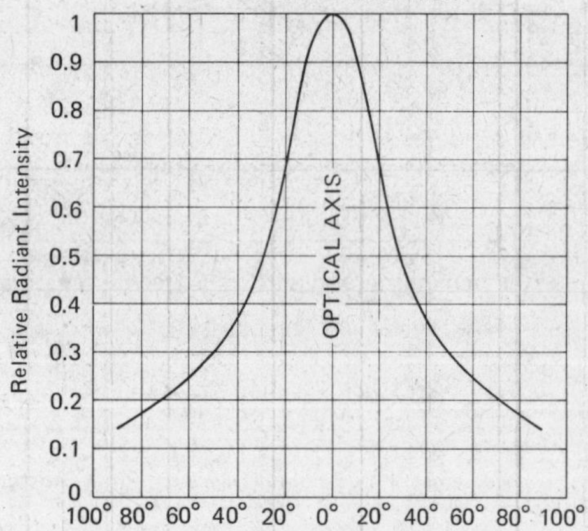

FIGURE 2

D2506, SEPTEMBER 1978

FORMERLY TIL209MC
FOR LIGHT-EMITTING DIODES SIMILAR IN SIZE TO LAMP STYLE T1

installation instructions

The bushing can be mounted in any panel having a thickness up to 2 mm (5/64 inch). To mount the bushing, drill a hole of diameter 5,2 mm (13/64 or 0.205 inch) and remove the extruded burr. Insert the bushing through the front of the panel. While holding the bushing in place, insert the LED from the rear of the panel until the LED snaps into place.

mechanical data

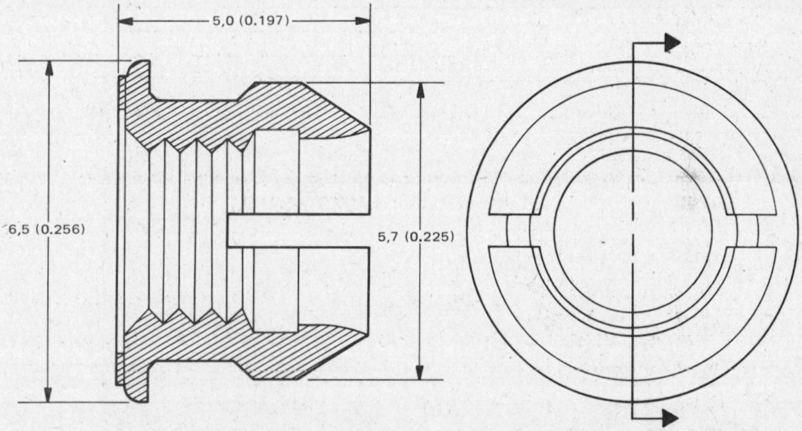

ALL DIMENSIONS ARE IN MILLIMETERS AND PARENTHETICALLY IN INCHES AND ARE NOMINAL

TYPICAL INSTALLATION

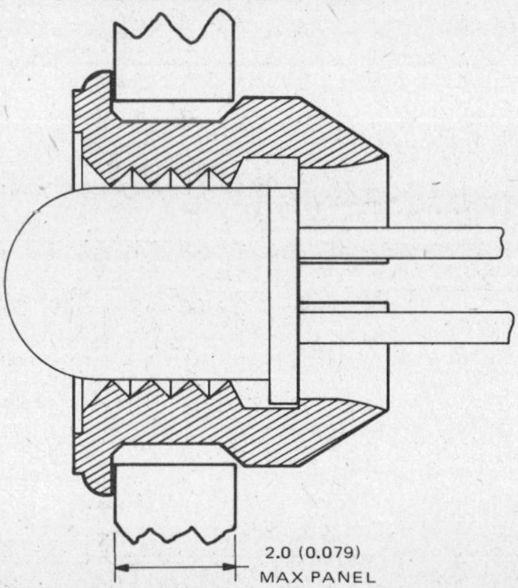

LED PANEL MOUNTING BUSHING WITH LOCK COLLAR

D2689, DECEMBER 1982

CAN BE USED AS A REPLACEMENT FOR TIL220MC AND TILM2 FOR LIGHT-EMITTING DIODES SIMILAR IN SIZE TO LAMP STYLE T1 3/4

installation instructions

This mounting bushing can be mounted in any panel having a thickness up to 3.2 mm (0.125 inch). To mount the bushing, drill a hole of diameter 6,35 mm (0.250 inch) and remove the extruded burr. Insert the bushing through the front of the panel. While holding the bushing in place, insert the LED from the rear of the panel until the LED snaps into place. The orientation of the flat side of the LED, which denotes the cathode lead, must be noted prior to insertion. After the LED is seated with its mounting flange snapped in the slot, push the lock collar over the rear side of the bushing until seated flush with the panel.

mechanical data

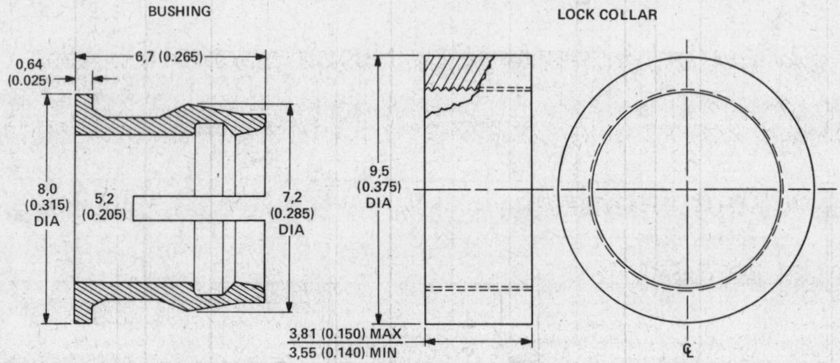

ALL DIMENSIONS ARE IN MILLIMETERS AND PARENTHETICALLY IN INCHES

TYPICAL INSTALLATION

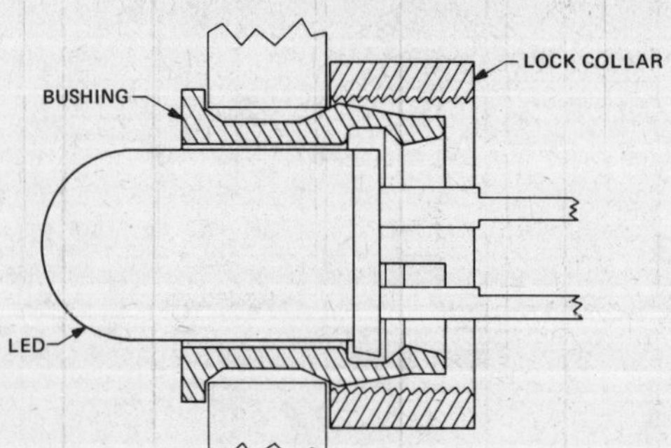

Copyright © 1982 by Texas Instruments Incorporated

TEXAS INSTRUMENTS
INCORPORATED
POST OFFICE BOX 225012 • DALLAS, TEXAS 75265

LED Displays

- Quick Reference Guide
- Hexadecimal Display with TTL Logic
- Seven-Segment Display with TTL Logic
- Seven-Segment (0.3 Inch and 0.5 Inch)
- Plus or Minus One (0.3 Inch and 0.5 Inch)
- Alphanumeric
- Multidigit
- Various Colors Available
  Red, High-Efficiency Red, Yellow, and Green

See Section 11 for High-Reliability LED Displays.

SINGLE-DIGIT DISPLAYS
QUICK REFERENCE GUIDE

DEVICE	TYPE OF CHARACTER(S)	CHARACTER HEIGHT mm (INCHES)	COLOR OF DISPLAY	PACKAGE	REMARKS
5082-7730 5082-7731 5082-7740	7-segment	7,6 (0.300)	Red	14- or 10-lead dual-in-line plastic	Direct replacements for Hewlett-Packard devices
TIL302 TIL302A TIL303 TIL303A	7 segment	6,9 (0.270)	Red	14-lead dual-in-line plastic	TIL302—left decimal. TIL303—right decimal
TIL304 TIL304A	Polarity and overflow unit	6,9 (0.270)	Red	14-lead dual-in-line plastic	Right decimal
TIL305	5 X 7 alphanumeric	7,6 (0.300)	Red	14-lead dual-in-line plastic	Left decimal
TIL306 TIL306A TIL307 TIL307A TIL308 TIL308A TIL309 TIL309A	7-segment	6,9 (0.270)	Red	16-lead dual-in-line plastic	TIL306 and TIL308—left decimal TIL307 and TIL309—right decimal
TIL311 TIL311A	Hexadecimal	6,9 (0.270)	Red	14-lead dual-in-line plastic	Logic includes latch, decoder, and driver Left and right decimals
TIL312 TIL313	7-segment	7,6 (0.300)	Red	14-lead dual-in-line plastic	TIL312 has common anode, right and left decimals TIL313 has common cathode, right decimal only
TIL314 TIL315	7-segment	7,6 (0.300)	Green	14-lead dual-in-line plastic	TIL314 has common anode, right and left decimals TIL315 has common cathode, right decimal only
TIL321A TIL322A	7-segment	12,7 (0.500)	Red	10-lead dual-in-line plastic	Right decimal, TIL321A is common-anode TIL322A is common-cathode
TIL323A TIL324A	7-segment	12,7 (0.500)	Green	10-lead dual-in-line plastic	TIL323A has common anode TIL324A has common cathode
TIL327	Polarity and overflow unit	7,6 (0.300)	Red	14-lead dual-in-line plastic	Plus/minus one with common anode Left decimal
TIL328	Polarity and overflow unit	7,6 (0.300)	Green	14-lead dual-in-line plastic	Plus/minus one with left decimal
TIL330A	Polarity and overflow unit	12,7 (0.500)	Red	10-lead dual-in-line plastic	Plus/minus one with common anode Left decimal
TIL331A	Polarity and overflow unit	12,7 (0.500)	Green	10-lead dual-in-line plastic	Plus/minus one with left decimal Common anode
TIL333 TIL334	7-segment	7,6 (0.300)	High-efficiency Red	14-lead dual-in-line plastic	TIL333 has common anode, right and left decimals TIL334 has common cathode, right decimal only
TIL335	Polarity and overflow unit				TIL335—Plus/minus one with left decimal

TEXAS INSTRUMENTS
INCORPORATED
POST OFFICE BOX 225012 • DALLAS, TEXAS 75265

SINGLE–DIGIT DISPLAYS
QUICK REFERENCE GUIDE (Continued)

DEVICE	TYPE OF CHARACTER(S)	CHARACTER HEIGHT mm (INCHES)	COLOR OF DISPLAY	PACKAGE	REMARKS
TIL339 TIL340	7-segment	7,6 (0.300)	Yellow	14-lead dual-in-line plastic	TIL339–Common anode, right and left decimals TIL340–Common cathode, right decimal TIL341–Plus/minus one with left decimal
TIL341	Polarity and overflow unit				
TIL345 TIL346	7-segment	12,7 (0.500)	Yellow	10-lead dual-in-line plastic	TIL345–Common anode TIL346–Common cathode TIL347–Plus/minus one common anode
TIL347	Polarity and overflow unit				
TIL348 TIL349	7-segment	12,7 (0.500)	High-efficiency Red	10-lead dual-in-line plastic	TIL348–Common anode TIl349–Common cathode TIL350–Plus/minus one common anode
TIL350	Polarity and overflow unit				
TIL729 TIL730	7-segment	12, (0.500)	Red	10-lead dual-in-line plastic	TIL729–Common anode–mitered segments TIL730–Common cathode–mitered segments

MULTIDIGIT DISPLAYS
QUICK REFERENCE GUIDE

DEVICE	TYPE OF CHARACTERS	CHARACTER HEIGHT mm (INCHES)	COLOR	NUMBER OF DIGITS	REMARKS
HDSP6504	Alpha-numeric	3,81 (0.150)	Red	4	ASCII 64-character set plus specials. Uses AC5947 driver
HDSP6508				8	
TIL393-6	7-segment	2,6 (0.102)†	Red	6	Mechanical dimensions are identical for all part numbers.
TIL393-8				8	
TIL393-9				9	
TIL804-8	7-segment	6,9 (0.270)	Red	8	Mechanical dimensions are identical for all part numbers.
TIL804-10				10	
TIL804-12				12	

†Height of magnified character. Additional magnifier available for 3,3 mm (0.130 in) high characters.

D2458, SEPTEMBER 1978—REVISED MARCH 1983

SOLID-STATE RED DISPLAYS

- 7,62-mm (0.300-inch) Character Height
- Wide Viewing Angle
- Designed to be Interchangeable with Hewlett-Packard 5082-7730, -7731, -7740
- Continuous Uniform Segments
- Categorized for Uniformity of Luminous Intensity among Units within Each Category
- High Contrast

absolute maximum ratings

Reverse Voltage at 25 °C Free-Air Temperature, Each Segment or Decimal Point .... 6 V
Peak Forward Current at (or below) 50 °C Free-Air Temperature,
Each Segment or Decimal Point .... 150 mA
Average Forward Current at (or below) 50 °C Free-Air Temperature (See Notes 1 and 2),
Each Segment or Decimal Point .... 25 mA
Operating Free-Air Temperature Range .... −40 °C to 85 °C
Storage Temperature Range .... −40 °C to 85 °C
Lead Temperature 1,6 mm (1/16 inch) below Seating Plane for 3 Seconds .... 260 °C

operating characteristics of each segment or decimal point at 25 °C free-air temperature

PARAMETER		TEST CONDITIONS	5082-7730 5082-7731			5082-7740			UNIT
			MIN	TYP	MAX	MIN	TYP	MAX	
I_V	Average Luminous Intensity per Segment (See Note 3)	I_F = 100 mA, Duty Cycle = 10%		610			610		μcd
		I_F = 20 mA	240	700		240	700		
	Segment-to-Segment Luminous Intensity Ratio	I_F = 20 mA		<1.5:1			<1.5:1		
λ_p	Wavelength at Peak Emission			655			655		nm
λ_d	Dominant Wavelength (See Note 4)			640			640		nm
$\Delta\lambda$	Spectral Bandwidth			20			20		nm
V_F	Static Forward Voltage			1.6	2		1.6	2	V
α_{VF}	Temperature Coefficient of Forward Voltage			−2			−2		mV/°C
I_R	Static Reverse Current	V_R = 3 V			100			100	μA

mechanical data

The display chips are mounted on a header and this assembly is then molded within an electrically nonconductive plastic case. To optimize device performance, materials are used that are limited to certain solvents for cleaning operations. It is recommended that only Freon TF†, isopropanol, or water be used.

5082-7730 5082-7731 5082-7740

COMMON ANODE, LEFT DECIMAL — COMMON ANODE, RIGHT DECIMAL — COMMON CATHODE, RIGHT DECIMAL

†Trademark of E.I. duPont de Nemours, Inc.

NOTES:
1. This average value applies for any 10-ms period.
2. Derate linearly to 10 mA at 85 °C free-air temperature at the rate of 0.25 mA/°C.
3. Luminous intensity is measured with a light sensor and filter combination that approximates the CIE (International Commission on Illumination) eye-response curve.
4. The dominant wavelength is derived from the CIE Chromaticity Diagram and is the single wavelength that defines the color of the emitted light.

Copyright © 1983 by Texas Instruments Incorporated

TEXAS INSTRUMENTS
INCORPORATED
POST OFFICE BOX 225012 • DALLAS, TEXAS 75265

5082-7730
5082-7731
5082-7740

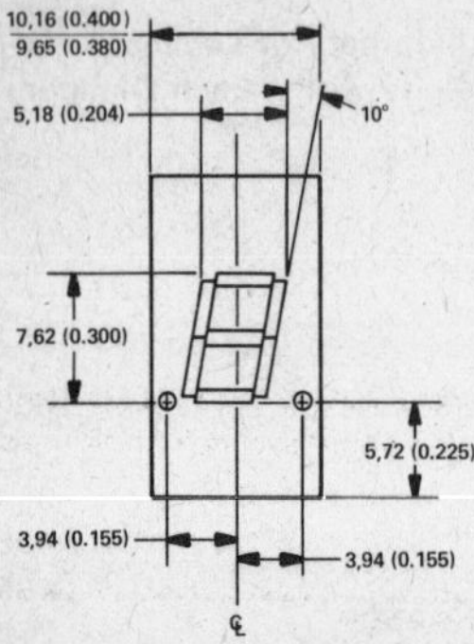

5082-7730
5082-7731

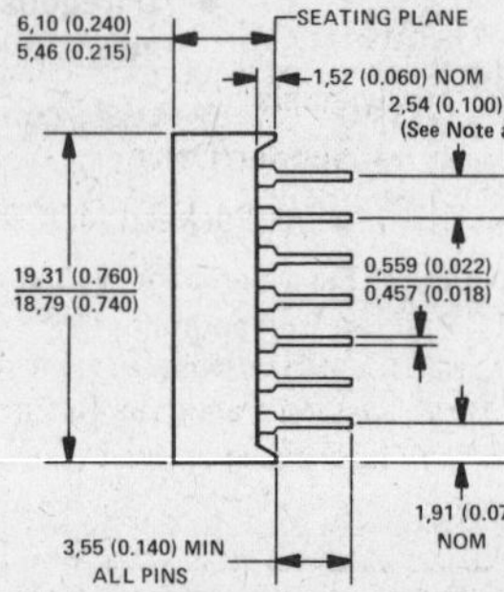

5082-7730
5082-7731

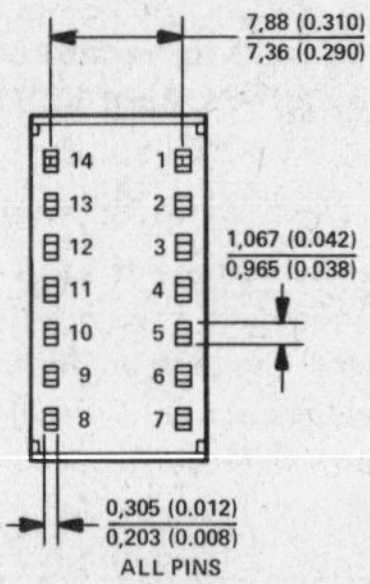

5082-7740

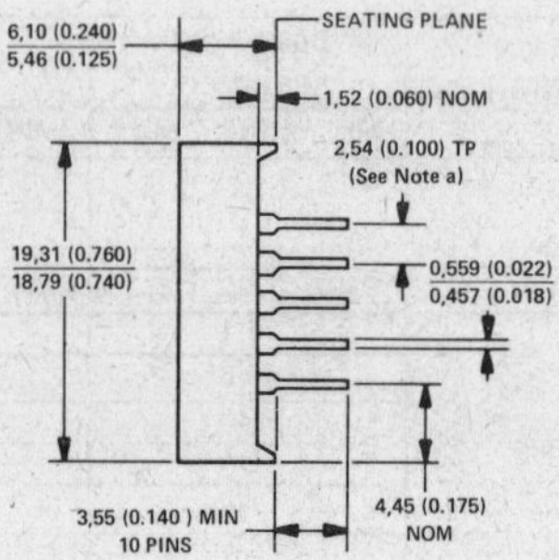

5082-7740

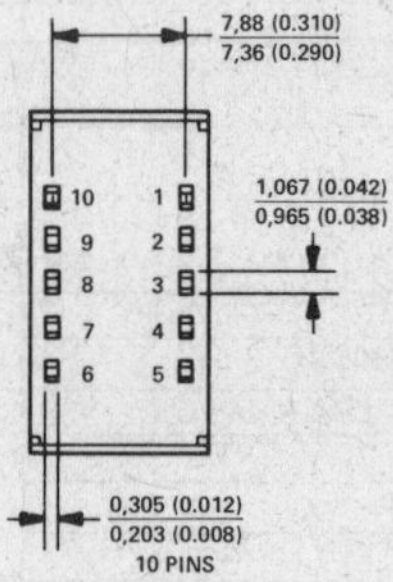

NOTES: a. Each pin centerline is located within 0,26 mm (0.010 inch) of its true longitudinal position.
b. All dimensions associated with segment and decimal point location are nominal.

ALL LINEAR DIMENSIONS ARE IN MILLIMETERS AND PARENTHETICALLY IN INCHES

5082-7730

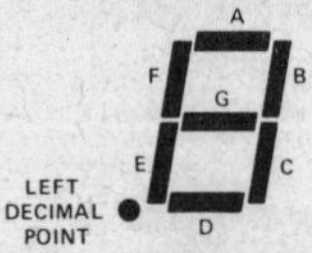

PIN 1–CATHODE A
PIN 2–CATHODE F
PIN 3–ANODE: DIGIT & DECIMAL
PIN 4–OMITTED
PIN 5–OMITTED
PIN 6–CATHODE LEFT DECIMAL
PIN 7–CATHODE E
PIN 8–CATHODE D
PIN 9–OMITTED
PIN 10–CATHODE C
PIN 11–CATHODE G
PIN 12–OMITTED
PIN 13–CATHODE B
PIN 14–ANODE: DIGIT & DECIMAL
PIN 3 IS INTERNALLY CONNECTED TO PIN 14

5082-7731

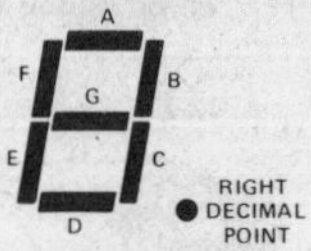

PIN 1–CATHODE A
PIN 2–CATHODE F
PIN 3–ANODE: DIGIT & DECIMAL
PIN 4–OMITTED
PIN 5–OMITTED
PIN 6 OMITTED
PIN 7–CATHODE E
PIN 8–CATHODE D
PIN 9–CATHODE RIGHT DECIMAL
PIN 10–CATHODE C
PIN 11–CATHODE G
PIN 12–OMITTED
PIN 13–CATHODE B
PIN 14–ANODE: DIGIT & DECIMAL
PIN 3 IS INTERNALLY CONNECTED TO PIN 14

5082-7740

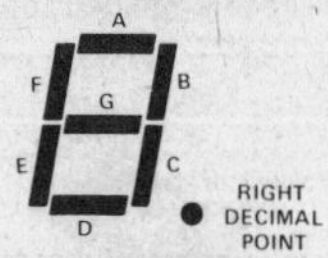

PIN 1–CATHODE: DIGIT & DECIMAL
PIN 2–ANODE F
PIN 3–ANODE G
PIN 4–ANODE E
PIN 5–ANODE D
PIN 6–CATHODE: DIGIT & DECIMAL
PIN 7–ANODE DECIMAL
PIN 8–ANODE C
PIN 9–ANODE B
PIN 10–ANODE A
PIN 1 IS INTERNALLY CONNECTED TO PIN 6

1282

TEXAS INSTRUMENTS
INCORPORATED
POST OFFICE BOX 225012 • DALLAS, TEXAS 75265

D2556, MAY 1980—REVISED APRIL 1983

SOLID-STATE 4- and 8-CHARACTER RED LED DISPLAYS

- 16-Segment Font Plus Colon
- 3,81-mm (0.150-inch) Character Height
- Displays ASCII 64-Character Set Plus Specials
- Compatible with AC5947 18-Segment Decoder/Driver
- Designed to Be Interchangeable with Hewlett Packard HDSP6504, HDSP6508

mechanical data

The displays are formed by placing a one-piece clear plastic lens on a printed-circuit board that contains the light-emitting-diode chips. To optimize device performance, materials are used that are limited to certain solvents for cleaning operations. It is recommended that only Freon† TF, isopropanol, or water be used.

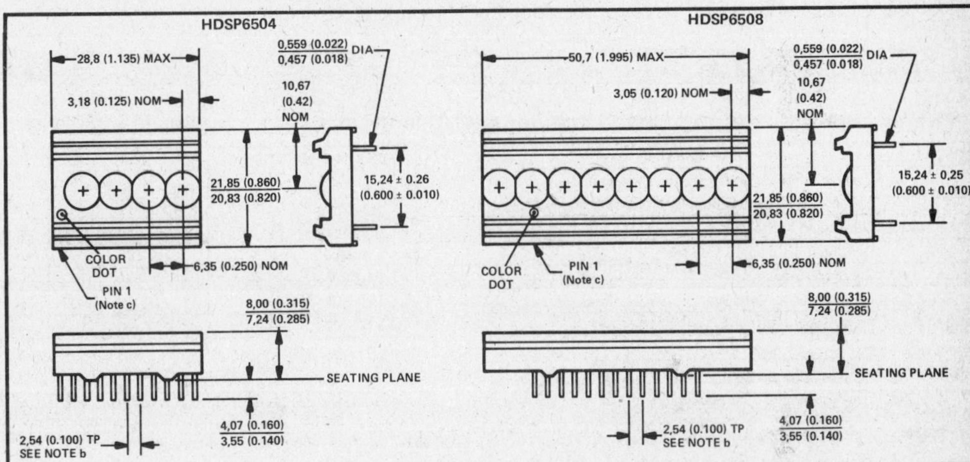

NOTES: a. All linear dimensions are in millimeters and parenthetically in inches.
b. Each pin centerline is located within 0,26 mm (0.010 inch) of its true longitudinal position.
c. Pin 1 is identified by a color dot adjacent to the lead.

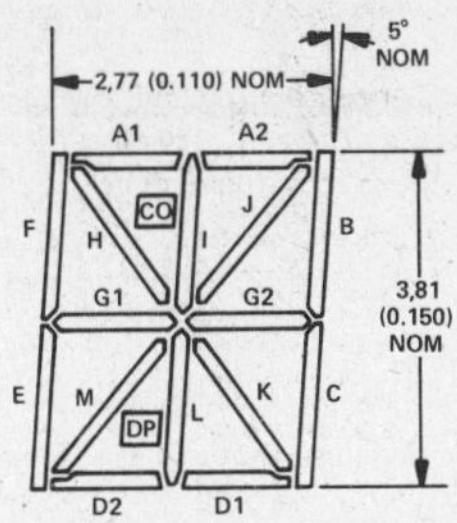

PIN NUMBER	HDSP6504		HDSP6508	
1	ANODE	SEGMENT G1	ANODE	SEGMENT G1
2	ANODE	SEGMENT DP	ANODE	SEGMENT DP
3	CATHODE	DIGIT 1	CATHODE	DIGIT 1
4	ANODE	SEGMENT D2	ANODE	SEGMENT D2
5	ANODE	SEGMENT L	ANODE	SEGMENT L
6	CATHODE	DIGIT 3	CATHODE	DIGIT 3
7	ANODE	SEGMENT E	ANODE	SEGMENT E
8	ANODE	SEGMENT M	ANODE	SEGMENT M
9	ANODE	SEGMENT K	ANODE	SEGMENT K
10	CATHODE	DIGIT 4	CATHODE	DIGIT 4
11	ANODE	SEGMENT D1	ANODE	SEGMENT D1
12	ANODE	SEGMENT J	CATHODE	DIGIT 6
13	ANODE	SEGMENT CO	CATHODE	DIGIT 8
14	ANODE	SEGMENT G2	CATHODE	DIGIT 7
15	ANODE	SEGMENT A2	CATHODE	DIGIT 5
16	ANODE	SEGMENT I	ANODE	SEGMENT J
17	CATHODE	DIGIT 2	ANODE	SEGMENT CO
18	ANODE	SEGMENT B	ANODE	SEGMENT G2
19	ANODE	SEGMENT A1	ANODE	SEGMENT A2
20	ANODE	SEGMENT C	ANODE	SEGMENT I
21	ANODE	SEGMENT H	CATHODE	DIGIT 2
22	ANODE	SEGMENT F	ANODE	SEGMENT B
23			ANODE	SEGMENT A1
24			ANODE	SEGMENT C
25			ANODE	SEGMENT H
26			ANODE	SEGMENT F

† Trademark of E. I. duPont de Nemours, Inc.

Copyright © 1983 by Texas Instruments Incorporated

TEXAS INSTRUMENTS
INCORPORATED
POST OFFICE BOX 225012 • DALLAS, TEXAS 75265

TYPES HDSP6504, HDSP6508
ALPHANUMERIC DISPLAYS

description

These displays are intended for use under pulsed conditions by enabling each character cathode sequentially and enabling the desired anodes in phase with the character-enabling pulse. The pulse rate is kept high enough so that the light from each character appears to the eye to be constant.

absolute maximum ratings

Reverse Voltage at 25°C Free-Air Temperature, Each Segment . . . . . 5 V
Peak Forward Current at (or below) 25°C Free-Air Temperature, Each Segment . . . . . 200 mA
Average Forward Current at (or below) 25°C Free-Air Temperature, Each Segment (See Note 1) . . . . . 7 mA
Average Power Dissipation at (or below) 25°C Free-Air Temperature, Each Character (See Note 2) . . . . . 138 mW
Operating Free-Air Temperature Range . . . . . −40°C to 85°C
Storage Temperature Range . . . . . −40°C to 100°C
Lead Temperature 1,6 mm (1/16 inch) below Seating Plane for 5 Seconds . . . . . 260°C

recommended operating conditions over operating free-air temperature range

Peak Forward Current, Each Segment . . . . . 7 mA
Average Forward Current, Each Segment (see Note 1) . . . . . 0.8 mA

operating characteristics of each segment at 25°C free-air temperature

	PARAMETER	TEST CONDITIONS	MIN	TYP	MAX	UNIT
I_v	Average Luminous Intensity per Character with All Segments Conducting Except DP and CO (See Note 3)		0.45	1.65		mcd
λ_p	Wavelength at Peak Emission	I_F = 30 mA per segment, See Note 4	640	655	680	nm
λ_d	Dominant Wavelength (See Note 5)			640		nm
$\Delta\lambda$	Spectral Bandwidth			20		nm
V_F	Static Forward Voltage	I_F = 30 mA		1.7	1.9	V
I_R	Static Reverse Current	V_R = 5 V		10	100	μA

NOTES:
1. This average applies for any 10-ms period.
2. Derate linearly to 62 mW at 85°C at the rate of 2.17 mW/°C.
3. Luminous intensity is measured with a light sensor and filter combination that approximates the CIE (International Comission on Illumination) eye-response curve. The luminous intensity ratio between segments within a digit is designed so that each segment will have the same luminance. Thus all the segments will appear to the eye to have equal brightness.
4. These parameters must be measured at t_w = 312 μs, duty cycle = 6.25%.
5. The dominant wavelength λ_d is derived from the CIE chromaticity diagram and is the single wavelength that defines the color of the emitted light.

TEXAS INSTRUMENTS
INCORPORATED
POST OFFICE BOX 225012 • DALLAS, TEXAS 75265

APPLICATIONS INFORMATION

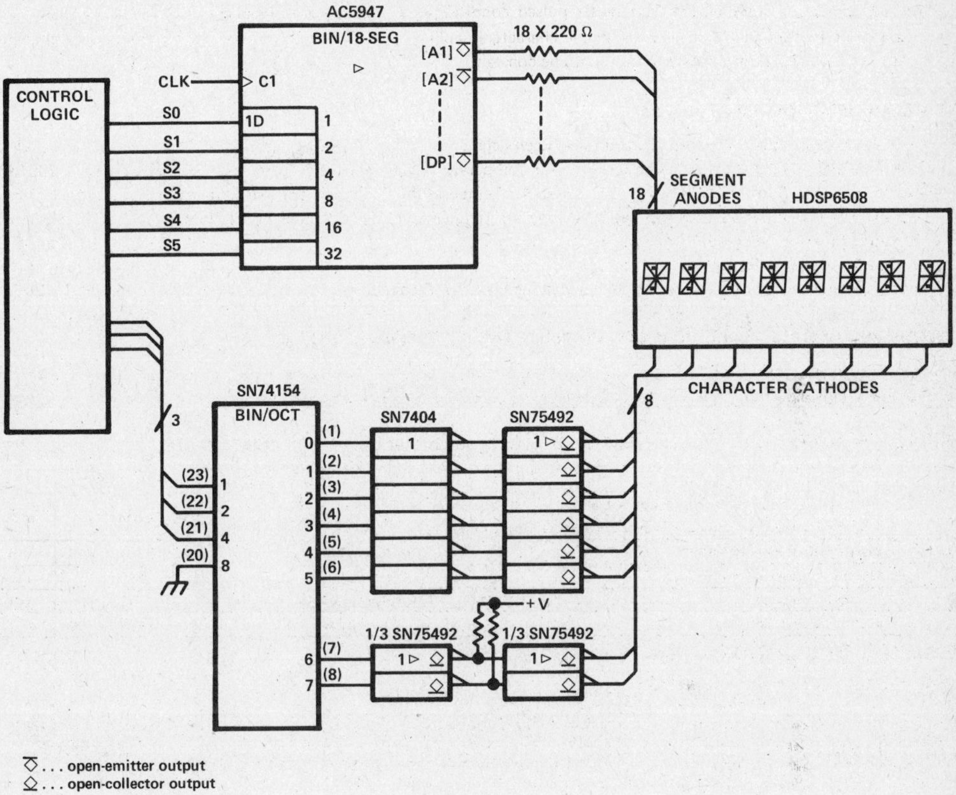

		S3 = 8, S2 = 4, S1 = 2, S0 = 1															
		0	1	2	3	4	5	6	7	8	9	10	11	12	13	14	15
S5 = 32 S4 = 16	0	@	A	B	C	D	E	F	G	H	I	J	K	L	M	N	O
	16	P	Q	R	S	T	U	V	W	X	Y	Z	[	\	]	↗	←
	32		!	"	#	$	%	&	'	(	)	*	+	,	−	.	/
	48	0	1	2	3	4	5	6	7	8	9	:	;	<	=	>	?

The numbers of the rows and columns are the total weights of the S inputs that must be active to select the various characters. To obtain the symbol ✱ , column 10 is selected by taking S3 (8) and S1 (2) both high (with S0 and S2 low) and row 32 is selected by taking S5 (32) high (with S4 low).

FIGURE 1–8-DIGIT DISPLAY CIRCUIT

D1021, APRIL 1971 – REVISED JUNE 1982

RED SOLID-STATE DISPLAYS

- 6,9-mm (0.270-Inch) Character Height
- High Luminous Intensity
- Low Power Requirements
- Each Unit Visually Checked for Uniformity of Elements
- Sign, Overflow, Left or Right Decimal Capability
- Wide Viewing Angle
- Compatible with Most TTL and DTL Circuits

mechanical data

These assemblies consist of display chips mounted on a header with either a red molded plastic body for the TIL302, TIL303, and TIL304 or a red plastic cap for the TIL302A, TIL303A, and TIL304A. Multiple displays may be mounted on 11,43-mm (0.450-inch) centers.

TIL302
TIL302A
TIL303
TIL303A

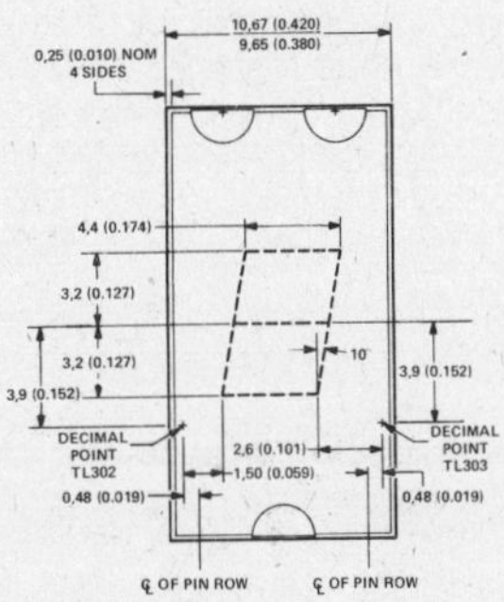

TIL304
TIL304A

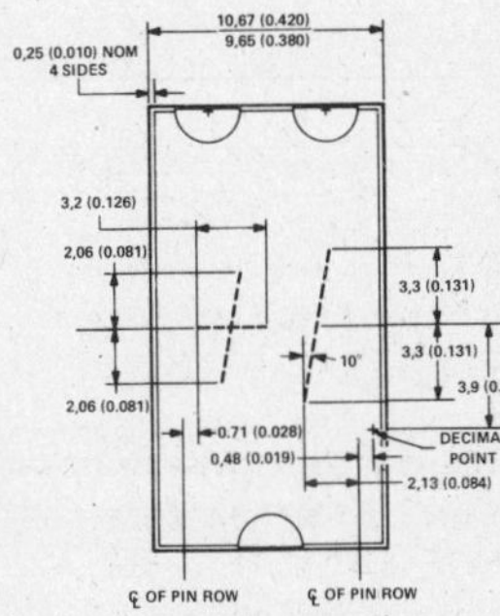

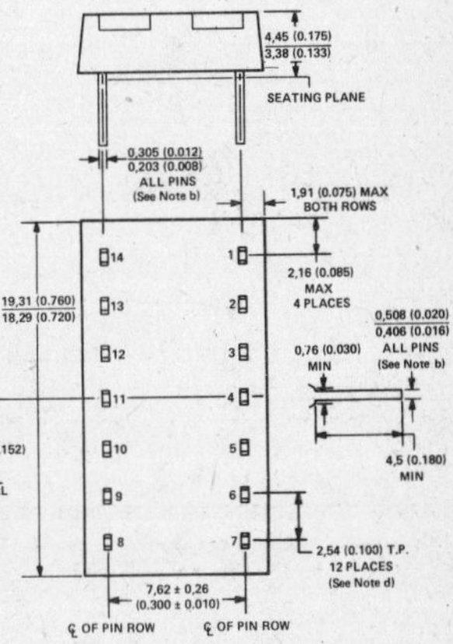

NOTES:
a. All linear dimensions are in millimeters and parenthetically in inches.
b. Lead dimensions are not controlled above the seating plane.
c. Centerlines of character segments and decimal points are shown as dashed lines. Associated dimensions are nominal.
d. The true-position pin spacing is 2,54 mm (0.100 inch) between centerlines. Each centerline is located within 0,26 mm (0.010 inch) of its true longitudinal position relative to pins 1 and 11.
e. On TIL302A, TIL303A, and TIL304A devices, the 3 mold indentations are not present.

TIL302, TIL302A

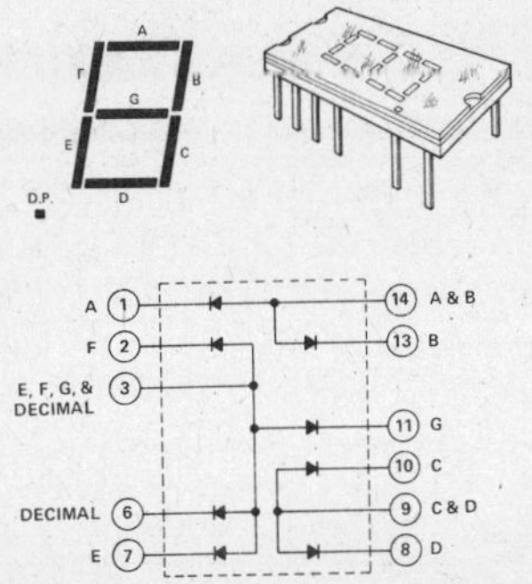

PINS 4, 5, AND 12 OMITTED

TIL303, TIL303A

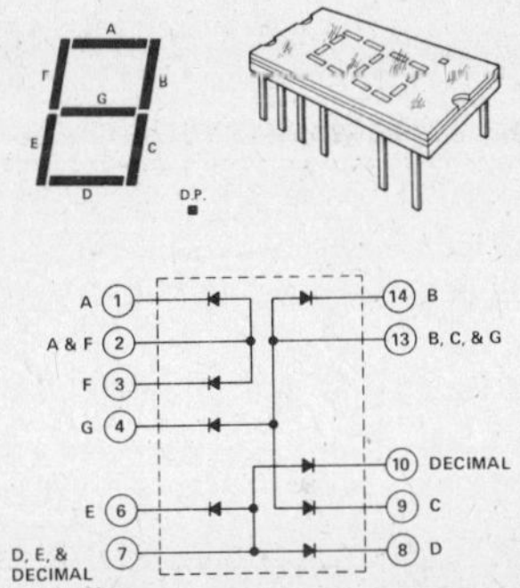

PINS 5, 11, AND 12 OMITTED

TIL304, TIL304A

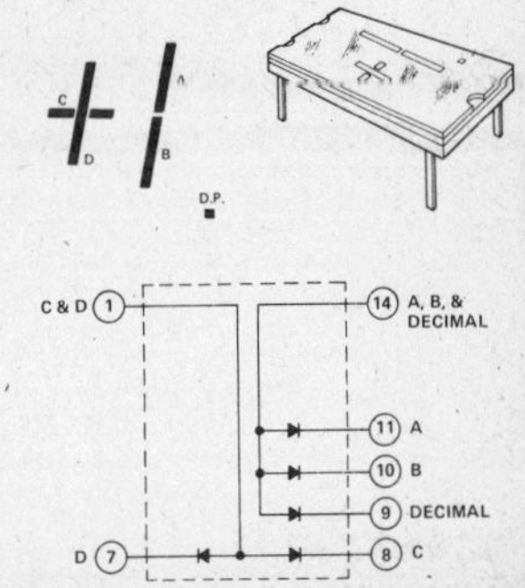

PINS 2, 3, 4, 5, 6, 12, AND 13 OMITTED

10
LED DISPLAYS

Copyright © 1982 by Texas Instruments Incorporated

absolute maximum ratings over operating free-air temperature range (unless otherwise noted)

Reverse Voltage at 25°C Free-Air Temperature:	
Each Segment	6 V
Decimal Point	3 V
Peak Forward Current, Each Segment or Decimal Point (See Note 1)	200 mA
Continuous Forward Current:	
Each Segment or Decimal Point	30 mA
Total for TIL302, TIL302A, TIL303, TIL303A	240 mA
Total for TIL304, TIL304A	150 mA
Operating Free-Air Temperature Range	0°C to 70°C
Storage Temperature Range	−25°C to 85°C

NOTE 1: This value applies for PRR ⩾ 60 Hz, duty cycle ⩽ 10%.

operating characteristics of each segment at 25°C free-air temperature (unless otherwise noted)

PARAMETER		TEST CONDITIONS	MIN	TYP	MAX	UNIT
I_v	Luminous Intensity (See Note 2)	I_F = 20 mA	100	275		µcd
λ_p	Wavelength at Peak Emission			660		nm
$\Delta\lambda$	Spectral Bandwidth			20		nm
V_F	Static Forward Voltage		3	3.4	3.8	V
α_{VF}	Average Temperature Coefficient of Static Forward Voltage	I_F = 20 mA, T_A = 0°C to 70°C		−2.7		mV/°C
I_R	Static Reverse Current	V_R = 6 V			100	µA
C	Anode-to-Cathode Capacitance	V_R = 0, f = 1 MHz		85		pF

operating characteristics of decimal point at 25°C free-air temperature (unless otherwise noted)

PARAMETER		TEST CONDITIONS	MIN	TYP	MAX	UNIT
I_v	Luminous Intensity (See Note 2	I_F = 20 mA	40	110		µcd
λ_p	Wavelength at Peak Emission			660		nm
$\Delta\lambda$	Spectral Bandwidth			20		nm
V_F	Static Forward Voltage		1.5	1.65	2	V
α_{VF}	Average Temperature Coefficient of Static Forward Voltage	I_F = 20 mA, T_A = 0°C to 70°C		−1.4		mV/°C
I_R	Static Reverse Current	V_R = 3 V			100	µA
C	Anode-to-Cathode Capacitance	V_R = 0, f = 1 MHz		120		pF

NOTE 2: Luminous intensity is measured with a light sensor and filter combination that approximates the CIE (International Commission on Illumination) eye-response curve.

TYPICAL CHARACTERISTICS

RELATIVE SPECTRAL CHARACTERISTICS

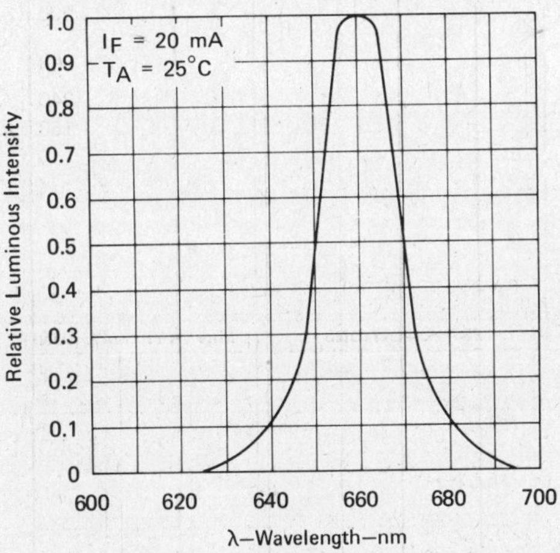

FIGURE 1

RELATIVE LUMINOUS INTENSITY
vs
FREE-AIR TEMPERATURE

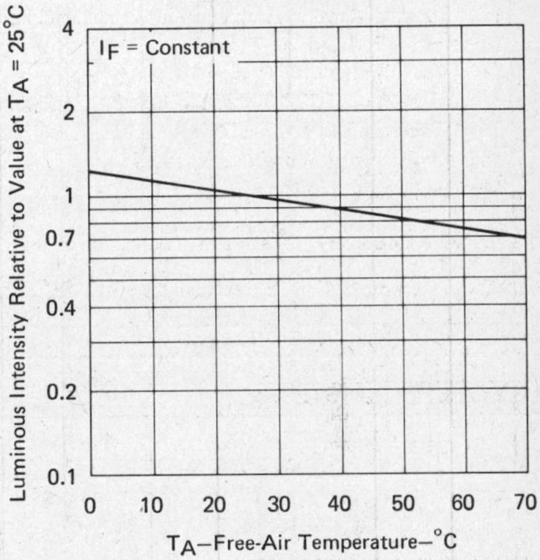

FIGURE 2

RELATIVE LUMINOUS INTENSITY
vs
FORWARD CURRENT

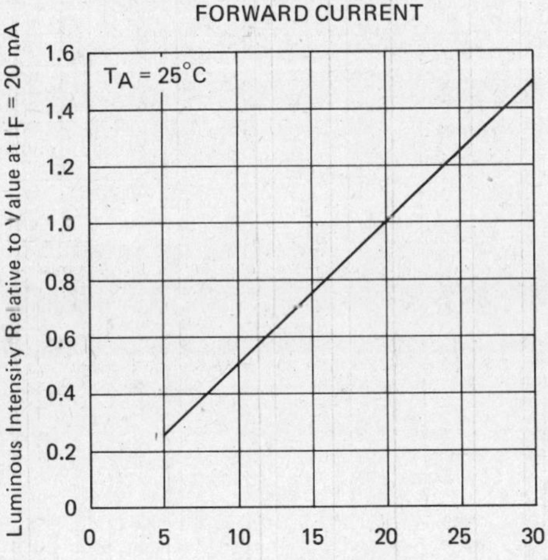

FIGURE 3

FORWARD CONDUCTION CHARACTERISTICS

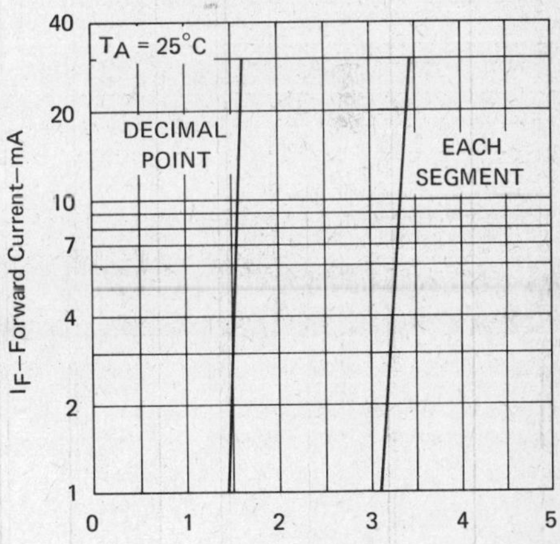

FIGURE 4

10
LED DISPLAYS

TYPICAL APPLICATION DATA

TIL302
TIL302A

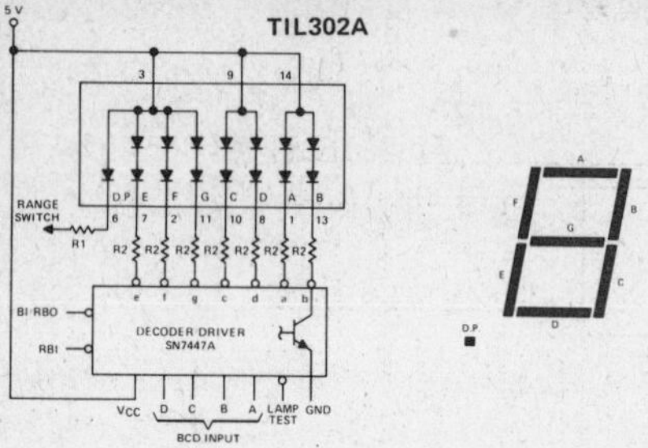

TIL303
TIL303A

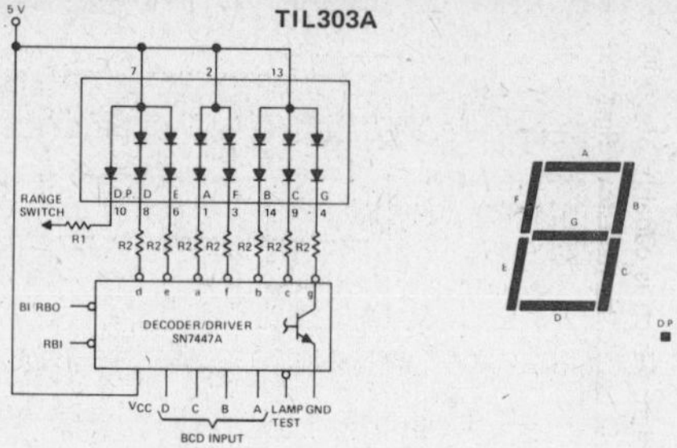

NOTE: R1 and R2 are selected for desired brightness.

FUNCTION TABLE
SN7447A

DECIMAL OR FUNCTION	INPUTS						BI/RBO†	SEGMENTS							NOTE
	LT	RBI	D	C	B	A		a	b	c	d	e	f	g	
0	H	H	L	L	L	L	H	ON	ON	ON	ON	ON	ON	OFF	1
1	H	X	L	L	L	H	H	OFF	ON	ON	OFF	OFF	OFF	OFF	1
2	H	X	L	L	H	L	H	ON	ON	OFF	ON	ON	OFF	ON	1
3	H	X	L	L	H	H	H	ON	ON	ON	ON	OFF	OFF	ON	1
4	H	X	L	H	L	L	H	OFF	ON	ON	OFF	OFF	ON	ON	1
5	H	X	L	H	L	H	H	ON	OFF	ON	ON	OFF	ON	ON	1
6	H	X	L	H	H	L	H	OFF	OFF	ON	ON	ON	ON	ON	1
7	H	X	L	H	H	H	H	ON	ON	ON	OFF	OFF	OFF	OFF	1
8	H	X	H	L	L	L	H	ON	ON	ON	ON	ON	ON	ON	1
9	H	X	H	L	L	H	H	ON	ON	ON	OFF	OFF	ON	ON	1
10	H	X	H	L	H	L	H	OFF	OFF	OFF	ON	ON	OFF	ON	1
11	H	X	H	L	H	H	H	OFF	OFF	ON	ON	OFF	OFF	ON	1
12	H	X	H	H	L	L	H	OFF	ON	OFF	OFF	OFF	ON	ON	1
13	H	X	H	H	L	H	H	ON	OFF	OFF	ON	OFF	ON	ON	1
14	H	X	H	H	H	L	H	OFF	OFF	OFF	ON	ON	ON	ON	1
15	H	X	H	H	H	H	H	OFF	OFF	OFF	OFF	OFF	OFF	OFF	1
BI	X	X	X	X	X	X	L	OFF	OFF	OFF	OFF	OFF	OFF	OFF	2
RBI	H	L	L	L	L	L	L	OFF	OFF	OFF	OFF	OFF	OFF	OFF	3
LT	L	X	X	X	X	X	H	ON	ON	ON	ON	ON	ON	ON	4

H = high level (logic 1 in positive logic), L = low level (logic 0 in positive logic), X = irrelevant.

†BI/RBO is wire-AND logic serving as blanking input (BI) and/or ripple-blanking output (RBO).

NOTES: 1. The blanking input (BI) must be open or held at a high logic level when output functions 0 through 15 are desired. The ripple-blanking input (RBI) must be open or high if blanking of a decimal zero is not desired.

2. When a low logic level is applied directly to the blanking input (BI), all segment outputs are off regardless of any other input.

3. When the ripple-blanking input (RBI) and inputs A, B, C, and D are at a low logic level with the lamp test input high, all segment outputs are off and the ripple-blanking output (RBO) of the decoder goes to a low level (response condition).

4. When the blanking input/ripple blanking output (BI/RBO) is open or held high and a low is applied to the lamp-test input, all segments are illuminated.

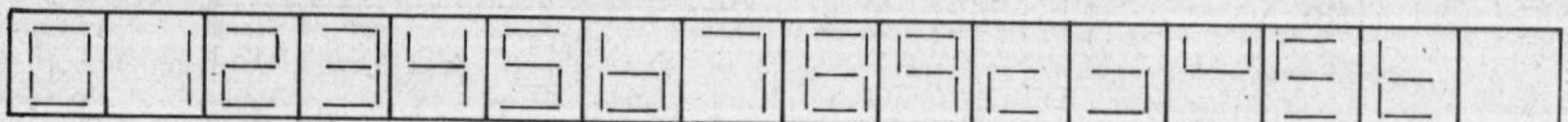

0 1 2 3 4 5 6 7 8 9 10 11 12 13 14 15

NUMERICAL DESIGNATIONS—RESULTANT DISPLAYS

TEXAS INSTRUMENTS
INCORPORATED
POST OFFICE BOX 225012 • DALLAS, TEXAS 75265

TYPICAL APPLICATION DATA

The TIL303, TIL303A, TIL304, and TIL304A are used in this application to make a three-digit display with sign, which is capable of 100% overrance ("1" plus three digits). The decimal point is located via an external range switch. The clear function will blank the overflow digit and reset the three digits to zero. Following resetting, input pulses will be counted, decoded, and displayed.

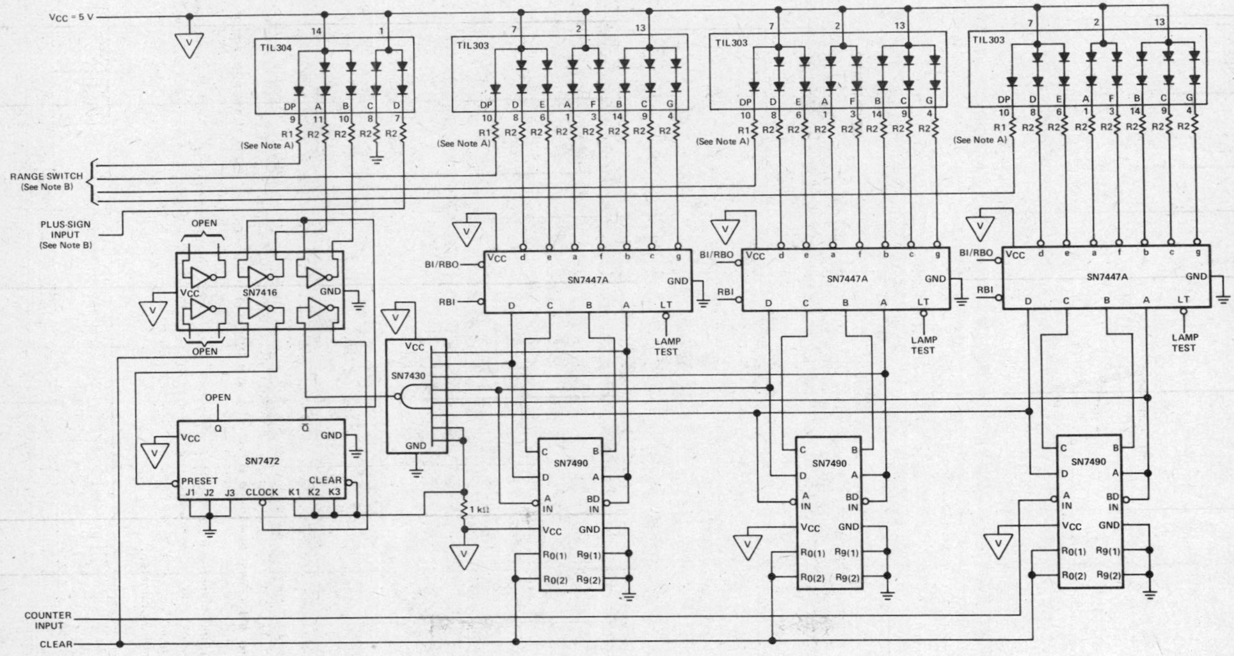

NOTES: A. R1 and R2 are selected for desired brightness.
B. Grounding of any of these lines will illuminate the associated function.

▽ . . . V_{CC} bus

TEXAS INSTRUMENTS
INCORPORATED
POST OFFICE BOX 225012 • DALLAS, TEXAS 75265

SOLID-STATE DISPLAY WITH RED TRANSPARENT PLASTIC ENCAPSULATION

- 7,62-mm (0.300-inch) Character Height
- High Luminous Intensity
- Low Power Requirements
- Wide Viewing Angle
- 5 X 7 Array with X-Y Select and Decimal
- Compatible with USASCII and EBCDIC Codes

mechanical data

This assembly consists of a display chip mounted on a printed circuit board with a red molded plastic body. Multiple displays may be mounted on 11,43-mm (0.450-inch) centers.

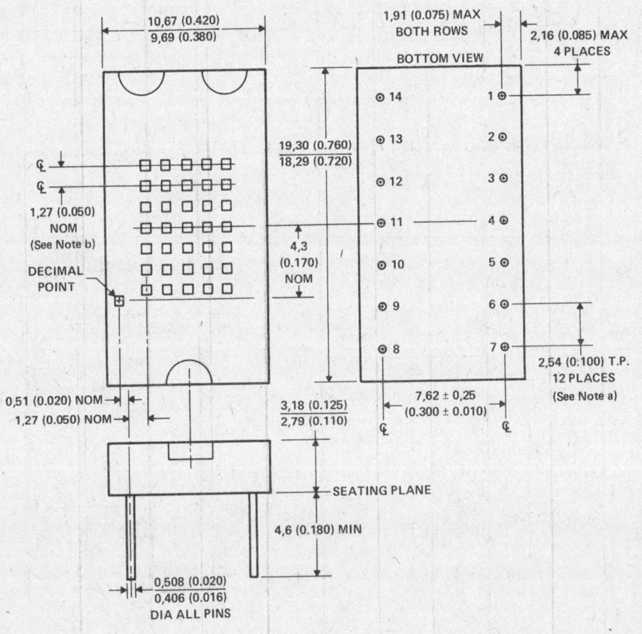

NOTES: a. The true-position spacing is 2,54 mm (0.100 inch) between lead centerlines. Each pin centerline is located within 0,25 mm (0.010 inch) of its true longitudinal position.

b. Vertical and horizontal spacing between centerlines of rows and columns nominally 1,27 mm (0.050 inch).

ALL DIMENSIONS ARE IN MILLIMETERS AND PARENTHETICALLY IN INCHES

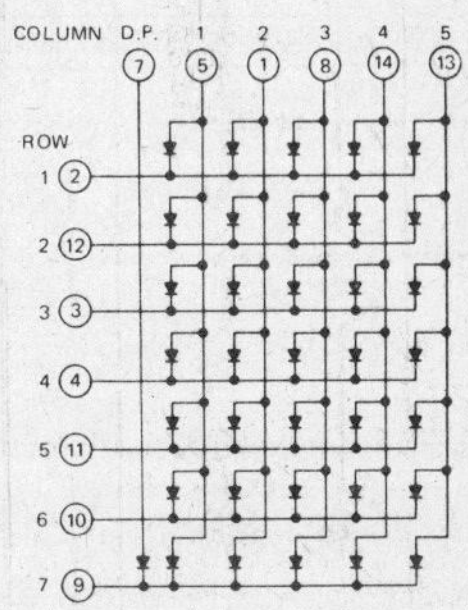

TOP VIEW ORIENTATION

CL-37

Copyright © 1983 by Texas Instruments Incorporated

absolute maximum ratings over operating free-air temperature range (unless otherwise noted)

Reverse Voltage at 25°C Free-Air Temperature	3 V
Peak Forward Current, Each Diode	100 mA
Average Forward Current (see Note 1):	
Each Diode	10 mA
Total	200 mA
Operating Free-Air Temperature Range	0° to 70°C
Storage Temperature Range	−25°C to 85°C

operating characteristics of each diode at 25°C free-air temperature (unless otherwise noted)

PARAMETER		TEST CONDITIONS	MIN	TYP	MAX	UNIT
I_V	Luminous Intensity (see Note 2)	I_F = 10 mA	40	110		µcd
λ_p	Wavelength at Peak Emission			660		nm
$\Delta\lambda$	Spectral Bandwidth			20		nm
V_F	Static Forward Voltage		1.5	1.65	2	V
α_{VF}	Average Temperature Coefficient of Static Forward Voltage	I_F = 10 mA, T_A = 0°C to 70°C		−1.4		mV/°C
I_R	Static Reverse Current	V_R = 3 V			10	µA
C	Anode-to-Cathode Capacitance	V_R = 0, f = 1 MHz		80		pF

NOTES: 1. This average value applies for any 1-ms period.
2. Luminous intensity is measured with a light sensor and filter combination that approximates the CIE (International Commission on Illumination) eye-response curve.

TYPICAL CHARACTERISTICS

RELATIVE LUMINOUS INTENSITY
vs
FREE-AIR TEMPERATURE

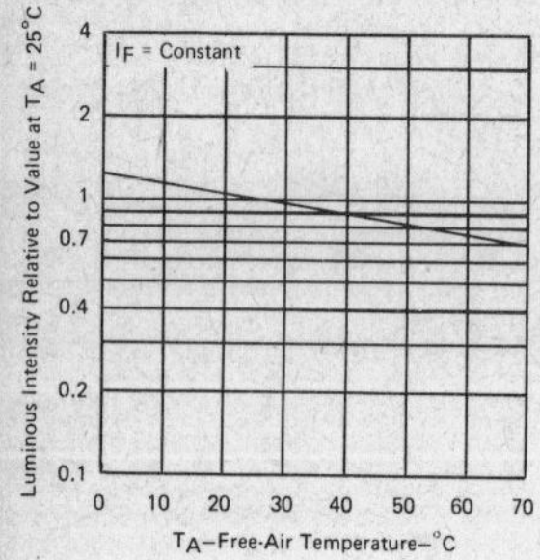

FIGURE 1

RELATIVE LUMINOUS INTENSITY
vs
FORWARD CURRENT

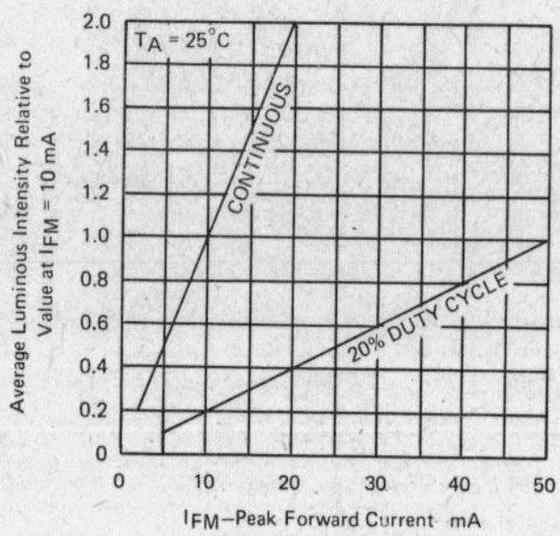

FIGURE 2

FORWARD CONDUCTION
CHARACTERISTICS

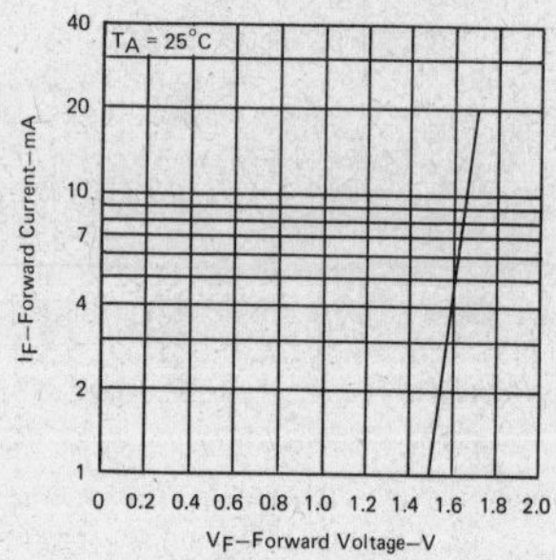

FIGURE 3

TEXAS INSTRUMENTS
INCORPORATED
POST OFFICE BOX 225012 • DALLAS, TEXAS 75265

SOLID-STATE DISPLAYS WITH INTEGRAL TTL MSI CIRCUIT CHIP FOR USE IN ALL SYSTEMS WHERE THE DATA TO BE DISPLAYED IS THE PULSE COUNT

- 6,9-mm (0.270-Inch) Character Height
- High Luminous Intensity
- TIL306 and TIL306A Have Left Decimal
- TIL307 and TIL307A Have Right Decimal
- Easy System Interface
- Wide Viewing Angle
- Internal TTL MSI Chip and Counter, Latch, Decoder, and Driver
- Constant-Current Drive for Light-Emitting Diodes

mechanical data

These assemblies consist of display chips and a TTL MSI chip mounted on a header with either a red molded plastic body for the TIL306 and TIL307 or a red plastic cap for the TIL306A and TIL307A. Multiple displays may be mounted on 11,43-mm (0.450-inch) centers.

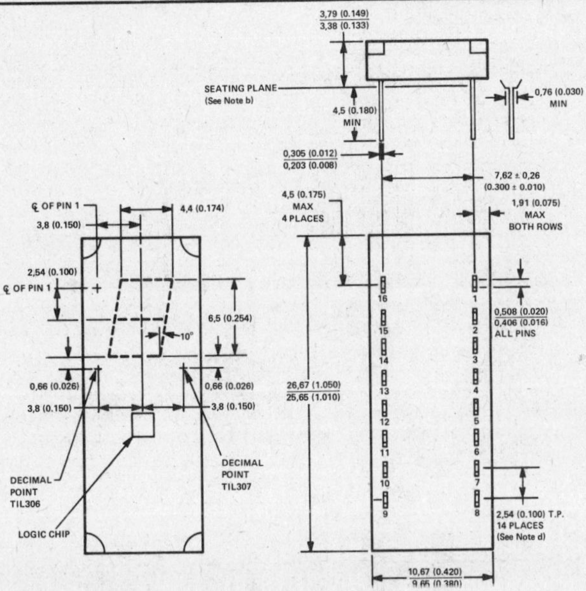

PIN ASSIGNMENTS

PIN 1	LATCH OUTPUT Q_B (BINARY WEIGHT 2)
PIN 2	LATCH OUTPUT Q_C (BINARY WEIGHT 4)
PIN 3	LATCH OUTPUT Q_D (BINARY WEIGHT 8)
PIN 4	LATCH OUTPUT Q_A (BINARY WEIGHT 1)
PIN 5	LATCH STROBE INPUT
PIN 6	RIPPLE-BLANKING INPUT
PIN 7	MAXIMUM-COUNT OUTPUT
PIN 8	GROUND
PIN 9	PARALLEL COUNT ENABLE INPUT
PIN 10	SERIAL COUNT ENABLE INPUT
PIN 11	RIPPLE-BLANKING OUTPUT
PIN 12	CLEAR INPUT
PIN 13	DECIMAL POINT INPUT
PIN 14	BLANKING INPUT
PIN 15	CLOCK INPUT
PIN 16	SUPPLY VOLTAGE, V_{CC}

NOTES: a. All linear dimensions are in millimeters and parenthetically in inches.
b. Lead dimensions are not controlled above the seating plane.
c. Centerlines of character segments and decimal points are shown as dashed lines. Associated dimensions are nominal.
d. The true-position pin spacing is 2,54 mm (0.100 inch) between centerlines. Each centerline is located with 0,26 mm (0.010 inch) of its true longitudinal position relative to pins 1 and 16.
e. On TIL306A and TIL307A devices, the 4 mold indentations are not present.

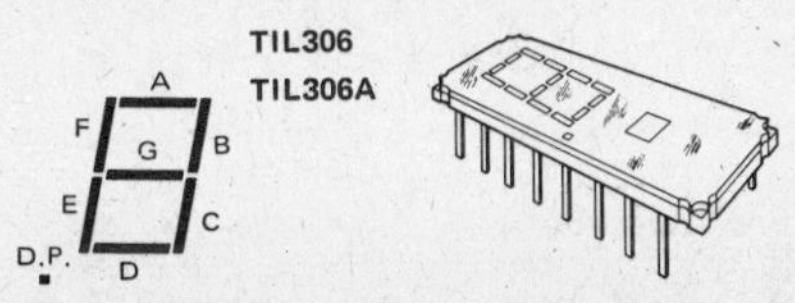

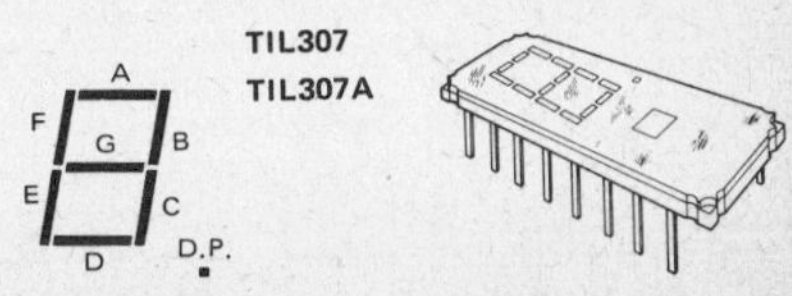

Copyright © 1982 by Texas Instruments Incorporated

functional block diagram

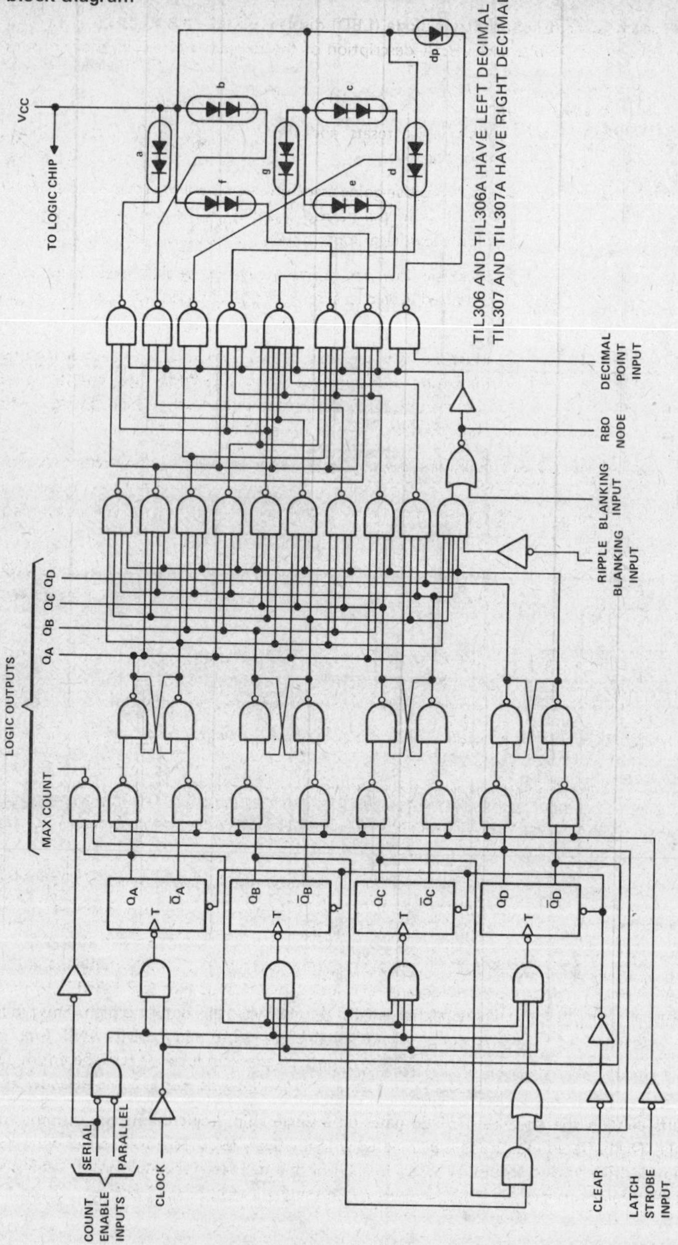

SYNCHRONOUS BCD COUNTER, 4-BIT LATCH, DECODER/DRIVER, SEVEN-SEGMENT LED DISPLAY WITH DECIMAL POINT

description

These internally-driven seven-segment light-emitting-diode (LED) displays contain a BCD counter, a four-bit latch, and a decoder/LED driver in a single 16-pin package. A description of the functions of the inputs and outputs of these devices follows:

FUNCTION	PIN NO.	DESCRIPTION
CLEAR INPUT	12	When low, resets and holds counter at 0. Must be high for normal counting.
CLOCK INPUT	15	Each positive-going transition will increment the counter provided that the circuit is in the normal counting mode (serial and parallel count enable inputs low, clear input high).
PARALLEL COUNT ENABLE INPUT (PCEI)	9	Must be low for normal counting mode. When high, counter will be inhibited. Logic level must not be changed when the clock is low.
SERIAL COUNT ENABLE INPUT (SCEI)	10	Must be low for normal counting mode, also must be low to enable maximum count output to go low. When high, counter will be inhibited and maximum count output will be driven high. Logic level must not be changed when the clock is low.
MAXIMUM COUNT OUTPUT	7	Will go low when the counter is at 9 and serial count enable input is low. Will return high when the counter changes to 0 and will remain high during counts 1 through 8. Will remain high (inhibited) as long as serial count enable input is high.
LATCH STROBE INPUT	5	When low, data in latches follow the data in the counter. When high, the data in the latches are held constant, and the counter may be operated independently.
LATCH OUTPUTS (Q_A, Q_B, Q_C, Q_D)	4, 1, 2, 3	The BCD data that drives the decoder can be stored in the 4-bit latch and is available at these outputs for driving other logic and/or processors. The binary weights of the outputs are: $Q_A = 1$, $Q_B = 2$, $Q_C = 4$, $Q_D = 8$.
DECIMAL POINT INPUT	13	Must be high to display decimal point. The decimal point is not displayed when this input is low or when the display is blanked.
BLANKING INPUT (BI)	14	When high, will blank (turn off) the entire display and force RBO low. Must be low for normal display. May be pulsed to implement intensity control of the display.
RIPPLE-BLANKING INPUT (RBI)	6	When the data in the latches is BCD 0, a low input will blank the entire display and force the RBO low. This input has no effect if the data in the latches is other than 0.
RIPPLE-BLANKING OUTPUT (RBO)	11	Supplies ripple-blanking information for the ripple-blanking input of the next decade. Provides a low if BI is high, or if RBI is low and the data in the latches is BCD 0; otherwise, this output is high. This pin has a resistive pull-up circuit suitable for performing a wire-AND function with any open-collector output. Whenever this pin is low the entire display will be blanked; therefore, this pin may be used as an active-low blanking input.

The TTL MSI circuits contain the equivalent of 86 gates on a single chip. Logic inputs and outputs are completely TTL/DTL compatible. The buffered inputs are implemented with relatively large resistors in series with the bases of the input transistors to lower drive-current requirements to one-half of that required for a standard Series 54/74 TTL input. The serial-carry input, actually two internal loads, is rated as one standard series 54/74 load.

TEXAS INSTRUMENTS
INCORPORATED
POST OFFICE BOX 225012 • DALLAS, TEXAS 75265

description (continued)

The logic outputs, except RBO, are active pull-up, and the latch outputs Q_A, Q_B, Q_C, and Q_D are each capable of driving three standard Series 54/74 loads at a low logic level or six loads at a high logic level while the maximum-count output is capable of driving five Series 54/74 loads at a low logic level or ten loads at a high logic level. The RBO node with passive pull-up serves as a ripple-blanking output with the capability to drive three Series 54/74 loads.

The LED driver outputs are designed specifically to maintain a relatively constant on-level current of approximately seven milliamperes through each LED segment and decimal point. All inputs are diode-clamped to minimize transmission-line effects, thereby simplifying system design. Maximum clock frequency is typically 18 megahertz and power dissipation is typically 600 milliwatts with all segments on.

The display format is as follows:

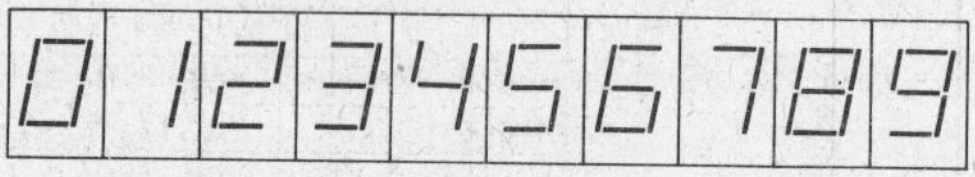

The displays may be interconnected to produce an n-digit display with the following features:

- Ripple-blanking input and output for blanking leading or trailing zeroes
- Floating-decimal-point logic capability
- Overriding blanking for suppressing entire display or pulse-modulation of LED brightness
- Dual count-enable inputs for parallel look-ahead and serial ripple logic to build high-speed fully synchronous, multidigit counter systems with no external logic, minimizing total propagation delay from the clock to the last latch output
- Provision for ripple-count cascading between packages
- Positive-edge-triggered synchronous BCD counter
- Parallel BCD data outputs available to drive logic processors or remote slaved displays simultaneously with data being displayed
- Latch strobe input allows counter to operate while a previous data point is displayed
- Reset-to-zero capability with clear input.

absolute maximum ratings over operating case temperature range (unless otherwise noted)

Supply Voltage, V_{CC} (See Note 1): Continuous . . . . 5.5 V
Nonrepetitive Peak, $t_w \leq 100$ ms . . . . 7 V
Input Voltage (See Note 1) . . . . 5.5 V
Operating Case Temperature Range (See Note 2) . . . . 0°C to 85°C
Storage Temperature Range . . . . −25°C to 85°C

NOTES: 1. Voltage values are with respect to network ground terminal.
2. Case temperature is the surface temperature of the plastic measured directly over the integrated circuit. Forced-air cooling may be required to maintain this temperature.

recommended operating conditions

			MIN	NOM	MAX	UNIT
Supply Voltage, V_{CC}			4.75	5	5.25	V
Normalized Fan-Out from Each Output, N (to Series 54/74 Integrated Circuits)	Low Logic Level	Q_A, Q_B, Q_C, Q_D, RBO			3	
		Maximum Count			5	
	High Logic Level	RBO			3	
		Q_A, Q_B, Q_C, Q_D			6	
		Maximum Count			10	
Clock Pulse Duration, $t_{w(clock)}$		High Logic Level	25			ns
		Low Logic Level	55			
Clear Pulse Duration, $t_{w(clear)}$			25			ns
Latch Strobe Pulse Duration, $t_{w(latch\ strobe)}$			45			ns
Setup Time, t_{su}		Serial Carry and Parallel Carry	30			ns
		Clear Inactive State	60			

TEXAS INSTRUMENTS
INCORPORATED
POST OFFICE BOX 225012 • DALLAS, TEXAS 75265

operating characteristics at 25°C case temperature

PARAMETER			TEST CONDITIONS	MIN	TYP‡	MAX	UNIT
I_v	Luminous Intensity (See Note 3)	Figure 8	V_{CC} = 5 V	700	1200		µcd
		Decimal Point		40	70		µcd
λ_p	Wavelength at Peak Emission		V_{CC} = 5 V, See Note 4		660		nm
$\Delta\lambda$	Spectral Bandwidth		V_{CC} = 5 V, See Note 4		20		nm
V_{IH}	High-Level Input Voltage			2			V
V_{IL}	Low-Level Input Voltage					0.8	V
V_{IK}	Input Clamp Voltage		V_{CC} = 4.75 V, I_I = −12 mA			−1.5	V
V_{OH}	High-Level Output Voltage	RBO	V_{CC} = 4.75 V, I_{OH} = −120 µA	2.4			V
		Q_A, Q_B, Q_C, Q_D	V_{CC} = 4.75 V, I_{OH} = −240 µA				
		Maximum Count	V_{CC} = 4.75 V, I_{OH} = −400 µA				
V_{OL}	Low-Level Output Voltage (See Note 5)	Q_A, Q_B, Q_C, Q_D, RBO	V_{CC} = 4.75 V, I_{OL} = 4.8 mA			0.4	V
		Maximum Count	V_{CC} = 4.75 V, I_{OL} = 8 mA				
I_I	Input Current at Maximum Input Voltage		V_{CC} = 5.25 V, V_I = 5.5 V			1	mA
I_{IH}	High-Level Input Current	Serial Carry	V_{CC} = 5.25 V, V_I = 2.4 V			40	µA
		RBO Node		−0.12	−0.5		mA
		Other Inputs				20	µA
I_{IL}	Low-Level Input Current	Serial Carry	V_{CC} = 5.25 V, V_I = 0.4 V			−1.6	mA
		RBO Node			−1.5	−2.4	
		Other Inputs				−0.8	
I_{OS}	Short-Circuit Output Current	Q_A, Q_B, Q_C, Q_D	V_{CC} = 5.25 V	−9		−27.5	mA
		Maximum Count		−15		−55	
I_{CC}	Supply Current		V_{CC} = 5.25 V, See Note 4		120	200	mA

‡All typical values are at V_{CC} = 5 V.

NOTES: 3. Luminous intensity is measured with a light sensor and filter combination that approximates the CIE (International Commission on Illumination) eye-response curve.
4. These parameters are measured with all LED segments and the decimal point on.
5. This parameter is measured with the display blanked.

switching characteristics, V_{CC} = 5 V, T_C = 25°C

PARAMETER§	FROM (INPUT)	TO (OUTPUT)	TEST CONDITIONS	MIN	TYP	MAX	UNIT
f_{max}				12	18		MHz
t_{PLH}	Serial Look-Ahead	Maximum Count	C_L = 15 pF, R_L = 560 Ω, See Figure 1		12		ns
t_{PHL}					23		
t_{PLH}	Clock	Maximum Count			26		ns
t_{PHL}					29		
t_{PLH}	Clock	Q_A, Q_B, Q_C, Q_D	C_L = 15 pF, R_L = 1.2 kΩ, See Figure 1		28		ns
t_{PHL}					38		
t_{PHL}	Clear	Q_A, Q_B, Q_C, Q_D			57		ns

§f_{max} ≡ Maximum clock frequency
t_{PLH} ≡ Propagation delay time, low-to-high-level output
t_{PHL} ≡ Propagation delay time, high-to-low-level output

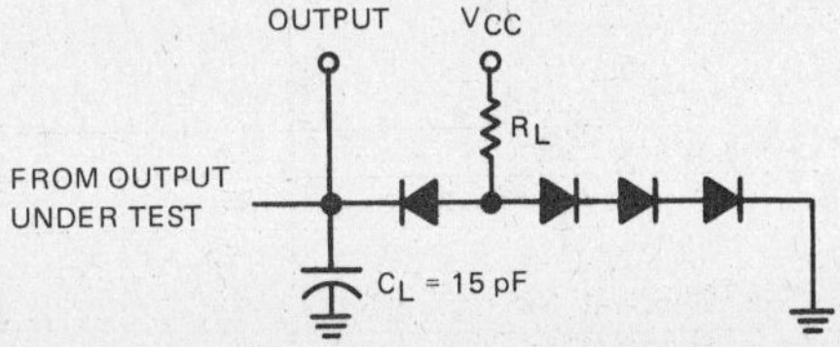

NOTES: A. C_L includes probe and jig capacitance.
B. All diodes are 1N3064.

LOAD CIRCUIT—FIGURE 1

TEXAS INSTRUMENTS
INCORPORATED
POST OFFICE BOX 225012 • DALLAS, TEXAS 75265

TYPICAL CHARACTERISTICS

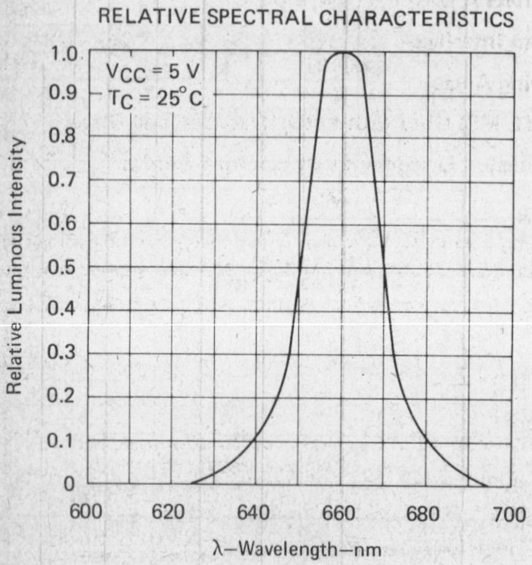

FIGURE 2

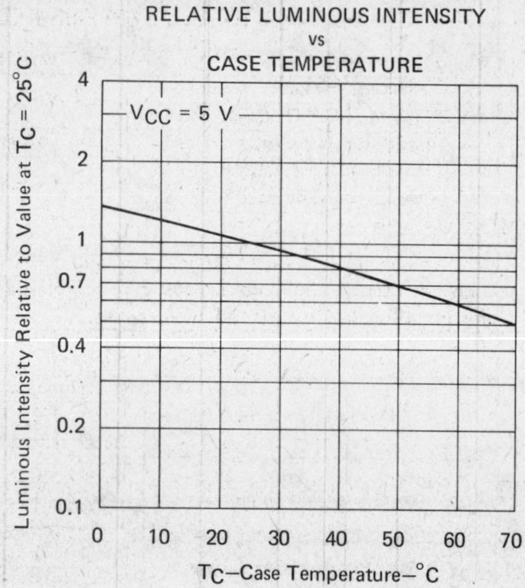

FIGURE 3

10 LED DISPLAYS

D1096, MARCH 1072 – REVISED JUNE 1982

SOLID-STATE DISPLAYS WITH INTEGRAL TTL MSI CIRCUIT CHIP FOR USE IN ALL SYSTEMS REQUIRING A DISPLAY OF BCD DATA

- 6,9-mm (0.270-Inch) Character Height
- TIL308 and TIL308A Have Left Decimal
- TIL309 and TIL309A Have Right Decimal
- Easy System Interface
- Wide Viewing Angle
- Internal TTL MSI Chip with Latch, Decoder, and Driver
- Constant-Current Drive for Light-Emitting Diodes

mechanical data

These assemblies consist of display chips and a TTL MSI chip mounted on a header with either a red molded plastic body for the TIL308 and TIL309 or a red plastic cap for the TIL308A and TIL309A. Multiple displays may be mounted on 11,43-mm (0.450-inch) centers.

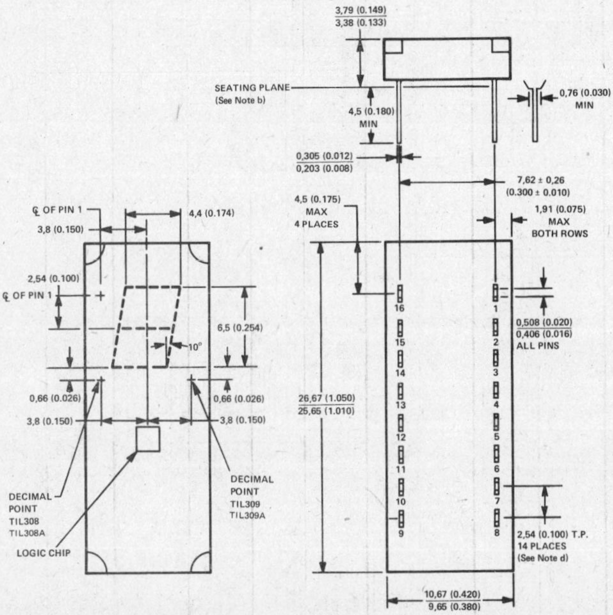

PIN ASSIGNMENTS FOR BOTH TYPES

PIN 1	LATCH OUTPUT Q_B (BINARY WEIGHT 2)
PIN 2	LATCH OUTPUT Q_C (BINARY WEIGHT 4)
PIN 3	LATCH OUTPUT Q_D (BINARY WEIGHT 8)
PIN 4	LATCH OUTPUT Q_A (BINARY WEIGHT 1)
PIN 5	LATCH STROBE INPUT
PIN 6	LATCH DATA INPUT C (BINARY WEIGHT 4)
PIN 7	LATCH DATA INPUT D (BINARY WEIGHT 8)
PIN 8	GROUND
PIN 9	NO INTERNAL CONNECTION
PIN 10	LATCH DATA INPUT B (BINARY WEIGHT 2)
PIN 11	BLANKING INPUT
PIN 12	LATCH DATA INPUT DP
PIN 13	LED TEST
PIN 14	LATCH OUTPUT DP
PIN 15	LATCH DATA INPUT A (BINARY WEIGHT 1)
PIN 16	SUPPLY VOLTAGE, V_{CC}

NOTES: a. All linear dimensions are in millimeters and parenthetically in inches.
b. Lead dimensions are not controlled above the seating plane.
c. Centerlines of character segments and decimal points are shown as dashed lines. Associated dimensions are nominal.
d. The true-position pin spacing is 2,54 mm (0.100 inch) between centerlines. Each centerline is located within 0,26 mm (0.010 inch) of its true longitudinal position relative to pins 1 and 16.
e. On TIL308A and TIL309A devices, the 4 mold indentations are not present.

TIL308
TIL308A

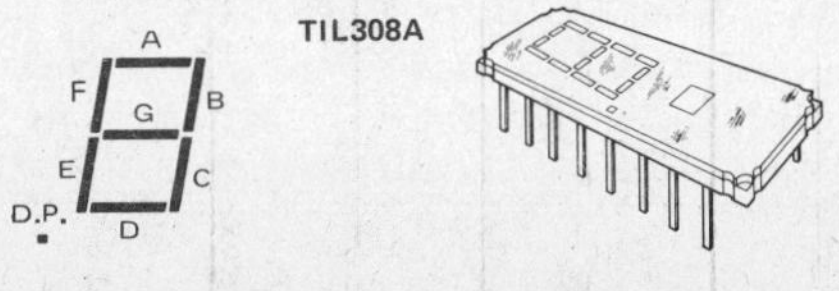

TIL309
TIL309A

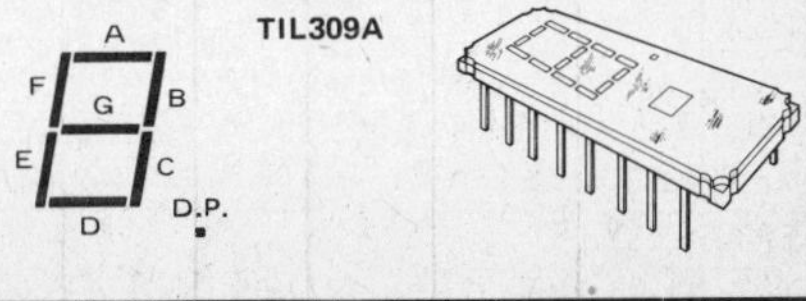

Copyright © 1982 by Texas Instruments Incorporated

TEXAS INSTRUMENTS
INCORPORATED
POST OFFICE BOX 225012 • DALLAS, TEXAS 75265

functional block diagram

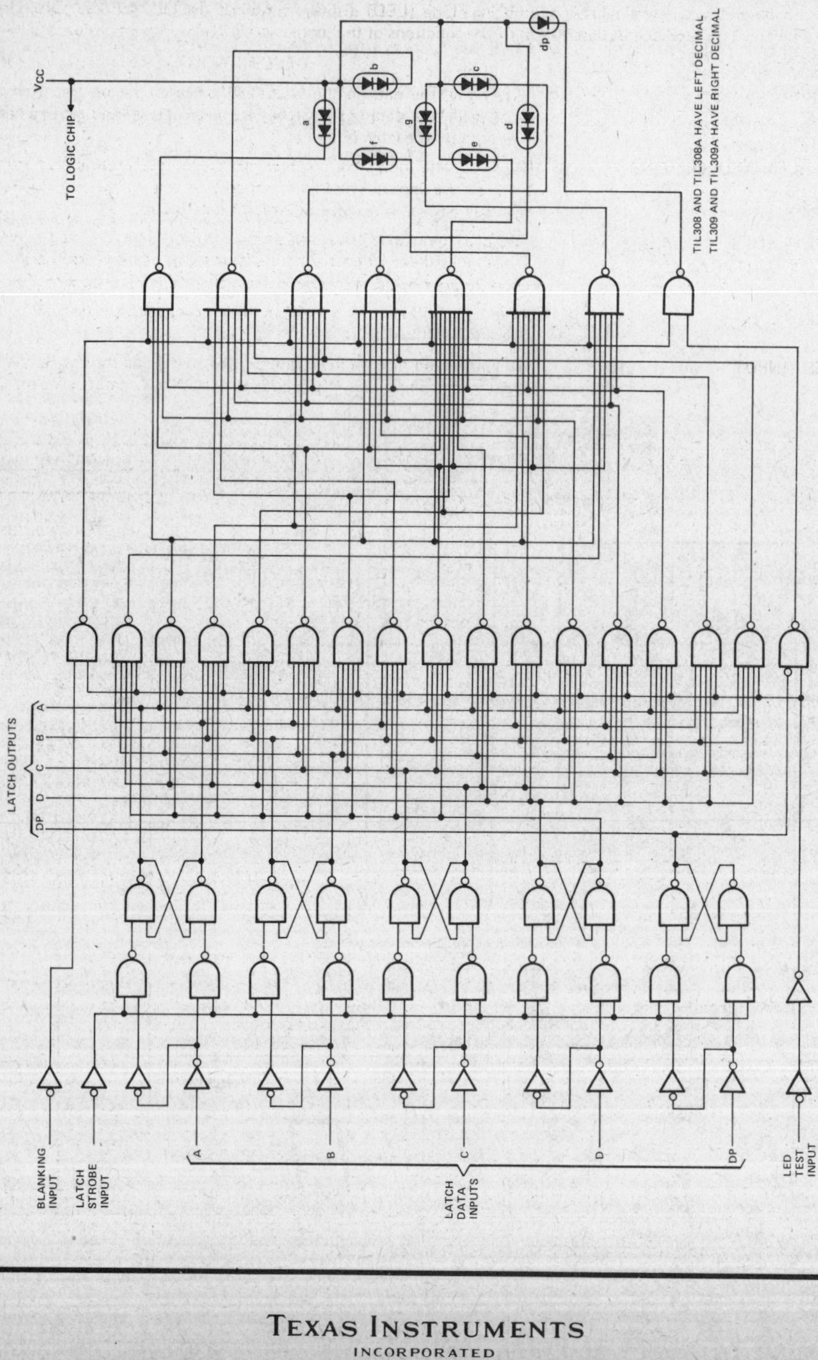

description

These internally-driven seven-segment light-emitting-diode (LED) displays contain a five-bit latch and a decoder/LED driver in a single 16-pin package. A description of the functions of the inputs and outputs of these devices follows:

FUNCTION	PIN NO.	DESCRIPTION
LATCH STROBE INPUT	5	When low, the data in the latches follow the data on the latch inputs. When high, the data in the latches are held constant and are unaffected by new data on the latch inputs.
LATCH DATA INPUTS A, B, C, D, DP	15, 10, 6, 7, 12	Data on these inputs are entered into the latches under the control of the latch strobe input. The binary weights of the inputs are: A = 1, B = 2, C = 4, D = 8. DP is decimal point latch data input.
LATCH OUTPUTS Q_A, Q_B, Q_C, Q_D, Q_{DP}	4, 1, 2, 3, 14	The BCD data that drives the decoder is stored in the five latches and is available at these outputs. The binary weights of the outputs are: $Q_A = 1$, $Q_B = 2$, $Q_C = 4$, $Q_D = 8$. Q_{DP} is decimal point latch output.
BLANKING INPUT	11	When low, will blank (turn off) the entire display. Must be high for normal operation of the display.
LED TEST INPUT	13	When low, will turn on the entire display, overriding the data in the latches and the blanking input. Must be high for normal operation of the display.

FUNCTION TABLE

FUNCTION	LATCH INPUTS						BLANKING	LED	LATCH OUTPUTS					DISPLAY	
	D	C	B	A	DP	STROBE	INPUT	TEST	Q_D	Q_C	Q_B	Q_A	Q_{DP}	TIL308	TIL309
0	L	L	L	L	L	L	H	H	L	L	L	L	L	0	0
1	L	L	L	H	H	L	H	H	L	L	L	H	H	.1	1.
2	L	L	H	L	L	L	H	H	L	L	H	L	L	2	2
3	L	L	H	H	H	L	H	H	L	L	H	H	H	.3	3.
4	L	H	L	L	L	L	H	H	L	H	L	L	L	4	4
5	L	H	L	H	H	L	H	H	L	H	L	H	H	.5	5.
6	L	H	H	L	L	L	H	H	L	H	H	L	L	6	6
7	L	H	H	H	H	L	H	H	L	H	H	H	H	.7	7.
8	H	L	L	L	L	L	H	H	H	L	L	L	L	8	8
9	H	L	L	H	H	L	H	H	H	L	L	H	H	.9	9.
A	H	L	H	L	L	L	H	H	H	L	H	L	L	A	A
MINUS SIGN	H	L	H	H	H	L	H	H	H	L	H	H	H	.-	-.
C	H	H	L	L	L	L	H	H	H	H	L	L	L	C	C
BLANK	H	H	L	H	H	L	H	H	H	H	L	H	H	.	.
E	H	H	H	L	L	L	H	H	H	H	H	L	L	E	E
F	H	H	H	H	H	L	H	H	H	H	H	H	H	.F	F.
BLANK	X	X	X	X	X	X	L	H	X	X	X	X	X		
LED TEST	X	X	X	X	X	X	X	L	X	X	X	X	X	.8	8.

H = high level, L = low level, X = irrelevant.
DP input has arbitrarily been shown activated (high) on every other line of the table.

description (continued)

The TTL MSI circuits contain the equivalent of 78 gates on a single chip. Logic inputs and outputs are completely TTL/DTL compatible. The buffered inputs are implemented with relatively large resistors in series with the bases of the input transistors to lower drive-current requirements to one-half of that required for a standard Series 54/74 TTL input.

Some of the additional features of these displays are as follows:

- Latched BCD and decimal point logic outputs provided to drive logic processors simultaneously with the displayed data
- Minimum number of inputs required . . . 4-line BCD plus decimal point
- Overriding blanking for suppressing entire display or for pulse-modulation of LED brightness
- LED test input to simultaneously turn on all display segments and decimal point
- Can be operated in a real-time mode or latched-update-only mode by use of the latch strobe input
- Displays numbers 0 thru 9 as well as A, C, E, F, or minus sign
- Can be blanked by entry of BCD 13 or by use of the blanking input
- Decimal point controlled independently with decimal-point latch
- Constant-current-source TTL-LED interface for optimum performance.

The latch outputs except Q_{DP} are active pull-up, and each one, except Q_{DP}, is capable of driving three standard Series 54/74 loads. The LED driver outputs are designed specifically to maintain a relatively constant on-level current of approximately seven milliamperes through each LED segment and decimal point. All inputs are diode-clamped to minimize transmission-line effects, thereby simplifying system design. Power dissipation is typically 575 milliwatts with all segments on.

absolute maximum ratings over operating case temperature range (unless otherwise noted)

Supply Voltage, V_{CC} (See Note 1): Continuous . . . . 5.5 V
Nonrepetitive Peak, $t_w \leq 100$ ms . . . . 7 V
Input Voltage (See Note 1) . . . . 5.5 V
Operating Case Temperature Range (See Note 2) . . . . 0°C to 85°C
Storage Temperature Range . . . . −25°C to 85°C

NOTES: 1. Voltage values are with respect to network ground terminal.
2. Case temperature is the surface temperature of the plastic measured directly over the integrated circuit. Forced-air cooling may be required to maintain this temperature.

recommended operating conditions

			MIN	NOM	MAX	UNIT
Latch Strobe Pulse Duration, t_w			4.75	5	5.25	V
Normalized Fan-out from each output, N (to Series 54/74 Integrated Circuits)	Low Logic Level	Q_{DP}			1	
		Q_A, Q_B, Q_C, Q_D			3	
	High Logic Level	Q_{DP}			3	
		Q_A, Q_B, Q_C, Q_D			6	
Latch Strobe Pulse Duration, t_w			45			ns
Setup Time, t_{su}			60			ns
Hold Time, t_h			0			ns

TEXAS INSTRUMENTS
INCORPORATED
POST OFFICE BOX 225012 • DALLAS, TEXAS 75265

6

operating characteristics at 25°C case temperature

PARAMETER			TEST CONDITIONS	MIN	TYP†	MAX	UNIT
I_v	Luminous Intensity (See Note 3)	Figure 8	V_{CC} = 5 V	700	1200		μcd
		Decimal Point		40	70		
λp	Wavelength at Peak Emission		V_{CC} = 5 V, See Note 4		660		nm
Δλ	Spectral Bandwidth		V_{CC} = 5 V, See Note 4		20		nm
V_{IH}	High-Level Input Voltage			2			V
V_{IL}	Low-Level Input Voltage					0.8	V
V_{IK}	Input Clamp Voltage		V_{CC} = 4.75 V, I_I = −12 mA			−1.5	V
V_{OH}	High-Level Output Voltage	Q_{DP}	V_{CC} = 4.75 V, I_{OH} = −120 μA	2.4			V
		Q_A, Q_B, Q_C, Q_D	V_{CC} = 4.75 V, I_{OH} = −240 μA				
V_{OL}	Low-Level Output Voltage (See Note 5)	Q_{DP}	V_{CC} = 4.75 V, I_{OL} = 1.6 mA			0.4	V
		Q_A, Q_B, Q_C, Q_D	V_{CC} = 4.75 V, I_{OL} = 4.8 mA				
I_I	Input Current at Maximum Input Voltage		V_{CC} = 5.25 V, V_I = 5.5 V			1	mA
I_{IH}	High-Level Input Current		V_{CC} = 5.25 V, V_I = 2.4 V			20	μA
I_{IL}	Low-Level Input Current		V_{CC} = 5.25 V, V_I = 0.4 V			−0.8	mA
I_{OS}	Short-Circuit Output Current	Q_A, Q_B, Q_C, Q_D	V_{CC} = 5.25 V	−9		−27.5	mA
		Q_{DP}		−1		−3.2	
I_{CC}	Supply Current		V_{CC} = 5.25 V, All Inputs at 0 V		115	180	mA

†All typical values are at V_{CC} = 5 V.

NOTES: 3. Luminous intensity is measured with a light sensor and filter combination that approximates the CIE (International Commission on Illumination) eye-response curve.
4. These parameters are measured with all LED segments and the decimal point on.
5. This parameter is measured with the display blanked.

switching characteristics, V_{CC} = 5 V, T_C = 25°C

PARAMETER	FROM (INPUT)	TO (OUTPUT)	TEST CONDITIONS	MIN	TYP	MAX	UNIT
t_{PLH}	A, B, C, D, DP	$Q_A, Q_B, Q_C, Q_D, Q_{DP}$	C_L = 15 pF, R_L = 1.2 kΩ, See Figure 1		35		ns
t_{PHL}					40		ns

t_{PLH} ≡ Propagation delay time, low-to-high-level output
t_{PHL} ≡ Propagation delay time, high-to-low-level output

PARAMETER MEASUREMENT INFORMATION

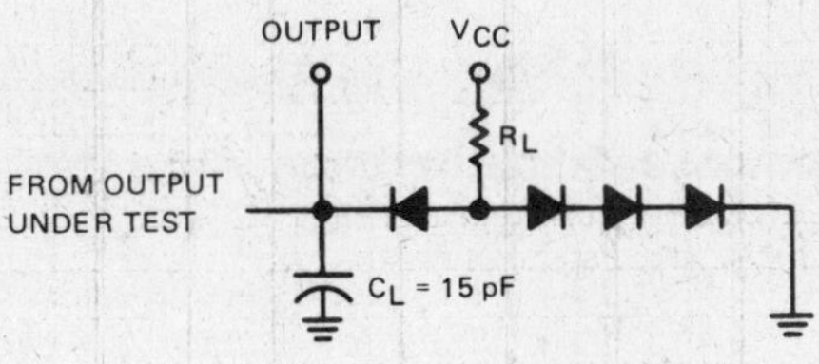

NOTES: A. C_L includes probe and jig capacitance.
B. All diodes are 1N3064.
C. Measurements made with latch strobe input grounded.

LOAD CIRCUIT—FIGURE 1

TYPICAL CHARACTERISTICS

RELATIVE SPECTRAL CHARACTERISTICS

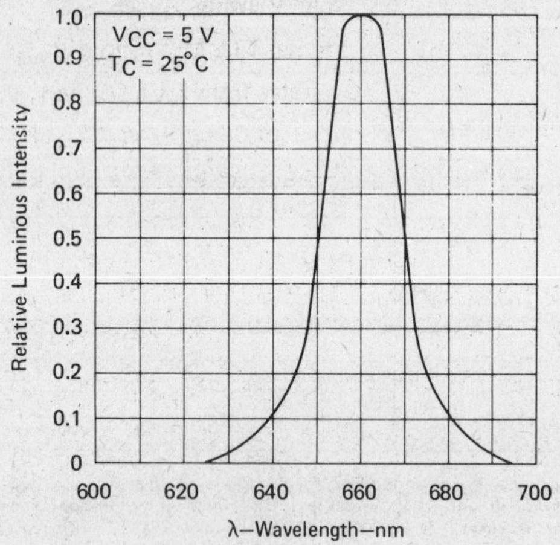

FIGURE 2

RELATIVE LUMINOUS INTENSITY
vs
CASE TEMPERATURE

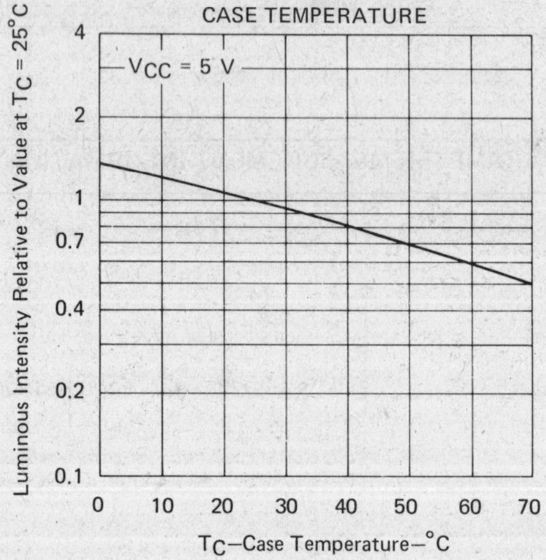

FIGURE 3

10

LED DISPLAYS

TEXAS INSTRUMENTS
INCORPORATED
POST OFFICE BOX 225012 • DALLAS, TEXAS 75265

SOLID-STATE HEXADECIMAL DISPLAY WITH INTEGRAL TTL CIRCUIT TO ACCEPT, STORE, AND DISPLAY 4-BIT BINARY DATA

- 7,62-mm (0.300-Inch) Character Height
- High Brightness
- Left-and-Right-Hand Decimals
- Separate LED and Logic Power Supplies May Be Used
- Wide Viewing Angle
- Internal TTL MSI Chip with Latch, Decoder, and Driver
- Operates from 5-Volt Supply
- Constant-Current Drive for Hexadecimal Characters
- Easy System Interface

mechanical data

These assemblies consist of display chips and a TTL MSI chip mounted on a header with either a red molded plastic body for the TIL311 or a red plastic cap for the TIL311A. Multiple displays may be mounted on 11,43-mm (0.450-inch) centers.

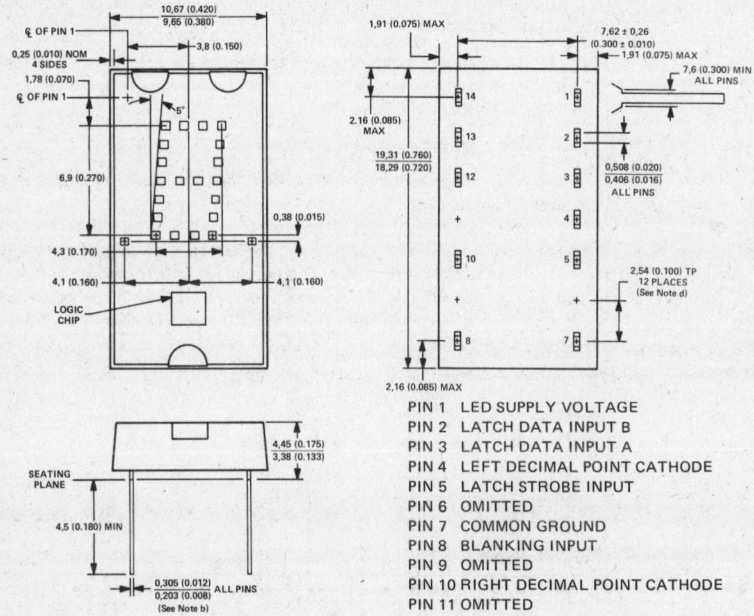

PIN 1 LED SUPPLY VOLTAGE
PIN 2 LATCH DATA INPUT B
PIN 3 LATCH DATA INPUT A
PIN 4 LEFT DECIMAL POINT CATHODE
PIN 5 LATCH STROBE INPUT
PIN 6 OMITTED
PIN 7 COMMON GROUND
PIN 8 BLANKING INPUT
PIN 9 OMITTED
PIN 10 RIGHT DECIMAL POINT CATHODE
PIN 11 OMITTED
PIN 12 LATCH DATA INPUT D
PIN 13 LATCH DATA INPUT C
PIN 14 LOGIC SUPPLY VOLTAGE, V_{CC}

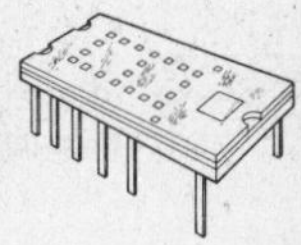

NOTES: a. All linear dimensions are in millimeters and parenthetically in inches.
b. Lead dimensions are not controlled above the seating plane.
c. Centerlines of character seqments and decimal points are shown as dashed lines. Associated dimensions are nominal.
d. The true-position pin spacing is 2,54 mm (0.100 inch) between centerlines. Each centerline is located within 0,26 mm (0.010 inch) of its true longitudinal position relative to pins 1 and 14.
e. On TIL311A devices, the 3 mold indentations are not present.

Copyright © 1982 by Texas Instruments Incorporated

description

This hexadecimal display contains a four-bit latch, decoder, driver, and 4 X 7 light-emitting-diode (LED) character with two externally-driven decimal points in a 14-pin package. A description of the functions of the inputs of this device follows.

FUNCTION	PIN NO.	DESCRIPTION
LATCH STROBE INPUT	5	When low, the data in the latches follow the data on the latch data inputs. When high, the data in the latches will not change. If the display is blanked and then restored while the enable input is high, the previous character will again be displayed.
BLANKING INPUT	8	When high, the display is blanked regardless of the levels of the other inputs. When low, a character is displayed as determined by the data in the latches. The blanking input may be pulsed for intensity modulation.
LATCH DATA INPUTS (A, B, C, D)	3, 2, 13, 12	Data on these inputs are entered into the latches when the enable input is low. The binary weights of these inputs are A = 1, B = 2, C = 4, D = 8.
DECIMAL POINT CATHODES	4, 10	These LEDs are not connected to the logic chip. If a decimal point is used, an external resistor or other current-limiting mechanism must be connected in series with it.
LED SUPPLY	1	This connection permits the user to save on regulated V_{CC} current by using a separate LED supply, or it may be externally connected to the logic supply (V_{CC}).
LOGIC SUPPLY (V_{CC})	14	Separate V_{CC} connection for the logic chip.
COMMON GROUND	7	This is the negative terminal for all logic and LED currents except for the decimal points.

The LED driver outputs are designed to maintain a relatively constant on-level current of approximately five milliamperes through each of the LED's forming the hexadecimal character. This current is virtually independent of the LED supply voltage within the recommended operating conditions. Drive current varies slightly with changes in logic supply voltage resulting in a change in luminous intensity as shown in Figure 2. This change will not be noticeable to the eye. The decimal point anodes are connected to the LED supply; the cathodes are connected to external pins. Since there is no current limiting built into the decimal point circuits, this must be provided externally if the decimal points are used.

The resultant displays for the values of the binary data in the latches are as shown below.

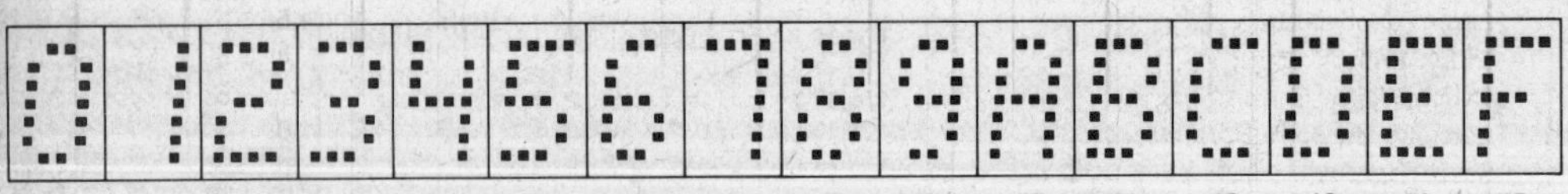

0 1 2 3 4 5 6 7 8 9 10 11 12 13 14 15

functional block diagram

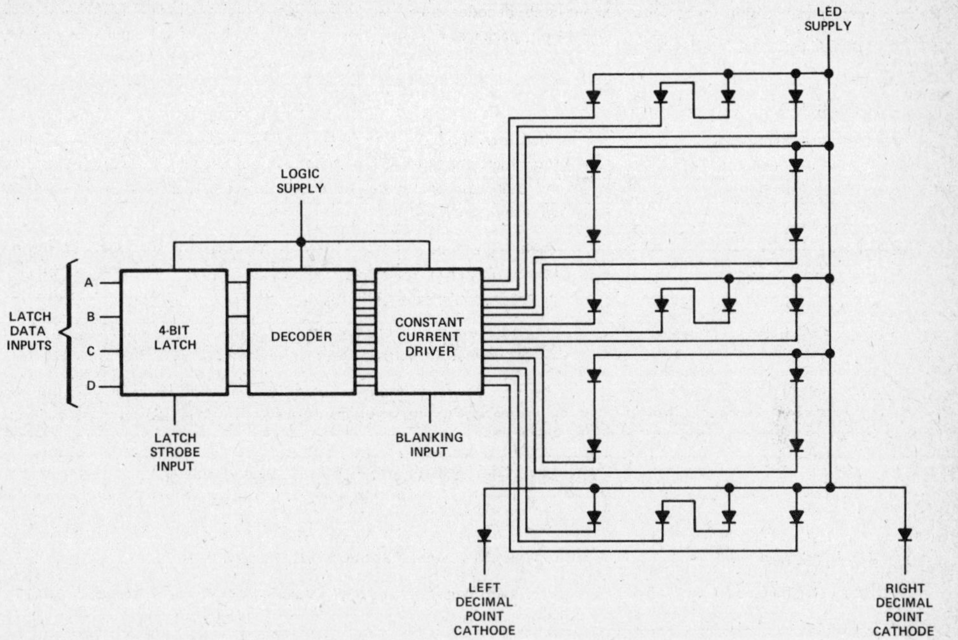

absolute maximum ratings over operating case temperature range (unless otherwise noted)

Logic Supply Voltage, V_{CC} (See Note 1)	7 V
LED Supply Voltage (See Note 1)	7 V
Input Voltage (Pins 2, 3, 5, 8, 12, 13; See Note 1)	5.5 V
Decimal Point Current	20 mA
Operating Case Temperature Range (See Note 2)	0°C to 85°C
Storage Temperature Range	−25°C to 85°C

NOTES: 1. Voltage values are with respect to common ground terminal.
2. Case temperature is the surface temperature of the plastic measured directly over the integrated circuit. Forced-air cooling may be required to maintain this temperature.

recommended operating conditions

	MIN	NOM	MAX	UNIT
Logic Supply Voltage, V_{CC}	4.5	5	5.5	V
LED Supply Voltage, V_{LED}	4	5	5.5	V
Decimal Point Current, $I_{F(DP)}$		5		mA
Latch Strobe Pulse Duration, t_w	40			ns
Setup Time, t_{su}	50			ns
Hold Time, t_h	40			ns

operating characteristics at 25°C case temperature

PARAMETER			TEST CONDITIONS	MIN	TYP	MAX	UNIT
I_V	Luminous Intensity (See Note 3)	Average Per Character LED	V_{CC} = 5 V, V_{LED} = 5 V, See Note 4	35	100		μcd
		Each decimal	$I_{F(DP)}$ = 5 mA	35	100		μcd
λ_p	Wavelength at Peak Emission		V_{CC} = 5 V, V_{LED} = 5 V, $I_{F(DP)}$ = 5 mA, See Note 5		660		nm
$\Delta\lambda$	Spectral Bandwidth				20		nm
V_{IH}	High-Level Input Voltage			2			V
V_{IL}	Low-Level Input Voltage					0.8	V
V_{IK}	Input Clamp Voltage		V_{CC} = 4.75 V, I_I = −12 mA			−1.5	V
I_I	Input Current at Maximum Input Voltage		V_{CC} = 5.5 V, V_I = 5.5 V			1	mA
I_{IH}	High-Level Input Current		V_{CC} = 5.5 V, V_I = 2.4 V			40	μA
I_{IL}	Low-Level Input Current		V_{CC} = 5.5 V, V_I = 0.4 V			−1.6	mA
I_{CC}	Logic Supply Current		V_{CC} = 5.5 V, V_{LED} = 5.5 V, $I_{F(DP)}$ = 5 mA, All inputs at 0 V		60	90	mA
I_{LED}	LED Supply Current				45	90	mA

NOTES: 3. Luminous intensity is measured with a light sensor and filter combination that approximates the CIE (International Commission on Illumination) eye-response curve.

4. This parameter is measured with A displayed, then again with E displayed.

5. These parameters are measured with .B. displayed.

TYPICAL CHARACTERISTICS

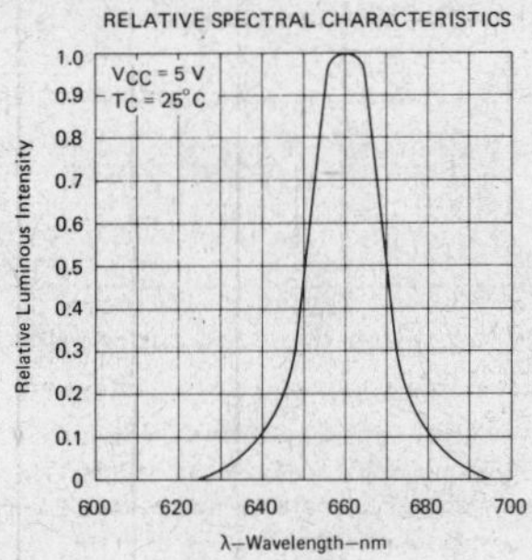

FIGURE 1

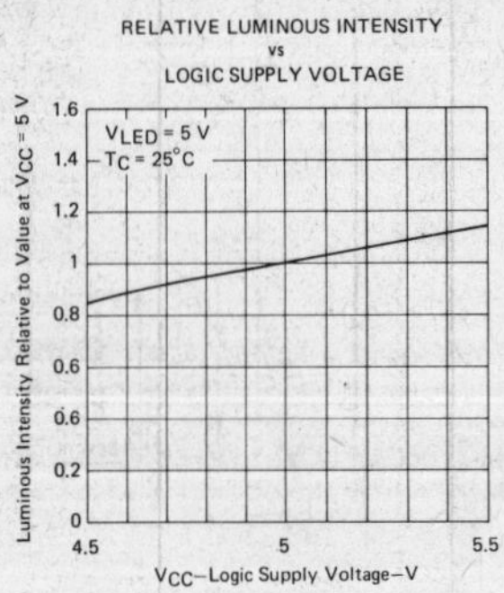

FIGURE 2

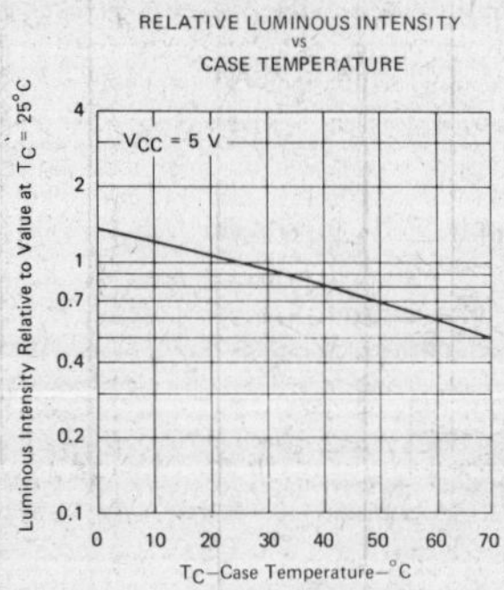

FIGURE 3

TEXAS INSTRUMENTS
INCORPORATED
POST OFFICE BOX 225012 • DALLAS, TEXAS 75265

10 LED DISPLAYS

D1924, SEPTEMBER 1981 – REVISED DECEMBER 1982

SOLID-STATE DISPLAYS WITH RED, GREEN, OR YELLOW CHARACTERS

- 7,62-mm (0.300-inch) Character Height
- Continuous Uniform Segments
- Wide Viewing Angle
- High Contrast
- Yellow and Green Displays are Categorized for Uniformity of Luminous Intensity and Wavelength among Units within Each Category

	SEVEN SEGMENTS WITH RIGHT AND LEFT DECIMALS, COMMON ANODE	SEVEN SEGMENTS WITH RIGHT DECIMAL, COMMON CATHODE	PULSE/MINUS ONE WITH LEFT DECIMAL
RED	TIL312	TIL313	TIL327
GREEN	TIL314	TIL315	TIL328
RED +	TIL333	TIL334	TIL335
YELLOW	TIL339	TIL340	TIL341

Red + stands for high-efficiency red.

mechanical data

NOTES: a. The true position spacing (T.P.) between centerlines is 2,54 mm (0.100 inch). Each pin centerline is within 0,26 mm (0.010 inch) of its true longitudinal position relative to pin 1.
b. All dimensions associated with segments and decimal points are nominal.
c. Left decimal point of TIL313, TIL315, TIL334, and TIL340 is not operational.

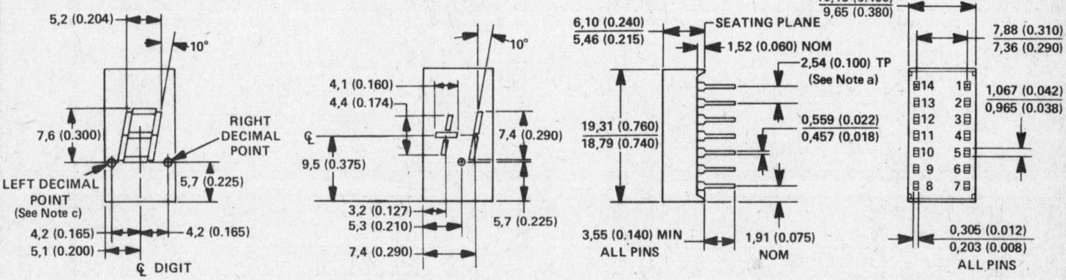

ALL DIMENSIONS ARE IN MILLIMETERS AND PARENTHETICALLY IN INCHES

TIL312, TIL314, TIL333, TIL339

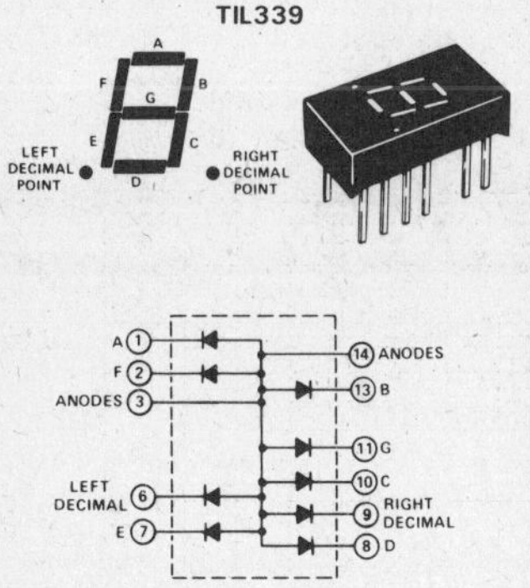

PINS 4, 5, AND 12 OMITTED

TIL313, TIL315, TIL334, TIL340

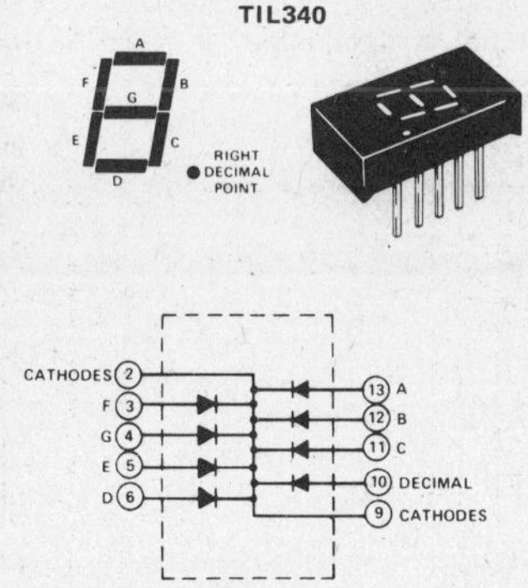

PINS 1, 7, 8, AND 14 OMITTED

TIL327, TIL328, TIL335, TIL341

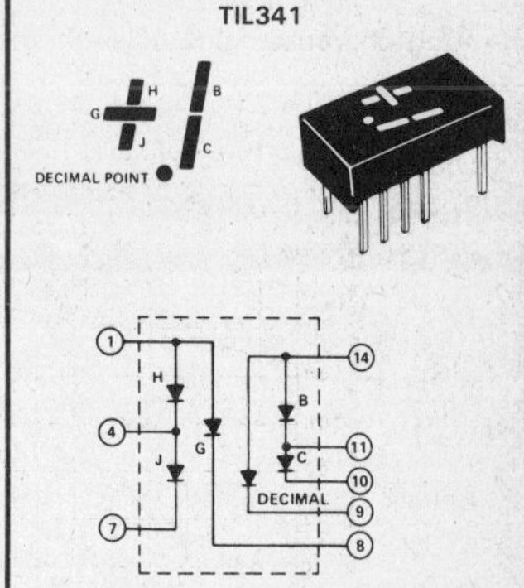

PINS 2, 3, 5, 6, 12, AND 13 OMITTED

Copyright © 1982 by Texas Instruments Incorporated

TEXAS INSTRUMENTS
INCORPORATED
POST OFFICE BOX 225012 • DALLAS, TEXAS 75265

mechanical data (continued)

The display chips are mounted on a lead frame and this assembly is then molded within an electrically nonconductive plastic case. To optimize device performance, materials are used that are limited to certain solvents for cleaning operations. It is recommended that only Freon TF†, isopropanol, or water be used.

†Trademark of E.I. duPont de Nemours, Inc.

absolute maximum ratings

Reverse Voltage at 25 °C Free-Air Temperature, Each Segment or Decimal Point . . . . . 5 V
Peak Forward Current at (or below) 25 °C Free-Air Temperature.
Each Segment or Decimal Point . . . . . 150 mA
Average Forward Current at (or below) 25 °C Free-Air Temperature (See Notes 1 and 2),
Each Segment or Decimal Point . . . . . 25 mA
Operating Free-Air Temperature Range . . . . . −40 °C to 85 °C
Storage Temperature Range . . . . . −40 °C to 85 °C
Lead Temperature 1.6 mm (1/16 Inch) Below Seating Plane for 5 Seconds . . . . . 260 °C

NOTES: 1. This average value applies for any 10-ms period.
2. Derate linearly to 10 mA at 85 °C free-air temperature at the rate of 0.25 mA/°C.

operating characteristics of each segment or decimal point at 25 °C free air temperature

PARAMETER			TEST CONDITIONS	RED TIL312, TIL313, TIL327			HIGH-EFFICIENCY RED TIL333, TIL334, TIL335			UNIT
				MIN	TYP	MAX	MIN	TYP	MAX	
I_V	Luminous Intensity (See Notes 3 and 4)	Average per Segment	I_F = 10 mA per segment	125			320			μcd
	Segment-to-Segment Luminous Intensity Ratio				1.5:1			1.5:1		
λ_p	Wavelength at Peak Emission				655			630		nm
$\Delta\lambda$	Spectral Bandwidth				20			40		nm
V_F	Static Forward Voltage		I_F = 20 mA		1.7	2		2.5	3	V
I_R	Static Reverse Current		V_R = 5 V			100			100	μA

operating characteristics of each segment or decimal point at 25 °C free air temperature

PARAMETER			TEST CONDITIONS	GREEN TIL314, TIL315, TIL328			YELLOW TIL339, TIL340, TIL341			UNIT
				MIN	TYP	MAX	MIN	TYP	MAX	
I_V	Luminous Intensity (See Notes 3 and 4)	Average per Segment	I_F = 10 mA per segment	125			320			μcd
	Segment-to-Segment Luminous Intensity Ratio				1.5:1			1.5:1		
λ_p	Wavelength at Peak Emission	(See Note 4)			565			585		nm
$\Delta\lambda$	Spectral Bandwidth				40			40		nm
V_F	Static Forward Voltage		I_F = 20 mA		2.5	3.5		2.5	3	V
I_R	Static Reverse Current		V_R = 5 V			100			100	μA

NOTE: 3. Luminous intensity is measured with a light sensor and filter combination that approximates the CIE (International Commission on Illumination) eye-response curve.
4. All displays are categorized for luminous intensity; yellow and green displays are also categorized for wavelength. The appropriate intensity (bin) letter and wavelength (bin) number are stamped on the top end of the package.

TEXAS INSTRUMENTS
INCORPORATED
POST OFFICE BOX 225012 • DALLAS, TEXAS 75265

TYPES TIL321A THRU TIL324A, TIL330A, TIL331A, TIL345 THRU TIL350 NUMERIC DISPLAYS

D2391, MARCH 1976–REVISED FEBRUARY 1983

SOLID-STATE DISPLAYS WITH RED, GREEN, OR YELLOW SEVEN SEGMENT DISPLAYS

- 12,7-mm (0.500-Inch) Character Height
- Continuous Uniform Segments
- High Contrast
- Categorized for Uniformity of Luminous Intensity and Wavelength among Units within Each Category for Yellow and Green Displays
- Low Power Requirements
- Wide Viewing Angle

	SEVEN SEGMENTS		PLUS/MINUS ONE
	COMMON ANODE	COMMON CATHODE	COMMON ANODE
RED	TIL321A	TIL322A	TIL330A
GREEN	TIL323A	TIL324A	TIL331A
YELLOW	TIL345	TIL346	TIL347
HIGH-EFFICIENCY RED	TIL348	TIL349	TIL350

mechanical data

NOTES: a. The true-position pin spacing is 2,54 mm (0.100 inch) between centerlines. Each pin centerline is located within 0,26 mm (0.010 inch) of its true longitudinal position relative to pins 3 and 8.

b. All dimensions associated with segments and decimal points are nominal.

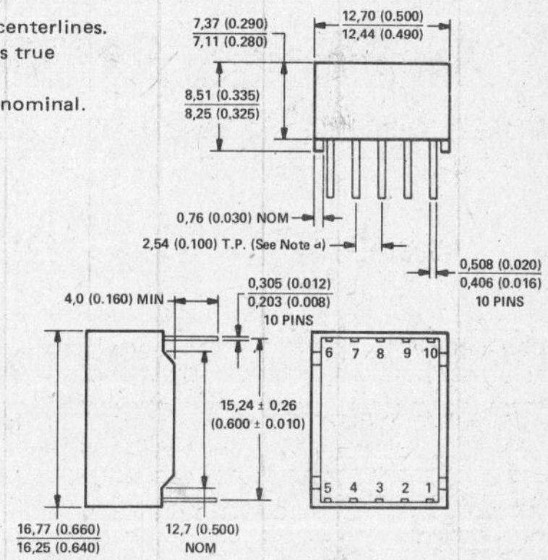

TIL321A THRU TIL324A, TIL345, TIL346, TIL348, TIL349

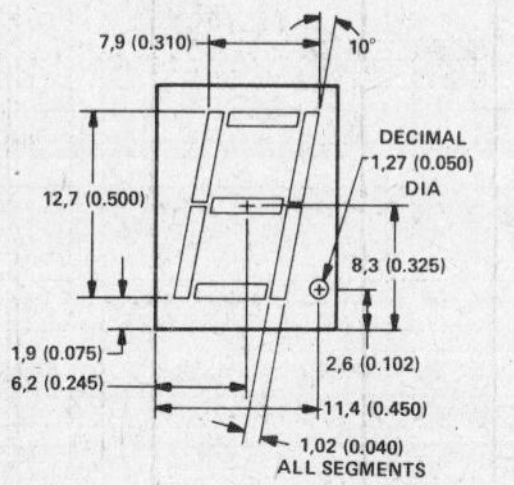

TIL330A, TIL331A, TIL347, TIL350

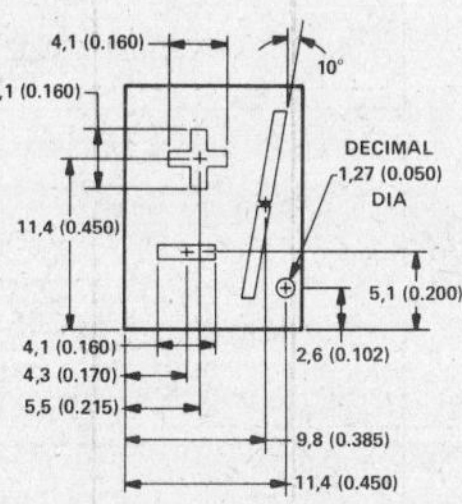

ALL LINEAR DIMENSIONS ARE IN MILLIMETERS AND PARENTHETICALLY IN INCHES

TIL321A, TIL323A, TIL345, TIL348

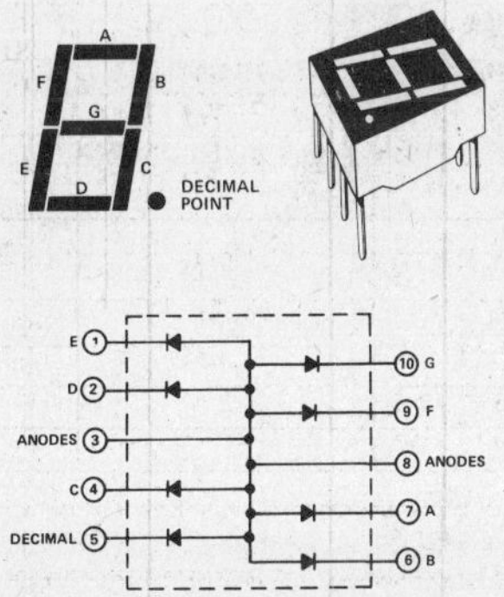

TIL322A, TIL324A, TIL346, TIL349

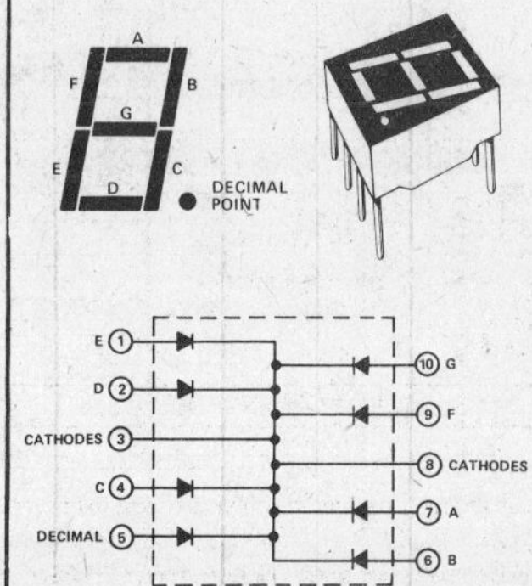

TIL330A, TIL331A, TIL347, TIL350

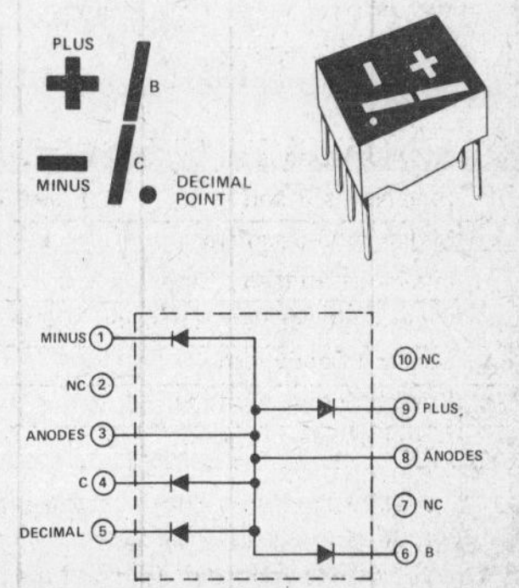

NC—No internal connection

Copyright © 1983 by Texas Instruments Incorporated

TEXAS INSTRUMENTS
INCORPORATED
POST OFFICE BOX 225012 • DALLAS, TEXAS 75265

mechanical data (continued)

The display chips are mounted on a lead frame, and this assembly is then molded within an electrically nonconductive plastic case. To optimize device performance, materials are used that are limited to certain solvents for cleaning operations. It is recommended that only Freon TF†, isopropanol, or water be used.

†Trademark of E.I. duPont de Nemours, Inc.

absolute maximum ratings

Reverse Voltage at 25°C Free-Air Temperature, Each Segment or Decimal Point .... 5 V
Peak Forward Current at (or below) 25°C Free-Air Temperature,
Each Segment or Decimal Point .... 150 mA
Average Forward Current at (or below) 25°C Free-Air Temperature (See Notes 1 and 2),
Each Segment or Decimal Point .... 25 mA
Operating Free-Air Temperature Range .... −40°C to 85°C
Storage Temperature Range .... −40°C to 85°C
Lead Temperature 1.6 mm (1/16 inch) Below Seating Plane for 5 Seconds .... 250°C

NOTES: 1. This average value applies for any 10-ms period.
2. Derate linearly to 10 mA at 85°C free-air temperature at the rate of 0.25 mA/°C.

operating characteristics of each segment or decimal point at 25°C free air temperature

PARAMETER			TEST CONDITIONS	RED TIL321A, TIL322A, TIL330			HIGH-EFFICIENCY RED TIL348, TIL349, TIL350			UNIT
				MIN	TYP	MAX	MIN	TYP	MAX	
I_V	Luminous Intensity (See Notes 3 and 4)	Average per Segment	I_F = 10 mA per segment	125			320			μcd
	Segment-to-Segment Luminous Intensity Ratio				1.5:1			1.5:1		
λ_p	Wavelength at Peak Emission				655			630		nm
$\Delta\lambda$	Spectral Bandwidth				20			40		nm
V_F	Static Forward Voltage		I_F = 20 mA		1.7	2		2.5	3	V
I_R	Static Reverse Current		V_R = 5 V			100			100	μA

PARAMETER			TEST CONDITIONS	GREEN TIL323A, TIL324A, TIL331A			YELLOW TIL345, TIL346, TIL347			UNIT
				MIN	TYP	MAX	MIN	TYP	MAX	
I_V	Luminous Intensity (See Notes 3 and 4)	Average per Segment	I_F = 10 mA per segment	125			320			μcd
	Segment-to-Segment Luminous Intensity Ratio				1.5:1			1.5:1		
λ_p	Wavelength at Peak Emission	(See Note 4)			565			585		nm
$\Delta\lambda$	Spectral Bandwidth				40			40		nm
V_F	Static Forward Voltage		I_F = 20 mA		2.5	3.5		2.5	3	V
I_R	Static Reverse Current		V_R = 5 V			100			100	μA

NOTES: 3. Luminous intensity is measured with a light sensor and filter combination that approximates the CIE (International Commission on Illumination) eye-response curve.
4. All displays are categorized for luminous intensity; yellow and green displays are also categorized for wavelength. The appropriate intensity (bin) letter and wavelength (bin) number are stamped on the top end of the package.

TEXAS INSTRUMENTS
INCORPORATED
POST OFFICE BOX 225012 • DALLAS, TEXAS 75265

D2134, DECEMBER 1975—REVISED DECEMBER 1982

SOLID-STATE RED DISPLAYS FOR CALCULATOR APPLICATIONS

- 2,6-mm (0.102-Inch) Magnified Image
- Seven-Segment Digits, Right-Hand Decimal Points
- Common-Cathode Configuration for Multiplex Applications
- 5,1-mm (0.200-Inch) Digit-to-Digit Spacing

TYPE	NUMBER OF DIGITS
TIL393-6	6
TIL393-8	8
TIL393-9	9

description

These multidigit displays are intended for use under pulsed conditions by enabling the common cathode of each digit sequentially and enabling the desired segment anode and/or right-hand decimal point anode in phase with the character-enabling pulse. The pulse rate is kept high enough so that the light from each character appears to the eye to be constant.

absolute maximum ratings over operating free-air temperature range

Reverse Voltage, Each Segment or Decimal Point . . . . . . 5 V
Peak Forward Current, Each Segment or Decimal Point . . . . . . 60 mA
Average Forward Current, Each Segment or Decimal Point (See Note 1) . . . . . . 5 mA
Operating Free-Air Temperature Range . . . . . . −40°C to 85°C
Storage Temperature Range . . . . . . −40°C to 85°C
Terminal Temperature for 5 Seconds . . . . . . 230°C

NOTE 1: This average value applies for any 10-ms period.

operating characteristics of each segment or decimal at 25°C free-air temperature

PARAMETER			TEST CONDITIONS	MIN	TYP	MAX	UNIT
$I_{v(pk)}$	Peak Luminous Intensity (See Note 2)	Segment	I_{FM} = 10 mA, t_w = 5 ms, PRR = 100 Hz	200	600		μcd
		Decimal		200	600		μcd
λ_p	Wavelength at Peak Emission				660		nm
$\Delta\lambda$	Spectral Bandwidth				20		nm
V_F	Static Forward Voltage				1.7	2.1	V

NOTE 2: Luminous intensity is measured with a light sensor and filter combination that approximates the CIE (International Commission on Illumination) eye-response curve.

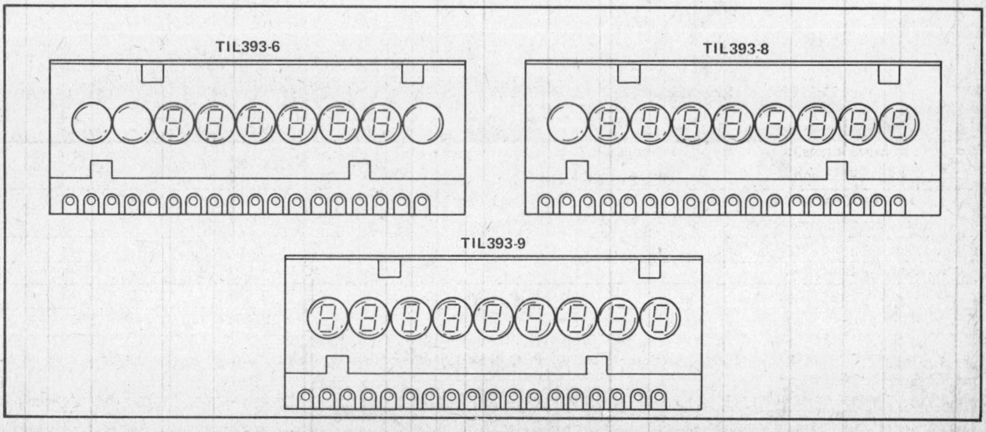

10
LED DISPLAYS

Copyright © 1982 by Texas Instruments Incorporated

mechanical data

The GaAsP monolithic chips (one for each digit) are mounted on a printed-circuit board. A clear plastic lens is attached to the p-c board providing protection for the chips and resulting in a magnified digit image of 2,6 mm (0.102 inch). The same lens is used for all three types.

The display may be mounted by use of a lead-frame assembly on 2,54-mm (0.100-inch) centers with the pins soldered into the p-c board holes, or by insertion into a p-c board edge connector. A rosin-core 60/40 tin/lead wire solder or a low-temperature deactivating flux with solid-core 60/40 solder can be used for hand-soldering operations.

Chlorinated hydrocarbon solvents must not be used for cleaning as the plastic lens may be damaged. Methanol, isopropanol, ethanol, Freon TP-35†, or Freon TE-35† solvents may be used with caution. Time must be allowed for the solvent to evaporate from beneath the display lens.

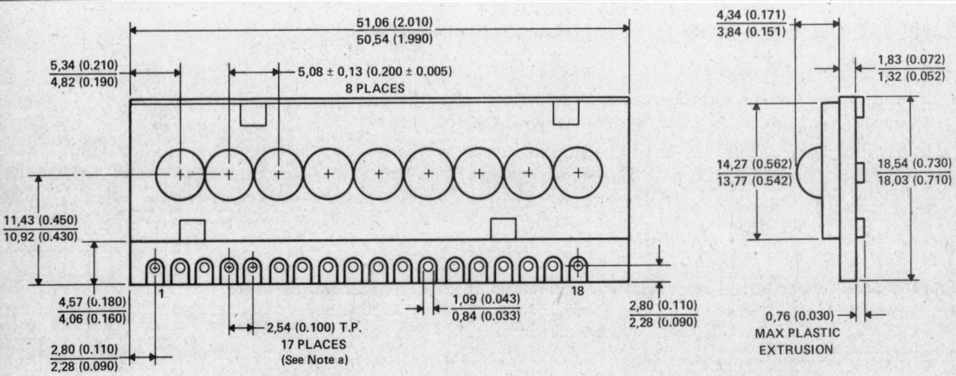

MAGNIFIED CHARACTER FONT

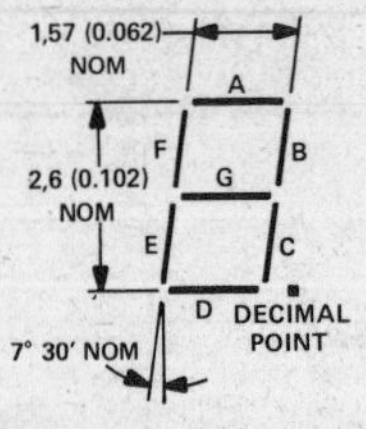

TOTAL VIEWING ANGLE (See Note b)

Horizontal	36° NOM
Vertical	32° NOM

TAB ASSIGNMENTS

TAB	FUNCTION
1	NO INTERNAL CONNECTION
2	DIGIT 1 CATHODE‡§
3	SEGMENT C ANODES
4	DIGIT 2 CATHODE§
5	DECIMAL POINT ANODES
6	DIGIT 3 CATHODE
7	SEGMENT A ANODES
8	DIGIT 4 CATHODE
9	SEGMENT E ANODES
10	DIGIT 5 CATHODE
11	SEGMENT D ANODES
12	DIGIT 6 CATHODE
13	SEGMENT G ANODES
14	DIGIT 7 CATHODE
15	SEGMENT B ANODES
16	DIGIT 8 CATHODE
17	SEGMENT F ANODES
18	DIGIT 9 CATHODE§

‡Make no external connection to tab 2 of TIL393-8.
§Make no external connection to tab 2, 4, or 18 of TIL393-6.

NOTES: a. The true-position tab spacing is 2,54 mm (0.100 inch) between centerlines. Each tab centerline is located within 0,26 mm (0.010 inch) of its true position relative to tab 1.
b. Total viewing angle that encompasses all the lines of sight along which the digit is legible within a given plane.

ALL DIMENSIONS ARE IN MILLIMETERS AND PARENTHETICALLY IN INCHES

†Trademark of E.I. duPont de Nemours, Inc.

TEXAS INSTRUMENTS
INCORPORATED
POST OFFICE BOX 225012 • DALLAS, TEXAS 75265

D2726, MARCH 1983

SOLID-STATE DISPLAYS WITH RED CHARACTERS

- **High Luminous Intensity . . . Typ I_V = 900 μcd at I_F = 10 mA**
- **12,7-mm (0.500-inch) Character Height**
- **TIL729 Has Common Anode**
- **TIL730 Has Common Cathode**
- **High Contrast Optimized with a Gray Package**
- **Right Hand Decimal Point**
- **Low Power Requirements**
- **Wide Viewing Angle**

mechanical data

NOTES: a. The true-position pin spacing is 2,54 mm (0.100 inch) between centerlines. Each pin centerline is located within 0,26 mm (0.010 inch) its true longitudinal position relative to pins 3 and 8.

b. All dimensions associated with segments and decimal points are nominal.

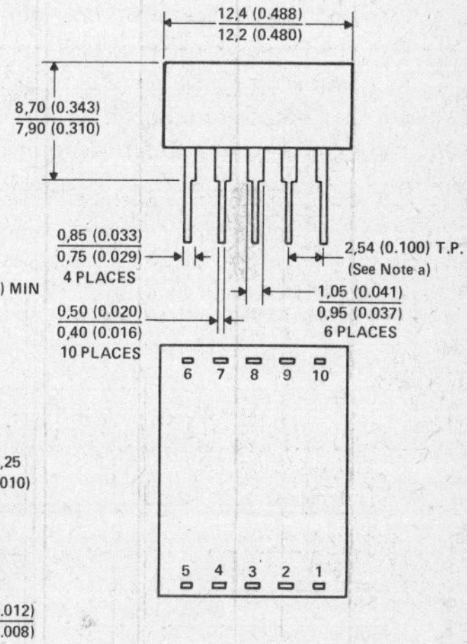

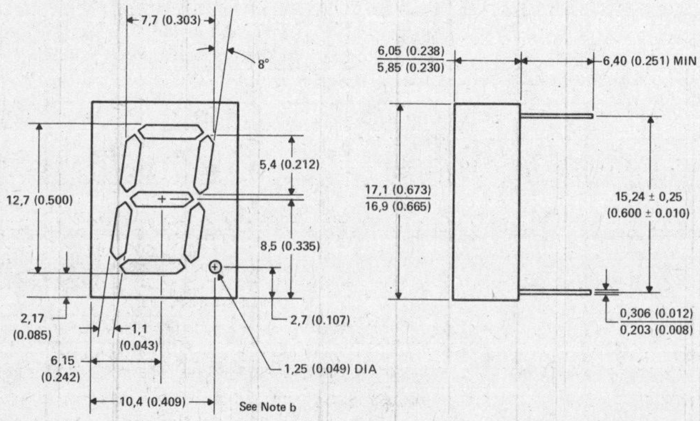

ALL LINEAR DIMENSIONS ARE IN MILLIMETERS AND PARENTHETICALLY IN INCHES

TIL729

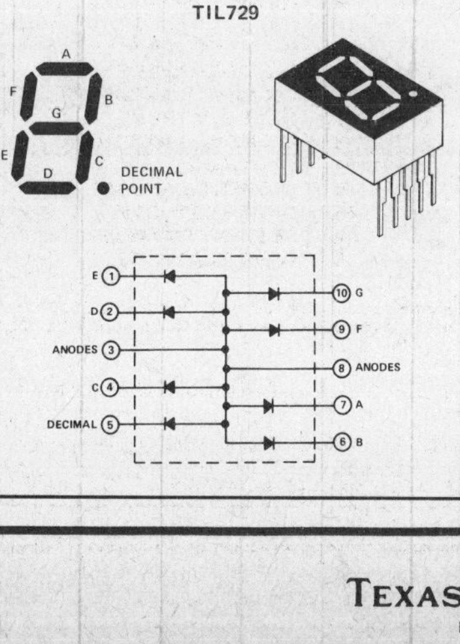

TIL730

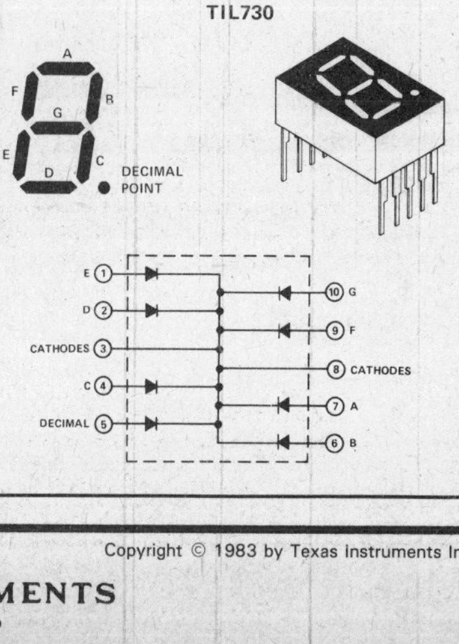

Copyright © 1983 by Texas Instruments Incorporated

mechanical data (continued)

The display chips are mounted on a lead frame, and this assembly is then molded within an electrically nonconductive plastic case. To optimize device performance, materials are used that are limited to certain solvents for cleaning operations. It is recommended that only Freon TF†, isopropanol, or water be used. For high contrast, the displays have a gray top surface.

†Trademark of E.I. duPont de Nemours, Inc.

absolute maximum ratings

Reverse Voltage at 25 °C Free-Air Temperature, Each Segment or Decimal Point .... 6 V
Peak Forward Current at (or below) 25 °C Free-Air Temperature,
Each Segment or Decimal Point .... 150 mA
Average Forward Current at (or below) 25 °C Free-Air Temperature (See Notes 1 and 2)
Each Segment or Decimal Point .... 20 mA
Operating Free-Air Temperature Range .... −40 °C to 85 °C
Storage Temperature Range .... −40 °C to 85 °C
Lead Temperature 1.6 mm (1/16 inch) Below Seating Plane for 5 Seconds .... 260 °C

NOTES: 1. This average value applies for any 10-ms period.
2. Derate linearly to 10 mA at 85 °C free-air temperature at the rate of 0.17 mA/°C.

operating characteristics of each segment or decimal point at 25 °C free-air temperature

PARAMETER			TEST CONDITIONS	MIN	TYP	MAX	UNIT
I_V	Luminous Intensity (See Note 3)	Segment	I_F = 10 mA	400	800		μcd
		Decimal Point		100	250		μcd
	Segment-to-Segment Luminous Intensity Ratio				1.5:1	2:1	
λ_p	Wavelength at Peak Emission			640	660	680	nm
$\Delta\lambda$	Spectral Bandwidth				25		nm
V_F	Static Forward Voltage				1.65	1.9	V
I_R	Static Reverse Current		V_R = 6 V		<10	100	μA

NOTE 3: Luminous intensity is measured with a light sensor and filter combination that approximates the CIE (International Commission on Illumination) eye-response curve.

TEXAS INSTRUMENTS
INCORPORATED
POST OFFICE BOX 225012 • DALLAS, TEXAS 75265

D2287, MARCH 1977—REVISED DECEMBER 1982

SOLID-STATE COMMON-CATHODE RED DISPLAY WITH RIGHT-HAND DECIMAL POINTS

- 6,9-mm (0.270-Inch) Character Height
- Multiplex Operation — Minimum Pin Connections
- High Luminous Intensity
- Wide Viewing Angle
- Viewing Distance up to 4.5 Meters (15 Feet)

applications

- Digital Frequency Read-Out
- Calculators
- Instrumentation Displays
- Data Terminals

description

This multidigit display is formed by mounting and bonding LED chips on a printed circuit board. Individual reflectors are used over the LED chips on each digit to form the segments. A diffuser placed over the reflectors results in a uniformly bright segment with a high contrast ratio.

The anodes of all like-positioned segments are connected together on the printed circuit board and brought out to a common pad connection. This type of configuration requires a minimum number of pad connections, but it requires that the display be used in a multiplexed mode. Each character is enabled sequentially by its cathode line and the desired segment and decimal anodes are enabled in phase with the cathode enabling pulse.

A peak current of 96 milliamperes is recommended for normal operating conditions at a duty cycle of 8.3% to obtain adequate display brightness. The pulse rate should be high enough so that the light from each character appears constant. A minimum pulse rate of 60 hertz can be used; however, rates of one kilohertz to ten kilohertz are recommended.

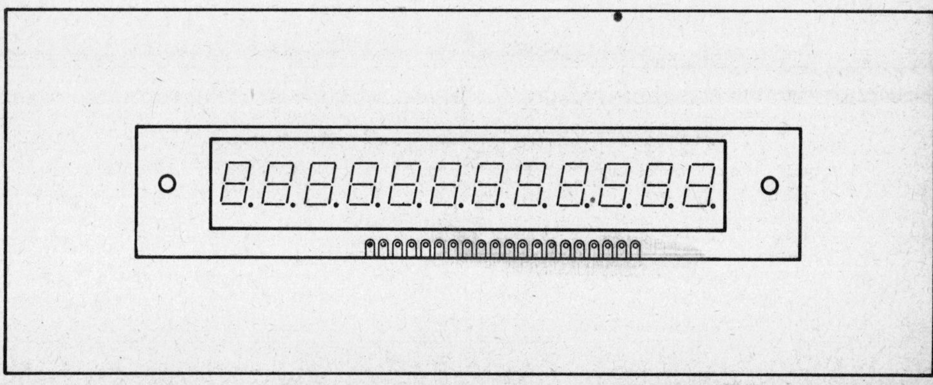

Copyright © 1982 Texas Instruments Incorporated

TEXAS INSTRUMENTS
INCORPORATED
POST OFFICE BOX 225012 • DALLAS, TEXAS 75265

mechanical data

The display may be mounted by soldering the pads directly to another printed-circuit board, by use of a lead-frame assembly on 0.100-inch centers with the pins soldered into the p-c board holes, or by insertion into a p-c board edge connector. A rosin-core 60/40 tin/lead solder, or a solid-core 60/40 solder with a low-temperature deactivating flux can be used for hand-soldering operations. Soldering temperature of each pad should not exceed 230°C for five seconds. Care should be exercised to keep the temperature of the plastic cover below 100°C as higher temperatures or direct contact of a hot soldering iron with the plastic could cause distortion or deformation of the character appearance.

Flux clean up using chlorinated hydrocarbon solvents should be avoided as they may damage the plastic parts. Methanol, isopropanol, ethanol, or Freon† TP-35 may be used with caution. Solvents can leave residues that may blur or obstruct the image.

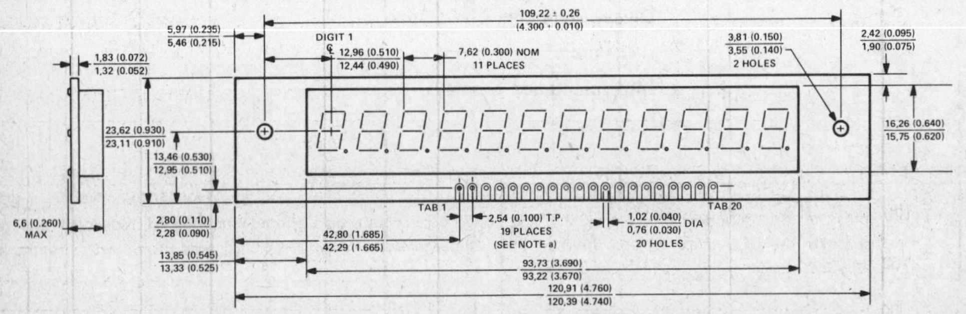

CHARACTER FONT

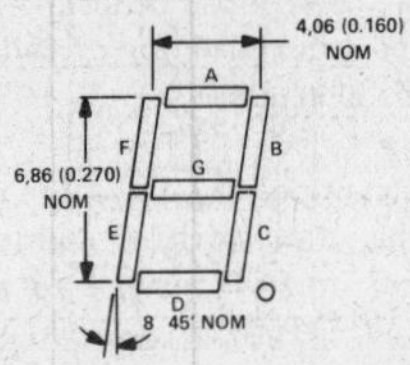

TAB ASSIGNMENTS

TAB	FUNCTION	TAB	FUNCTION
1	DIGIT 1 CATHODES	11	DIGIT 11 CATHODES
2	DIGIT 4 CATHODES	12	SEGMENT A ANODES
3	DIGIT 3 CATHODES	13	SEGMENT B ANODES
4	DIGIT 2 CATHODES	14	SEGMENT C ANODES
5	DIGIT 5 CATHODES	15	SEGMENT D ANODES
6	DIGIT 6 CATHODES	16	DECIMAL POINT ANODES
7	DIGIT 7 CATHODES	17	SEGMENT G ANODES
8	DIGIT 8 CATHODES	18	SEGMENT F ANODES
9	DIGIT 9 CATHODES	19	SEGMENT E ANODES
10	DIGIT 10 CATHODES	20	DIGIT 12 CATHODES

NOTE: a. The true-position (T.P.) pin spacing is 2,54 mm (0.100 inch) between centerlines. Each pin centerline is located within 0,13 mm (0.005 inch) of its true longitudinal position relative to Tab 1.

ALL DIMENSIONS ARE IN MILLIMETERS AND PARENTHETICALLY IN INCHES. INCH DIMENSIONS GOVERN.

product options

Texas Instruments Incorporated can supply multidigit displays that are variations of the basic 12-digit TIL804. Options include fewer digits or decimal points than 12 each and a choice of location of the omitted digits or decimal points. Fewer than eight digits are not recommended in order to be effective from the standpoints of cost and physical size. For custom arrangements contact your TI field office or Optoelectronics Marketing at the following address:

Texas Instruments Incorporated
Optoelectronics Marketing
P.O. Box 225012
Dallas, Texas 75265
Phone: (214) 995-3821

†Trademark of E.I. du Pont de Nemours, Inc.

absolute maximum ratings over operating free-air temperature range (unless otherwise noted)

Reverse Voltage, Each Segment or Decimal Point . . . . 5 V
Peak Forward Current at (or below) 25°C Free-Air Temperature (See Note 1),
Each Segment or Decimal Point . . . . 150 mA
Average Forward Current at (or below) 25°C Free-Air Temperature (See Notes 1 and 2),
Each Segment or Decimal Point . . . . 25 mA
Operating Free-Air Temperature Range . . . . −40°C to 70°C
Storage Temperature Range . . . . −40°C to 70°C

NOTES: 1. For operation above 25°C free-air temperature, refer to Figures 1 and 2.
2. This average value applies for any 10 millisecond period.

PEAK FORWARD CURRENT PER SEGMENT OR DECIMAL POINT vs FREE-AIR TEMPERATURE

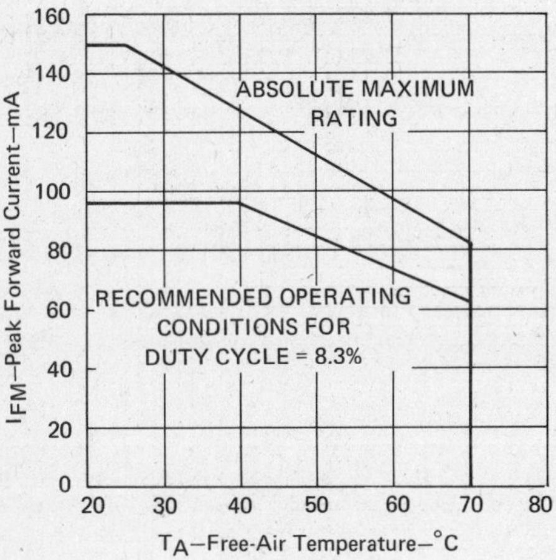

FIGURE 1

AVERAGE FORWARD CURRENT PER SEGMENT OR DECIMAL POINT vs FREE-AIR TEMPERATURE

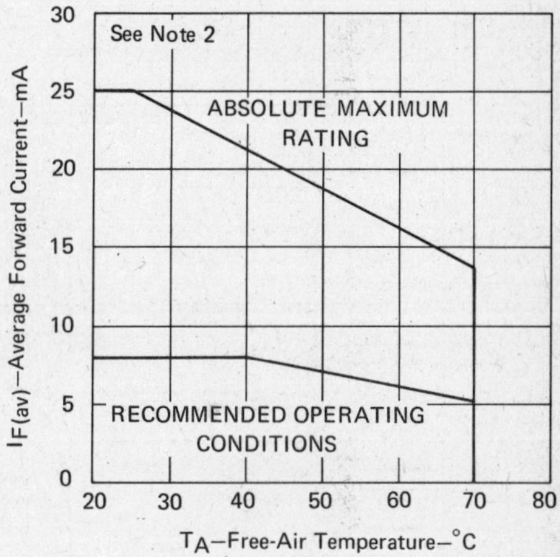

FIGURE 2

operating characteristics of each segment or decimal at 25°C free-air temperature

PARAMETER			TEST CONDITIONS	MIN	TYP	MAX	UNIT
I_V	Luminous Intensity (See Note 3)	Segment	I_F = 10 mA	100	150		µcd
		Decimal			100		
λ_p	Wavelength at Peak Emission				655		nm
$\Delta\lambda$	Spectral Bandwidth				20		nm
V_F	Static Forward Voltage		I_F = 20 mA		1.7	2	V

NOTE 3: Luminous intensity is measured with a light sensor and filter combination that approximates the CIE (International Commission on Illumination) eye-response curve.

TEXAS INSTRUMENTS
INCORPORATED
POST OFFICE BOX 225012 • DALLAS, TEXAS 75265

TYPICAL APPLICATION DATA

Each digit of the display is connected in a common-cathode configuration and the anodes of like-positioned segments of all digits are connected together for multiplex operation. Normal operation of each digit is 8.3% (1/12) duty cycle or less, except on custom devices with fewer digits. For example, a device with eight digits might be operated at 12.5% (1/8) duty cycle.

Figure 3, below, shows a typical interface circuit between the TIL804 and a TMS 1200 microcomputer. The typical conditions shown are intended as a guide only. These conditions will give a bright display easily read under high ambient light conditions as would be found in an office or laboratory; that is, 25 to 50 foot candles. If a brighter display is required, the average and peak currents through the segments could be increased.

Note that the display is to be operated under multiplexed conditions only.

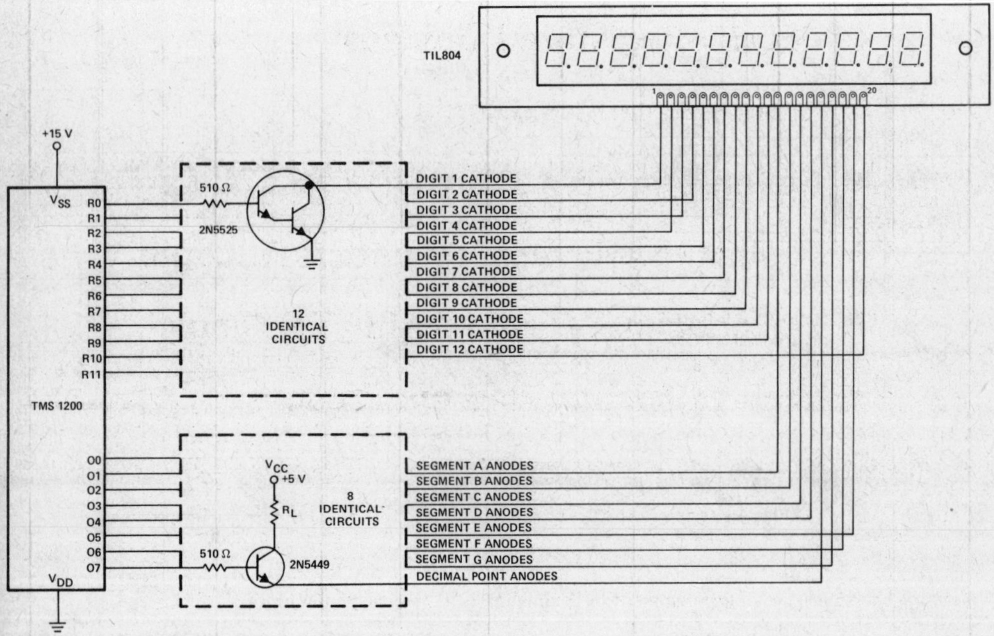

$$R_L = \frac{V_{DD} - V_{CE(sat)\,(2N5449)} - V_{CE(sat)\,(2N5525)} - V_{F(LED)}}{I_{FM}}$$

TYPICAL CONDITIONS

$V_{CC} = 5$ V
$V_{CE(sat)(2N5449)} = 0.2$ V
$V_{CE(sat)}$ $(2N5525) = 1.0$ V
$V_{F(LED)} = 1.8$ V
$I_{FM} = 96$ mA
Duty Cycle = 8.3%
$R_L = 21\ \Omega$

FIGURE 3

TEXAS INSTRUMENTS
INCORPORATED
POST OFFICE BOX 225012 • DALLAS, TEXAS 75265

High-Reliability LED Displays

- Quick Reference Guide
- JEDEC-Registered Devices
- Seven-Segment Display
- Seven-Segment Display with TTL Logic
- Alphanumeric Display with TTL Logic
- Hexadecimal Display with TTL Logic
- High-Efficiency Red and Yellow

HIGH-RELIABILITY LED DISPLAYS
QUICK REFERENCE GUIDE

JEDEC PART NO.	TI PART NO.	TYPE OF CHARACTER	CHARACTER HEIGHT mm (INCHES)	COLOR OF DISPLAY	PACKAGE	REMARKS
4N41	TIL501	7-segment	6,9 (0.270)	Red	14-lead hermetically sealed dual-in-line	Electrically and mechanically interchangeable with TIL302
4N56	TIL505	Hexadecimal	7,6 (0.300)	Red	14-lead hermetically sealed dual-in-line	Self-contained four-bit latch, decoder, and driver with 4 x 7 font
4N57	TIL506	7-segment	7,6 (0.300)	Red	8-lead hermetically sealed	Internal TTL MSI chip with decoder and driver. Left decimal
4N58	TIL507	5 x 7 alphanumeric	7,6 (0.300)	Red	16-lead hermetically sealed dual-in-line	Integral D-type flip-flop column drivers and series limiting resistors. Left decimal.
—	TIL509	Hexadecimal	7,6 (0.300)	Yellow	14-lead hermetically sealed dual in-line	Self-contained four-bit latch, decoder, and driver with 4 x 7 font
—	TIL510	5 x 7 alphanumeric	7,6 (0.300)	Yellow	16-lead hermetically sealed dual-in-line	Integral D-type flip-flop column drivers and series limiting resistors. Left decimal.

TEXAS INSTRUMENTS
INCORPORATED
POST OFFICE BOX 225012 • DALLAS, TEXAS 75265

HERMETICALLY SEALED SOLID-STATE VISIBLE DISPLAY (FORMERLY TIL501)

- Electrically and Mechanically Interchangeable with TIL302
- 6,9-mm (0.270-Inch) Character Height
- High Luminous Intensity
- Low Power Requirements
- Withstands Severe Environmental Conditions
- Left-Hand Decimal
- Wide Viewing Angle
- Compatible with Most TTL Circuits
- Each Unit Checked for Uniformity of Elements

*mechanical data

The display is mounted on a ceramic header, which is then hermetically sealed to a glass cover. Multiple displays may be mounted on 11,4-mm (0.450-inch) centers.

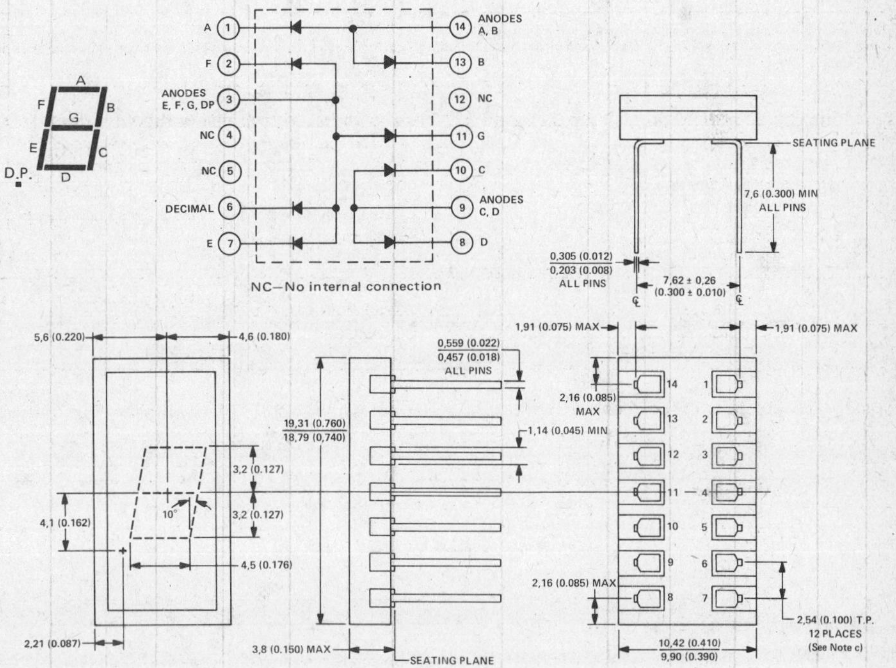

NOTES: a. All linear dimensions are in millimeters and parenthetically in inches.

b. Centerlines of character segments and decimal points are shown as dashed lines. Associated dimensions are nominal.

c. The true-position pin spacing is 2,54 mm (0.100 inch) between centerlines. Each centerline is located within 0,26 (0.010 inch) of its true longitudinal position relative to pins 4 and 11.

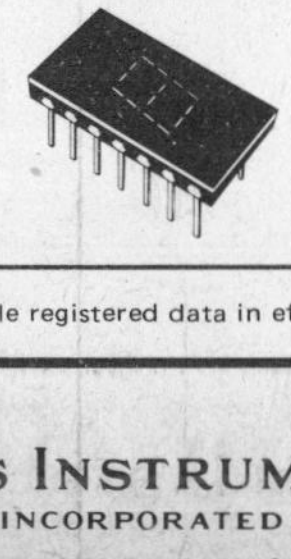

*JEDEC registered data. This data sheet contains all applicable registered data in effect at the time of publication.

Copyright © 1982 by Texas Instruments Incorporated

TEXAS INSTRUMENTS
INCORPORATED
POST OFFICE BOX 225012 • DALLAS, TEXAS 75265

*absolute maximum ratings

Reverse Voltage at 25°C Free-Air Temperature:
Each Segment . . . 6 V
Decimal Point . . . 3 V
Peak Forward Current at (or below) 70°C Free-Air Temperature, (See Note 1)
Each Segment or Decimal Point . . . 200 mA
Average Forward Current at (or below) 70°C Free-Air Temperature (See Notes 2 and 3):
Each Segment or Decimal Point . . . 30 mA
Total . . . 240 mA
Operating Free-Air Temperature Range . . . −55°C to 100°C
Storage Temperature Range . . . −65°C to 125°C
Lead Temperature 1,6 mm (1/16 Inch) Below the Seating Plane for 10 Seconds . . . 260°C

NOTES: 1. Derate linearly to 100°C free-air temperature at the rate of 6.67 mA/°C.
2. These average values apply for any 10-ms period.
3. Derate linearly to 100°C free-air temperature at the rates of 1 mA/°C for each segment or decimal point and 8 mA/°C for the total device.

*operating characteristics of each segment at 25°C free-air temperature (unless otherwise noted)

PARAMETER		TEST CONDITIONS	MIN	TYP	MAX	UNIT
I_V	Luminous Intensity (See Note 4)	I_F = 20 mA	200	700		μcd
λ_p	Wavelength at Peak Emission		640	660	680	nm
$\Delta\lambda$	Spectral Bandwidth			20		nm
V_F	Static Forward Voltage		3	3.4	3.8	V
α_{VF}	Average Temperature Coefficient of Static Forward Voltage	I_F = 20 mA, T_A = 0°C to 100°C		−2.7		mV/°C
I_R	Static Reverse Current	V_R = 6 V			100	μA
C	Anode-to-Cathode Capacitance	V_R = 0, f = 1 MHz		85		pF

*operating characteristics of decimal point at 25°C free-air temperature (unless otherwise noted)

PARAMETER		TEST CONDITIONS	MIN	TYP	MAX	UNIT
I_V	Luminous Intensity (See Note 4)	I_F = 20 mA	100	350		μcd
λ_p	Wavelength at Peak Emission		640	660	680	nm
$\Delta\lambda$	Spectral Bandwidth			20		nm
V_F	Static Forward Voltage		1.5	1.65	2	V
α_{VF}	Average Temperature Coefficient of Static Forward Voltage	I_F = 20 mA, T_A = 0°C to 100°C		−1.4		mV/°C
I_R	Static Reverse Current	V_R = 3 V			100	μA
C	Anode-to-Cathode Capacitance	V_R = 0, f = 1 MHz		120		pF

NOTE 4: Luminous intensity is measured with a light sensor and filter combination that approximates the CIE (International Commission on Illumination) eye-response curve.

*JEDEC registered data

TEXAS INSTRUMENTS
INCORPORATED
POST OFFICE BOX 225012 • DALLAS, TEXAS 75265

TYPICAL CHARACTERISTICS

RELATIVE SPECTRAL CHARACTERISTICS

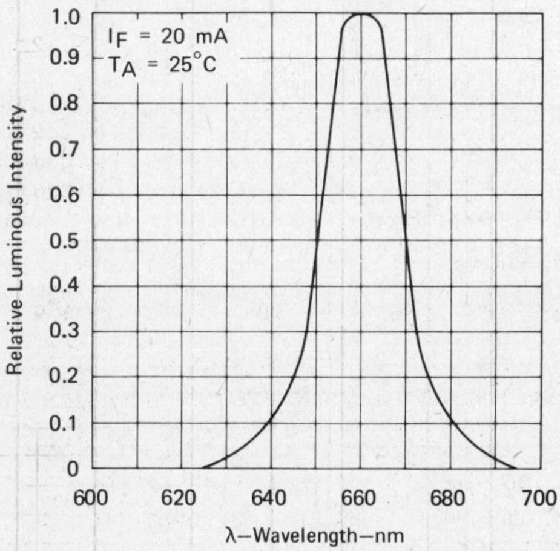

FIGURE 1

RELATIVE LUMINOUS INTENSITY
vs
FREE-AIR TEMPERATURE

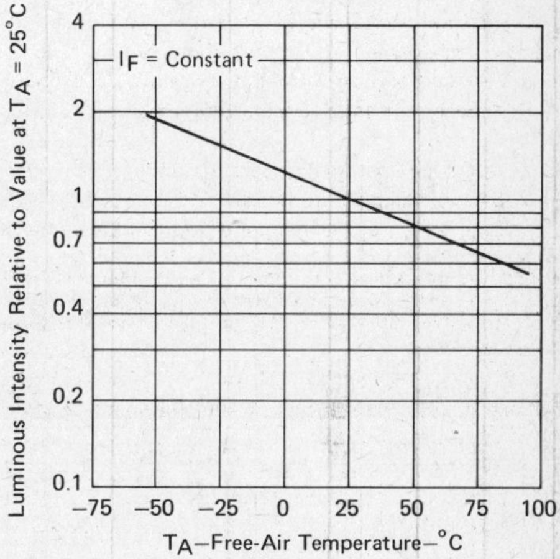

FIGURE 2

RELATIVE LUMINOUS INTENSITY
vs
FORWARD CURRENT

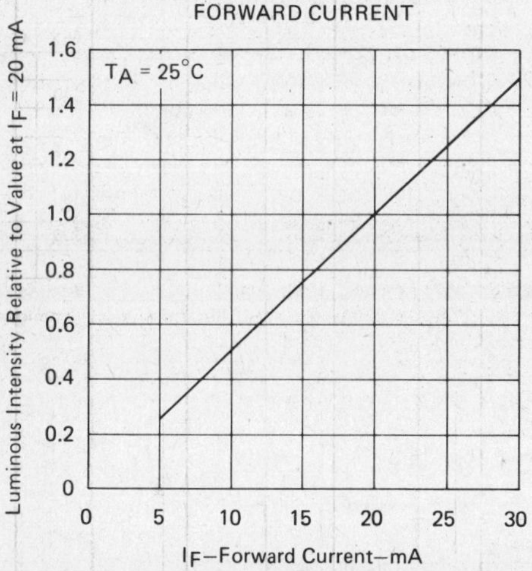

FIGURE 3

FORWARD CONDUCTION CHARACTERISTICS

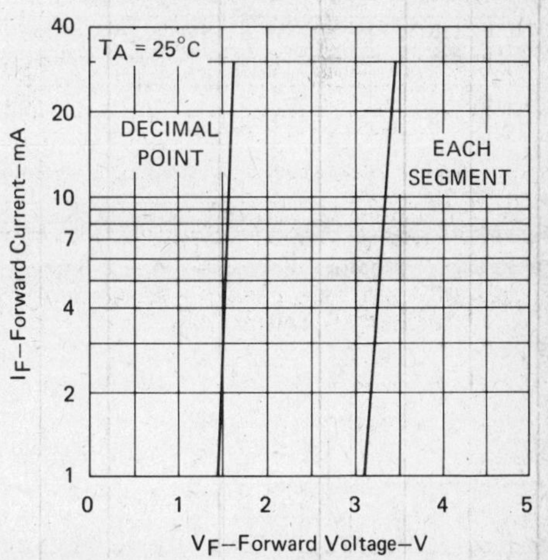

FIGURE 4

11

HI-REL DISPLAYS

TYPICAL APPLICATION DATA

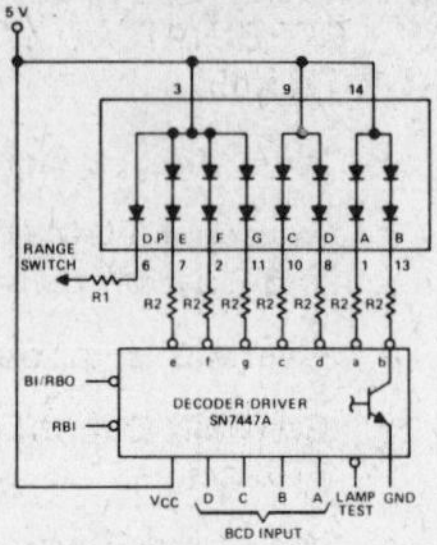

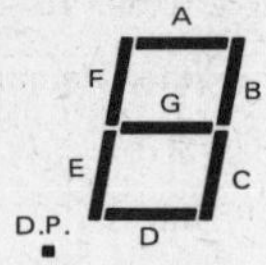

NOTES: A. R1 and R2 are selected for desired brightness.

B. SN74L47 may be used in place of SN7447A in applications where segment forward current will not exceed 20 mA, or SN74LS47 may be used for current up to 24 mA. An alternate font is available in the SN74247 and SN74LS247. For use below 0°C and/or above 70°C, substitute parts from the 54 Family.

FUNCTION TABLE
SN7447A, SN74L47, SN74LS47

DECIMAL OR FUNCTION	INPUTS						BI/RBO†	SEGMENTS							NOTE
	LT	RBI	D	C	B	A		a	b	c	d	e	f	g	
0	H	H	L	L	L	L	H	ON	ON	ON	ON	ON	ON	OFF	1
1	H	X	L	L	L	H	H	OFF	ON	ON	OFF	OFF	OFF	OFF	1
2	H	X	L	L	H	L	H	ON	ON	OFF	ON	ON	OFF	ON	1
3	H	X	L	L	H	H	H	ON	ON	ON	ON	OFF	OFF	ON	1
4	H	X	L	H	L	L	H	OFF	ON	ON	OFF	OFF	ON	ON	1
5	H	X	L	H	L	H	H	ON	OFF	ON	ON	OFF	ON	ON	1
6	H	X	L	H	H	L	H	OFF‡	OFF	ON	ON	ON	ON	ON	1
7	H	X	L	H	H	H	H	ON	ON	ON	OFF	OFF	OFF	OFF	1
8	H	X	H	L	L	L	H	ON	ON	ON	ON	ON	ON	ON	1
9	H	X	H	L	L	H	H	ON	ON	ON	OFF‡	OFF	ON	ON	1
10	H	X	H	L	H	L	H	OFF	OFF	OFF	ON	ON	OFF	ON	1
11	H	X	H	L	H	H	H	OFF	OFF	ON	ON	OFF	OFF	ON	1
12	H	X	H	H	L	L	H	OFF	ON	OFF	OFF	OFF	ON	ON	1
13	H	X	H	H	L	H	H	ON	OFF	OFF	ON	OFF	ON	ON	1
14	H	X	H	H	H	L	H	OFF	OFF	OFF	ON	ON	ON	ON	1
15	H	X	H	H	H	H	H	OFF	OFF	OFF	OFF	OFF	OFF	OFF	1
BI	X	X	X	X	X	X	L	OFF	OFF	OFF	OFF	OFF	OFF	OFF	2
RBI	H	L	L	L	L	L	L	OFF	OFF	OFF	OFF	OFF	OFF	OFF	3
LT	L	X	X	X	X	X	H	ON	ON	ON	ON	ON	ON	ON	4

H = high level (logic 1 in positive logic), L = low level (logic 0 in positive logic), X = irrelevant.

†BI/RBO is wire-AND logic serving as blanking input (BI) and/or ripple-blanking output (RBO).
‡These segments would be on if the SN74247 or SN74LS247 were used.

NOTES: 1. The blanking input (BI) must be open or held at a high logic level when output functions 0 through 15 are desired. The ripple-blanking input (RBI) must be open or high if blanking of a decimal zero is not desired.

2. When a low logic level is applied directly to the blanking input (BI), all segment outputs are off regardless of any other input.

3. When the ripple-blanking input (RBI) and inputs A, B, C, and D are at a low logic level with the lamp test input high, all segment outputs are off and the ripple-blanking output (RBO) of the decoder goes to a low level (response condition).

4. When the blanking input/ripple blanking output (BI/RBO) is open or held high and a low is applied to the lamp-test input, all segments are illuminated.

SN7447A, SN74L47, SN74LS47

ALTERNATE FONT
SN74247, SN74LS247

0 1 2 3 4 5 6 7 8 9 10 11 12 13 14 15 6 9

NUMERICAL DESIGNATIONS—RESULTANT DISPLAYS

RECOMMENDED DECODE/DRIVE WITH BCD INPUTS

D1940, SEPTEMBER 1982

HERMETICALLY SEALED SOLID-STATE HEXADECIMAL DISPLAY WITH INTEGRAL TTL CIRCUIT TO ACCEPT, STORE, AND DISPLAY 4-BIT BINARY DATA (FORMERLY TIL505)

- Electrically Interchangeable with TIL311
- 7,62-mm (0.300-Inch) Character Height
- Left- and Right-Hand Decimals
- Separate LED and Logic Power Supplies May Be Used
- Easy System Interface
- Wide Viewing Angle
- Internal TTL MSI Chip with Latch, Decoder, and Driver
- Operates from 5-Volt or 6-Volt Supply
- Constant-Current Drive for Hexadecimal Characters
- Withstands Severe Environmental Conditions

*mechanical data

The display and TTL MSI chip are mounted on a ceramic header, which is then hermetically sealed to a glass window. Multiple displays may be mounted on 12,7-mm (0.500-inch) centers.

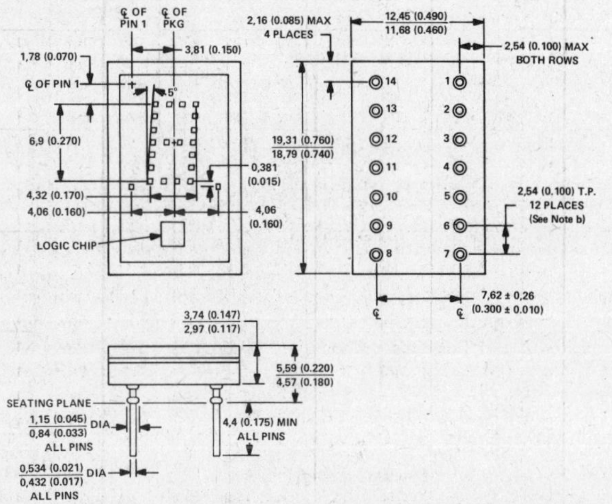

PIN 1 LED SUPPLY VOLTAGE
PIN 2 LATCH DATA INPUT B
PIN 3 LATCH DATA INPUT A
PIN 4 LEFT DECIMAL POINT CATHODE
PIN 5 NO INTERNAL CONNECTION†
PIN 6 LATCH STROBE INPUT†
PIN 7 COMMON GROUND
PIN 8 BLANKING INPUT
PIN 9 NO INTERNAL CONNECTION
PIN 10 RIGHT DECIMAL POINT CATHODE
PIN 11 NO INTERNAL CONNECTION
PIN 12 LATCH DATA INPUT D
PIN 13 LATCH DATA INPUT C
PIN 14 LOGIC SUPPLY VOLTAGE, V_{CC}

†To make pin connections compatible with the TIL311, connect pins 5 and 6 together.

NOTES:
a. The true-position pin spacing is 2,54 mm (0.100 inch) between centerlines. Each pin centerline is located within 0,26 mm (0.010 inch) of its true longitudinal position relative to pins 1 and 14.
b. Lead dimensions are not controlled above the seating plane.
c. Dimensions associated with position of LED's are between centerlines and are nominal.
d. All dimensions are in millimeters and parenthetically in inches.

*JEDEC registered data. This data sheet contains all applicable registered data in effect at the time of publication.

Copyright © 1982 by Texas Instruments Incorporated

TEXAS INSTRUMENTS
INCORPORATED
POST OFFICE BOX 225012 • DALLAS, TEXAS 75265

TYPE 4N56
HEXADECIMAL DISPLAY WITH LOGIC

*description

This hexadecimal display contains a four-bit latch, decoder, driver, and 4 X 7 light-emitting-diode (LED) character with two externally driven decimal points in a 14-pin package. A description of the functions of the inputs of this device follows.

FUNCTION	PIN NO.	DESCRIPTION
LATCH STROBE INPUT	6	When low, the data in the latches follow the data on the latch data inputs. When high, the data in the latches will not change. If the display is blanked and then restored while the enable input is high, the previous character will again be displayed.
BLANKING INPUT	8	When high, the display is blanked regardless of the levels of the other inputs. When low, a character is displayed as determined by the data in the latches. The blanking input may be pulsed for intensity modulation.
LATCH DATA INPUTS (A, B, C, D)	3, 2, 13, 12	Data on these inputs are entered into the latches when the enable input is low. The binary weights of these inputs are A = 1, B = 2, C = 4, D = 8.
DECIMAL POINT CATHODES	4, 10	These LEDs are not connected to the logic chip. If a decimal point is used, an external resistor or other current-limiting mechanism must be connected in series with it.
LED SUPPLY	1	This connection permits the user to save on regulated V_{CC} current by using a separate LED supply, or it may be externally connected to the logic supply (V_{CC}).
LOGIC SUPPLY (V_{CC})	14	Separate V_{CC} connection for the logic chip.
COMMON GROUND	7	This is the negative terminal for all logic and LED currents except for the decimal points.

The LED driver outputs are designed to maintain a relatively constant on-level current of approximately five milliamperes through each of the LED's forming the hexadecimal character. This current is virtually independent of the LED supply voltage within the recommended operating conditions. Drive current varies with changes in logic supply voltage resulting in a change in luminous intensity as shown in Figure 2. The decimal point anodes are connected to the LED supply; the cathodes are connected to external pins. Since there is no current limiting built into the decimal point circuits, this must be provided externally if the decimal points are used.

The TTL MSI chip is specially designed with a wider supply voltage range than standard Series 54/74 circuits so that it will operate from either a five-volt or a six-volt power supply.

The resultant displays for the values of the binary data in the latches are as shown below.

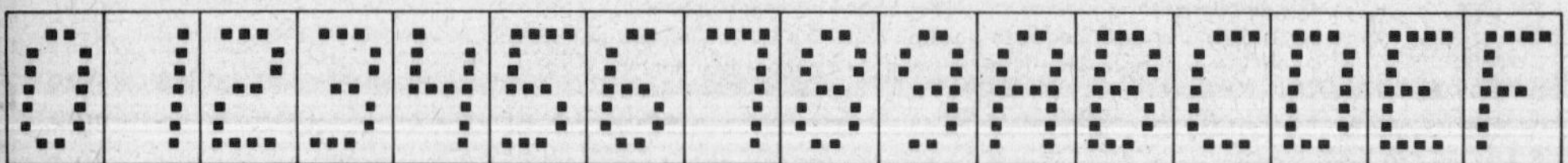

0 1 2 3 4 5 6 7 8 9 10 11 12 13 14 15

*JEDEC registered data.

TEXAS INSTRUMENTS
INCORPORATED
POST OFFICE BOX 225012 • DALLAS, TEXAS 75265

***functional block diagram**

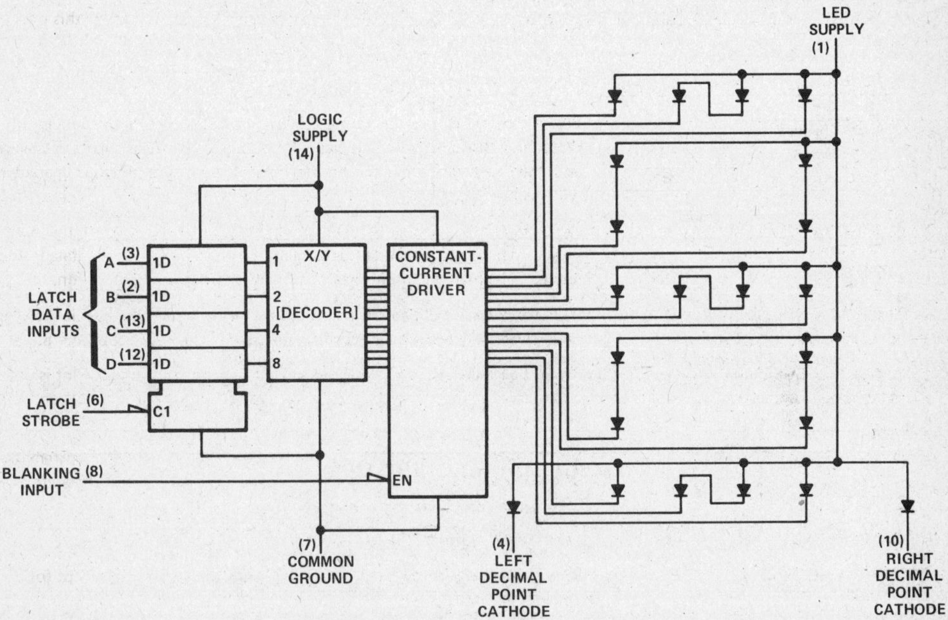

***absolute maximum ratings over operating free-air temperature range (unless otherwise noted)**

Logic Supply Voltage, V_{CC} (See Note 1)	7 V
LED Supply Voltage (See Note 1)	7 V
Input Voltage (Pins 2, 3, 6, 8, 12, 13; See Note 1)	5.5 V
Decimal Point Current	20 mA
Operating Free-Air Temperature Range	–55°C to 100°C
Storage Temperature Range	–65°C to 125°C

NOTE 1: Voltage values are with respect to common ground terminal.

***recommended operating conditions**

	MIN	NOM	MAX	UNIT
Logic Supply Voltage, V_{CC}	4.5	5	6.5	V
LED Supply Voltage, V_{LED}	4	5	7	V
High-Level Input Voltage, V_{IH}	2			V
Low-Level Input Voltage, V_{IL}			0.8	V
Decimal Point Current, $I_{F(DP)}$		5		mA
Latch Strobe Pulse Duration, t_w	40			ns
Data Setup Time Before Latch Strobe Goes High, t_{su}	50			ns
Data Hold Time After Latch Strobe Goes High, t_h	40			ns

NOTES: 2. The minimum setup time is the interval immediately preceding the positive-going transition of the latch strobe input during which interval the data to be displayed must be maintained at the latch data inputs to ensure its recognition.

3. The minimum hold time is the interval immediately following the positive-going transition of the latch strobe input during which interval the data to be displayed must be maintained at the latch data inputs to ensure its continued recognition.

*JEDEC registered data.

*operating characteristics at 25°C free-air temperature

PARAMETER			TEST CONDITIONS	MIN	TYP	MAX	UNIT
I_v	Luminous Intensity (See Note 4)	Average Per Character LED	V_{CC} = 5 V, V_{LED} = 5 V, See Note 5	35	100		µcd
		Each decimal	$I_{F(DP)}$ = 5 mA	35	100		µcd
λ_p	Wavelength at Peak Emission		V_{CC} = 5 V, V_{LED} = 5 V,	640	660	680	nm
$\Delta\lambda$	Spectral Bandwidth		$I_{F(DP)}$ = 5 mA, See Note 6		20		nm
V_{IK}	Input Clamp Voltage		V_{CC} = 4.75 V, I_I = −12 mA			−1.5	V
I_I	Input Current at Maximum Input Voltage		V_{CC} = 5.5 V, V_I = 5.5 V			1	mA
I_{IH}	High-Level Input Current		V_{CC} = 5.5 V, V_I = 2.4 V			40	µA
I_{IL}	Low-Level Input Current		V_{CC} = 5.5 V, V_I = 0.4 V			−1.6	mA
I_{CC}	Logic Supply Current		V_{CC} = 5.5 V, V_{LED} = 5.5 V,		60	90	mA
I_{LED}	LED Supply Current		$I_{F(DP)}$ = 5 mA, All inputs at 0 V		45	90	mA

NOTES: 4. Luminous intensity is measured with a light sensor and filter combination that approximates the CIE (International Commission on Illumination) eye-response curve.
5. This parameter is measured with A displayed, then again with E displayed.
6. These parameters are measured with .B. displayed.

*JEDEC registered data.

TYPICAL CHARACTERISTICS

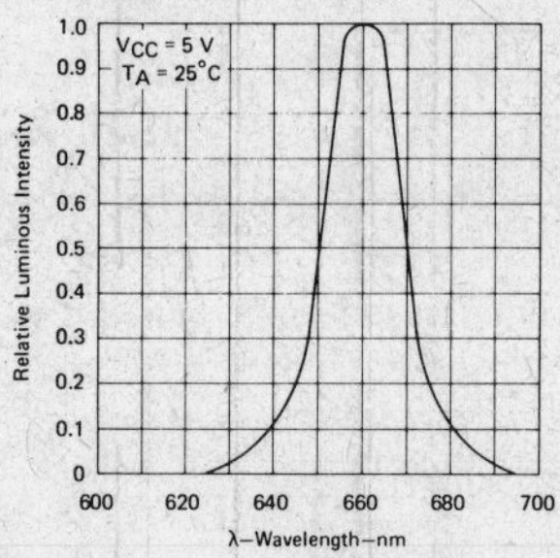

FIGURE 1

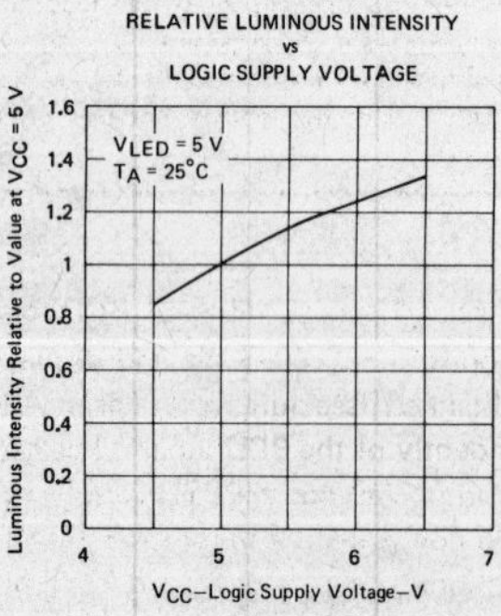

FIGURE 2

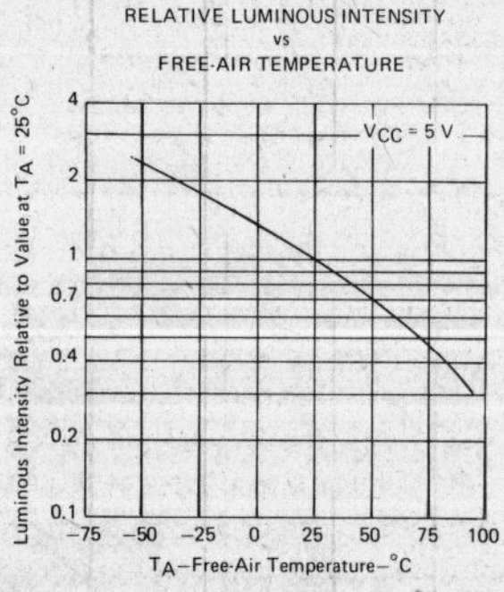

FIGURE 3

D1941, JULY 1981 — REVISED SEPTEMBER 1982

HERMETICALLY SEALED SOLID-STATE SEVEN-SEGMENT DISPLAY WITH TTL DECODER/DRIVER (FORMERLY TIL506)

- Withstands Military Environmental Conditions
- 7,62-mm (0.300-Inch) Character Height
- Internal TTL MSI Chip with Decoder and Driver
- BCD Four-Line Input
- Wide Viewing Angle
- High Luminous Intensity
- Left-Hand Decimal
- Constant-Current Drive for Light-Emitting Diodes
- Compatible with Most TTL Circuits

*mechanical data

The display and TTL logic chip are mounted on a ceramic header, which is then hermetically sealed to a glass window. Multiple displays may be mounted on 15.9-mm (0.625-inch) centers.

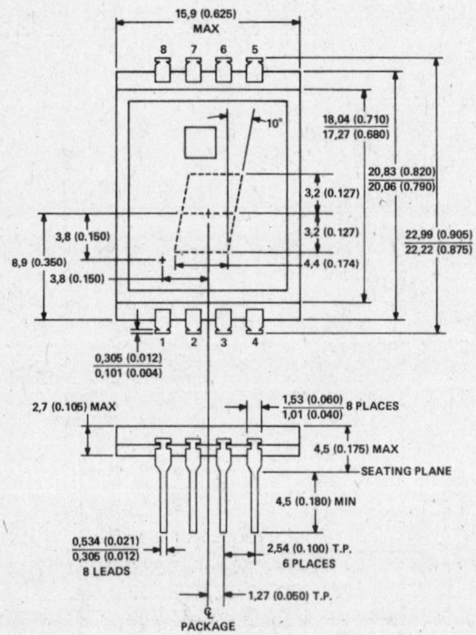

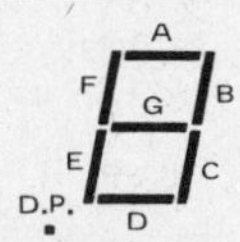

PIN 1 DATA INPUT $\overline{A}$
(BINARY WEIGHT 1)
PIN 2 DECIMAL POINT INPUT
PIN 3 LED SUPPLY VOLTAGE, V_{LED}
PIN 4 DATA INPUT $\overline{D}$
(BINARY WEIGHT 8)
PIN 5 DATA INPUT $\overline{C}$
(BINARY WEIGHT 4)
PIN 6 GROUND
PIN 7 LOGIC SUPPLY VOLTAGE, V_{CC}
PIN 8 DATA INPUT $\overline{B}$
(BINARY WEIGHT 2)

NOTES: a. All dimensions are in millimeters and parenthetically in inches.
b. The true-position pin spacing is 2,54 mm (0.100 inch) between centerlines. Each pin centerline is located within 0,26 mm (0.010 inch) of its true longitudinal position relative to the package centerline.
c. Centerlines of character segments and decimal points are shown as dashed lines. Associated dimensions are nominal.

*description

The 4N57 contains a seven-segment numeric display with left-hand decimal and a TTL MSI BCD-to-seven-segment decoder and driver. It accepts four-line binary-coded-decimal (BCD) input in negative logic and displays the decimal number in a seven-segment format. Invalid inputs are automatically blanked (see function table). A low-logic-level voltage (⩽0.8 V) at the decimal point input turns on the decimal independently of the BCD inputs. The decimal point, as well as each segment, is driven by a constant current from the logic chip. Varying the LED supply voltage will not significantly affect the brightness of the display. The brightness may be controlled by pulse-width modulation of the BCD inputs alternating between a valid code and an invalid code (e.g., all inputs low).

*JEDEC registered data. This data sheet contains all applicable registered data in effect at the time of publication.

Copyright © 1982 by Texas Instruments Incorporated

FUNCTION TABLE

FUNCTION	DATA INPUTS $\overline{D}$	$\overline{C}$	$\overline{B}$	$\overline{A}$	$\overline{DP}$	DISPLAY
0	H	H	H	H	H	0
1	H	H	H	L	L	.1
2	H	H	L	H	H	2
3	H	H	L	L	L	.3
4	H	L	H	H	H	4
5	H	L	H	L	L	.5
6	H	L	L	H	H	6
7	H	L	L	L	L	.7
8	L	H	H	H	H	8
9	L	H	H	L	L	.9
BLANK	L	H	L	H	H	
BLANK	L	H	L	L	L	.
BLANK	L	L	H	H	H	
BLANK	L	L	H	L	L	.
BLANK	L	L	L	H	H	
BLANK	L	L	L	L	L	.

H = high logic level, L = low logic level
$\overline{DP}$ input has arbitrarily been shown activated (low) on every other line of the table.

***functional block diagram**

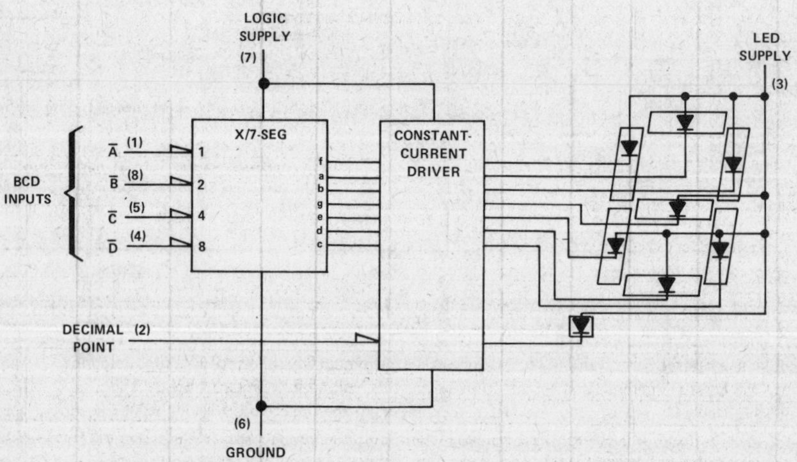

*JEDEC registered data.

*absolute maximum ratings over operating free-air temperature range (unless otherwise noted)

Logic Supply Voltage, V_{CC} (See Note 1) . . . . 7 V
LED Supply Voltage, V_{LED}, at (or below) 70°C Free-Air Temperature (See Note 2) . . . . 5.5 V
Data Input Voltage . . . . 5.5 V
Operating Free-Air Temperature Range . . . . −55°C to 100°C
Storage Temperature Range . . . . −65°C to 125°C

NOTES: 1. Voltage values are with respect to the ground terminal.
2. For operation above 70°C free-air temperature, refer to LED Supply Voltage Derating Curve, Figure 1.

*recommended operating conditions

	MIN	NOM	MAX	UNIT
Logic Supply Voltage, V_{CC}	4.5	5	5.5	V
LED Supply Voltage, V_{LED} (See Figure 1)	4	4.6	5	V
High-Level Input Voltage, V_{IH}	2			V
Low-Level Input Voltage, V_{IL}			0.8	V
Operating Free-Air Temperature, T_A	−55		100	°C

LED SUPPLY VOLTAGE DERATING CURVE

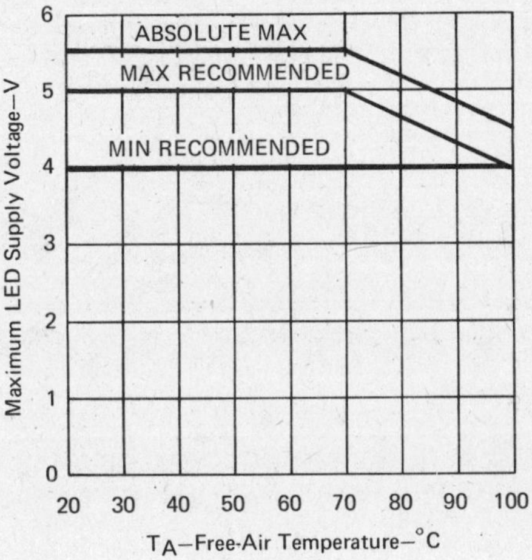

FIGURE 1

*operating characteristics at 25°C free-air temperature

PARAMETER			TEST CONDITIONS		MIN	TYP	MAX	UNIT
I_v	Luminous Intensity (See Note 3)	Figure 8	V_{CC} = 5 V, See Note 4	V_{LED} = 4.6 V,	700			µcd
		Decimal Point			40			
λ_p	Wavelength at Peak Emission				640	660	680	nm
$\Delta\lambda$	Spectral Bandwidth					20		nm
V_{IK}	Input Clamp Voltage		V_{CC} = 4.5 V,	I_I = −12 mA			−1.5	V
I_I	Input Current at Maximum Input Voltage		V_{CC} = 5.5 V,	V_I = 5.5 V			1	mA
I_{IH}	High-Level Input Current		V_{CC} = 5.5 V,	V_I = 2.4 V			20	µA
I_{IL}	Low-Level Input Current		V_{CC} = 5.5 V,	V_I = 0.4 V			−0.8	mA
I_{CC}	Logic Supply Current		V_{CC} = 5.5 V, DP at 5 V,	V_{LED} = 5 V, Other inputs at 0 V			75	mA
I_{LED}	LED Supply Current						160	mA

NOTES: 3. Luminous intensity is measured with a light sensor and filter combination that approximates the CIE (International Commission on Illumination) eye-response curve.
4. These parameters were measured with all LED segments and the decimal point on.

*JEDEC registered data.

TEXAS INSTRUMENTS
INCORPORATED
POST OFFICE BOX 225012 • DALLAS, TEXAS 75265

D1957, JULY 1981—REVISED FEBRUARY 1983

HERMETICALLY SEALED SOLID-STATE DISPLAY WITH INTEGRAL TTL COLUMN DRIVERS (FORMERLY TIL507)

- **Withstands Military Environmental Conditions**
- **7,6-mm (0.300-Inch) Character Height**
- **Integral D-Type Flip-Flop Column Drivers and Series Limiting Resistors**
- **Wide Viewing Angle**
- **Compatible with Most TTL Circuits**
- **High Luminous Intensity**
- **Left Decimal**

*mechanical data

The display and TTL logic chip are mounted on a ceramic header which is then hermetically sealed to a glass window. Multiple displays may be mounted on 12,2-mm (0.480-inch) centers.

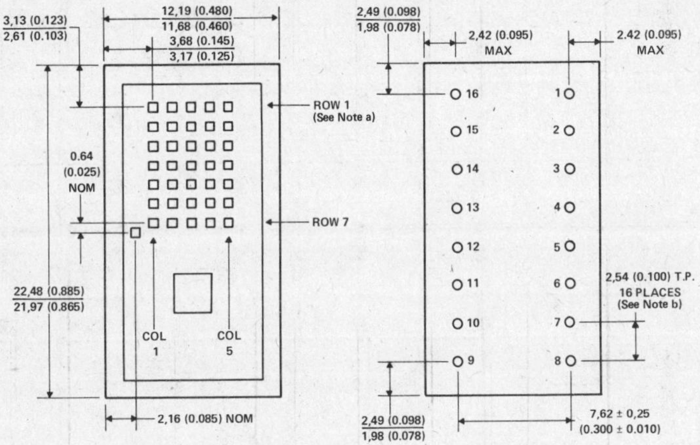

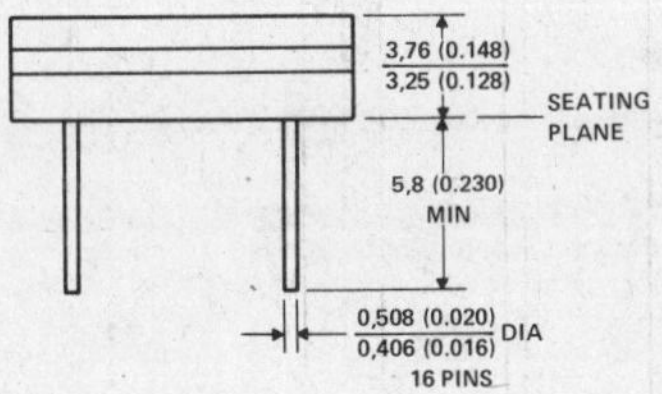

PIN 1	ROW 1
PIN 2	ROW 3
PIN 3	ROW 5
PIN 4	ROW 7 & D.P. ANODE
PIN 5	COLUMN 1
PIN 6	D.P. COLUMN
PIN 7	CLOCK
PIN 8	GROUND
PIN 9	V_{CC}
PIN 10	COLUMN 3
PIN 11	COLUMN 5
PIN 12	COLUMN 4
PIN 13	COLUMN 2
PIN 14	ROW 6
PIN 15	ROW 4
PIN 16	ROW 2

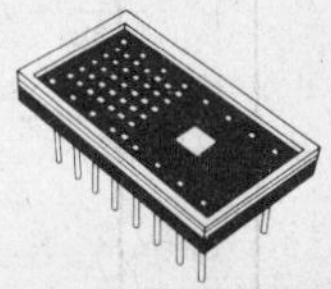

NOTES:
a. Vertical and horizontal spacing between centerline of rows and columns is nominally 0,13 mm (0.055 inch).
b. The true-position pin spacing is 2,54 mm (0.100 inch) between centerlines. Each centerline is located within 0,26 mm (0.010 inch) of its true longitudinal position relative to pins 1 and 16.

ALL DIMENSIONS ARE IN MILLIMETERS AND PARENTHETICALLY IN INCHES.

*JEDEC registered data. This data sheet contains all applicable registered data in effect at the time of publication.

11
HI-REL DISPLAYS

Copyright © 1983 by Texas Instruments Incorporated

*description

The 4N58 is a 5 X 7 matrix of light-emitting diodes plus a decimal point. The device includes an IC logic chip similar to SN54174 containing six D-type flip-flops that can transfer data from a character generator to the five columns of the matrix and the decimal point. The chip also includes six cathode column drivers with series limiting resistors.

The rows are strobed by sequentially applying a positive voltage to each row input. As each row is strobed the data set up at column inputs are transferred to the column drivers on the rising edge of each clock pulse. A high column input causes the LED to turn on. After the minimum hold time requirement has been satisfied, the column data inputs may change whether the clock is high or low.

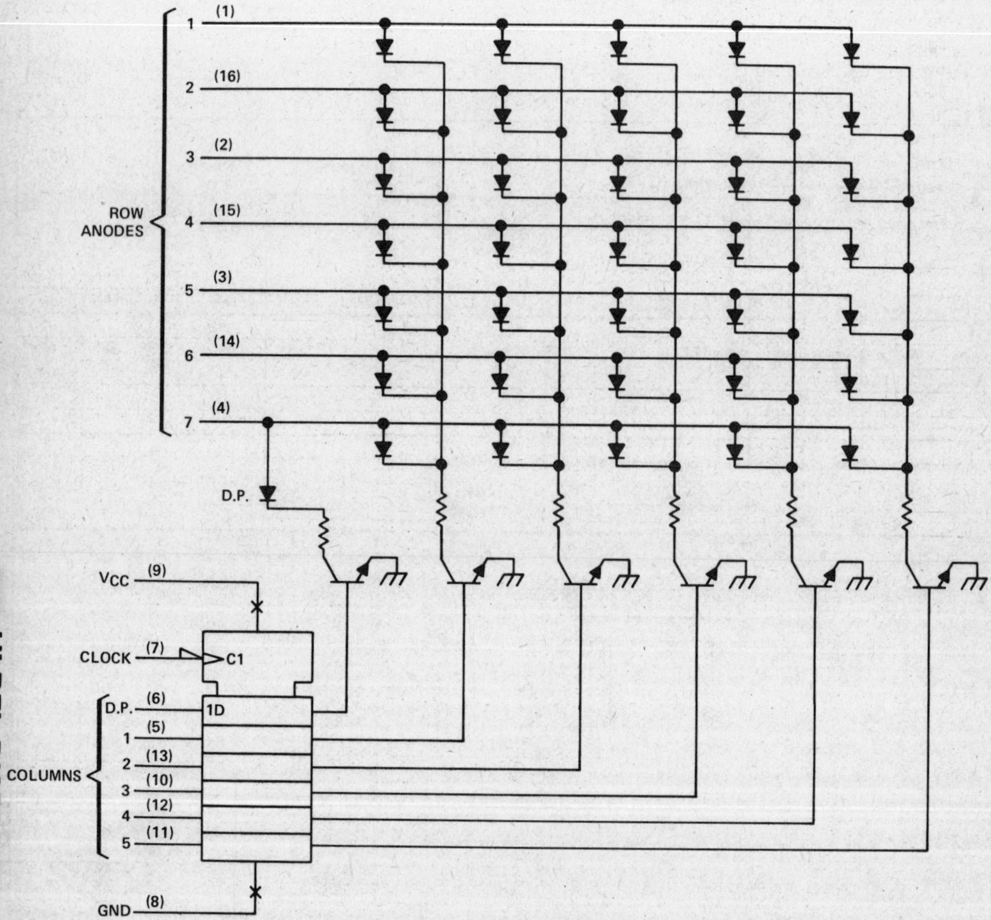

*JEDEC registered data.

*absolute maximum ratings over operating free-air temperature range (unless otherwise noted)

Logic Supply Voltage, V_{CC} (See Note 1)	7 V
Row Anode Voltage, V_{row}	5.5 V
Input Voltage (Column and Clock)	5.5 V
Operating Free-Air Temperature Range	−55°C to 100°C
Storage Temperature Range	−65°C to 125°C

NOTE 1: Voltage values are with respect to network ground terminal.

*recommended operating conditions

	MIN	NOM	MAX	UNIT
Logic Supply Voltage	4.5	5	5.5	V
High-Level Row Anode Voltage, V_{row}	3.5†	4	5	V
High-Level Input Voltage, V_{IH}	2			V
Low-Level Input Voltage, V_{IL}			0.8	V
Clock Frequency, f_{clock}		3		MHz
Duration of Clock Pulse, t_w	200			ns
Data Setup Time, t_{su}	50			ns
Data Hold Time, t_h	5			ns
Operating Free-Air Temperature, T_A	−55		100	°C

†Voltage may be reduced to 0 V to control intensity of the display.

*operating characteristics at 25°C free-air temperature

PARAMETER			TEST CONDITIONS	MIN	TYP	MAX	UNIT
I_v	Luminous Intensity (See Note 2)		V_{CC} = 5 V, I_F = 10 mA	40	110		µcd
λ_p	Wavelength at Peak Emission		V_{CC} = 5 V, V_{row} = 4 V	640	660	680	nm
$\Delta\lambda$	Spectral Bandwidth		V_{CC} = 5 V, V_{row} = 4 V		20		nm
V_{IK}	Input Clamp Voltage		V_{CC} = 4.5 V, I_I = −12 mA			−1.5	V
I_{IH}	High-Level Input Current		V_{CC} = 5.5 V, V_I = 2.4 V			150	µA
I_{IL}	Low-Level Input Current		V_{CC} = 5.5 V, V_I = 0.4 V			−1	mA
I_{row}	Row Input Current	Row 1 thru Row 6	See Note 3		500	800	mA
I_{row}	Row Input Current	Row 7	See Note 3		600	1000	mA
I_{CC}	Logic Supply Current		See Note 3		45	65	mA

NOTES: 2. Luminous intensity is measured with a light sensor and filter combination that approximates the CIE (International Commission on Illumination) eye-response curves.

3. Maximum values of row input current and logic supply current are stated for V_{CC} = 5.5 V, V_{row} = 5 V. Typical values are stated for V_{CC} = 5 V, V_{row} = 4 V. All column inputs are high.

*JEDEC registered data.

YELLOW HEXADECIMAL DISPLAY WITH LOGIC

D2688, SEPTEMBER 1982

HERMETICALLY SEALED SOLID-STATE HEXADECIMAL DISPLAY WITH INTEGRAL TTL CIRCUIT TO ACCEPT, STORE, AND DISPLAY 4-BIT BINARY DATA

- Electrically Interchangeable with TIL505, 4N56 and TIL311
- 7,62-mm (0.300-Inch) Character Height
- Left- and Right-Hand Decimals
- Separate LED and Logic Power Supplies May Be Used
- High Luminous Intensity
- Wide Viewing Angle
- Internal TTL MSI Chip with Latch, Decoder, and Driver
- Operates from 5-Volt or 6-Volt Supply
- Constant-Current Drive for Hexadecimal Characters
- Withstands Severe Environmental Conditions

mechanical data

The display and TTL MSI chip are mounted on a ceramic header, which is then hermetically sealed to a glass window. Multiple displays may be mounted on 12,7-mm (0.500-inch) centers.

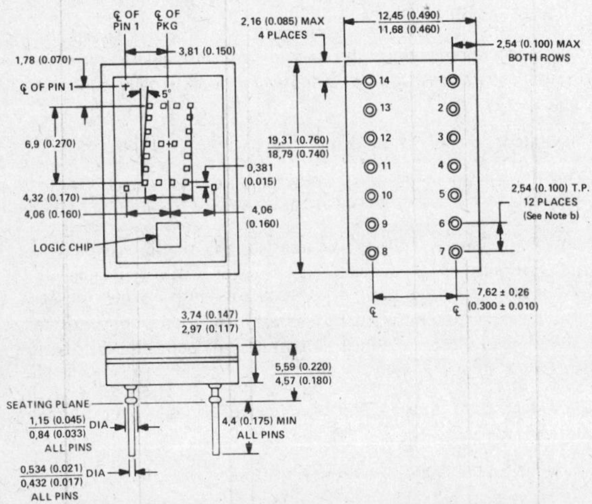

PIN 1 LED SUPPLY VOLTAGE
PIN 2 LATCH DATA INPUT B
PIN 3 LATCH DATA INPUT A
PIN 4 LEFT DECIMAL POINT CATHODE
PIN 5 NO INTERNAL CONNECTION†
PIN 6 LATCH STROBE INPUT†
PIN 7 COMMON GROUND
PIN 8 BLANKING INPUT
PIN 9 NO INTERNAL CONNECTION
PIN 10 RIGHT DECIMAL POINT CATHODE
PIN 11 NO INTERNAL CONNECTION
PIN 12 LATCH DATA INPUT D
PIN 13 LATCH DATA INPUT C
PIN 14 LOGIC SUPPLY VOLTAGE, V_{CC}

† To make pin connections compatible with the TIL311, connect pins 5 and 6 together.

NOTES:
a. The true-position pin spacing is 2,54 mm (0.100 inch) between centerlines. Each pin centerline is located within 0,26 mm (0.010 inch) of its true longitudinal position relative to pins 1 and 14.
b. Lead dimensions are not controlled above the seating plane.
c. Dimensions associated with position of LED's are between centerlines and are nominal.
d. All dimensions are in millimeters and parenthetically in inches.

PRODUCT PREVIEW

This document contains information on a product under development. Texas Instruments reserves the right to change or discontinue this product without notice.

YELLOW HEXADECIMAL DISPLAY WITH LOGIC

description

The TIL509 is a hexadecimal display containing a four-bit latch, decoder, driver, and 4 X 7 yellow light-emitting diode (LED) character with two externally driven decimal points in a 14-pin package. Electrically this device is identical to the TIL311 and 4N56.

FUNCTION	PIN NO.	DESCRIPTION
LATCH STROBE INPUT	6	When low, the data in the latches follow the data on the latch data inputs. When high, the data in the latches will not change. If the display is blanked and then restored while the enable input is high, the previous character will again be displayed.
BLANKING INPUT	8	When high, the display is blanked regardless of the levels of the other inputs. When low, a character is displayed as determined by the data in the latches. The blanking input may be pulsed for intensity modulation.
LATCH DATA INPUTS (A, B, C, D)	3, 2, 13, 12	Data on these inputs are entered into the latches when the enable input is low. The binary weights of these inputs are A = 1, B = 2, C = 4, D = 8.
DECIMAL POINT CATHODES	4, 10	These LEDs are not connected to the logic chip. If a decimal point is used, an external resistor or other current-limiting mechanism must be connected in series with it.
LED SUPPLY	1	This connection permits the user to save on regulated V_{CC} current by using a separate LED supply, or it may be externally connected to the logic supply (V_{CC}).
LOGIC SUPPLY (V_{CC})	14	Separate V_{CC} connection for the logic chip.
COMMON GROUND	7	This is the negative terminal for all logic and LED currents except for the decimal points.

The LED driver outputs are designed to maintain a relatively constant on-level current of approximately five milliamperes through each of the LED's forming the hexadecimal character. This current is virtually independent of the LED supply voltage within the recommended operating conditions. Drive current varies with changes in logic supply voltage resulting in a change in luminous intensity as shown in Figure 2. The decimal point anodes are connected to the LED supply; the cathodes are connected to external pins. Since there is no current limiting built into the decimal point circuits, this must be provided externally if the decimal points are used.

The TTL MSI chip is specially designed with a wider supply voltage range than standard Series 54/74 circuits so that it will operate from either a five-volt or a six-volt power supply.

The resultant displays for the values of the binary data in the latches are as shown below.

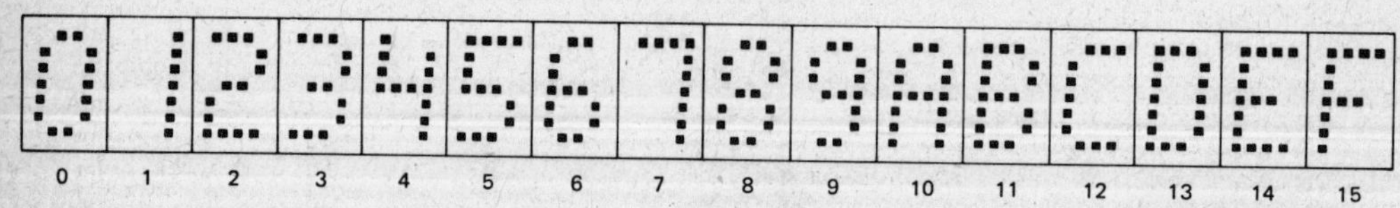

TEXAS INSTRUMENTS
INCORPORATED
POST OFFICE BOX 225012 • DALLAS, TEXAS 75265

***functional block diagram**

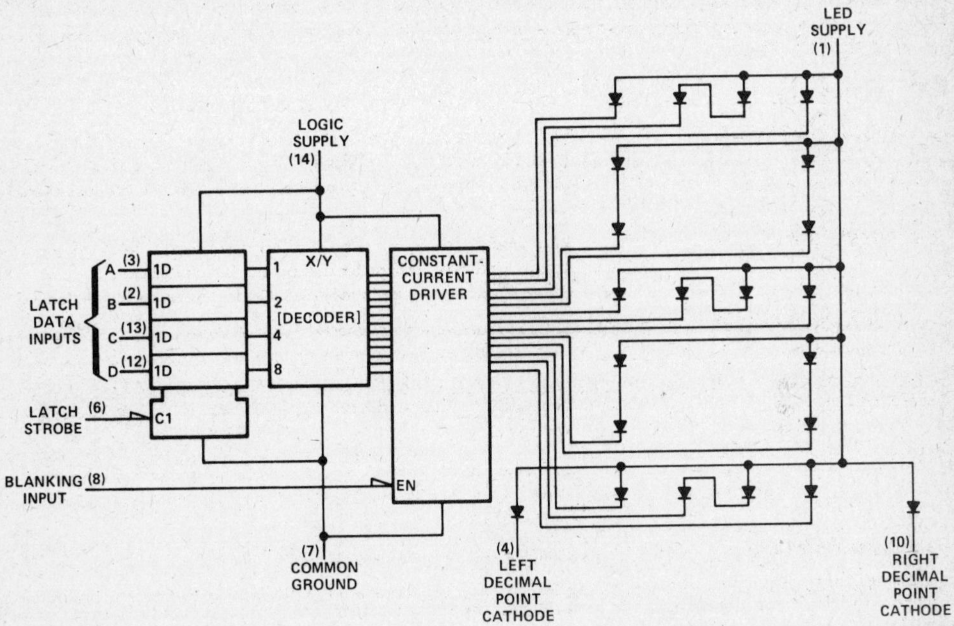

absolute maximum ratings over operating free-air temperature range (unless otherwise noted)

Logic Supply Voltage, V_{CC} (See Note 1)	7 V
LED Supply Voltage (See Note 1)	7 V
Input Voltage (Pins 2, 3, 6, 8, 12, 13; See Note 1)	5.5 V
Decimal Point Current	20 mA
Operating Free-Air Temperature Range	−55°C to 100°C
Storage Temperature Range	−65°C to 125°C

NOTE 1: Voltage values are with respect to common ground terminal.

recommended operating conditions

	MIN	NOM	MAX	UNIT
Logic Supply Voltage, V_{CC}	4.5	5	6.5	V
LED Supply Voltage, V_{LED}	4	5	7	V
High-Level Input Voltage, V_{IH}	2			V
Low-Level Input Voltage, V_{IL}			0.8	V
Decimal Point Current, $I_{F(DP)}$		5		mA
Latch Strobe Pulse Duration, t_w	40			ns
Data Setup Time Before Latch Strobe Goes High, t_{su}	50			ns
Data Hold Time After Latch Strobe Goes High, t_h	40			ns

NOTES: 2. The minimum setup time is the interval immediately preceding the positive-going transition of the latch strobe input during which interval the data to be displayed must be maintained at the latch data inputs to ensure its recognition.

3. The minimum hold time is the interval immediately following the positive-going transition of the latch strobe input during which interval the data to be displayed must be maintained at the latch data inputs to ensure its continued recognition.

operating characteristics at 25°C free-air temperature

	PARAMETER		TEST CONDITIONS	MIN	TYP	MAX	UNIT
I_v	Luminous Intensity (See Note 4)	Average Per Character LED	V_{CC} = 5 V, V_{LED} = 5 V, See Note 5	300			μcd
		Each decimal	$I_{F(DP)}$ = 5 mA	300			μcd
λ_p	Wavelength at Peak Emission		V_{CC} = 5 V, V_{LED} = 5 V, $I_{F(DP)}$ = 5 mA, See Note 6	570	580	590	nm
$\Delta\lambda$	Spectral Bandwidth				40		nm
V_{IK}	Input Clamp Voltage		V_{CC} = 4.75 V, I_I = −12 mA			−1.5	V
I_I	Input Current at Maximum Input Voltage		V_{CC} = 5.5 V, V_I = 5.5 V			1	mA
I_{IH}	High-Level Input Current		V_{CC} = 5.5 V, V_I = 2.4 V			40	μA
I_{IL}	Low-Level Input Current		V_{CC} = 5.5 V, V_I = 0.4 V			−1.6	mA
I_{CC}	Logic Supply Current		V_{CC} = 5.5 V, V_{LED} = 5.5 V, $I_{F(DP)}$ = 5 mA, All inputs at 0 V		60	90	mA
I_{LED}	LED Supply Current				45	90	mA

NOTES: 4. Luminous intensity is measured with a light sensor and filter combination that approximates the CIE (International Commission on Illumination) eye-response curve.

5. This parameter is measured with A displayed, then again with E displayed.

6. These parameters are measured with B displayed.

HERMETICALLY SEALED SOLID-STATE YELLOW DISPLAY WITH INTEGRAL TTL COLUMN DRIVERS (YELLOW VERSION OF 4N58)

- Withstands Military Environmental Conditions
- 7,6-mm (0.300-Inch) Character Height
- Integral D-Type Flip-Flop Column Drivers and Series Limiting Resistors
- Wide Viewing Angle
- Compatible with Most TTL Circuits
- High Luminous Intensity
- Left Decimal

mechanical data

The display and TTL logic chip are mounted on a ceramic header which is then hermetically sealed to a glass window. Multiple displays may be mounted on 12,2-mm (0.480-inch) centers.

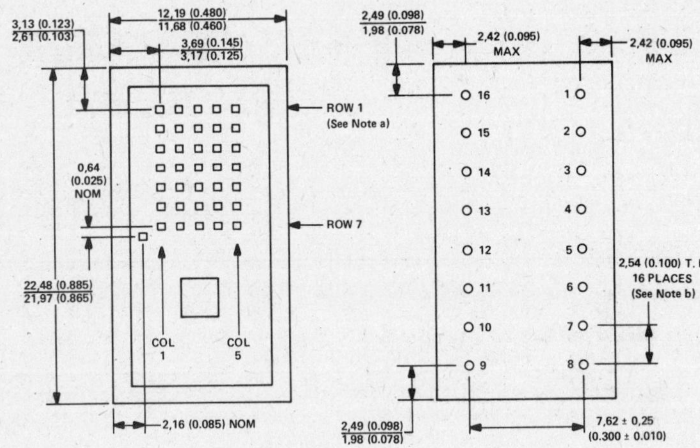

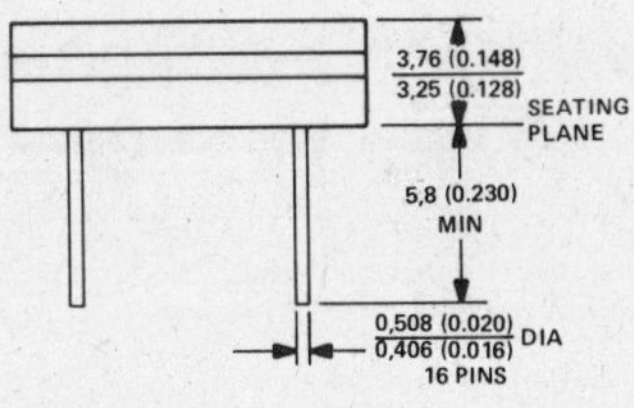

PIN 1 ROW 1
PIN 2 ROW 3
PIN 3 ROW 5
PIN 4 ROW 7 & D.P. ANODE
PIN 5 COLUMN 1
PIN 6 D.P. COLUMN
PIN 7 CLOCK
PIN 8 GROUND
PIN 9 V_{CC}
PIN 10 COLUMN 3
PIN 11 COLUMN 5
PIN 12 COLUMN 4
PIN 13 COLUMN 2
PIN 14 ROW 6
PIN 15 ROW 4
PIN 16 ROW 2

NOTES: a. Vertical and horizontal spacing between centerline of rows and columns is nominally 0,13 mm (0.005 inch).
b. The true-position pin spacing is 2,54 mm (0.100 inch) between centerlines. Each centerline is located within 0,26 mm (0.010 inch) of its true longitudinal position relative to pins 1 and 16.

ALL DIMENSIONS ARE IN MILLIMETERS AND PARENTHETICALLY IN INCHES.

11 HI-REL DISPLAYS

PRODUCT PREVIEW

This document contains information on a product under development. Texas Instruments reserves the right to change or discontinue this product without notice.

Copyright © 1982 by Texas Instruments Incorporated

TEXAS INSTRUMENTS
INCORPORATED
POST OFFICE BOX 225012 • DALLAS, TEXAS 75265

description

The TIL510 is a 5 X 7 matrix of yellow light-emitting diodes plus a decimal point. The device includes an IC logic chip similar to SN54174 containing six D-type flip-flops that can transfer data from a character generator to the five columns of the matrix and the decimal point. The chip also includes six cathode column drivers with series limiting resistors. This device is electrically and functionally identical to the TIL507 and 4N58.

absolute maximum ratings over operating free-air temperature range (unless otherwise noted)

Logic Supply Voltage, V_{CC} (See Note 1)	7 V
Row Anode Voltage, V_{row}	5.5 V
Input Voltage (Column and Clock)	5.5 V
Operating Free-Air Temperature Range	−55°C to 100°C
Storage Temperature Range	−65°C to 125°C

NOTE 1: Voltage values are with respect to network ground terminal.

recommended operating conditions

	MIN	NOM	MAX	UNIT
Logic Supply Voltage	4.5	5	5.5	V
High-Level Row Anode Voltage, V_{row}	3.5†	4	5	V
High-Level Input Voltage, V_{IH}	2			V
Low-Level Input Voltage, V_{IL}			0.8	V
Clock Frequency, f_{clock}		3		MHz
Duration of Clock Pulse, t_w	200			ns
Data Setup Time, t_{su}	50			ns
Data Hold Time, t_h	5			ns
Operating Free-Air Temperature, T_A	−55		100	°C

†Voltage may be reduced to 0 V to control intensity of the display.

operating characteristics at 25°C free-air temperature

PARAMETER			TEST CONDITIONS	MIN	TYP	MAX	UNIT
I_v	Luminous Intensity (See Note 2)		$V_{CC} = 5$ V, $I_F = 10$ mA	300			µcd
λ_p	Wavelength at Peak Emission		$V_{CC} = 5$ V, $V_{row} = 4$ V	570	580	590	nm
$\Delta\lambda$	Spectral Bandwidth		$V_{CC} = 5$ V, $V_{row} = 4$ V		40		nm
V_{IK}	Input Clamp Voltage		$V_{CC} = 4.5$ V, $I_I = -12$ mA			−1.5	V
I_{IH}	High-Level Input Current		$V_{CC} = 5.5$ V, $V_I = 2.4$ V			150	µA
I_{IL}	Low-Level Input Current		$V_{CC} = 5.5$ V, $V_I = 0.4$ V			−1	mA
I_{row}	Row Input Current	Row 1 thru Row 6	See Note 3		500	800	mA
I_{row}	Row Input Current	Row 7	See Note 3		600	1000	mA
I_{CC}	Logic Supply Current		See Note 3		45	65	mA

NOTES: 2. Luminous intensity is measured with a light sensor and filter combination that approximates the CIE (International Commission on Illumination) eye-response curve.

3. Maximum values of row input current and logic supply current are stated for $V_{CC} = 5.5$ V, $V_{row} = 5$ V. Typical values are stated for $V_{CC} = 5$ V, $V_{row} = 4$ V. All column inputs are high.

Fiber-Optic Components and Amplifiers

- Quick Reference Guide
- Silicon Photodiode and Phototransistor
- Silicon Integrated Analog Receivers
- High-Speed Transimpedance Amplifiers
- Gallium Aluminum Arsenide Infrared Emitters
- Hermetically Sealed Packages

SILICON PHOTODETECTORS QUICK REFERENCE GUIDE

DEVICE	DETECTOR TYPE	RADIANT RESPONSIVITY (A/W)	RISETIME (ns) @ 5 V	FEATURES
TIED458	Phototransistor	120	10,000	High responsivity
TIED459	PIN Photodiode	0.42	10	High speed

SILICON INTEGRATED ANALOG RECEIVERS QUICK REFERENCE GUIDE

DEVICE	RADIANT RESPONSIVITY (mV/μW)	EQUIVALENT INPUT NOISE RADIANT POWER (μW)	TRANSITION TIME (ns)	FEATURES
TIED460	60	0.007	80 for t_W = 500 ns	Single +5 V supply, Converts optical input to voltage output
TIED461	26	0.015	35 for t_W = 250 ns	
TIED462	12	0.04	18 for t_W = 100 ns	
TIED463	4.8	0.13	10 for t_W = 50 ns	

TRANSIMPEDANCE AMPLIFIERS QUICK REFERENCE GUIDE

DEVICE	BANDWIDTH (MHz)	FORWARD TRANSFER IMPEDANCE (kΩ)	EQUIVALENT INPUT NOISE CURRENT (pA/$\sqrt{Hz}$)	FEATURES
TIEF150	100	1	8.5	Converts photodetector current to voltage output
TIEF151	50	4	4.5	
TIEF152	20	12	3	

GALLIUM ALUMINUM ARSENIDE INFRARED-EMITTING DIODES QUICK REFERENCE GUIDE

DEVICE	RADIANT POWER OUTPUT (μW)* @ 50 mA	RADIANT PULSE RISETIME (ns)	HALF-INTENSITY BEAM ANGLE	λ_p (nm)	FEATURES
TIES494	45	12	20°	820	Microlens metal-case packaging
TIES495	75	12	20°	820	
TIES496	110	12	20°	820	

*Radiant power transmitted through a 0,2-mm (0.008-inch) diameter mechanical aperture into a numerical aperture of 0.25.
All values shown are typical.

DETECTOR DESIGNED FOR FIBER-OPTIC APPLICATIONS

- **High Radiant Responsivity . . . 120 A/W**
- **Large Effective Detector Area with Internal 0.8-mm-Diameter Spherical Microlens**
- **Optically Compatible with TIES494, TIES495, and TIES496 Infrared-Emitting Diodes**
- **Mechanically Compatible with Various Commercial Fiber-Optic Connector Receptacles**

mechanical data

The device is in a hermetically sealed welded case with a flat glass window in the case top. The coin header is gold plated; the metal window can is nickel plated.

LEADS 1 AND 2 ARE ELECTRICALLY INSULATED FROM THE CASE

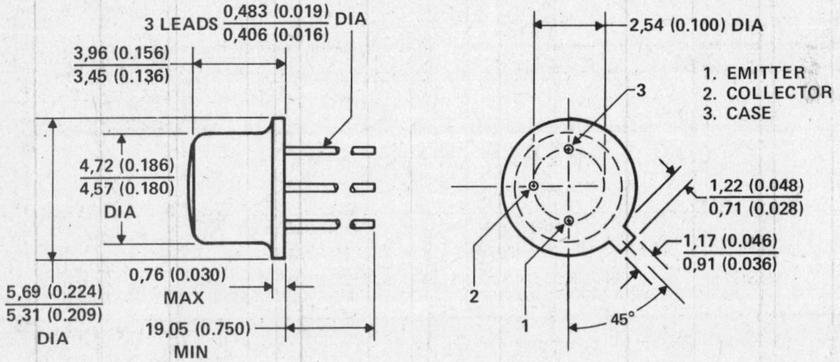

The maximum extension of the glass above the top of the case is 0,25 mm (0.010 inch).
ALL DIMENSIONS ARE IN MILLIMETERS AND PARENTHETICALLY IN INCHES

CL-77

absolute maximum ratings at 25 °C free-air temperature (unless otherwise noted)

Collector-Emitter Voltage	30 V
Emitter-Collector Voltage	7 V
Continuous Collector Current	20 mA
Continuous Device Dissipation at (or below) 25 °C Free-Air Temperature (See Note 1)	100 mW
Operating Free-Air Temperature Range	−40 °C to 85 °C
Storage Temperature Range	−40 °C to 100 °C
Lead Temperature 1,6 mm (1/16 inch) from Case for 10 seconds	240 °C

NOTE 1. Derate linearly to 40 mW at 85 °C free-air temperature at the rate of 1.0 mW/°C.

Copyright © 1983 by Texas Instruments Incorporated

operating characteristics at 25 °C free-air temperature

	PARAMETER	TEST CONDITIONS			MIN	TYP	MAX	UNIT
$V_{(BR)CEO}$	Collector-Emitter Breakdown Voltage	I_C = 100 μA, See Note 2	P_I = 0,		30			V
$V_{(BR)ECO}$	Emitter-Collector Breakdown Voltage	I_E = 100 μA,	P_I = 0		7			V
I_D	Dark Current	V_{CE} = 10 V,	P_I = 0			5	40	nA
R_e	Radiant Responsivity (see Note 3)	V_{CC} = 5 V, See Figure 1	P_I = 1 μW,		60	120		A/W
t_r	Rise Time (see Note 4)	V_{CC} = 5 V, See Figure 2		R_L = 100 Ω		10		ns
				R_L = 1 kΩ		25		
				R_L = 10 kΩ		150		

NOTES: 2. The radiant power input P_I is the radiant power measured with a large-area detector from the output end of a 10-m fiber-optic cable assembly incorporating a graded-index optical fiber with a 100-μm core diameter, a 140-μm cladding diameter, and a numerical aperture of ≤0.30. A TIES495 GaAlAs infrared-emitting diode with λ_p = 820 nm is used at the input end of the fiber-optic cable assembly. The TIES495 has a typical half-intensity beam angle of 20°.

3. Radiant responsivity is defined as the change of detector current output divided by the change of radiant power input P_I.
4. Rise time is the time required for a change in detector current output from 10% to 90% of its peak value for a step change of incident radiant power. The fall time is approximately equal to the rise time.

PARAMETER MEASUREMENT INFORMATION

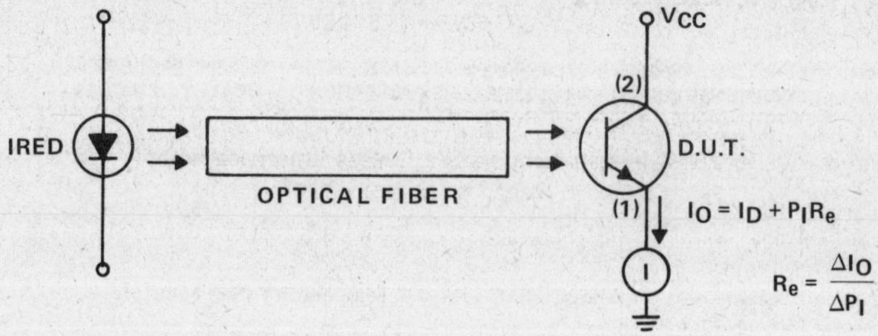

FIGURE 1— TEST CIRCUIT FOR RADIANT RESPONSIVITY

TEXAS INSTRUMENTS
INCORPORATED
POST OFFICE BOX 225012 • DALLAS, TEXAS 75265

PARAMETER MEASUREMENT INFORMATION

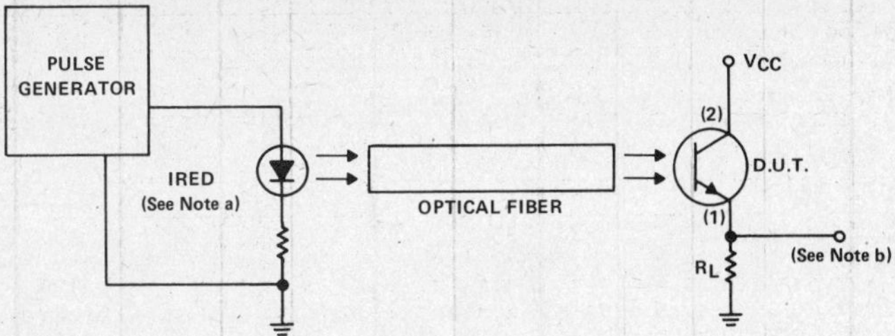

TEST CIRCUIT

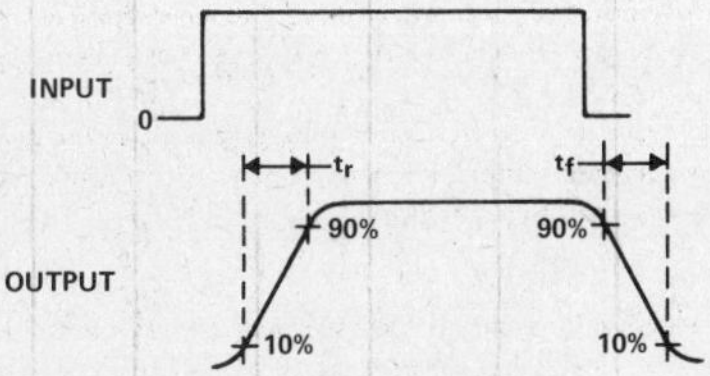

VOLTAGE WAVEFORMS

FIGURE 2 – SWITCHING TIMES

NOTES: a. Radiant power input is supplied by a pulsed GaAlAs infrared emitting diode with the following operating characteristics: λ_p = 820 nm, tr ≤ 50 ns.

b. The output waveform is monitored on an oscilloscope with the following characteristics: t_r ≤ 50 ns, Z_{in} ≥ 1 MΩ, C_{in} ≤ 20 pF.

TYPICAL CHARACTERISTICS

DARK CURRENT
vs
FREE-AIR TEMPERATURE

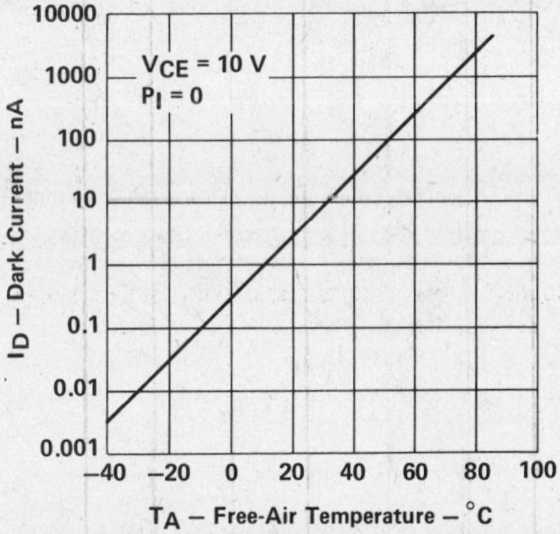

FIGURE 3

D2679, APRIL 1983

DETECTOR DESIGNED FOR FIBER-OPTIC APPLICATIONS

- High-Resistivity Silicon PIN Photodiode for High Performance at Low Voltage
- Rise Time . . . 10 ns at $V_R = 5$ V
- Low Capacitance . . . 3 pF at $V_R = 5$ V
- Large Effective Detector Area with Internal 1,2-mm-Diameter Spherical Microlens
- Optically Compatible with TIES494, TIES495, TIES496 Infrared-Emitting Diodes
- Mechanically Compatible with Commercial Fiber Optic Connector Receptacles

mechanical data

The device is in a hermetically-sealed welded case with a flat glass window in the case top. The coin header is gold plated; the metal window can is nickel plated.

LEADS 1 AND 2 ARE ELECTRICALLY INSULATED FROM THE CASE

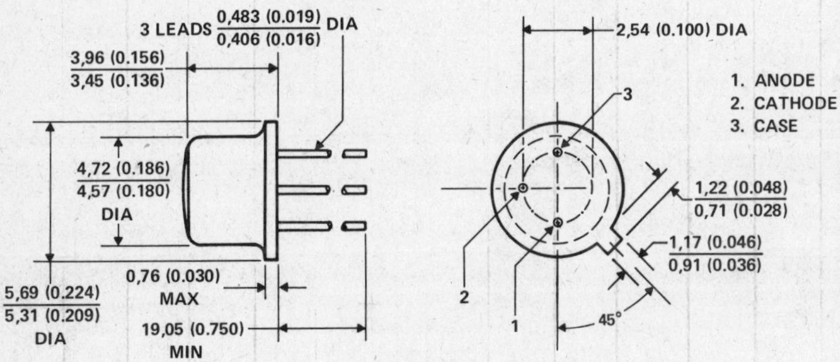

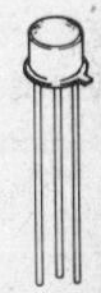

The maximum extension of the glass above the top of the case is 0,25 mm (0.010 inch).
ALL LINEAR DIMENSIONS ARE IN MILLIMETERS AND PARENTHETICALLY IN INCHES.

CL-77

absolute maximum ratings at 25°C free-air temperature (unless otherwise noted)

Reverse Voltage	50 V
Continuous Power Dissipation at (or below) 25°C Free-Air Temperature (see Note 1)	50 mW
Operating Free-Air Temperature Range	−40°C to 85°C
Storage Temperature Range	−40°C to 100°C
Lead Temperature 1,6 mm (1/16 inch) from Case for 10 Seconds	240°C

NOTE 1: Derate linearly to 20 mW at 85°C free-air temperature at the rate of 0.5 mW/°C.

Copyright © 1983 by Texas Instruments Incorporated

TEXAS INSTRUMENTS
INCORPORATED
POST OFFICE BOX 225012 • DALLAS, TEXAS 75265

operating characteristics at 25°C free-air temperature

PARAMETER		TEST CONDITIONS		MIN	TYP	MAX	UNIT
$V_{(BR)}$	Breakdown Voltage	I_R = 100 μA, See Note 2	P_I = 0,	50	100		V
I_D	Dark Current	V_R = 25 V,	P_I = 0		4	20	nA
C_T	Total Capacitance	V_R = 5 V,	P_I = 0		3		pF
R_e	Radiant Responsivity (see Note 3)	V_R = 5 V,	See Figure 1	0.32	0.42	0.6	A/W
t_r	Rise Time (see Note 4)	V_R = 5 V	See Figure 2		10	16	ns
		V_R = 12 V			7		
		V_R = 25 V			5		

NOTES: 2. The radiant power input P_I is the radiant power measured with a large-area detector from the output end of a 10-m fiber-optic cable assembly incorporating a graded-index optical fiber with a 100-μm core diameter, a 140-μm cladding diameter, and a numerical aperture of ≤0.30. A TIES495 GaAlAs infrared-emitting diode with λ_p = 820 nm is used at the input end of the fiber-optic cable assembly. The TIES495 has a typical half-intensity beam angle of 20°.

3. Radiant responsivity is defined as the change of detector current output divided by the change of radiant power input P_I.

4. Rise time is the time required for a change in detector current output from 10% to 90% of its peak value for a step change of incident radiant power. The electrical bandwidth (in MHz) at which the detector current output is reduced to $1/\sqrt{2}$ of the maximum low-frequency value is approximately $350/t_r$ (t_r in ns). The optical bandwidth at which the detector current output is reduced to 1/2 of the maximum low-frequency value is approximately $610/t_r$. The fall time is approximately equal to the rise time.

PARAMETER MEASUREMENT INFORMATION

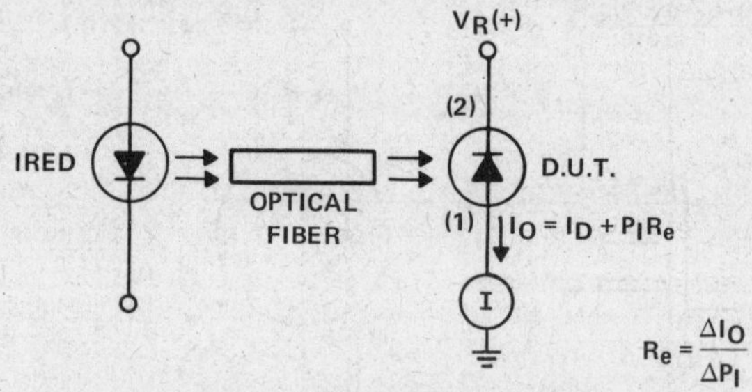

FIGURE 1—TEST CIRCUIT FOR RADIANT RESPONSIVITY

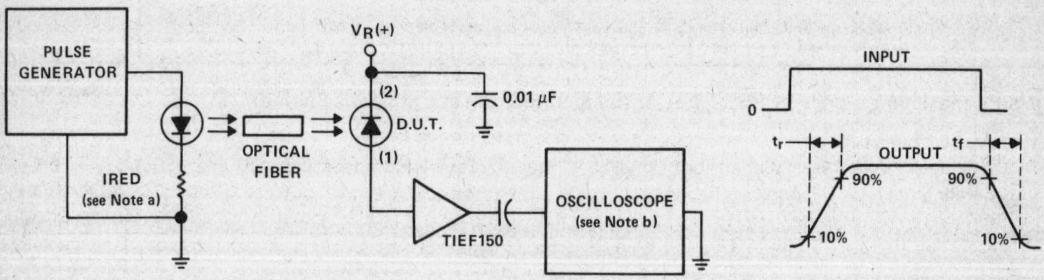

TEST CIRCUITS　　　WAVEFORMS

NOTES: a. Radiant power input is supplied by a pulsed GaAlAs infrared emitting diode with the following operating characteristics: λ_p = 850 nm, $t_w \geq 100$ ns, $t_r \leq 5$ ns.

b. The output waveform is monitored on an oscilloscope with the following characteristics: Z_{in} = 50 Ω, $t_r \leq 2$ ns. The measured rise time is corrected for the combined rise times of the optical source, the TIEF150 transimpedance amplifier, and the oscilloscope.

FIGURE 2—SWITCHING TIMES

TYPICAL CHARACTERISTICS

DARK CURRENT
vs
REVERSE VOLTAGE

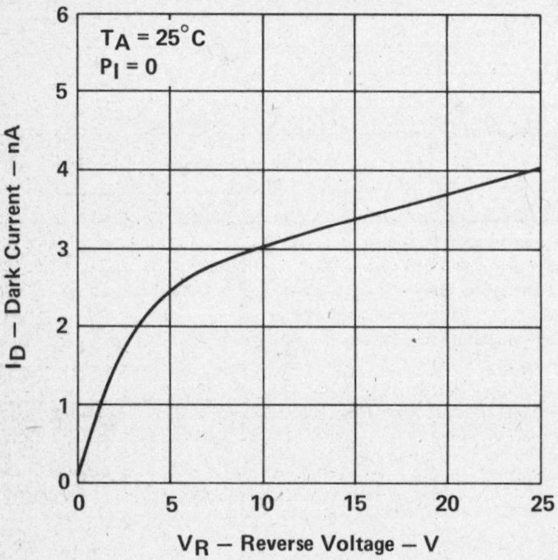

FIGURE 3

DARK CURRENT
vs
FREE-AIR TEMPERATURE

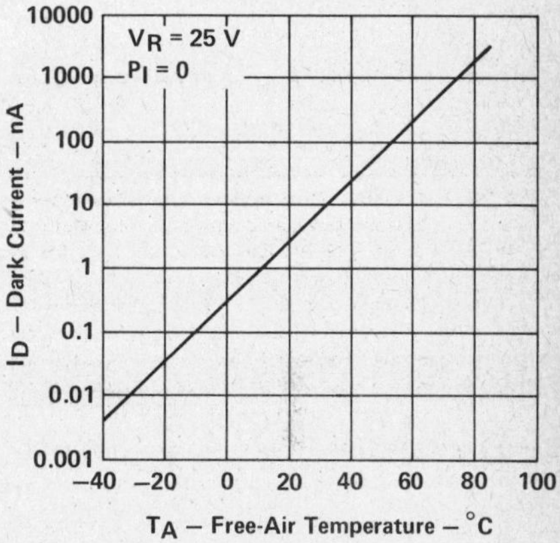

FIGURE 4

TOTAL CAPACITANCE
vs
REVERSE VOLTAGE

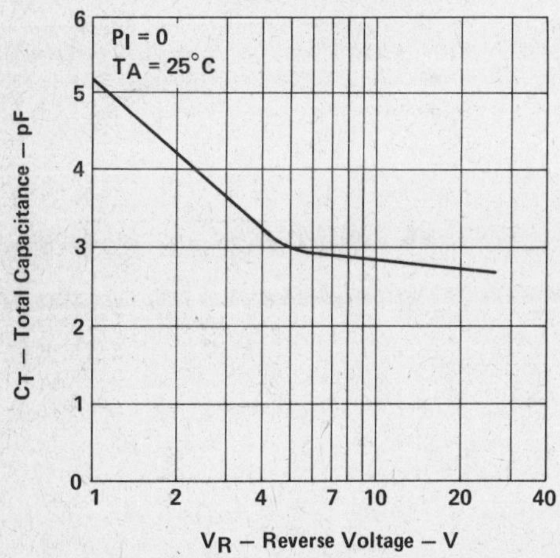

FIGURE 5

RELATIVE RADIANT RESPONSIVITY
vs
WAVELENGTH

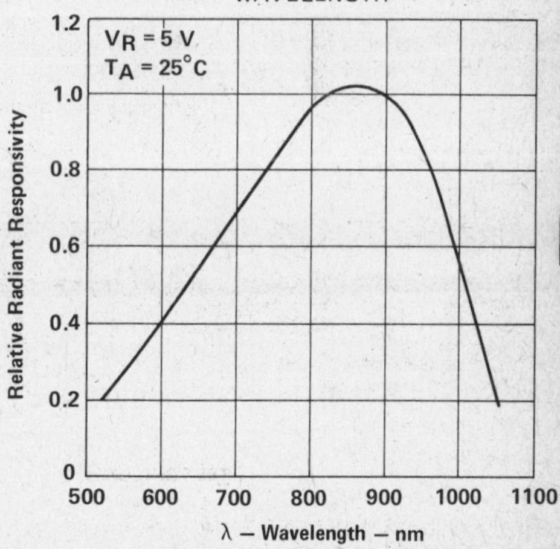

FIGURE 6

D2680, MARCH 1983

INTEGRATED ANALOG RECEIVERS FOR FIBER-OPTIC APPLICATIONS

- Monolithic Integrated Circuit Containing Both Photodetector and Transimpedance Preamplifier
- Converts Optical Input to Voltage Output
- Quasi-Differential Output for AC-Coupled Systems
- Fast Pulse Response Time . . . 10 ns for TIED463
- High Radiant Responsivity . . . 60 mV/μW for TIED460
- Large Effective Detector Area with Internal 0.7-mm-Diameter Spherical Microlens
- Optically Compatible with TIES494, TIES495, and TIES496 Infrared-Emitting Diodes
- Mechanically Compatible with Commercial Fiber Optic Connector Receptacles

mechanical data

The devices are in a hermetically sealed welded case with a flat glass window in the case top. The coin header is gold plated; the metal window can is nickel plated.

ALL LEADS ARE ELECTRICALLY INSULATED FROM THE CASE

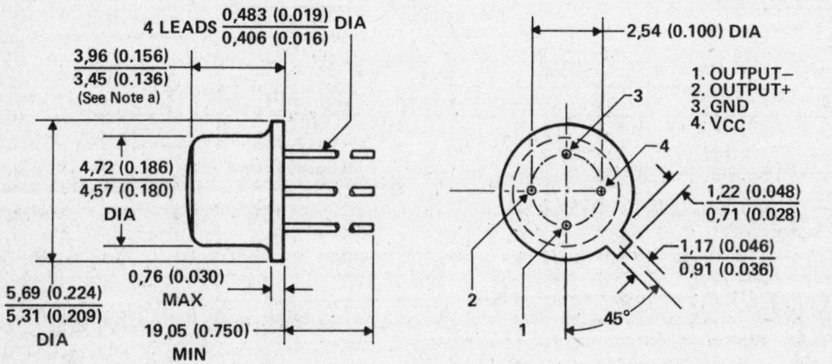

NOTE a: The maximum extension of the glass above the top of the case is 0,25 mm (0.010 inch).

ALL LINEAR DIMENSIONS ARE IN MILLIMETERS AND PARENTHETICALLY IN INCHES

schematic

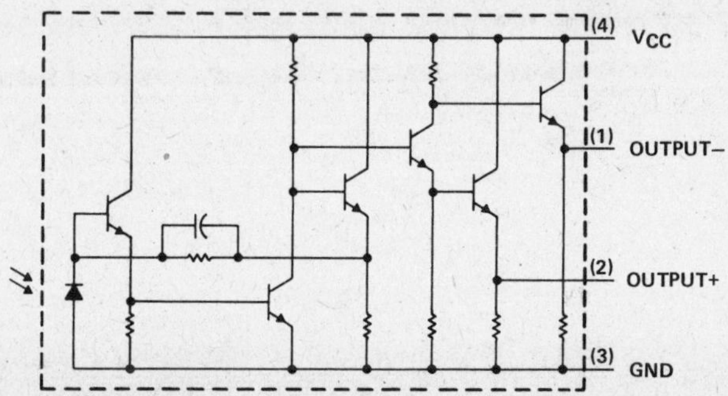

Copyright © 1983 by Texas Instruments Incorporated

TEXAS INSTRUMENTS
INCORPORATED
POST OFFICE BOX 225012 • DALLAS, TEXAS 75265

absolute maximum ratings at 25°C free-air temperature (unless otherwise noted)

Supply Voltage, V_{CC}	7.5 V
Operating Free-Air Temperature Range	−40°C to 85°C
Storage Temperature Range	−40°C to 100°C
Lead Temperature 1,6 mm (1/16 inch) from Case for 10 Seconds	240°C

operating characteristics at 25°C free-air temperature, V_{CC} = 5 V

	PARAMETER	TEST CONDITIONS		TIED460			TIED461			UNIT
				MIN	TYP	MAX	MIN	TYP	MAX	
$R_{e(s)}$	Steady-State Responsivity*		See Figure 1	40	60	100	18	26	45	mV/μW
$R_{e(p)}$	Pulsed Radiant Responsivity†	t_w = 500 ns			58					mV/μW
		t_w = 250 ns						24		
P_n	Equivalent Input Noise Radiant Power‡				0.007			0.015		μW
V_{OQ+}	Quiescent DC Output Voltage (Noninverting Output)	P_I = 0§,	See Figure 1	0.48	0.60	0.72	0.52	0.64	0.76	V
V_{OQ-}	Quiescent DC Output Voltage (Inverting Output)	P_I = 0,	See Figure 1	2.7	3.0	3.3	2.7	3.0	3.3	V
V_n	RMS Output Noise Voltage	P_I = 0,	See Figure 2		0.4	0.6		0.4	0.6	mV
z_o	Output Impedance¶	P_I = 0,	f = 20 kHz		200			200		Ω
$t_{t(p)}$	Pulsed Transition Time (20% to 80%)#	t_w = 500 ns,	See Figure 3		80	120				ns
		t_w = 250 ns,	See Figure 3					35	50	
I_{CC}	Supply Current	P_I = 0,	See Figure 1	3	4.0	5	3.3	4.4	5.5	mA

*Radiant responsivity is defined as the absolute change of output voltage divided by the change of radiant power input P_I. The steady-state radiant responsivity $R_{e(s)}$ applies for incident radiant power pulse durations of greater than 2 μs. The pulse radiant responsivity for short radiant power pulse durations is discussed below. The maximum output voltage change should be less than 0.7 V to maintain linear operation and to minimize pulse duration distortion. Bypass capacitors between V_{CC} (Pin 4) and GND (Pin 3) are required; a 1-μF tantalum capacitor in parallel with a 0.01-μF ceramic disc capacitor is recommended to be placed as close as possible to the device pins.

†The output voltage response to a step change of incident radiant power consists of two components: a fast component due to radiation absorbed in the photodiode high-field junction depletion region, and a slow component due to radiation absorbed in the photodiode nondepleted regions. The amplitude of the fast component is characterized by the pulse radiant responsivity $R_{e(p)}$ for short incident radiant power pulse durations.

‡Equivalent input noise radiant power P_n equals the RMS output noise voltage V_n divided by the steady-state radiant responsivity $R_{e(s)}$.

§The radiant power input P_I is the radiant power measured with a large-area detector from the output end of a 10-m fiber-optic cable assembly that incorporates a graded-index optical fiber with a 100-μm core diameter, a 140-μm cladding diameter, and a numerical aperture of ≤ 0.30. A TIES495 820-nm GaAlAs infrared-emitting diode is used at the input end of the fiber optic cable assembly. The TIES495 has a typical half-intensity beam angle of 20 degrees.

¶Capacitive coupling is required for load resistances smaller than 10,000 ohms to minimize disturbance of the quiescent dc output voltages.

#The pulsed transition time $t_{t(p)}$ is the time required for the output voltage to change from 20% to 80% of its peak value for short incident radiant power pulse durations. The transition time of the trailing edge is approximately equal to that of the leading edge. The output voltage pulse duration is typically 6 ns less than the incident radiant power pulse duration due to differences of propagation delay times for the leading and trailing transitions. The overall 10% to 90% rise time is typically 250 ns for pulse durations greater than 2 μs.

absolute maximum ratings at 25°C free-air temperature (unless otherwise noted)

Supply Voltage, V_{CC} .... 7.5 V
Operating Free-Air Temperature Range .... −40°C to 85°C
Storage Temperature Range .... −40°C to 100°C
Lead Temperature 1,6 mm (1/16 inch) from Case for 10 Seconds .... 240°C

operating characteristics at 25°C free-air temperature, V_{CC} = 5 V

PARAMETER		TEST CONDITIONS		TIED462			TIED463			UNIT
				MIN	TYP	MAX	MIN	TYP	MAX	
$R_{e(s)}$	Steady-State Responsivity*		See Figure 1	8	12	20	3.2	4.8	8	mV/μW
$R_{e(p)}$	Pulsed Radiant Responsivity†	t_w = 100 ns			10					mV/μW
		t_w = 50 ns						3.5		
P_n	Equivalent Input Noise Radiant Power‡				0.04			0.13		μW
V_{OQ+}	Quiescent Output Voltage (Noninverting Output)	P_I = 0§,	See Figure 1	0.56	0.68	0.8	0.62	0.74	0.86	V
V_{OQ-}	Quiescent Output Voltage (Inverting Output)	P_I = 0,	See Figure 1	2.6	2.9	3.2	2.6	2.9	3.2	V
V_n	RMS Output Noise Voltage	P_I = 0,	See Figure 2		0.5	0.7		0.6	0.9	mV
z_o	Output Impedance¶	P_I = 0,	f = 20 kHz		200			200		Ω
$t_{t(p)}$	Pulsed Transition Time (20% to 80%)#	t_w = 100 ns	See Figure 3		18	28				ns
		t_w = 50 ns,	See Figure 3					10	15	
I_{CC}	Supply Current	P_I = 0,	See Figure 1	3.6	4.8	6	4.2	5.6	7	mA

*Radiant responsivity is defined as the absolute change of output voltage divided by the change of radiant power input P_I. The steady-state radiant responsivity $R_{e(s)}$ applies for incident radiant power pulse durations of greater than 2 μs. The pulse radiant responsivity for short radiant power pulse durations is discussed below. The maximum output voltage change should be less than 0.7 V to maintain linear operation and to minimize pulse duration distortion. Bypass capacitors between V_{CC} (Pin 4) and GND (Pin 3) are required; a 1-μF tantalum capacitor in parallel with a 0.01-μF ceramic disc capacitor is recommended to be placed as close as possible to the device pins.

†The output voltage response to a step change of incident radiant power consists of two components: a fast component due to radiation absorbed in the photodiode high-field junction depletion region, and a slow component due to radiation absorbed in the photodiode nondepleted regions. The amplitude of the fast component is characterized by the pulse radiant responsivity $R_{e(p)}$ for short incident radiant power pulse durations.

‡Equivalent input noise radiant power P_n equals the RMS output noise voltage V_n divided by the steady-state radiant responsivity $R_{e(s)}$.

§The radiant power input P_I is the radiant power measured with a large-area detector from the output end of a 10-m fiber-optic cable assembly that incorporates a graded-index optical fiber with a 100-μm core diameter, a 140-μm cladding diameter, and a numerical aperture of ≤ 0.30. A TIES495 820-nm GaAlAs infrared-emitting diode is used at the input end of the fiber optic cable assembly. The TIES495 has a typical half-intensity beam angle of 20 degrees.

¶Capacitive coupling is required for load resistances smaller than 10,000 ohms to minimize disturbance of the quiescent dc output voltages.

#The pulsed transition time $t_{t(p)}$ is the time required for the output voltage to change from 20% to 80% of its peak value for short incident radiant power pulse durations. The transition time of the trailing edge is approximately equal to that of the leading edge. The output voltage pulse duration is typically 6 ns less than the incident radiant power pulse duration due to differences of propagation delay times for the leading and trailing transitions. The overall 10% to 90% rise time is typically 250 ns for pulse durations greater than 2 μs.

PARAMETER MEASUREMENT INFORMATION

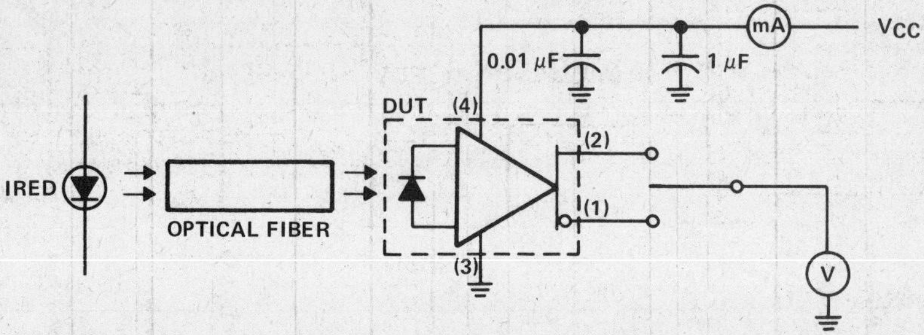

FIGURE 1—TEST CIRCUIT FOR STEADY-STATE PARAMETERS

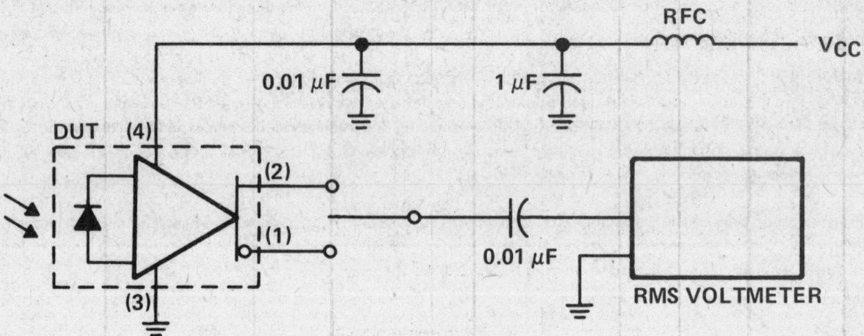

FIGURE 2—TEST CIRCUIT FOR NOISE MEASUREMENTS

PARAMETER MEASUREMENT INFORMATION

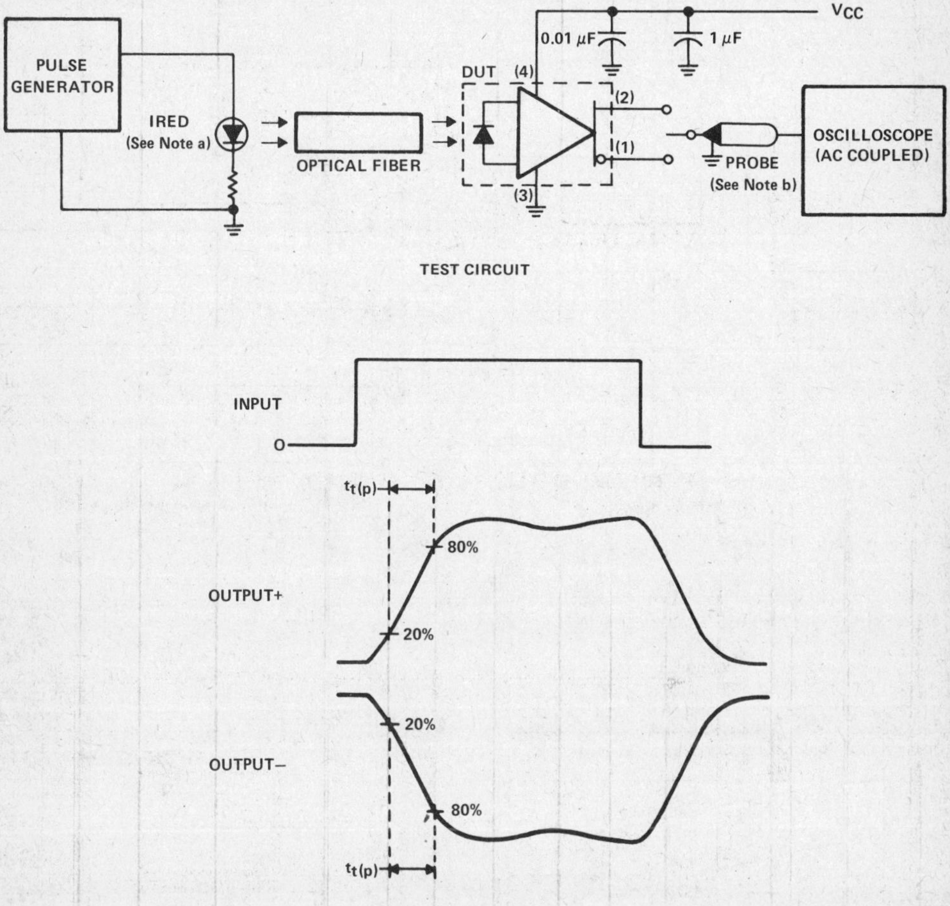

FIGURE 3–SWITCHING TIMES

NOTES: a. Radiant power input is supplied by a pulsed GaAlAs infrared-emitting diode with the following characteristics λ_p = 850 nm, $t_r \leqslant 5$ ns.

b. The input impedance of the probe is at least 100 kΩ. The combined rise time of the probe and the oscilloscope is equal to or less than 2 ns.

TYPICAL CHARACTERISTICS

RELATIVE RADIANT RESPONSIVITY
vs
FREE-AIR TEMPERATURE

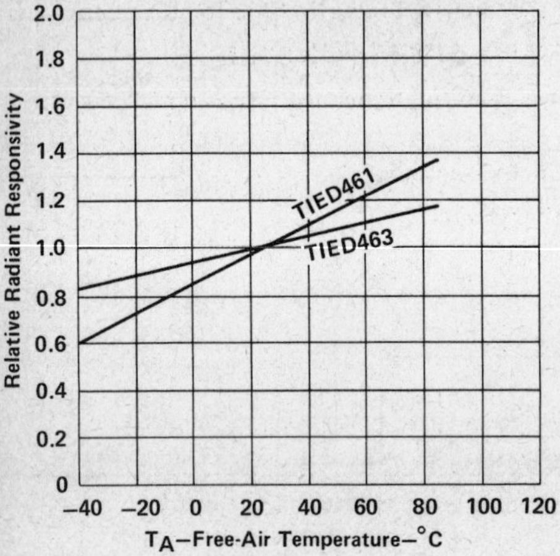

FIGURE 4

RELATIVE PULSE RESPONSE TIME
vs
FREE-AIR TEMPERATURE

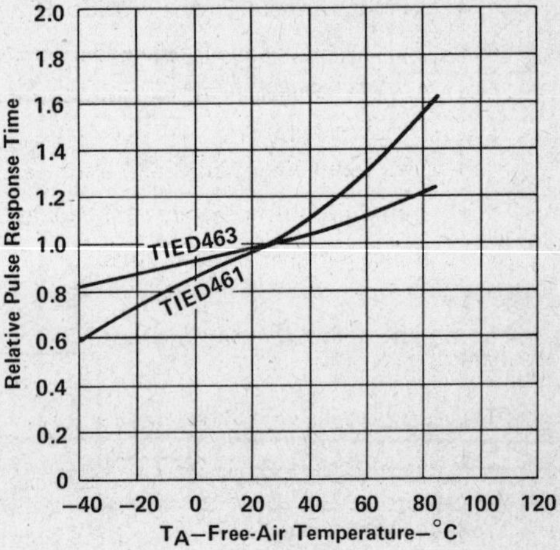

FIGURE 5

SUPPLY CURRENT
vs
FREE-AIR TEMPERATURE

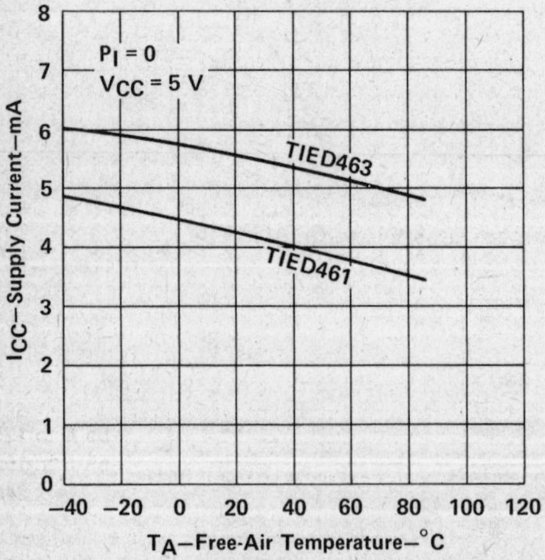

FIGURE 6

QUIESCENT DC OUTPUT VOLTAGE
vs
FREE-AIR TEMPERATURE

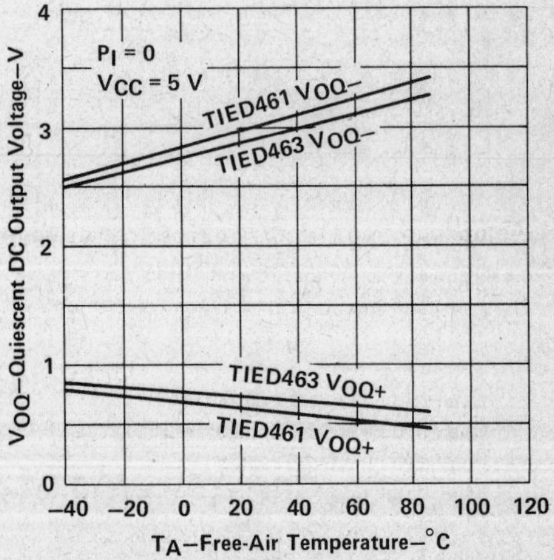

FIGURE 7

FIBER OPTIC COMPONENTS AND AMPLIFIERS

12

D1954, NOVEMBER 1974—REVISED DECEMBER 1977

OPTOELECTRONIC INTERFACE CIRCUITS FOR APPLICATIONS SUCH AS LASER RANGEFINDERS AND OPTICAL COMMUNICATIONS

(FORMERLY TIXL150, TIXL151, TIXL152)

- Designed for Current Sources such as Photodiodes and Photomultiplier Tubes
- Transimpedance Circuit Provides Output Voltage Linearly Proportional to Input Current
- Typical Frequency Responses from DC to 100 MHz, 50 MHz, and 20 MHz
- Typical Equivalent Input Noise Current Spectral Densities of 8.5 pA/$\sqrt{Hz}$, 4.5 pA/$\sqrt{Hz}$, and 3 pA/$\sqrt{Hz}$
- Low Input Impedance for Tolerance of High Input Capacitance
- Low Output Impedance for Loads as Small as 50 Ohms‡
- Single Supply of 4 to 6 Volts

mechanical data

The device is in a hermetically sealed welded case similar to but shorter than JEDEC TO-12.

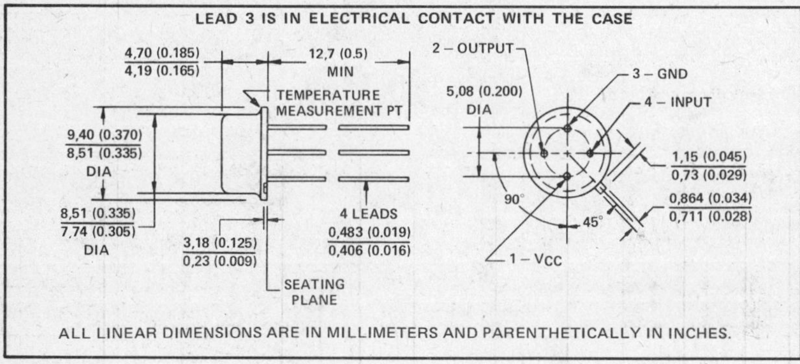

absolute maximum ratings at 25°C free-air temperature (unless otherwise noted)

Supply voltage V_{CC}		8 V
Continuous Input Current Range:	TIEF150	−5 mA to 2 mA
	TIEF151	−1.2 mA to 2 mA
	TIEF152	−0.5 mA to 2 mA
External Load Conductance		20 mmho‡
Operating Free-Air Temperature Range		−55°C to 125°C
Storage Temperature Range		−65°C to 150°C
Lead Temperature 1/16 Inch from Case for 10 Seconds		240°C

†Equivalent input noise current is defined as broadband rms output voltage divided by z_f and by the square root of noise bandwidth. The noise bandwidth is $\pi/2$ times the signal bandwidth measured between the frequencies at which response is down 3 dB with a high-frequency rolloff of 6 dB/octave.

‡Capacitive coupling is required for load resistances smaller than 1000 ohms to minimize disturbance of the amplifier bias.

electrical characteristics at 25°C free-air temperature, V_{CC} = 5.8 V

PARAMETER		TEST CONDITIONS§	TIEF150			TIEF151			TIEF152			UNIT
			MIN	TYP	MAX	MIN	TYP	MAX	MIN	TYP	MAX	
I_n	Equivalent Input Noise Current†	R_L = 50 Ω, See Note 1		8.5	10		4.5	7		3	5.5	pA/$\sqrt{Hz}$
z_f	Forward Transfer Impedance	R_L = 50 Ω, f = 20 kHz	0.8	1.0		2.8	4		8	12		kΩ
z_i	Input Impedance	R_L = 50 Ω, f = 20 kHz		35	70		100	140		300	500	Ω
z_o	Output Impedance	I_{in} = 0, f = 20 kHz		0.5	5		2	10		4	12	Ω
V_o	Maximum RMS Output Voltage	R_L = 50 Ω, f = 20 kHz	100			100			100			mV
B	Bandwidth (−3 dB)	R_L = 50 Ω	90	100		40	50		12	20		MHz
V_{IQ}	Quiescent Input Voltage	Input open		0.7			0.7			0.7		V
V_{OQ}	Quiescent Output Voltage	Input open		0.8			0.8			0.8		V
I_{CC}	Supply Current	Input open		4	6		4	6		4	7	mA

†Equivalent input noise current is defined as broadband rms output voltage divided by z_f and by the square root of noise bandwidth. The noise bandwidth is $\pi/2$ times the signal bandwidth measured between the frequencies at which response is down 3 dB with a high-frequency roll-off of 6 dB/octave.

§Output coupling capacitance = 1 μF, V_{CC} bypass capacitance = 0.01 μF.

NOTE 1: Equivalent input noise current is determined using a post-amplifier with response down 3 dB at 10 kHz and 150 MHz. Therefore, the overall signal bandwidth is equal to the bandwidth of the device under test.

schematic

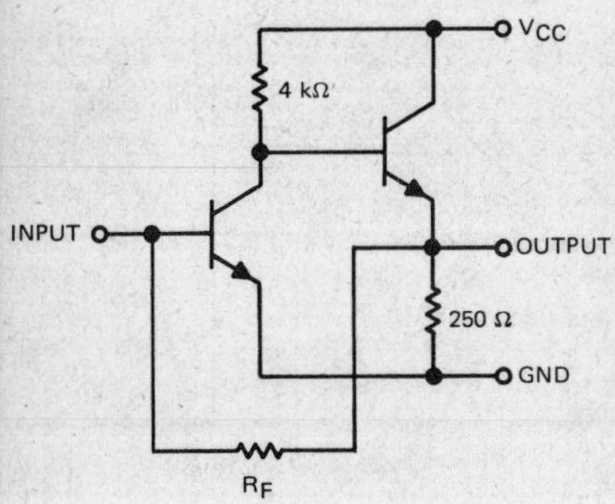

VALUE APPROXIMATELY EQUAL TO z_f

Resistor values shown are nominal

typical application

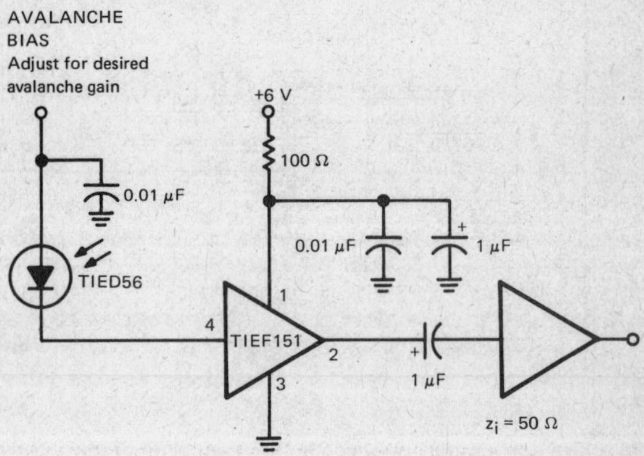

TYPICAL PERFORMANCE FOR M = 100, λ = 0.9 μm

R_e = 2.3 X 10^5 V/W

NEP = 2 X 10^{-13} W/$\sqrt{Hz}$

f_{lower} = 3 kHz

f_{upper} = 50 MHz

TEXAS INSTRUMENTS
INCORPORATED
POST OFFICE BOX 225012 • DALLAS, TEXAS 75265

D2681, APRIL 1983

SOURCE DESIGNED FOR FIBER-OPTIC APPLICATIONS

- High-Efficiency GaAlAs Infrared-Emitting Diode
- 820-nm Peak-Emission Wavelength
- Radiant Rise Time . . . 12 ns Typical
- Internal 0.5-mm-Diameter Spherical Microlens for Efficient Optical Coupling
- Optically Compatible with TIED458 Phototransistor, TIED459 Photodiode, and TIED460, TIED461, TIED462, TIED463 Integrated Analog Receivers
- Mechanically Compatible with Commercial Fiber Optic Connector Receptacles

mechanical data

The devices are in a hermetically sealed welded case with flat glass window in the case top. A coin header is used for increased thermal dissipation capability. The coin header is gold plated. The metal window can is nickel plated.

LEADS 1 AND 2 ARE ELECTRICALLY INSULATED FROM THE CASE

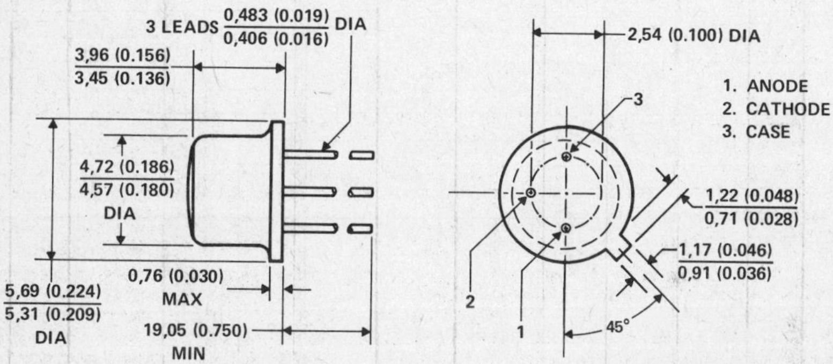

The maximum extension of the glass above the top of the case is 0,25 mm (0.010 inch).
ALL DIMENSIONS ARE IN MILLIMETERS AND PARENTHETICALLY IN INCHES

CL-77

Copyright © 1983 by Texas Instruments Incorporated

absolute maximum ratings at 25°C free-air temperature (unless otherwise noted)

Reverse Voltage .... 2 V
Continuous Forward Current at (or below) 25°C Free-Air Temperature (See Note 1) .... 60 mA
Peak Forward Current at (or below) 25°C Free-Air Temperature (See Notes 2 and 3) .... 100 mA
Operating Free-Air Temperature Range (See Notes 2 and 3) .... −40°C to 85°C
Storage Temperature Range .... −40°C to 100°C
Lead Temperature 1,6 mm (1/16 inch) from Case for 10 seconds .... 240°C

NOTES: 1. Derate linearly to 24 mA at 85°C free-air temperature at the rate of 0.60 mA/°C.
2. Derate linearly to 40 mA at 85°C free-air temperature at the rate of 1.0 mA/°C.
3. This value applies for $t_w \leq 100$ µs, duty cycle ≤ 50%.

operating characteristics at 25°C case temperature

PARAMETER			TEST CONDITIONS	MIN	TYP	MAX	UNIT
P_O	Radiant Power Output (see Note 4)	TIES494	I_F = 50 mA, See Figure 1	30	45	120	µW
		TIES495		50	75	160	
		TIES496		80	110	240	
P_O	Pulsed Radiant Power Output (see Note 4)	TIES494	I_{FM} = 100 mA, See Note 5		90		µW
		TIES495			150		
		TIES496			220		
λ_p	Wavelength at Peak Emission		I_F = 50 mA	790	820	860	nm
$\Delta\lambda$	Spectral Bandwidth		I_F = 50 mA		40		nm
θ_{HI}	Half-Intensity Beam Angle		I_F = 50 mA		20°		
V_F	Static Forward Voltage		I_F = 50 mA		1.6	2	V
V_F	Forward Voltage		I_F = 100 mA, See Note 5		1.8		V
C	Capacitance		V_F = 0		200		pF
t_r	Radiant Pulse Rise Time (see Note 6)		I_{FM} = 50 mA, $t_w \geq$ 100 ns, See Figure 2		12	20	ns

NOTES: 4. The radiant power output, P_O, is the radiant power transmitted through a 0,2-mm (0.008-inch) diameter mechanical aperture into a numerical aperture of 0.25. The radiant power coupled into a graded-index optical fiber with a 100-µm core diameter, a 140-µm cladding diameter, and a numerical aperture of 0.3 is typically 24% of P_O. The radiant power coupled into a graded-index optical fiber with a 50-µm core diameter, a 125-µm cladding diameter, and a numerical aperture of 0.2, is typically 4.5% of P_O.
5. These parameters must be measured using pulse techniques. $t_w \leq 100$ µs, duty cycle ≤50%.
6. Radiant pulse rise time is the time required for a change in radiant intensity from 10% to 90% of its peak value for a step change in forward diode current. The typical electrical bandwidth (in MHz) at which the radiant power output is reduced to $1/\sqrt{2}$ of the maximum low-frequency value is approximately $350/t_r$ (t_r in ns) or 29 MHz. The typical optical bandwidth at which the radiant power output is reduced to 1/2 of the maximum low-frequency value is approximately $610/t_r$ or 50 MHz. The radiant pulse fall time is typically equal to or less than the radiant pulse rise time.

temperature coefficients

PARAMETER		TEST CONDITIONS	MIN	TYP	MAX	UNIT
α_{PO}	Temperature Coefficient of Radiant Power Output	I_F = 50 mA		−0.5		%/°C
$\alpha_{\lambda p}$	Temperature Coefficient of Peak-Emission Wavelength			0.25		nm/°C
$\alpha_{\Delta\lambda}$	Temperature Coefficient of Spectral Bandwidth			0.08		nm/°C
α_{VF}	Temperature Coefficient of Static Forward Voltage			−1.3		mV/°C

TEXAS INSTRUMENTS
INCORPORATED
POST OFFICE BOX 225012 • DALLAS, TEXAS 75265

PARAMETER MEASUREMENT INFORMATION

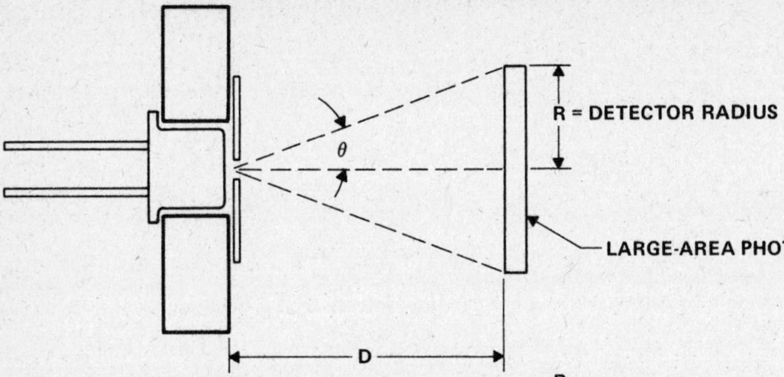

$$\text{NUMERICAL APERTURE} = \sin\theta = \frac{R}{\sqrt{R^2 + D^2}}$$

FIGURE 1—RADIANT POWER OUTPUT MEASUREMENT

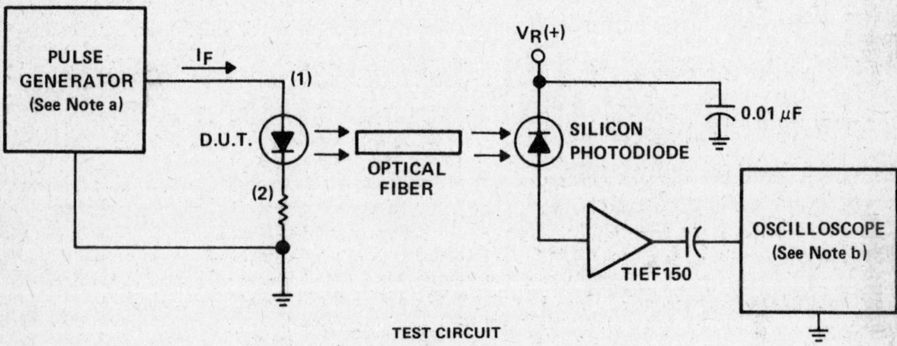

TEST CIRCUIT

Adjust amplitude of input pulse for I_{FM} = 50 mA

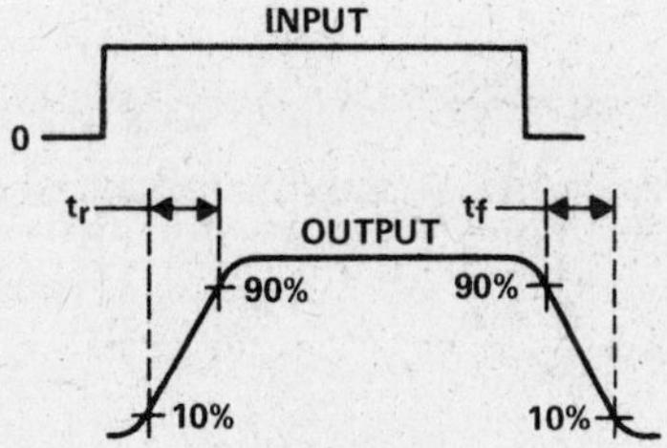

WAVEFORMS

NOTES: a. The input current waveform is supplied by a pulse generator with the following characteristics: Z_o = 50 Ω, $t_w \geq$ 100 ns, $t_r \leq$ 2ns.

b. The output waveform is monitored on an oscilloscope with the following characteristics: Z_{in} = 50 Ω, $t_r \leq$ 2 ns. The measured rise time is corrected for the combined rise times (≤6 ns) of the receiver circuit and the oscilloscope.

FIGURE 2—SWITCHING TIMES

FIBER OPTIC COMPONENTS AND AMPLIFIERS

12

High-Reliability Index

- LED Displays
- Optocouplers (Isolators)
- Infrared Emitters
- Infrared Detectors

High-Reliability Optoelectronic Products

Texas Instruments offers a large selection of high-reliability optoelectronic devices consisting of displays, optocouplers, infrared emitters, and infrared detectors/sensors. See referenced data sheets listed below for complete specifications on each device.

HI-REL LED DISPLAYS

- **In moisture-resistant ceramic packages**
- **TTL compatibility and reliability**

Available in

- **High-intensity red or yellow**
- **7-segment red**
- **Hexadecimal red or yellow**
- **Alphanumeric red or yellow**

For applications involving military or adverse environmental conditions, TI offers a variety of high-reliability moisture-resistant displays. These red and yellow LED displays offer high luminous intensity coupled with a wide viewing angle and low power requirements.

JEDEC PART NO.	TI PART NO.	TYPE OF CHARACTER	CHARACTER HEIGHT mm (INCHES)	COLOR OF DISPLAY	PACKAGE	REMARKS	PAGE
4N41	TIL501*	7-segment	6,9 (0.270)	Red	14-lead hermetically sealed dual-in-line	Electrically and mechanically interchangeable with TIL302	11-3
4N56	TIL505*	Hexadecimal	7,6 (0.300)	Red	14-lead hermetically sealed dual-in-line	Self-contained four-bit latch, decoder, and driver with 4 × 7 font	11-7
4N57	TIL506*	7-segment	7,6 (0.300)	Red	8-lead hermetically sealed	Internal TTL MSI chip with decoder and driver. Left decimal	11-11
4N58	TIL507*	5 × 7 alphanumeric	7,6 (0.300)	Red	16-lead hermetically sealed dual-in-line	Integral D-type flip-flop column drivers and series limiting resistors. Left decimal.	11-15
–	TIL509	Hexadecimal	7,6 (0.300)	Yellow	14-lead hermetically sealed dual in-line	Self-contained four-bit latch, decoder, and driver with 4 × 7 font similar to 4N56	11-19
–	TIL510	5 × 7 alphanumeric	7,6 (0.300)	Yellow	16-lead hermetically sealed dual-in-line	Integral D-type flip-flop column drivers and series limiting resistors. Left decimal similar to 4N58	11-23

*These part numbers have been replaced by the JEDEC part numbers shown.

HI-REL OPTOCOUPLERS (OPTOISOLATORS)

- Hermetically sealed TO-72 and TO-78 metal-can packages
- JAN, JANTX, and JANTXV versions available (4N22 and 4N47 series)
- Stable over wide temperature ranges
- High current transfer ratios (CTR)
- High-voltage electrical isolation . . . 1-kV rating

PART NUMBER		METAL-CAN PACKAGE	CTR (MIN %) @	I_F (mA)	PAGE
"SUPER-COUPLERS"*					
3N261		TO-72	50	1	7-3
3N262		TO-72	100	1	7-3
3N263		TO-72	200	1	7-3
4N47	JAN,JANTX, and JANTXV per MIL-S-19500/548	TO-78	50	1	7-21, 7-27
4N48	JAN, JANTX, and JANTXV per MIL-S-19500/548	TO-78	100	1	7-21, 7-27
4N49	JAN, JANTX, and JANTXV per MIL-S-19500/548	TO-78	200	1	7-9, 7-13
OPTOCOUPLERS					
4N22	JAN, JANTX, and JANTXV per MIL-S-19500/486A	TO-78	25	10	7-9, 7-13
4N23	JAN, JANTX, and JANTXV per MIL-S-19500/486A	TO-78	60	10	7-9, 7-13
4N24	JAN, JANTX, and JANTXV per MIL-S-19500/486A	TO-78	100	10	7-9, 7-13
TIL102		TO-78	25	10	7-33
TIL103		TO-78	100	10	7-33
TIL120		TO-72	25	10	7-51
TIL121		TO-72	50	10	7-51

*"Super-couplers" are optocouplers having high CTR (current transfer ratio) at low I_F.

TEXAS INSTRUMENTS
INCORPORATED
POST OFFICE BOX 225012 • DALLAS, TEXAS 75265

HI-REL EMITTERS AND DETECTORS

- **Hermetically sealed packages**
- **Wide temperature storage and operating range**
- **Spectrally and mechanically matched IR pairs**
- **Pill packages and TO-18 packages**

The devices listed below are subjected to the processing and lot acceptance in accordance to the sequence of tests in MIL-S-19500 for JANTX types. These are not to be construed to be JANTX-qualified parts. A detail specification is available upon request through your Field Sales Office or Optoelectronics Marketing at the following address:

Texas Instruments Incorporated
Optoelectronics Marketing
P.O. Box 225012, MS 12
Dallas, Texas 75265
Phone: (214) 995-3821

HI-REL INFRARED EMITTERS AND SENSORS

PART NUMBER	DESCRIPTION	METAL-CAN PACKAGE	PAGE
TIL24HR2	IR Emitter	Pill	3-3, 3-7
TIL31BHR2	IR Emitter	TO-18	3-9, 3-11
TIL81HR2	Phototransistor	TO-18	5-9, 5-13
TIL604HR2	Phototransistor	Pill	5-31, 5-39

Quality and Reliability

- Quality/Reliability Program
- Device Reliability Data

TEXAS INSTRUMENTS QUALITY/RELIABILITY PROGRAM FOR OPTOELECTRONICS

Texas Instruments has an extensive commitment to produce optoelectronic products with the highest quality and reliability performance possible. TI monitors the entire semiconductor process, from the earliest stages of crystal formation through completion of the final device. These monitored processes which follow rigid Quality Standards are illustrated in Table I. As an added emphasis on our quality trust, TI incorporates Quality Reviews with some of our major customers to monitor their incoming inspection reports with our reporting system. These customers' inputs are reviewed on a monthly basis by the top management of Texas Instruments and are used to constantly update our standard within the industry. Our continuing goal is to be the Number 1 supplier in the industry, and we have set up our QA Program to meet this challenge.

The broad spectrum of industrial and/or military applications demand our products to be operative under adverse conditions and prolonged usage. Refer to Table II for our overall testing capability. Table III defines the military standard test capabilities available at TI.

Extensive facilities are used in our failure analysis laboratory to analyze in-house and field failures of our products. Table IV illustrates our Failure Analysis Procedures and our test facilities.

In summary, this chapter includes the following tables:

Table I	General Standard Device Flow
Table II	Overall Test Capability
Table III	Military Standard Test Capability
Table IV	Failure Analysis Capability

Reliability data on our products is consolidated every 3 months and is available upon request by contacting your local TI Field Sales Office or by contacting TI direct.

Table I. General Standard Device Flow

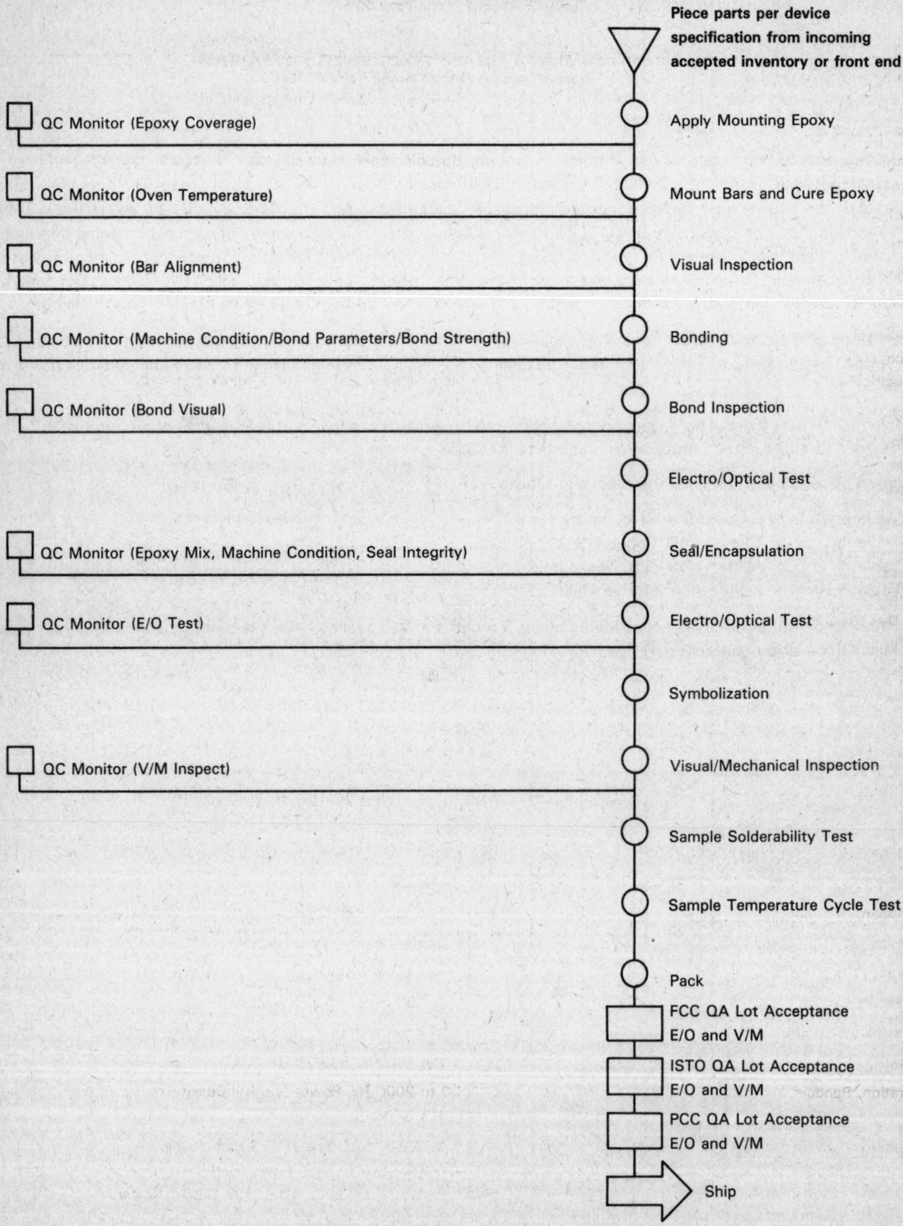

Table II. Overall Test Capability

Test	Capability
Acceleration, Sustained (Centrifuge)	50 to 50,000 G (standard) 50,000 to 125,000 G (nonstandard)*
Bond Strength	0 to 25 grams
Altitude (Barometric Pressure, Reduced)	150,000 ft simulated altitude
Dew Point	−65°C to 150°C
Electrostatic Susceptibility, MIL-STD-883, Method 3015	
Flammability	800°C to 1100°C
Moisture Resistance	+2°C to +96°C, 40% to 100% RH
Particle Detection	
Acoustical (PIND)	≥0.1 microgram
Electrical	Intermittency ≥ 1 μs with 100-mV amplitude
Pressure Cooker	0 to 15 psig of steam pressure
Radiographic Inspection (X-Ray)	
Film	Resolution to 0.001 inch, 150 kV, 5 mA
Real Time	360° rotation — Resolution to 0.001 inch
Salt Atmosphere/Spray	25°C to 45°C, up to 20% salt solution
Seal	
Gross Leak	
Bubble	$\geq 1 \times 10^{-5}$ atm cm^3/s
Dye Penetrant	$\geq 5 \times 10^{-6}$ atm cm^3/s
Weight Gain	$> 2 \times 10^{-6}$ atm cm^3/s
Radioactive Tracer Gas	$\geq 1 \times 10^{-10}$ atm cm^3/s
Symbolization (Resistance to Solvents)	
Shock (Mechanical)	To limits of: MIL-STD-202, Method 213 MIL-STD-750 MIL-STD-810, Method 516 MIL-STD-883
Solderability, Meniscograph	MIL-STD-883, Method 2022
Solderability/Soldering	Up to 280°C
Temperature Cycling	−185°C to 300°C
Terminal Strength (Lead Integrity)	Lead Fatigue, Tension, Torque
Thermal Shock	−196°C to 200°C
Ultrasonics	0 to 100 psi at 40 kHz or 25 kHz
Ultraviolet Exposure	To 12.5 mW/cm^2
Vibration, Fatigue	10 to 100 Hz, 5 to 70 G
Vibration, Random	20 to 2000 Hz, Power Spectral Density 1.3 G^2/Hz
Vibration, Variable	5 to 2000 Hz as limited by 1 inch double amplitude and 60 inches/second velocity. 0 to 70 G (standard), 70 to 100 G (nonstandard)*

*Limited fixture availability.

QUALITY AND RELIABILITY

14

Table III. Military Standard Test Capability

TEST CATEGORY	MIL-STD-202	MIL-STD-750	MIL-STD-883
Altitude	All Conditions **except** G	All Conditions **except** G	All Conditions **except** G
Bond Strength		Conditions A or B	Conditions A , C, or D
Dew Point		All Conditions	All Conditions
Flammability	All Conditions		
Immersion	All Conditions	All Conditions	All Conditions
Insulation Resistance	All Conditions	All Conditions	All Conditions
Meniscograph Solderability			All Conditions
Moisture Resistance	All Conditions	All Conditions	All Conditions
Resistance to Solvents (Symbolization)	All Conditions	All Conditions	All Conditions
Salt Atmosphere		All Conditions	All Conditions
Salt Spray	All Conditions	All Conditions	
Seal	All Conditions	All Conditions	All Conditions
Solderability	All Conditions	All Conditions	All Conditions
Soldering Heat	All Conditions	All Conditions	
Temperature Cycling	All Conditions **except** Method 107, Conditions D & E	All Conditions **except** Method 1051, Conditions D & E	All Conditions
Temperature Storage			Conditions A thru F
Terminal Strength (Lead Integrity)	All Conditions	All Conditions	All Conditions
Axial Lead Tensile Test		All Conditions	
Thermal Shock (Glass Strain)		All Conditions	All Conditions
Acceleration, Sustained (Centrifuge)	All Conditions	All Conditions	All Conditions Method 2001, Conditions G, H, and I, may require special fixturing.†
Particle Impact Noise Detection (PIND)		All Conditions	All Conditions
Forward Instability Shock (FIST)		All Conditions	
Backward Instability Shock (BIST)		All Conditions	
Shock (Mechanical)*	All Conditions	All Conditions	All Conditions Method 2002, Conditions F and G, may require special fixturing.†
Vibration, Fatigue		All Conditions	All Conditions
Vibration, Noise		All Conditions	All Conditions

†Call Physical Test supervisor for available fixtures
*Also perform mechanical shock per MIL-STD-810B, Method 516.

QUALITY AND RELIABILITY 14

TEXAS INSTRUMENTS
INCORPORATED
POST OFFICE BOX 225012 • DALLAS, TEXAS 75265

Table III. Military Standard Test Capability (Continued)

TEST CATEGORY	MIL-STD-202	MIL-STD-750	MIL-STD-883
Vibration, Random‡	All Conditions		
Vibration, Variable Frequency‡	All Conditions	All Conditions	All Conditions
X-Ray, Film§	All Conditions	All Conditions	All Conditions
X-Ray, Real Time (TV X-Ray)§	All Conditions	All Conditions	All Conditions

‡ Also perform random vibration and vibration, variable frequency per MIL-STD-810B, Method 514.1, procedures I, II, III, IV, V, VI, and VII. Omit paragraph 4.5.1.1, Resonant Search, and paragraph 4.5.1.2, Resonant Dwell.

§ Radiographic inspection is performed in accordance with many government and customer specifications. Before any new radiographic specification is acceptable for use as a test standard within the Semiconductor Group, it must be approved by Environmental Test Services. For questions pertaining to a particular specification, contact the Radiographic Group supervisor or cost center manager—phone (214) 995-3397 or 995-3931.

Table IV. Failure Analysis Capabilities

I. Nondestructive Techniques
- A. Hermeticity evaluation
- B. X-ray interpretation of bonding and die mount.
- C. Electrical characterization
  1. Breakdown, leakage, and functional tests run at temperature extremes.
  2. Polaroid documentation of curve traces and/or oscilloscope traces

II. Destructive Techniques
- A. Decapsulation/Delid of devices
- B. Probe and isolation of electrical defects
- C. Layer-by-layer removal of device levels by selective etching
- D. Microsection analysis
  1. Sections taken at shallow to 90° angles — sample sizes to 1.5 inches
  2. Selective staining to delineate diffusions, dielectrics, etc.
  3. Thickness measurements by SEM or optical microscopy
- E. Optical microphotography — magnifications to 5000X
- F. Infrared microscopy — transmission and reflection
- G. Nanometrics
- H. Planar plasma etching
- I. Scanning electron microscopy — SEM
  1. Routine magnification to 50,000X
  2. 50-Å resolution
  3. Back-scattered electron detector
  4. Military product lot acceptance of metallization
  5. Voltage contrast
  6. Specimen current amplifier
- J. Electron microprobe
  1. Chemical detection of elements with atomic number greater than 11.
  2. Typical 4- to 5-μm beam penetration
  3. Spot size typically 1000 to 2000 Å
- K. Auger spectroscope
- L. Ion microprobe mass analysis
- M. Gas and/or plastic composition analysis

QUALITY AND RELIABILITY

14

LS600 RELIABILITY DATA

INTRODUCTION

Texas Instruments designs and builds quality and reliability into all the products that it offers in the electronic marketplace. The quality control organization is uniquely responsible for coordinating the total effort and for providing direct action necessary to assure that quality and reliability objectives are met. Accordingly, quality control reaches from raw material inputs to evaluation of finished goods as evidenced by the many inspections and tests shown on the typical light sensor flow diagram in Figure 1.

The reliability data shown in this report is based on extensive tests performed by Texas Instruments to assure continued leadership in optical sensor quality and reliability. More than 42,200 units have been subjected to life test with an accumulation of over 39,000,000 device hours. The data is complete, representing all devices produced during the years 1966 through 1982. The tests were performed on ungraded, unburned-in devices and are typical of TI sensor products.

OPERATING LIFE TESTS

The 25°C life tests were performed with incident light intensity adjusted for power dissipation of each device of 50 milliwatts at 10 volts V_{CE}. Readings of dark current (I_D) and light current (I_L) were made at 0, 250, 500, and 1000 hours. Failure criteria were 0.2 μA maximum for I_D and 20% degradation of limits for I_L. A total of 3210 sensors were tested to these criteria with 6 failures. These samples were taken from lots whose total count exceeded 1,050,000 sensors. Data from these tests are shown in Figure 2.

The 55°C life tests were performed with incident light intensity adjusted for power dissipation on each device of 50 milliwatts at V_{CE} = 10 V. Readings of dark current (I_D) and light current (I_L) were made at 0, 168, and 1000 hours. failure criteria were 0.2 μA maximum for I_D and ±40% change in I_L within original specification limits. A total of 16,810 units were tested to these criteria with 55 failures. These samples were taken from lots whose total count exceeded 9,000,000 sensors. Data from these tests are shown in Figure 3.

The long-term reliability of the LS600 sensor is demonstrated by the plots shown in Figure 6 and Figure 7. The data is completely representative of all tests conducted during the reporting period. The projected degradation limits are based upon the exponential distribution of failure. Extended tests performed on small samples confirm that the degradation is within the limits as shown.

ENVIRONMENTAL TESTS

The tests listed in Figure 8 were performed on samples of the product with the catastrophic or degradation failures as shown. It must be pointed out that test conditions shown are not the ultimate strength of the product but represent the test requirements imposed by our customers. The ultimate strength of these devices is much higher in most cases. Inquiries concerning response to particular requirements should be addressed to your TI sales representative.

HIGH-TEMPERATURE REVERSE BIAS

Devices are stored in dark ovens at 150°C with 45 volts applied for 1000 hours. Readings of dark current (I_D), breakdown voltage ($V_{(BR)CEO}$), and light current (I_L) were made at 0, 168, and 1000 hours. Failure criteria were 0.2 μA maximum for I_D and 60% degradation within original limits for I_L. A total of 17,723 units were tested to these criteria with 75 failures. These samples were taken from lots whose total count exceeded 9,000,000 sensors. Data from these tests are shown in Figure 4.

STORAGE LIFE TESTS

Devices were stored in ovens at 150°C for 500 and 1000 hours (depending upon requirements). Readings of dark current (I_D) and light current (I_L) were made at 0, 168, and 1000 hours. Failure criteria were 0.2 μA maximum for I_D and 20% degradation of limits for I_L. A total of 5200 units were tested to these criteria with 6 failures. These samples were taken from lots whose total count exceeded 1,000,000 sensors. Data from these tests are shown in Figure 5.

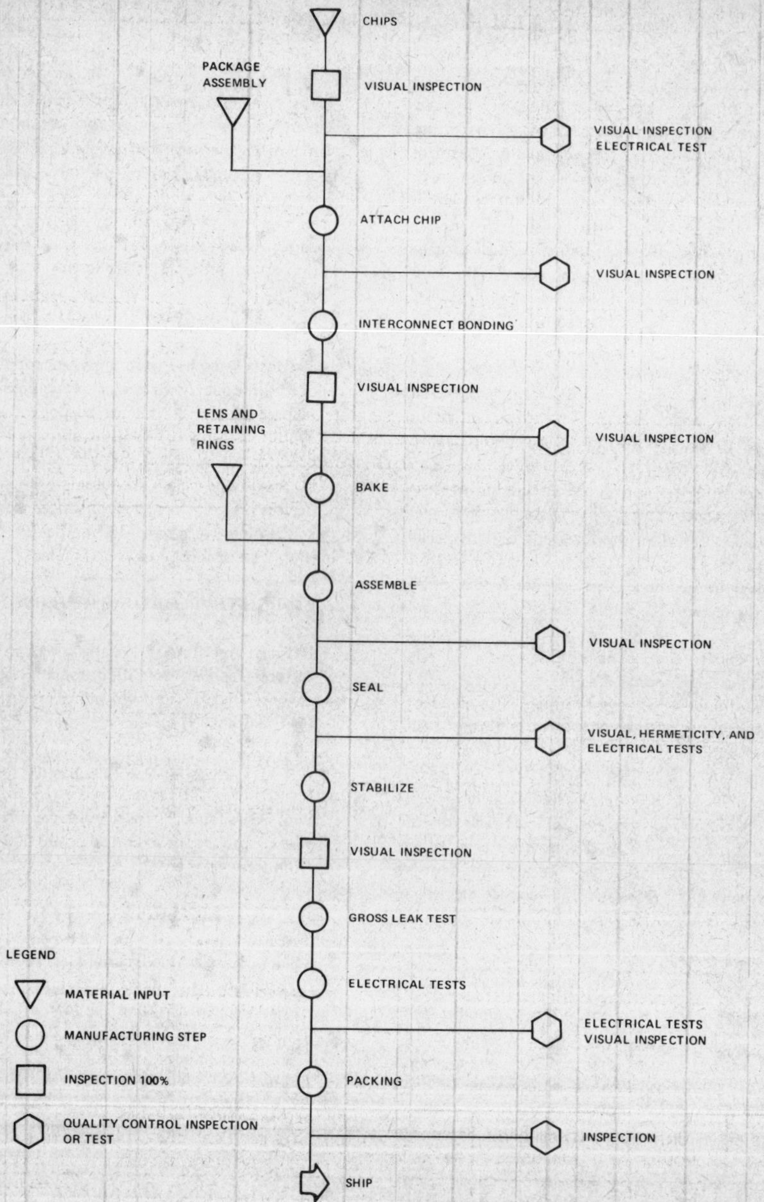

FIGURE 1. Typical Light Sensor Flow Diagram

TEXAS INSTRUMENTS
INCORPORATED
POST OFFICE BOX 225012 • DALLAS, TEXAS 75265

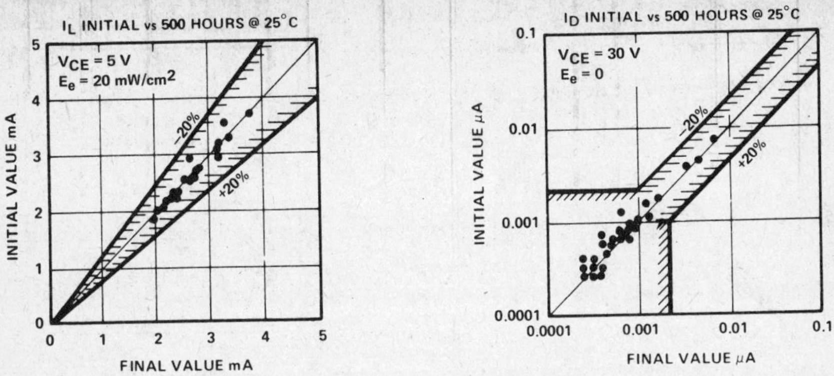

UNITS TESTED	UNIT HOURS	CATASTROPHIC FAILURES	DEGRADATION FAILURES			
			TOTAL	FAILURE RATE IN %/1,000 HOURS		MEAN TIME BETWEEN FAILURES
				60% CONFIDENCE	90%CONFIDENCE	
3210	2,847,000	0	6	0.20	0.33	390,000 HOURS

FIGURE 2. Operating Life at 25°C

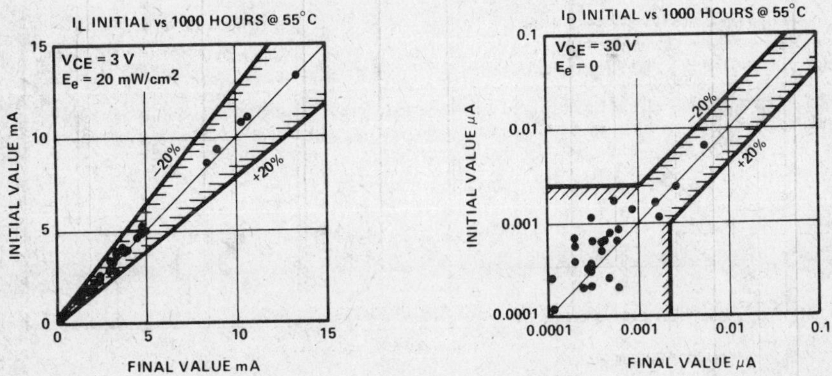

UNITS TESTED	UNIT HOURS	CATASTROPHIC FAILURES	DEGRADATION FAILURES			
			TOTAL	FAILURE RATE IN%/1,000 HOURS		MEAN TIME BETWEEN FAILURES
				60% CONFIDENCE	90%CONFIDENCE	
16,810	16,810,000	0	55	0.35	0.39	286,000 HOURS

FIGURE 3. Operating Life at 55°C.

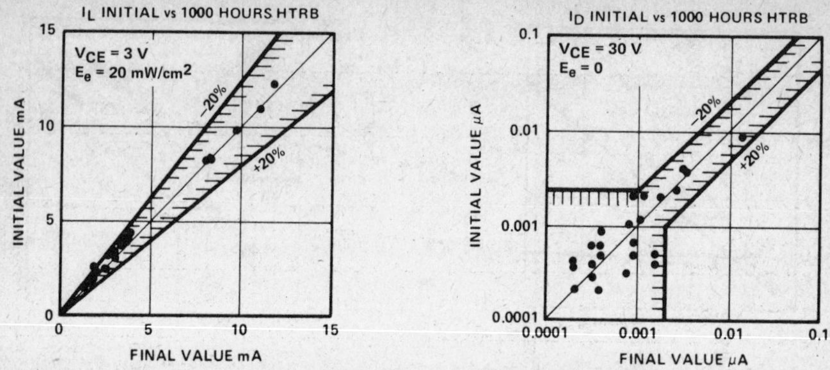

UNITS TESTED	UNIT HOURS	CATASTROPHIC FAILURES	DEGRADATION FAILURES			
			TOTAL	FAILURE RATE IN %/1,000 HOURS		MEAN TIME BETWEEN FAILURES
				60% CONFIDENCE	90%CONFIDENCE	
17,723	17,725,000	0	75	0.44	0.48	227,000 HOURS

FIGURE 4. High-Temperature Reverse Bias

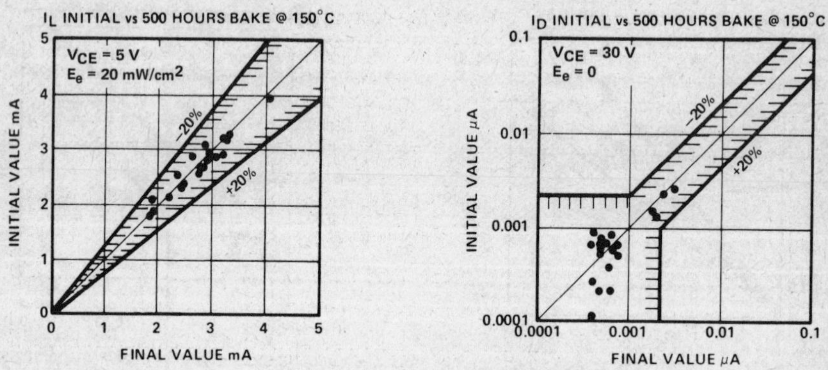

UNITS TESTED	UNIT HOURS	CATASTROPHIC FAILURES	DEGRADATION FAILURES			
			TOTAL	FAILURE RATE IN%/1,000 HOURS		MEAN TIME BETWEEN FAILURES
				60% CONFIDENCE	90%CONFIDENCE	
5300	4,143,200	0	5	0.12	0.19	828,640 HOURS

FIGURE 5. High-Temperature Storage

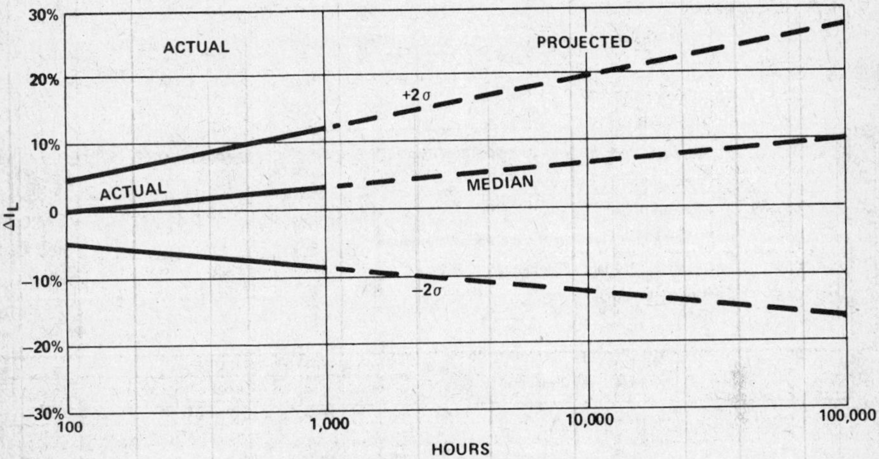

FIGURE 6. % ΔI_L vs Operating Life at 25°C

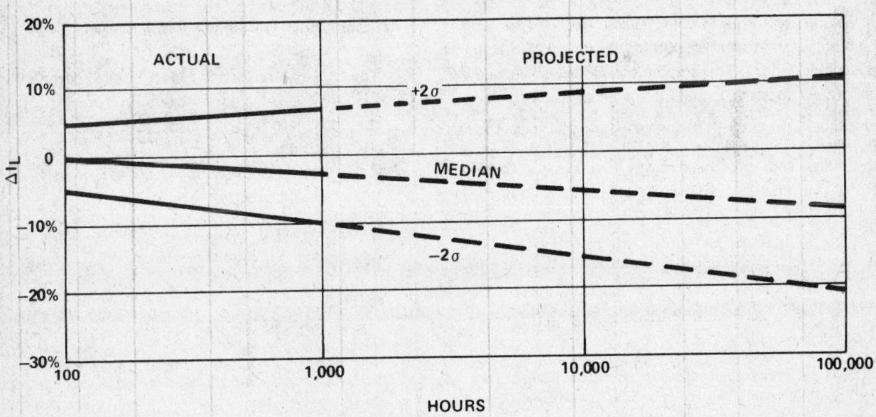

FIGURE 7. % ΔI_L vs Operating Life at 55°C

MIL-STD-750 Test Method	Test	Quantity Tested	Failures (Catastrophic or Degradation)
2026	Solderability: 240°C, 3 Minutes	126	0
1051	Temperature Cycle: 5 Cycles, 30 Min., −65 to +125°C	126	0
1051	Temperature Cycle −40°C to 100°C, 5 Cycles, 30 Minutes	17,400	8
1056	Thermal Shock: 5 Cycles	126	0
1021	Moisture Resistance	126	0
2016	Shock, Impact: 1000 g, 5 Each Axis, 0.5 millisecond	126	0
2056	Vibration, Variable Frequency: 10g	24,606	7
2046	Vibration Fatigue: 10g	126	0
2006	Constant Acceleration: 10kg, 1 Min.	126	0
1001	Barometric Pressure: 15mmHg, 45 V	126	0
1071	Hermetic Seal: Test Condition E	22,750	0

FIGURE 8. Environmental Test Results

TIL604HR HIGH-RELIABILITY PHOTOTRANSISTOR

Texas Instruments has always been known as a producer of high-quality products, and the LS600 series phototransistor is no exception as evidenced by the testing of more than 31,000 units with an accumulation of over 30,000,000 hours without a catastrophic failure. This small pill package, developed by Texas Instruments, has an excellent record for reliability over more than 10 years in military and aero-space applications. Utilizing the expertise, techniques, and processes developed during these years of building the LS600 phototransistors to high-reliability customer specifications, Texas Instruments now offers the TIL604HR2 as a standard high-reliability device to customers requiring extra reliability in their applications.

The phototransistors and the complementary TIL24HR2 infrared emitters are now available as standard product items. For more information, contact your nearest TI sales representative, or Optoelectronics Department Product Marketing.

TEXAS INSTRUMENTS
INCORPORATED
POST OFFICE BOX 225012 • DALLAS, TEXAS 75265

TIL23, TIL24 RELIABILITY DATA

INTRODUCTION

Texas Instruments has long been noted as a quality producer of semiconductor components. The TIL23 and TIL24 solid-state infrared-emitting diodes (IRED's) are high-quality, reliable additions to its line of optoelectronic products. They have been designed as highly reliable, long-life products capable of meeting demanding military and commercial needs. Quality control of these products begins with incoming inspection of raw materials and is continued throughout the manufacturing process as shown in assembly-test flow diagram (Figure 1). Conscientious quality control practiced by the manufacturing organization and monitored at critical steps by the quality control organization ensures that the designed reliability will be achieved in the finished product.

Since this product was announced in 1970, some five million device hours of reliability testing have been accumulated on ungraded, unburned-in samples, and additional data is continuously being accumulated. This report summarizes, in graphical form, data on the operating life of TIL23 and TIL24 at 10, 30, and 50 mA at 25°C and 50 mA at 55°C. Results of various mechanical and temperature stress tests are also presented.

OPERATING LIFE TESTS

Room temperature (25°C) life tests were performed at three different current levels: 10 mA, 30 mA and 50 mA. Readings of power output were made with a solar cell in a short-circuit current mode at 0, 168, 500 and 1,000 hours. Forward voltage was read at these intervals and no significant changes were observed. Extended operating life tests at 25°C (4,000 hours) on 300 units have substantiated the extrapolated degradation rates shown in Figures 2, 3, 4 and 5.

Since 1976 3,792,000 device hours have been accumulated on samples (see Figures 6 and 7) operated for 1000 hours. Failure criteria were degradation of output power of more than 50% or a change in V_F of more than 5%. Readings were taken at 0, 168, and 1000 hours. The samples were taken from lots whose total count exceeded three million LED's.

STORAGE LIFE TESTS

High-temperature (85°C) storage tests were performed for 1000 hours on 3312 devices (see Figure 8). Only one unit (ΔV_F = 19%) exceeded the failure criteria of 50% degradation of output power or 5% change in V_F.

ENVIRONMENTAL TESTS

The tests listed in Figure 10 were performed on samples of the product with the catastrophic failures as shown. It should be noted that the test conditions shown are not the ultimate strength of the product but represent the test requirements imposed by our customers. The ultimate strength of these devices is much higher in most cases. Inquiries concerning response to particular requirements should be addressed to your TI sales representative.

TIL24HR2, TIL31BHR2. . .HIGH-RELIABILITY INFRARED EMITTERS

Texas Instruments now offers the TIL24HR2 and TIL31BHR2 as standard product items to customers requiring extra reliability in their applications. The TIL24HR2 and TIL31BHR2 are used to provide dependable and reliable infrared sources in military and aerospace applications. The TIL24HR2 and TIL31BHR2 infrared emitters and the complementary TIL81HR2 and TIL604HR2 phototransistors are now available as standard product items. For more information, contact your nearest TI sales representative or Optoelectronic Department Product Marketing.

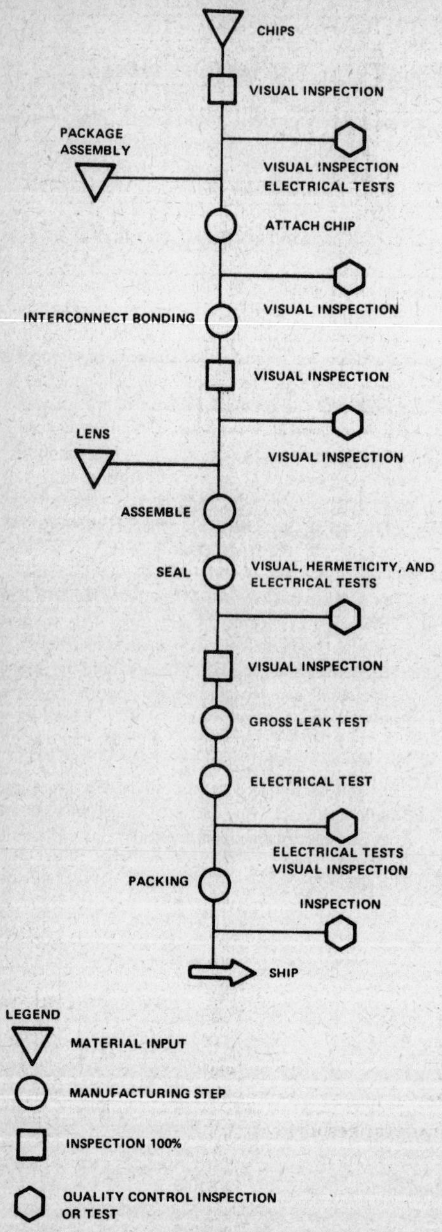

FIGURE 1. TIL23 AND TIL24 INFRARED-EMITTER FLOW DIAGRAM

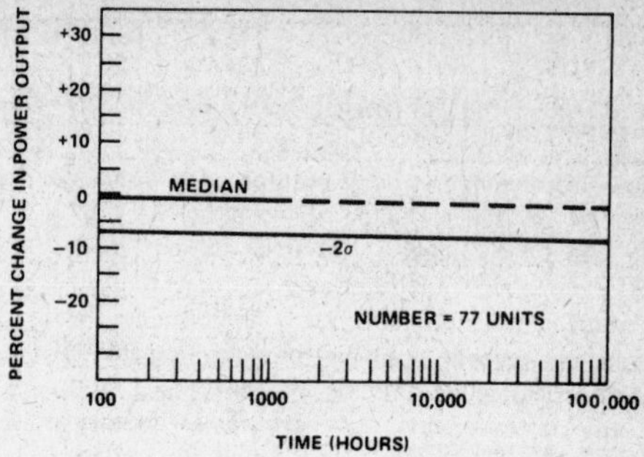

FIGURE 2. CHANGE IN POWER OUTPUT AS A FUNCTION OF OPERATING TIME AT I_F = 10 mA, 25°C

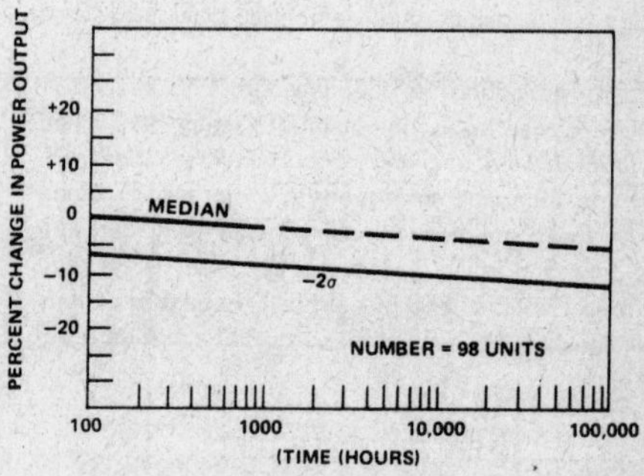

FIGURE 3. CHANGE IN POWER OUTPUT AS A FUNCTION OF OPERATING TIME AT I_F = 30 mA, 25°C

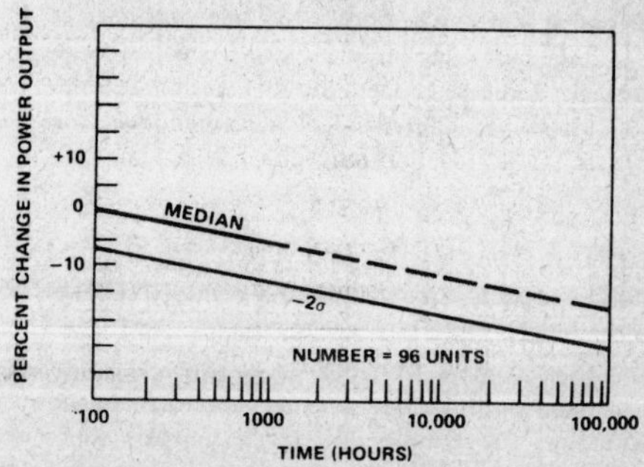

FIGURE 4. CHANGE IN POWER OUTPUT AS A FUNCTION OF OPERATING TIME AT I_F = 50 mA, 25°C

TEXAS INSTRUMENTS
INCORPORATED
POST OFFICE BOX 225012 • DALLAS, TEXAS 75265

UNITS TESTED	UNIT HOURS	CATASTROPHIC FAILURES	DEGRADATION FAILURES			
			TOTAL	FAILURE RATE IN %/1,000 HOURS		MEAN TIME BETWEEN FAILURES
				60% CONFIDENCE	90% CONFIDENCE	
3360	3,360,000	0	1	0.09	0.21	1,111,000 HOURS

FIGURE 5. OPERATING LIFE AT 25°C AND 50 mA

UNITS TESTED	UNIT HOURS	CATASTROPHIC FAILURES	DEGRADATION FAILURES			
			TOTAL	FAILURE RATE IN %/1,000 HOURS		MEAN TIME BETWEEN FAILURES
				60% CONFIDENCE	90% CONFIDENCE	
432	432,000	0	2	0.72	1.23	140,000 HOURS

FIGURE 6. OPERATING LIFE AT 25°C AND 75 mA

UNITS TESTED	UNIT HOURS	CATASTROPHIC FAILURES	DEGRADATION FAILURES			
			TOTAL	FAILURE RATE IN %/1,000 HOURS		MEAN TIME BETWEEN FAILURES
				60% CONFIDENCE	90% CONFIDENCE	
3360	3,360,000	0	3	0.17	0.20	588,000 HOURS

FIGURE 7. OPERATING LIFE AT 55°C AND 50 mA

UNITS TESTED	UNIT HOURS	CATASTROPHIC FAILURES	DEGRADATION FAILURES			
			TOTAL	FAILURE RATE IN %/1,000 HOURS		MEAN TIME BETWEEN FAILURES
				60% CONFIDENCE	90% CONFIDENCE	
3312	3,312,000	0	1	0.06	0.12	1,666,000 HOURS

FIGURE 8. STORAGE LIFE AT 85°C

MIL-STD-750 Test Method	Test	Quantity Tested	Failures (Catastrophic or Degradation)
1051	Temperature Cycle: 5 Cycles, 30 Minutes, −40°C to +100°C	828	0
1051	Temperature Cycle: 5 Cycles, 30 Minutes, −65°C to +150°C	50	0
1056	Thermal Shock: 5 Cycles, Condition A	50	0
1021	Moisture Resistance	50	0
2016	Shock, Impact: 1500 g, Z_1 Axis, 0.5 milliseconds	1896	2
2056	Vibration, Variable Frequency: 20 g	1656	4
2006	Constant Acceleration: 20 kg, 1 Min. Z_1	146	0
1071	Hermetic Seal: Test Condition E	390	1

FIGURE 9. ENVIRONMENTAL TEST RESULTS

Applications

- Application Summary
- Optoelectronic and Robotic Applications
- Low-Voltage Monitor
- Indicator of Analog Quantities
- Fluid-Level Indicator
- Voltage-Level Indicator
- Pulse Generated by Interrupting Light Beam
- Multiplexing Displays
- TIL311 Hexadecimal LED Display
- Counting Circuits Using TIL306 and TIL308 Displays
- Optocouplers in Circuits
- Interfacing Using Optocouplers
- Bar-Code Scanner

APPLICATIONS SUMMARY

OPTOELECTRONICS IN ROBOTIC APPLICATIONS

A robot system is a good example of a system that can apply many optoelectronic devices. The block diagram of Figure 1 illustrates functions that must be considered when designing a robot system. These functions include displays to provide information to the operator for the operation of the system as well as indicators to alert the operator of any system problems. Optical sensors provide information to the computer about the position, velocity, and acceleration of the moving parts of the robot. Vision can be included to provide information to the computer about the location of parts and/or the condition of the parts that the robot must handle. Tactile feedback of some form may be necessary so that the robot does not apply excessive force to the parts it must move. The data from the sensors is transmitted to the computer to provide the necessary control.

LEDs (green or red) are good choices to alert operators to problems, to use as on/off indicators, and to display binary system status. Seven-segment displays (single or multidigit), hexadecimal displays, and alphanumeric displays are available to provide communication from the computer to the operator about more detailed status of the robot. Sensor/detector arrays of various types can be used with encoder discs and reflective surfaces as shown in Figure 1 to provide position information, velocity information, and to serve as limit switches to prevent excessive range of motion. Optoisolators (optocouplers) can be used with fiber-optic links to transmit data both to the robot and to the control system. Such couplers provide electrical isolation between the high-power circuits necessary to drive the robot and the low-power circuits used in the computer or microprocessor.

An interfacing scheme for a simple pick and place robot is shown in Figure 2. This system uses a TMS9900-101M microcomputer coupled to a terminal, memory expansion, and I/O interface. Robot inputs are from sensors represented in Figure 2 by switches SW0-SW11. The output from the CPU to the robot is to solenoids SOL0-SOL11, which represent relays or stepping motors.

Figure 3 shows the output coupling from the 9901 (Programmable Systems Interface) IC to the solenoids. The TIL119 optocoupler provides isolation from the higher voltage required to operate the solenoids as well as isolation from the noise generated by the magnetic field when the solenoid is turned on and off. The values of resistors and choice of transistors will depend on the size of the solenoid and can be calculated using standard design procedures.

Figure 4 is the input coupling needed from the robot to the interface IC. The optocoupler again provides isolation between the robot's high-power circuits and the low-power 9901 interface IC. R1 and R2 are chosen for the logic level and forward current desired. The switch SW0 can be the output from a sensor detector array such as TIL143. The phototransistor of the TIL143 can be connected directly in place of the TIL119.

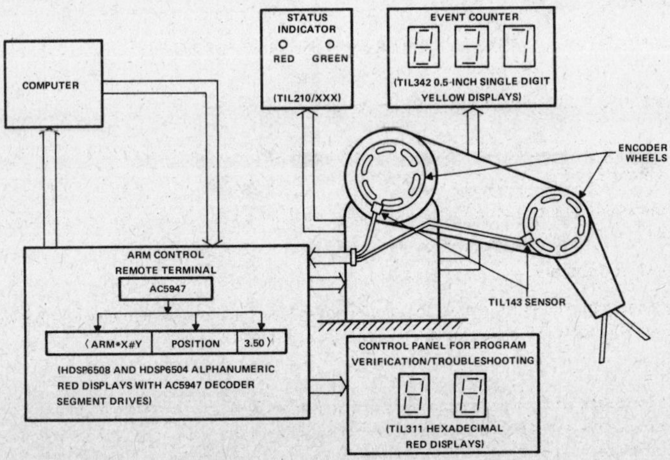

Figure 1. Robot Arm Application

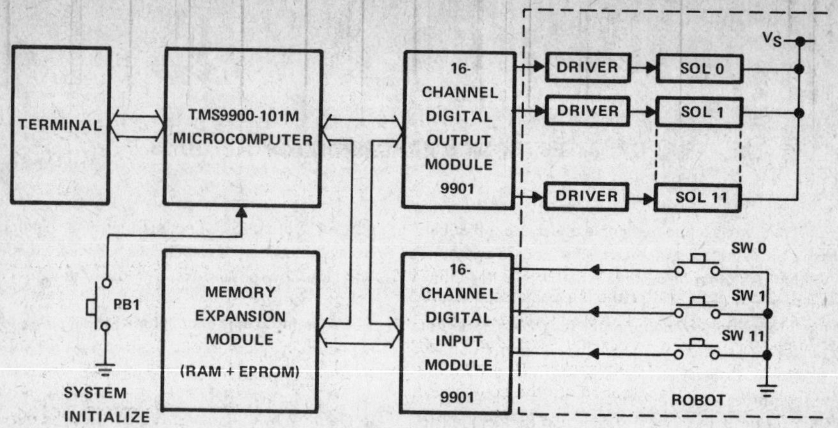

Figure 2. Interface for Pick and Place Robot

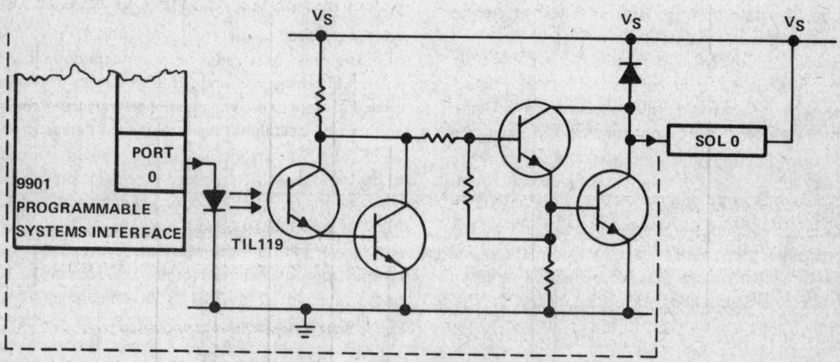

Figure 3. Output Coupling from 9901 to Solenoid

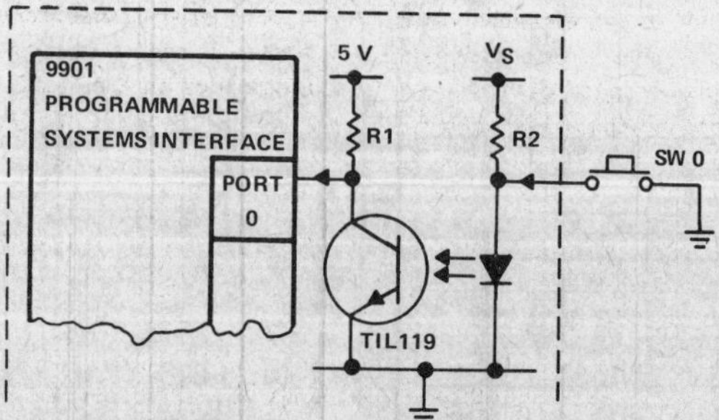

Figure 4. Input Coupling from Limit Switch to 9901

LOW-VOLTAGE MONITOR

description

Figure 1a gives a circuit that can be used to monitor the voltage of a battery and warns the operator when the battery voltage is below a preset level by turning on an LED. In the circuit shown, the values are set for a 12-volt automobile battery. The preset level is 10 volts. The important devices in this application are the comparator, LM239, and the LED, TIL209A. The comparator is a device that senses two different voltages and provides an output that is either low voltage or high voltage, depending on the relative size of the two input voltages. One of the inputs is called the noninverting input and is designated by a plus sign, while the other is the inverting input and is designated by a minus sign. If V+ is more positive than V−, the output is high. If V− is more positive than V+, the output is low. Figure 1b shows how the output switches with V− set at +5 volts. If V+ is less than +5 volts, the output voltage is low. If V+ is greater than +5 volts, the output voltage is high.

In the circuit shown in Figure 1a, the reference voltage at the inverting (−) input of the LM239 comparator is set by a 5-volt zener diode. R1 and R2 are used as a voltage divider to provide one-half the battery voltage to the noninverting (+) input. When the battery voltage decreases to less than 10 volts, the voltage at the noninverting input goes below 5 volts, which is less than the reference voltage. This causes the output of the amplifier to switch from a high level to a low level, which turns on the LED. Current through the LED is limited by R4. By using eight such circuits and selecting appropriate values of R1 and R2 for each, a battery status display could be made to indicate the battery voltage in one-volt steps from 7 volts to 15 volts. If the number of levels to be detected is more than two, it probably is less expensive and possibly more reliable to use complete packaged analog level detectors because of the single integrated circuit and the reduced number of components.

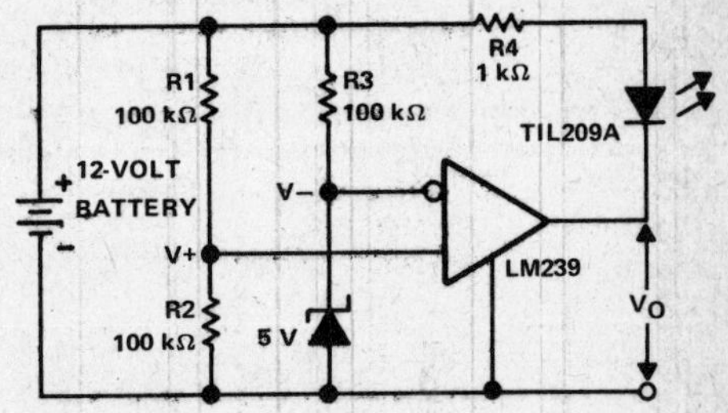

a. SCHEMATIC OF CIRCUIT FOR LOW-VOLTAGE INDICATOR

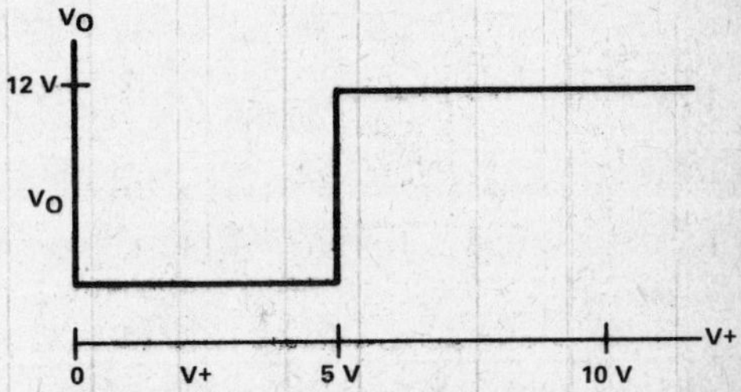

b. TRANSFER CURVE FOR LM239 WITH V − EQUAL TO +5 V

(Source: L.B. Masten and B.R. Masten, *Understanding Optronics*, Texas Instruments Incorporated, 1981.)

Figure 1. Low-Voltage Monitor

Application

Any battery operated circuit that is critical can use such a monitor circuit to signal the fact that preventive maintenance must be performed.

INDICATOR OF ANALOG QUANTITIES

The circuit shown in Figure 1 uses TI's TL500C and TL502C analog-to-digital (A/D) converter integrated circuits to provide an accurate display of an analog input voltage. With the input circuit shown, the display provides a readout proportional to the intensity of the light incident on the phototransistor. Since the TIL81 responds to radiation in the range of 500 to 1100 nanometers, while the human eye responds to radiation in the range of 400 to 700 nanometers, it is responding more to infrared than to the visible spectrum.

If input circuit number 1 is used as an alternate input circuit, the display is an accurate thermometer that can be calibrated to display either degrees Celsius or degrees Fahrenheit.

The alternative input circuit number 2 can be used to provide an output that is proportional to milliamperes of input current. At the same time, or it might be an independent requirement, electrical isolation is obtained between the input transducer and the display unit.

Two parts are required for the complete integrated circuit A/D converter, an analog processor and a digital signal controller. The analog processor is controlled by digital signals from the controller to provide the basic function for a dual-slope integrating A/D converter.

The analog processor (TL500C and TL501C) contains the necessary analog switches and decoding circuits, a reference voltage generator, a buffer, an integrator, and a comparator. The easiest way to complete the A/D converter is to use the matching digital processors (TL502C, TL503C) but the analog processor can also be controlled by discrete logic circuits or a microprocessor that has been programmed with the proper routine.

Each TL502C and TL503C includes oscillator, counter, control logic, and digit enable circuits. The TL502C provides multiplexed outputs for seven-segment displays, while the TL503C has multiplexed BCD outputs to couple directly to other computer circuits, or to displays requiring BCD code.

The TL500C and TL501C analog processors are designed to automatically compensate for internal zero offsets, to integrate a differential voltage at the analog inputs, to integrate a voltage at the reference input in the opposite direction, and to provide an indication of zero-voltage crossing.

The TL500C provides 4 1/2-digit readout accurately when used with a precision external reference voltage. The TL501C provides 100-ppm linearity error and 3 1/2-digit accuracy capability. These devices are manufactured using TI's advanced technology to produce JFET, MOSFET, and bipolar devices on the same chip. The TL500C and TL501C are intended for operation over the temperature range of 0 °C to 70 °C.

When the analog processor and digital processor are used together they provide an A/D converter that has automatic zero-offset compensation, true differential inputs, high-impedance inputs, and outputs that can drive up to 4 1/2 digits of display. As a result, the A/D converter is a very versatile circuit combination for converting analog inputs from high-impedance sensors of light intensity, pressure, temperature, moisture, and position and can be used to provide display and control signals for weight scales, industrial controllers, thermometers, light level indicators, etc.

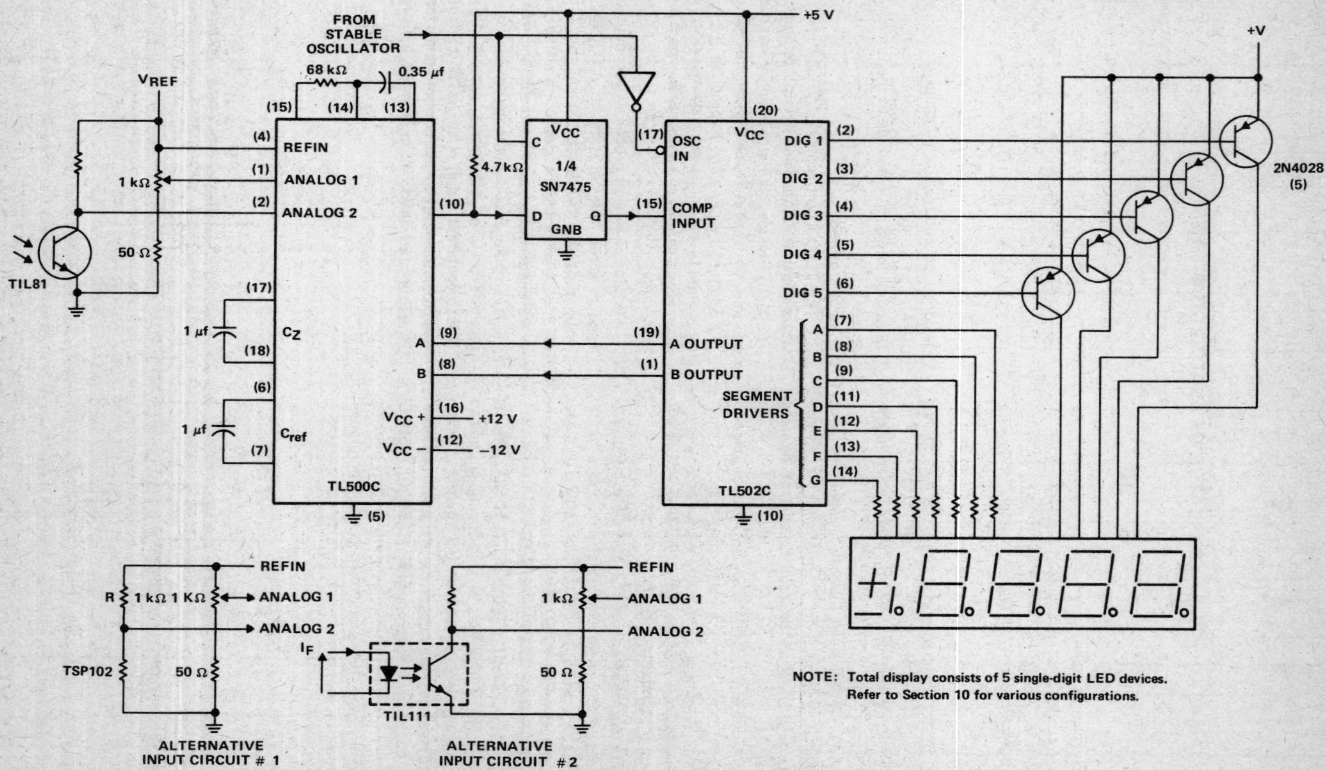

NOTE: Total display consists of 5 single-digit LED devices.
Refer to Section 10 for various configurations.

Figure 1.

FLUID LEVEL CONTROLLER

description

Figure 1 shows a typical circuit that can be used to maintain fluid between two levels. The timing diagram is shown in Figure 2. When the fluid drops below level B, detector Q2 provides a current that produces a positive-going voltage pulse to the noninverting input of comparator 2. Comparator 2 produces a high-logic-level voltage at the set input of the RS flip-flop consisting of two cross-coupled NOR gates. This sets the output of the flip-flop to a high-logic-level voltage. When the flip-flop output goes to the high-logic-level voltage, the buffer amplifier passes a control signal to turn on the pump and the fluid level begins to rise. As soon as the level goes above B, detector Q2 is turned off, the comparator switches back, and the set input to the flip-flop goes low. However, the flip-flop output remains at a high-logic-level voltage because the reset input is also at a low-logic-level voltage. When the fluid level reaches level A, detector Q1, which has been producing photocurrent, turns off and provides a negative-going voltage pulse at the inverting input of comparator 1. As a result, the output voltage level of comparator 1 goes to a high-logic-level voltage. This resets the RS flip-flop and the flip-flop output goes to a low-logic-level voltage and the resulting control signal turns off the pump. When the fluid level drops below level A, detector Q1 again produces photocurrent to raise the voltage at the inverting input of comparator 1 above the reference. The output of comparator 1 and the reset input of the flip-flop go to a low-logic-level voltage. However, because the set input is low also, the flip-flop output remains at the low-logic-level voltage. Therefore, the pump remains off until the fluid level drops below level B again.

The parallel RC circuit connecting the emitters of Q1 and Q2 to the comparators acts as a differentiating circuit that couples sharp changes to the comparators, but filters out slow changes due to varying ambient light conditions.

applications

Variations on this control circuit can be made to keep something that moves within certain boundary conditions. An elevator is a good example.

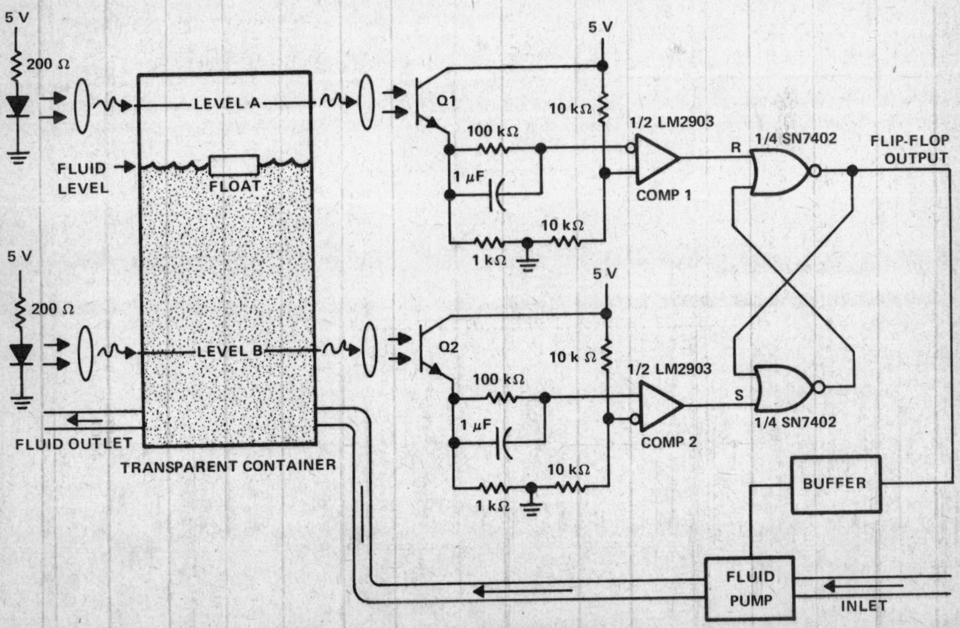

Figure 1. Fluid Level Controller

APPLICATIONS 15

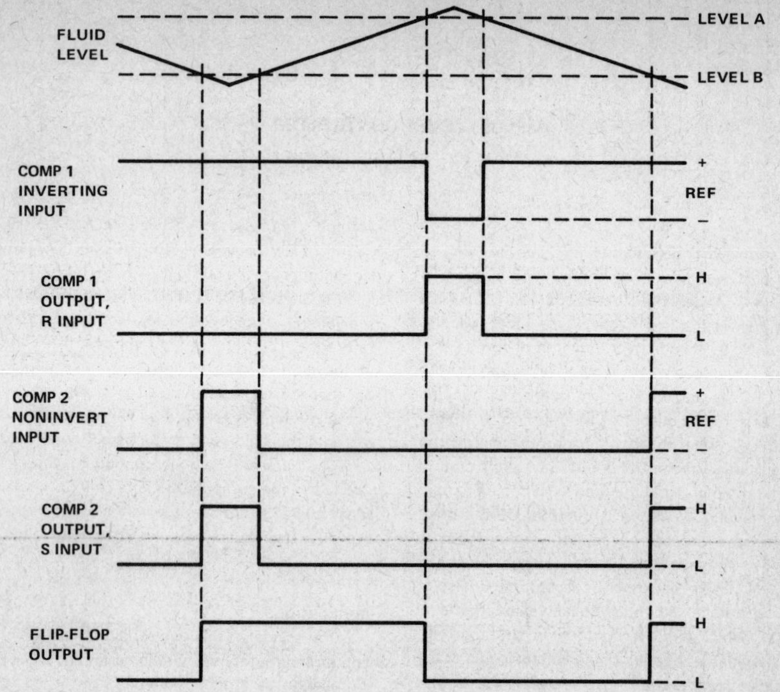

Figure 2. Timing Diagram

VOLTAGE-LEVEL INDICATOR

description

The circuit in Figure 1 provides a visual indication of the input analog voltage level. The circuit uses a type TL489C 5-step analog level detector, which has a high input impedance at pin 8 and open-collector outputs capable of sinking up to 40 milliamperes. This makes it suitable for driving a linear array of 5 LEDs to indicate the level is 5 steps. The voltage at the analog input should be scaled by the input potentiometer R2 and the series resistor R1 to ensure that the voltage at pin 8 is in the range of zero to approximately one volt and should never exceed eight volts. The TL489C operates from a standard 5-volt V_{CC} and the LEDs are supplied from V+. If a logarithmic display is desired the TL487C 5-step logarithm analog level detector can be substituted for the TL489C. LEDs 1, 2, 3, 4, and 5 will turn on sequentially as the input voltage at pin 8 of the TL489C increases by 200-millivolt steps. LED 1 will also flash periodically when the input voltage at pin 8 is less than 200 millivolts. The resistor value for R3 is selected by:

$$R3 = \frac{V+ - V_{LED} - V_{OL}}{I_{LED}}$$

$$R3 = \frac{V+ - V_{LED} - 0.5\ V}{I_{LED}}$$

where $V_{OL} = 0.5$ V.

calculations

For an I_{LED} current of 10 mA, $V_{LED} = 1.6$ V and with V+ = 12 V, the resistor value would be:

$$R3 = \frac{(12 - 1.6 - 0.5)\ V}{10\ mA} = \frac{9.9\ V}{10\ mA}$$

9.9 V is essentially 10 V, therefore,

$$R3 = \frac{10}{10 \times 10^{-3}}$$

$$R3 = 1\ k\Omega$$

further discussion

When the analog input is less than 200 millivolts, output 2 of TL489C is off. Under these conditions, C1 charges to V+ through R4, the LED, and R3. R4 is chosen to make

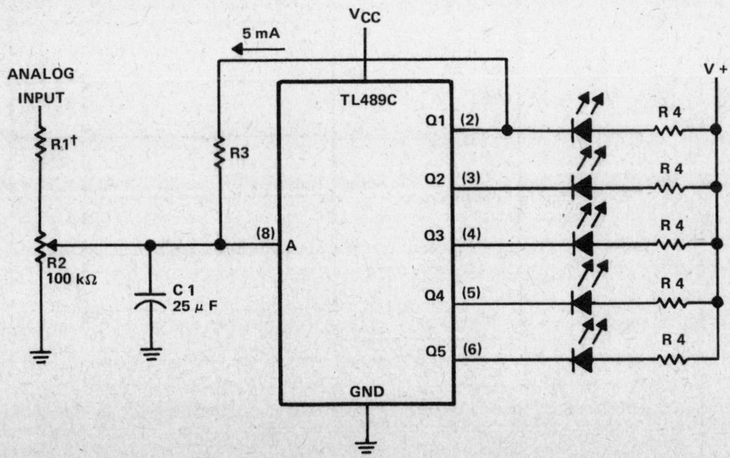

† R1 is chosen to ensure that the voltage across R2 is less than 8 volts. Normally it will be set to 1 volt.

Figure 1. 5-Step Voltage-Level Indicator

APPLICATIONS

15

the charging current equal to 5 milliamperes when the voltage across C1 is zero. Therefore,

$$5\ \text{mA} = \frac{V+ - V_{LED}}{R3 + R4}$$

$$R3 + R4 = \frac{12 - 1.6}{5\ \text{mA}}$$

$$R4 = \frac{12 - 1.6}{5\ \text{mA}} - R3$$

$$R4 = \frac{10.4}{5\ \text{mA}} - 1\ \text{k}\Omega$$

$$R4 = 2.08\ \text{k}\Omega - 1\ \text{k}\Omega$$

Subtracting and using a standard value,

$$R4 = 1\ \text{k}\Omega$$

When C1 charges above 200 millivolts, the TL489C output 2 is pulled down to $V_{CE(sat)}$ and discharges C1 below the input threshold and the TL489C output 2 is turned off. As a result, C1 starts to charge again and the cycle repeats causing the LED in output 2 to flash. When the input is well above 200 millivolts, output 2 is always on at $V_{CE(sat)}$ and C1 remains discharged.

a 10-step indicator

A TL490C can be used to provide a 10-step display of the voltage level as shown in Figure 2. The TL490C is a 10-step adjustable analog level detector that is also capable of sinking up to 40 milliamperes at each output. The voltage range at the input pin should range from 0 to 2 volts.

Circuits of this type are useful as liquid-level indicators, pressure indicators, and temperature indicators. They may also be used with a set of active filters to provide a visual indication of harmonic content of audio signals.

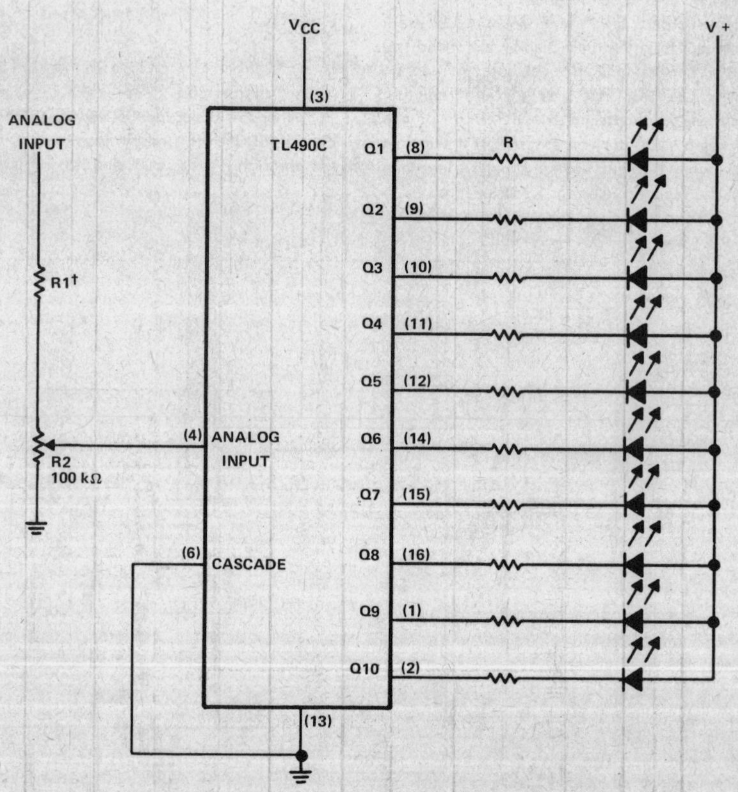

Figure 2. 10-Step Voltage-Level Indicator

†R1 is chosen to ensure that the voltage across R2 is less than 8 volts. Normally it will be set to 1 V.

APPLICATIONS 15

PULSE GENERATED BY INTERRUPTING A LIGHT BEAM

description

The circuit of Figure 1 is designed to put out a pulse when an object on the conveyor belt blocks the light source. The light source is a tungsten lamp (at 2870 K), which will keep the TIL81 phototransistor turned on. This produces a high-logic-level voltage at pin 1 of the SN7414 Schmitt-trigger inverter and a TTL-compatible low logic level at pin 5 of the monostable multivibrator SN74121. When an object blocks the light, the TIL81 turns off causing the Schmitt-trigger inverter to trigger the one-shot SN74121. The R_{ext} and C_{ext} shown in Figure 1 are selected to give an output pulse of 100-microsecond duration when B (pin 5) is triggered by a positive-going pulse. Other values of R and C for pulse durations from 50 nanosconds to 10 milliseconds are shown in Figures 3 and 4. The SN7414 Schmitt-trigger provides pulses with steep transistions from slowly varying waveforms that are input from the TIL81.

application

As shown in Figure 1, the circuit can be used in many manufacturing operations where the primary function can be triggered by interrupting a light beam counting objects as they move down a conveyor, triggering a drilling or insertion action after a part has moved into place, or opening sorting bins or closing sorting bins after a part has moved in place.

Opening doors when people walk onto a surface pad, detecting when an object moves into a light beam, and sensing opaque surfaces over transparent surfaces are a few more typical types of applications.

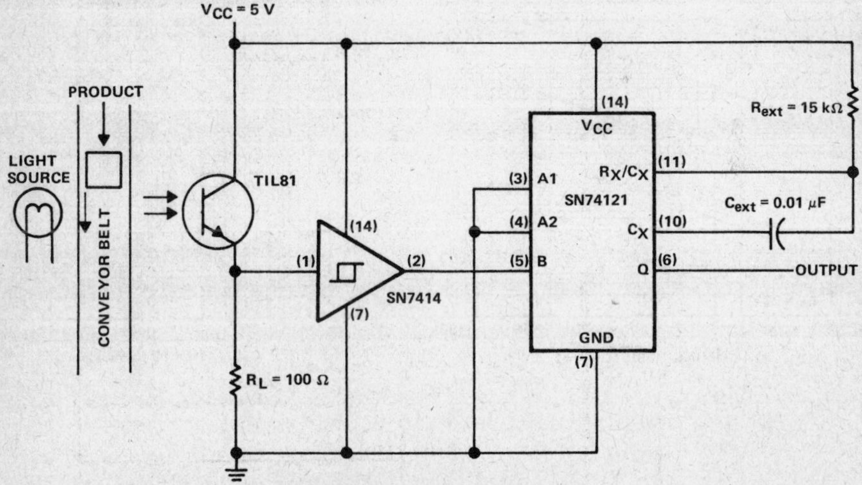

Figure 1. Circuit Schematic

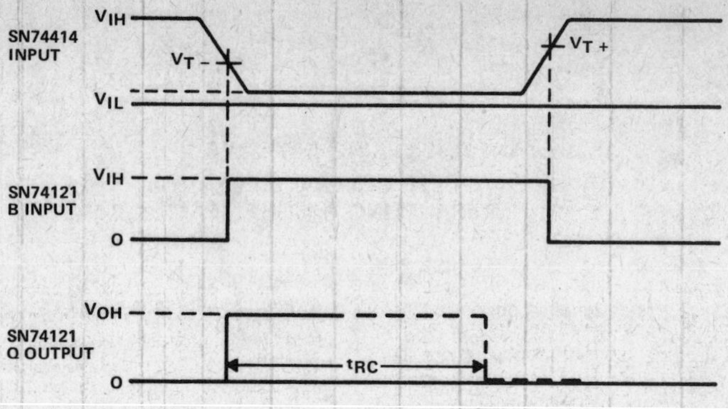

t_{RC} DEPENDS ON VALUE OF R_{ext} AND C_{ext} CHOSEN FROM FIGURE 3 AND 4

Figure 2. Waveforms

OUTPUT PULSE WIDTH
vs
TIMING RESISTOR VALUE

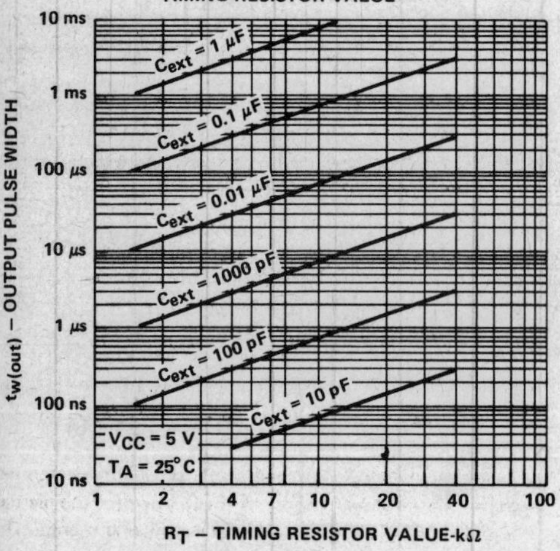

Figure 3.

OUTPUT PULSE WIDTH
vs
EXTERNAL CAPACITANCE

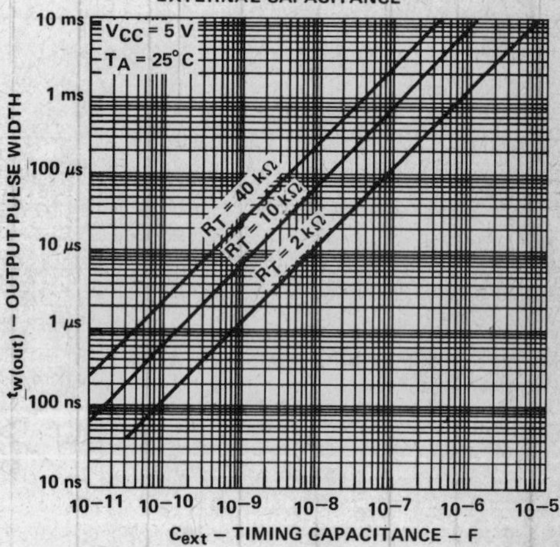

Figure 4.

MULTIPLEXING DISPLAYS

seven-segment displays

To display numbers and symbols an array of display elements is required. Two common configurations are shown in Figure 1. Figure 1a shows the seven-segment display that can be used to display the decimal numerals and some alphabetical characters by turning on appropriate segment patterns. Figure 1b shows a 5 × 7 dot matrix that can be used to display any alphanumeric symbol by turning on the appropriate dot pattern. The pattern required for each

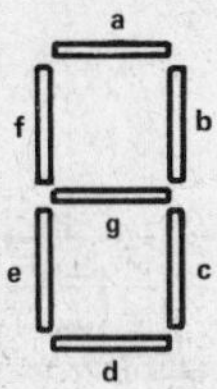

a. SEVEN-SEGMENT LED DISPLAY

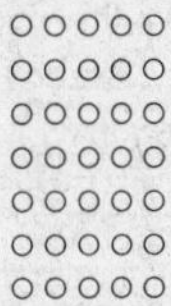

b. 5 X 7 DOT MATRIX DISPLAY

Figure 1. Display Matrices

number, character, or symbol to be displayed must be stored in a read-only memory or a display decoder in order to properly display a desired character. The interface to a seven-segment display is the BCD-to-seven-segment decoder driver like the SN7446 shown in Figure 2a. The input to the decoder is the BCD code for the number to be displayed. The RBI and BI signals can be taken low to turn off all segments, regardless of the input code. When BI is high, the LT (amp test) input can be brought low to turn on all segments to perform a lamp test operation. The BI/RBO can serve as an output for ripple blanking to other decoders. When RBI is brought low, RBO as an output will go low for rippling a blanking signal to other display decoders. The segment drivers A through H are connected to the LED's of the display to control which LED's are turned on.

The entire circuit and display is available as a single device, the TIL321A, shown in Figure 2b. This device has the 4-bit BCD code input, a decimal point input, and depends on a non-BCD code to provide blanking. Devices also exist that include a register as well as a decoder/driver and display in the same unit. The TIL308 shown in Figure 2c is one of those. It stores the four BCD inputs in a quadruple S-R flip-flop whose outputs are available from the device. There is a latch strobe input that, when low, stores the BCD code in the 4-bit register. There is a blanking input, BI, that, when low turns off all segments, and an LED test input that, when low, turns on all segments. If the LED test and the BI inputs are both high, the display shows the number whose code is latched in the device data register. Such a register simplifies the I/O requirements of the microcomputer since it can be treated as a complete stsorage location. It may be connected to either the data bus or any special system I/O bus.

The interface to a 5 × 7 or other dot matrix is handled in much the same way as the seven-segment device. The simplest device of this type is the TIL311, which displays hexadecimal characters using LEDs arranged on a 4 × 7 dot matrix pattern as shown in Figure 3a. It includes a 4-bit data register with a latch strobe input that causes the 4-bit input data to be entered while the strobe is low. As long as the strobe stays high, the information displayed and stored will not change. Thus, one could treat the strobe as a rising-edge latch signal. The overall structure of the TIL311 is shown in Figure 3. There is a blanking input that, when high, causes the display to be blanked. There is a left and right decimal point input available.

The control of a 5 × 7 dot matrix display device like the TIL305 requires a ROM or EPROM in which the display pattern for each character to be displayed is stored. The basic circuit structure is shown in Figure 4 for an individual interface to a TIL305. The TTL signals from the seven input lines (ASCII code inputs) are connected to the inputs I1 through I7. The current-drive capability is provided by SN75491 drivers acting as sink drivers from the output lines 01 through 07 and as source drivers for the column lines on the TIL305. At the time a column line is driven with current, the column select code CA through CE must simultaneously be applied to the column select lines of the EPROM. The EPROM outputs the seven row signals for a

APPLICATIONS 15

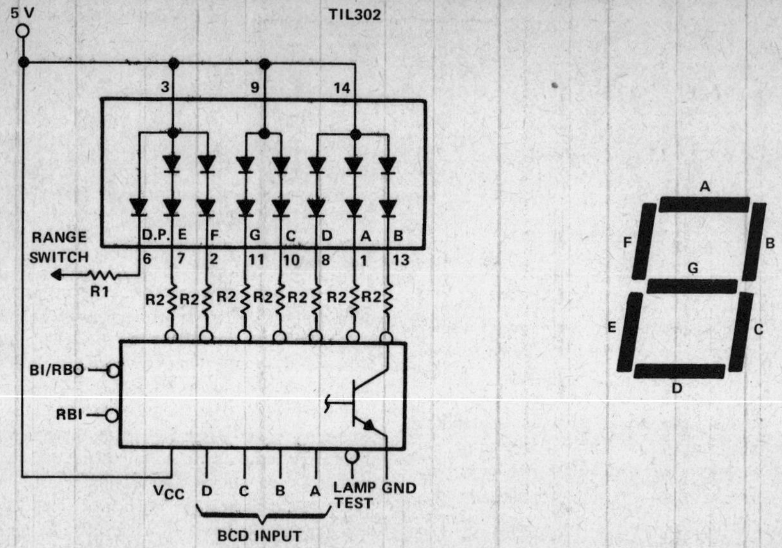

a. SEVEN-SEGMENT DISPLAY INTERFACE

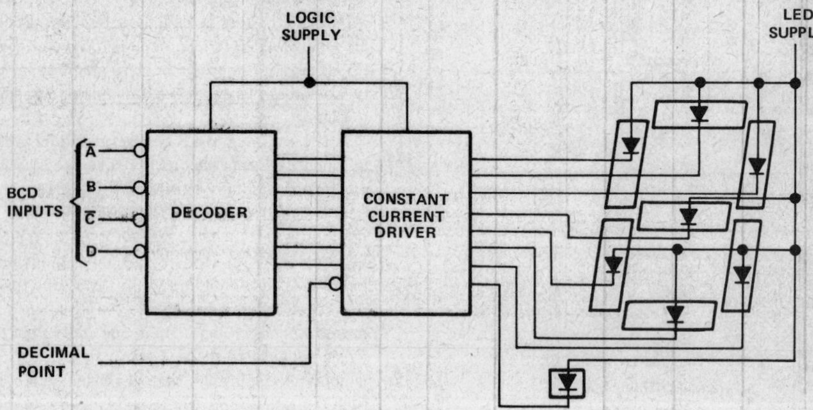

b. TIL321A

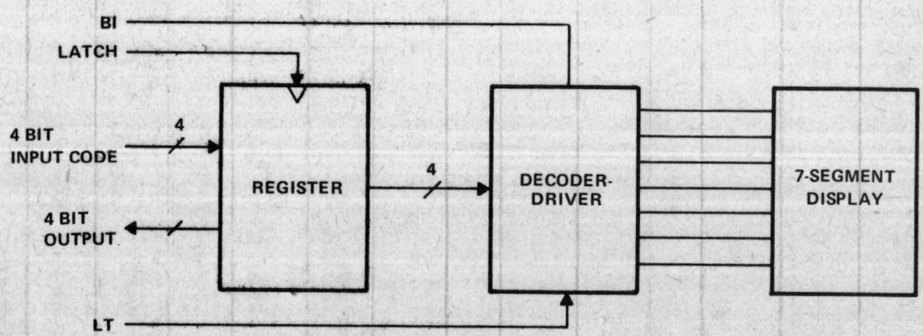

c. TIL308

Figure 2. Seven-Segment Displays

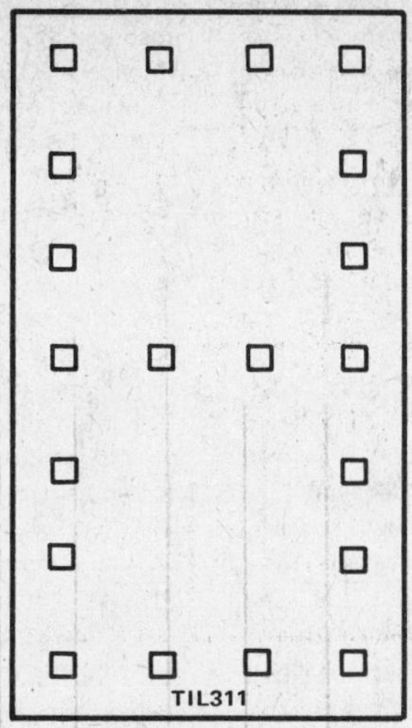

a. MATRIX 4 X 7 PATTERN.

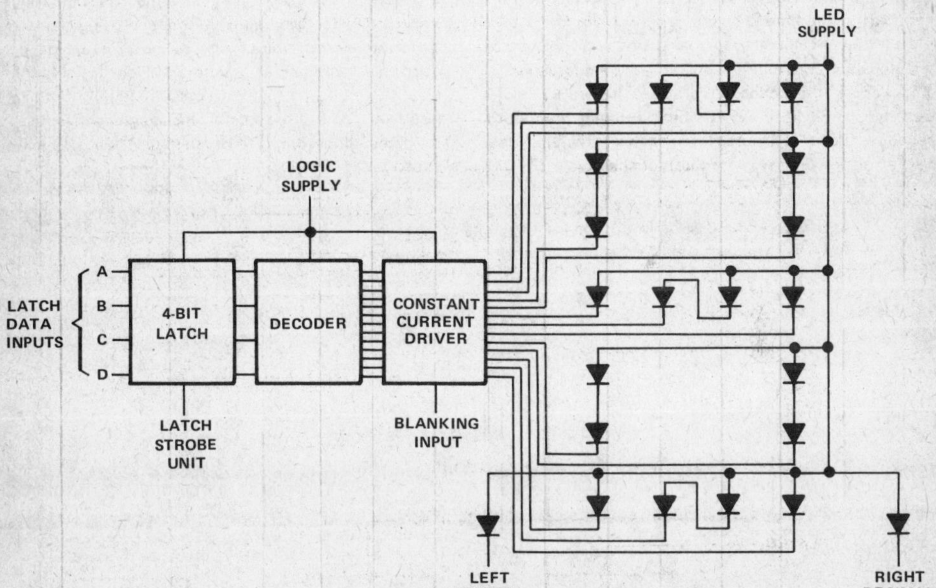

b. SCHEMATIC

Figure 3. 4 X 7 Matrix

selected character for a selected column. Thus, the circuitry must scan through the columns at an appropriate rate by using either a ring counter or a counter-decoder combination.

In the example of Figure 4, an SN75496 5-bit ring counter is set up so that only one bit position will have a 1 at any given time. This is achieved with the wired-NAND control (SN7416) on the serial input. For example, if all outputs A through E (all bits) in the register are 0, the serial input line will be at the 1 level, and a 1 will be shifted into the first bit position. This 1 (in the A position) causes the serial input line to go low (to a 0), which will be shifted in to fill the lower bits with zeroes. The original 1 will propagate through to E with each rising edge of the clock. When the 1 is at E, a 1 will again be generated at the serial input to insert a new 1 into A when the 1 is shifted out of E. Thus, there is only one 1 in the shift register at any time. Only one column of the EPROM is addressed at any time, and only one column of LED drivers is turned on at any time. Also in the example of Figure 4, the unijunction oscillator is set to provide a clock pulse sequence at a frequency of about 1,000 pulses per second. A new column is selected and turned on for about a millisecond, and a column is on 20 percent of the time.

The circuit of Figure 4 provides only a single-character display position. If a multiple-position character display is required, it is not reasonable to provide a separate EPROM for each display unit. In other words, it is not feasible to repeat the circuit of Figure 4 for each character in the multiple position display. A circuit that shares the EPROM resource must be used. This means that the display must provide a RAM for storage of the character codes to be displayed and a sequence controller that will sequence through the codes stored in RAM while the different TIL305s are activated. The basic structure is shown in Figure 5 for a 16-character display.

There must be 16-location RAM, and each location must store a 6-bit ASCII code. There must be a modulo-16 counter that determines which RAM code and character position is to be used at any given time. The TIL305 that is activated is selected by the output of a 4-to-16 decoder. The decoder turns on one group of the SN75493 sink drivers for the selected character position. The sink drivers for all other character positions are turned off, and the associated TIL305s for those positions remain off. The modulo-16 counter is incremented by the trailing edge of output E from the central 5-bit ring counter, since that marks the beginning of the new column 1 display. There must be a provision for writing into the RAM from the processor and a write control signal, W', that will switch the RAM address from the modulo-16 counter to the processor address lines. The SN74LS245 for each bit provides the switching required. This connection allows the information being displayed to be controlled by processor memory write or output operations. The overall structure of Figure 5 is somewhat complicated, but it can be cost effective in both power dissipation and parts costs. A similar approach can be used for time multiplexing of 7-segment displays to save power consumption.

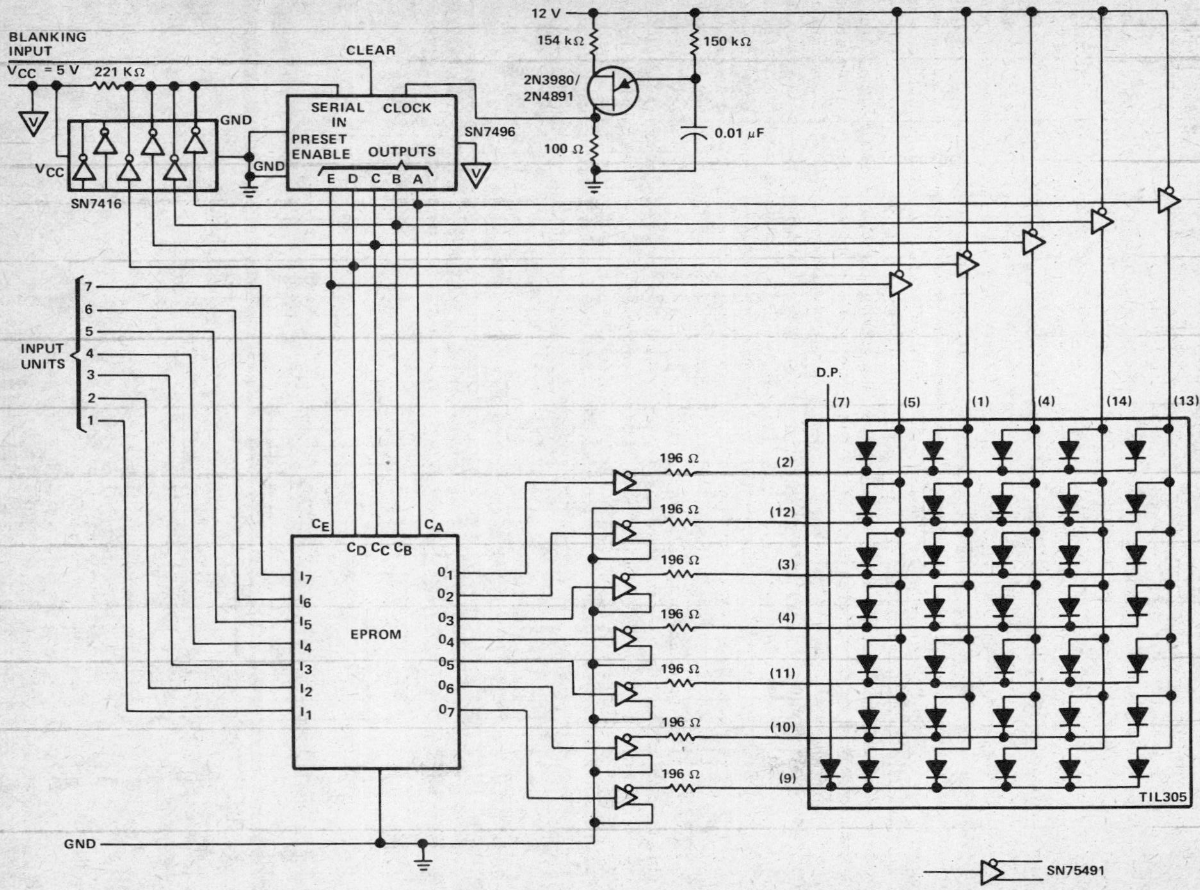

Figure 4. 5 X 7 Matrix

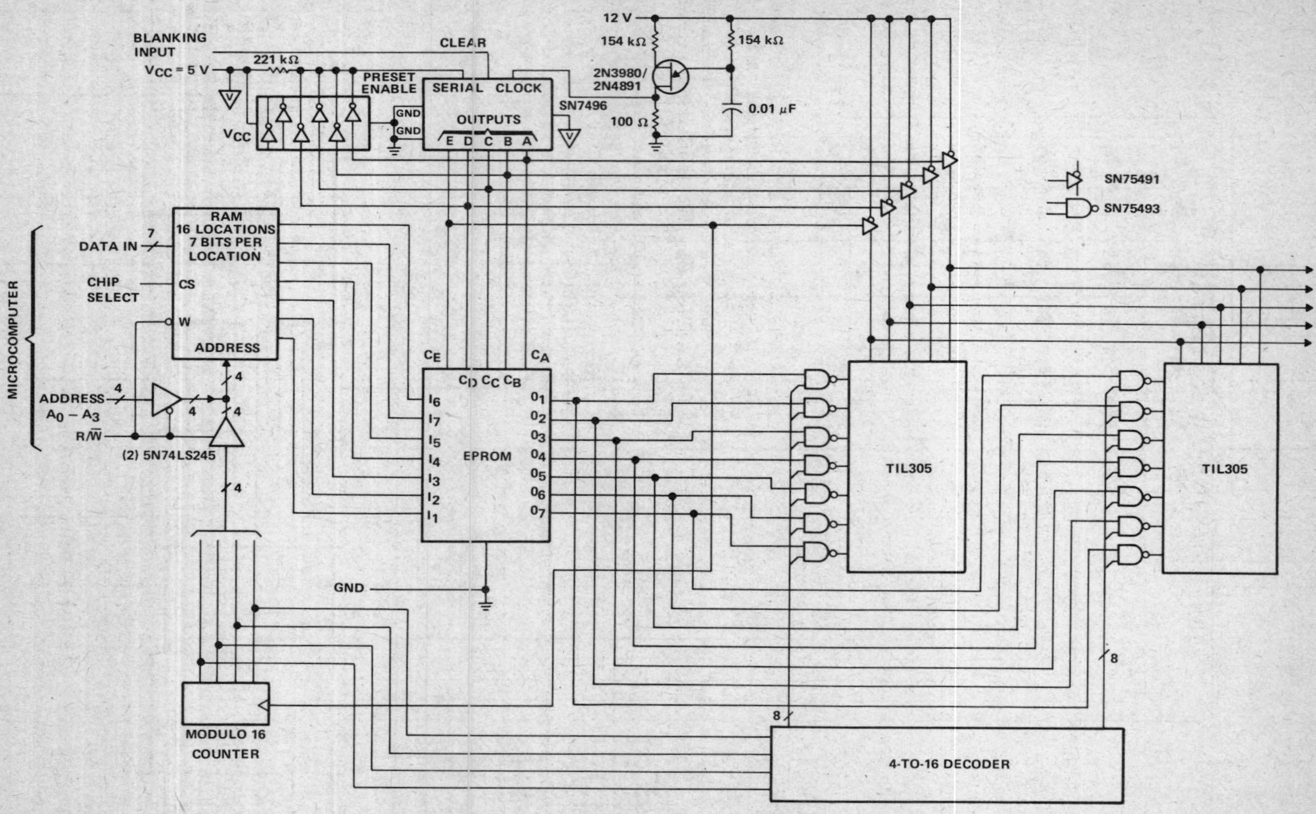

Figure 5. 16-Character Display

TIL311 HEXADECIMAL LED DISPLAY

The TIL311 is designed to store and display decimal and hexadecimal data. The device consists of an MSI logic chip to perform logic and storage functions plus a light emitting diode (LED) display in a single 14-pin dual in-line package.

It accepts parallel 8-4-2-1 data on four input lines and displays the corresponding decimal or hexadecimal character on a 4-by-7 dot matrix. Figure 1 illustrates the hexadecimal character representation for the decimal numbers 0 through 15. The logic levels are designed to be TTL compatible: a high level is 2 V to 5 V, a low level is 0 V to 0.8 V.

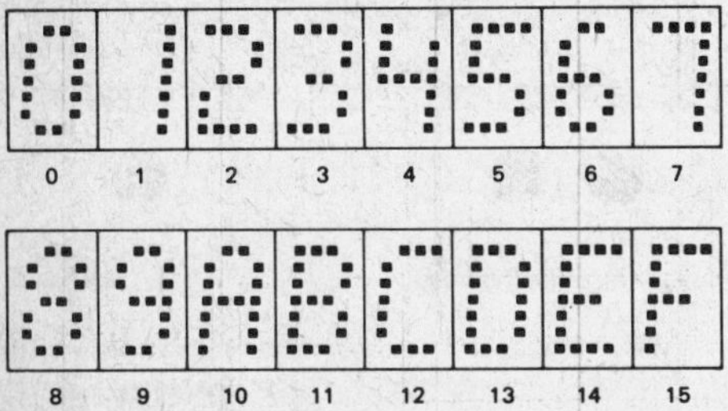

FIGURE 1. TIL311 Hexadecimal Character Configuration

The block diagram in Figure 2 shows the major sections of the TIL311: latches, decoder, current driver, and LED display. The inputs are DATA, LATCH STROBE, BLANKING, and DP. DATA is parallel 8-4-2-1 coded data. When LATCH STROBE is low, the data in the latches follow the data inputs. When LATCH STROBE goes high, the data on the input lines at strobe time is stored in the latches.

The 4-bit code is decoded and the required diodes are turned on via the constant-current drivers to display the proper character.

The LED display contains two decimal points: one to the left and one to the right of the character. A low input to one of the DP inputs will turn that decimal point on.

BLANKING must be low to display the character. When BLANKING goes high, the character is turned off regardless of the inputs. The BLANKING input does not change the data stored in the latches. BLANKING may be pulsed to intensity-modulate the display. The apparent brightness of the display is proportional to the duty cycle of the modulating signal, assuming a frequency high enough to avoid visible flicker. For example, at 1 kHz, a 50% duty

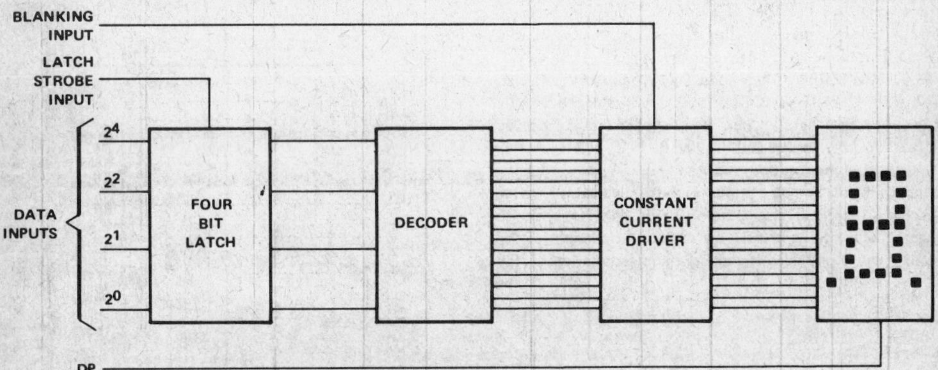

FIGURE 2. TIL311 Hexadecimal Display Block Diagram

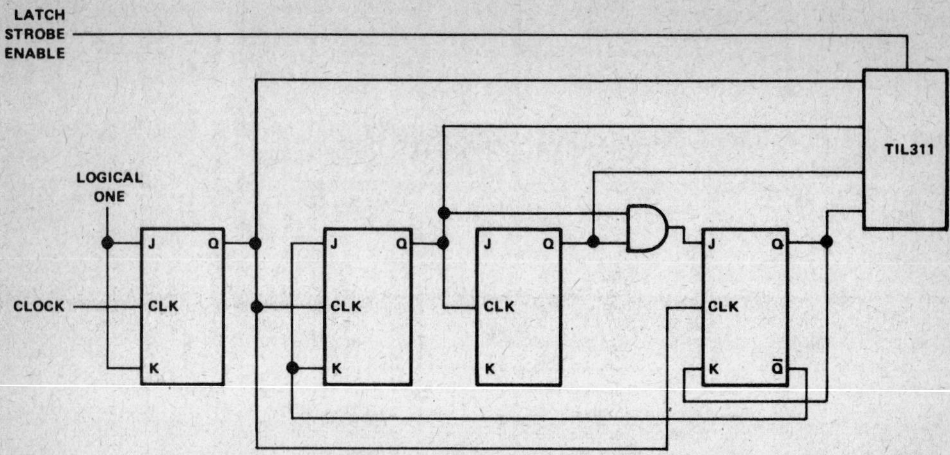

FIGURE 3. TIL311 Used As Counter Display

FIGURE 4. Discrete Light Display for a 16-Bit Register

cycle would cause an apparent brightness of 50% of the steady-state brightness.

Figure 3 illustrates the use of the TIL311 as a decimal display. The JK flip-flops are connected as a count-by-ten counter and represent one decade position in a multi-decade counter. The four Q outputs of the four flip-flops furnish the data inputs to the TIL311. Normally LATCH STROBE will be held high so that the display does not follow the counting. When counting is complete for a given time base, LATCH STROBE is pulsed with a negative-going pulse. The new data is then transferred from the decade counter into the latches and displayed.

Another application for the TIL311 is to display register information on computer control panels and service panels. Figure 4 illustrates the use of discrete lights to display the contents of a 16-bit register. The length of the display can easily lead to errors in interpretation of the data. Figure 5 illustrates the use of the TIL311 to display the same data in the same 16-bit register. The 16 register positions are divided into four 4-bit groups. The four bits in each group provide the inputs to each of four TIL311 displays. The resulting four hexadecimal character display provides a more concise interpretation of the register data.

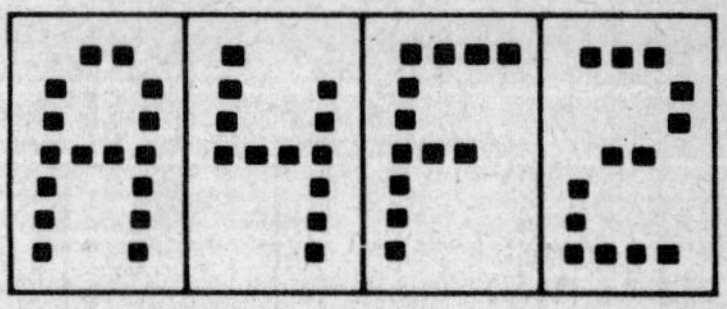

FIGURE 5. Hexadecimal Display for a 16-Bit Register

COUNTING CIRCUITS
USING TIL306 AND TIL308 LEDs

Digital instruments have experienced a constant evolution since 1960. Counters that once occupied several inches of rack space in a 19-inch rack have been replaced by units the size of a text book with performance characteristics surpassing the older models. A major contribution to these changes is the continued advances in solid-state devices: integrated circuits have replaced the tubes and transistors and light-emitting diodes (LEDs) have replaced the incandescent displays.

Texas Instruments has introduced a new product that simplifies further the design of systems utilizing counters or digital read-outs. By combining an IC chip to perform the logic function and an LED display in a single 16-pin dual in-line package, Texas Instruments has provided the designer a device that reduces the complexity of his system without reducing flexibility of design. Two of these devices are the TIL306 and TIL308. The TIL306 and TIL308 have decimal points to the left side of the character. The TIL307 and TIL309 have decimal points to the right side of the character, but are otherwise identical to the TIL306 and TIL308, respectively. They can be combined to count, store, and display data in multiple decade positions.

CIRCUIT DESCRIPTION

The TIL306, as shown in Figure 1, consists of four major sections: counter, latches, decoder/driver, and LED display.

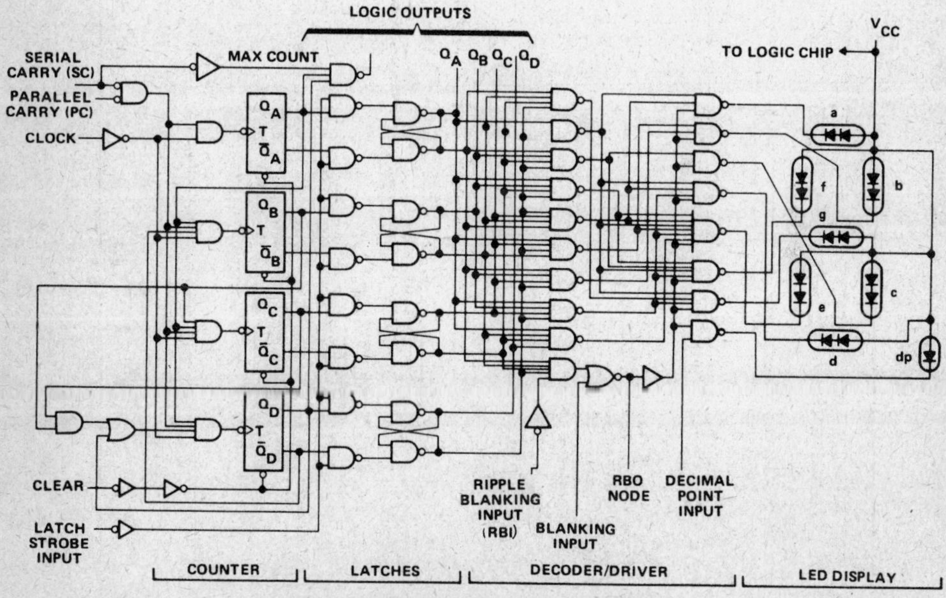

FIGURE 1. Functional Block Diagram of TIL306

The counter is connected as a synchronous counter. This configuration takes advantage of the minimal propogation delay to give maximum speed capability. Inputs to the counter are CLEAR, CLOCK, SERIAL CARRY, and PARALLEL CARRY. The counter and its inputs generate an output, MAX COUNT. Additional connections are LATCH STROBE, BLANKING, RIPPLE BLANKING, RBO, DECIMAL POINT and LOGIC OUTPUTS. All inputs and outputs are designed to be TTL compatible. A high level is a minimum of 2 V and a low level is a maximum of 0.8 V. A low input to CLEAR will reset the counter to zero independently of any other input. As long as the input remains low the counter remains at zero. A high is required to allow the counter to count.

The CLOCK input is the signal to be counted. With an input the counter will advance from 0 to 9. At a count of 9 the counter automatically resets to 0 with the next pulse. The counter changes state on the positive-going edge of the clock pulse. The clock pulse to the counter is controlled by SERIAL CARRY and PARALLEL CARRY.

The MAX COUNT output goes low when the counter reaches a count of 9, and then goes high when the counter progresses to 0 on the next clock input. This output can be connected to the CLOCK input of the next decade position for asynchronous operation or to the SERIAL CARRY input of the next decade position for synchronous operation.

A high on SERIAL CARRY inhibits the counter and forces MAX COUNT to go high regardless of the state of the counter stages. When SERIAL CARRY and PARALLEL CARRY go low, the CLOCK is enabled to the counter stages and the MAX COUNT gate is allowed to sense the status of the counter. The logic level of SERIAL CARRY must not be allowed to change while CLOCK is low or erroneous counts may result.

PARALLEL CARRY permits look ahead carry inputs from lower order decade positions. A high input inhibits the clock to the counter stages. When PARALLEL CARRY and SERIAL CARRY go low the clock to the counter stages is enabled. The logic level of PARALLEL CARRY must not be allowed to change while CLOCK is low or erroneous counts may result.

LATCH STROBE transfers the data in the counter stages to the latch storage to be displayed. With LATCH STROBE low, the latch flip-flops follow the states of the counter flip-flops. When LATCH STROBE goes high, the counter data is stored in the latch flip-flops. The counter can continue to count while the previous information is stored in the latches.

The DECIMAL POINT input controls the display of the decimal point. A high is required to turn on the LED decimal point display.

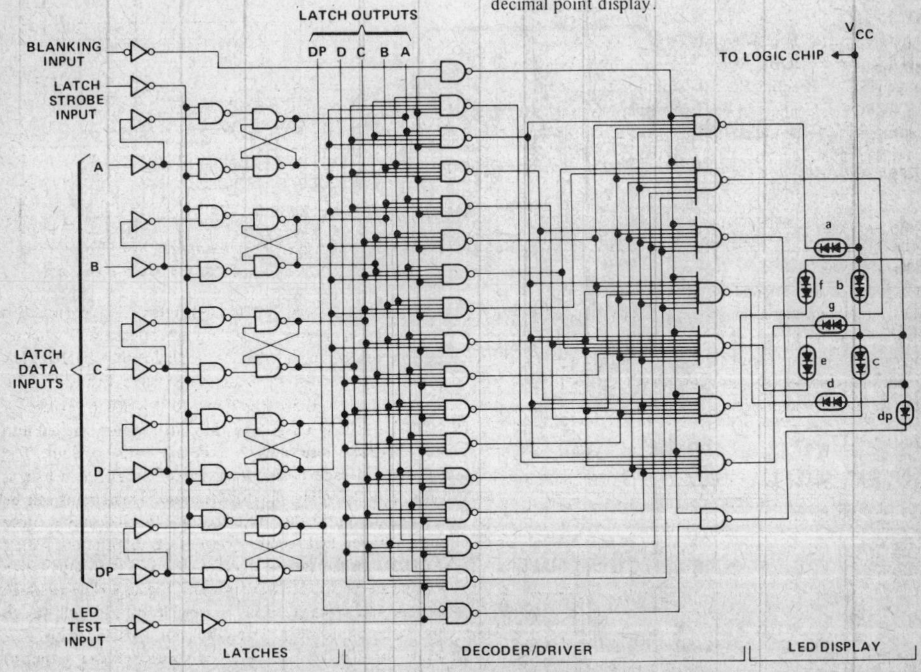

FIGURE 2. Functional Block Diagram of TIL308

A high on BLANKING inhibits the driver and gates and blanks the LED display. For normal operation, the BLANKING input must be low.

A low on RIPPLE BLANKING blanks the display if the latch flip-flops contain a count of zero. This combination also forces the RBO NODE to go low. By connecting the RBO NODE of one decade position to the RIPPLE BLANKING input of the next decade position, zero suppression can be achieved. This is discussed in detail in a later portion of this report, Counter Circuit Description. The RBO NODE has a resistor pullup, which allows this output to be used as an input. A low level applied to RBO will blank the LED display independently of other input.

The TIL308 looks physically identical to the TIL306. However, the TIL306 contains a counter section: the TIL308 does not. The TIL308 accepts 8-4-2-1 BCD code from external sources, stores it in latches, and displays the stored character by means of an LED display. As shown in Figure 2, the TIL308 consists of the three major sections: latch, decoder/driver, and LED display.

The inputs and outputs, designed to be TTL compatible, consist of DATA INPUTS, DATA OUTPUTS, LATCH STROBE, BLANKING, and LED TEST.

The BCD data and decimal point on the DATA INPUT lines are transferred into the latch flip-flops when LATCH STROBE is low. The BCD data and decimal point data stored in the latches are available at DATA OUTPUT. With LATCH STROBE high the DATA INPUT lines can change without effecting the data stored in the latches.

BLANKING must be high to display the data stored in the latches. When BLANKING goes low, the decoder drivers are inhibited and LED display is turned off. The data stored in the latches are not effected by BLANKING.

LED TEST can be used to test the LED display. A low to LED TEST will override all other signals and turn all of the LEDs on. LED TEST does not change the status of the latches.

With the basic operation of the circuits outlined, two typical interconnection methods are shown in Figure 3 and 4. Figure 3 shows the TIL306 connected in the synchronous mode. Figure 4 shows the TIL306 in the asynchronous mode. The asynchronous mode will be used in the following example of a counter.

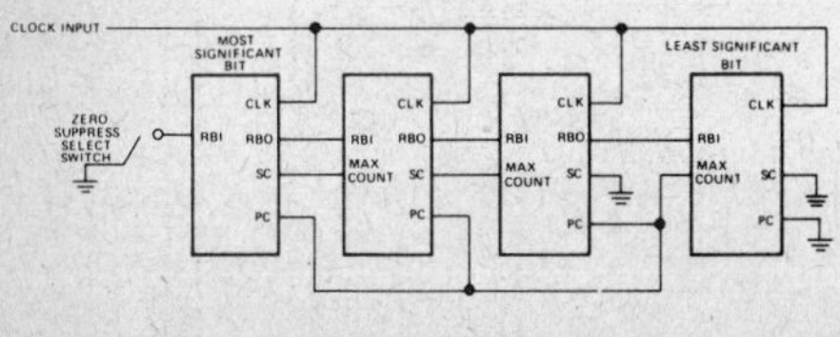

FIGURE 3. TIL306 Interconnections for Synchronous-Count Mode and High-Order-Zero Suppression.

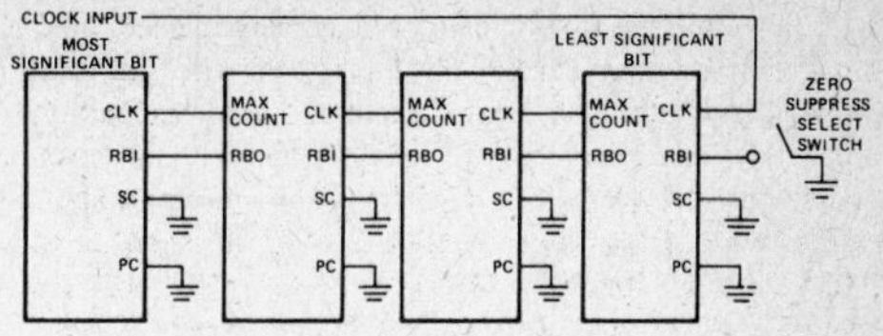

FIGURE 4. TIL306 Interconnections for Asynchronous-Counting Mode and Low-Order-Zero Suppression.

COUNTER CIRCUIT DESCRIPTION

The counter is a major constituent in digital instruments. Digital voltmeters, frequency counters, event counters, and period counters all have a circuit in common, very much like the one shown in Figure 4.

The circuit to be discussed in detail in this report incorporates both the TIL306 and the TIL308. One of the limiting factors of the TIL306 is that the counter typically does not count faster than 18 MHz. Combining the TIL306 with a TIL308 and feeding the TIL308 from a high-speed counter expands the system to a much higher frequency. Figure 5 shows a BCD counter capable of working at 100 MHz. The circuit consists of two SN74S112 Schottky

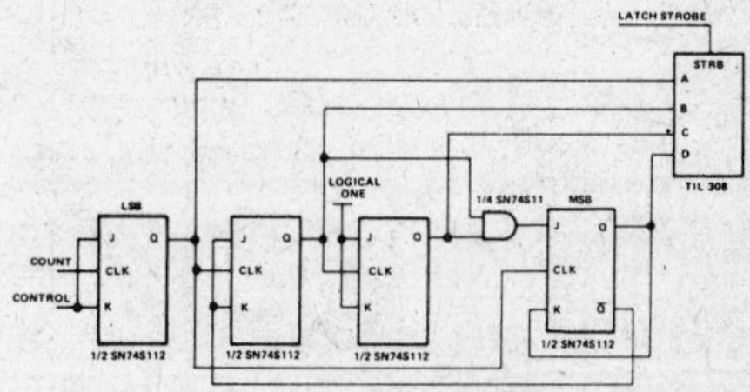

FIGURE 5. 100 MHz Decade Counter Using Texas Intruments Schottky TTL Logic and A TIL308 Display.

TTL circuits and one SN74S11 Schottky TTL circuit. This configuration results in an asynchronous BCD counter capable of dividing a 100-MHz signal down to 10 MHz. The speed is a result of Texas Instruments Schottky TTL devices that allow flip-flops to toggle in excess of 100 MHz. The Q outputs of the four flip-flops are fed into one TIL308, resulting in a decade with readout. The following decade position consists of a TIL306, which is capable of handling the 10 MHz rate. This circuit can be expanded even further by preceeding the Schottky counter stage with an ECL counter stage. ECL IC flip-flops with a 400-*MHz* toggle rate and discrete built ECL flip-flops with a toggle rate of 800 MHz are possible. Figure 6 shows a block diagram of a stage which is capable of counting up to 800 MHz. Since ECL levels do not coincide with TTL levels, an ECL-TTL converter is necessary. The output of the converter will drive the TIL308 without any interference caused by switching speed problems.

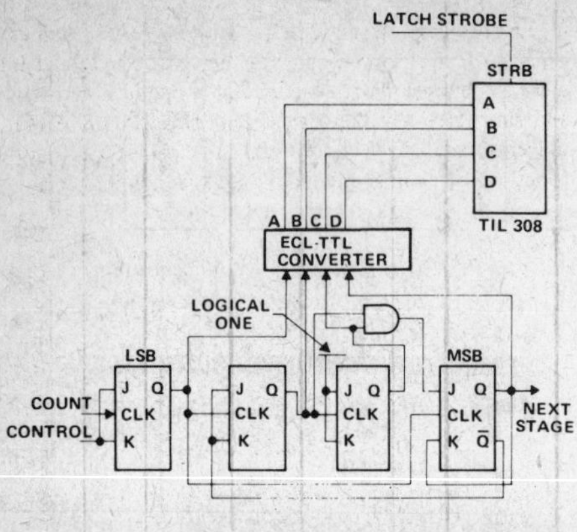

FIGURE 6. 800-MHz Decade Counter Using ECL Logic and A TIL308 Display.

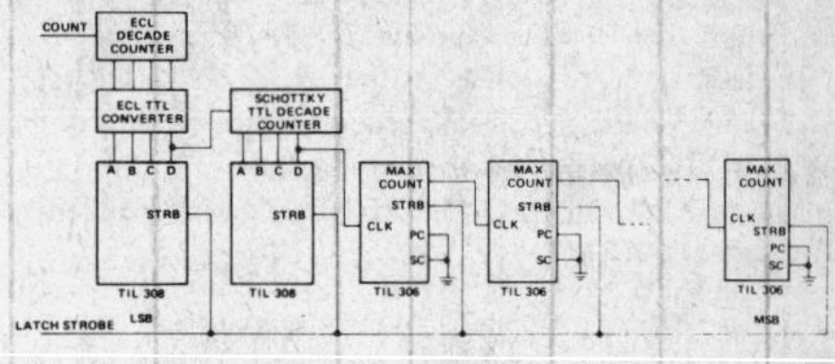

FIGURE 7. Nine-Digit Counter

A *B*

FIGURE 8. Two Counters with Identical Performance. Counter (A) Uses TIL306 Devices; Counter (B) Does not. Note how many less Components are Needed in the Counter Using TIL306 Devices.

Figure 7 is a block diagram representation of a nine-digit readout, consisting of an ECL decade counter with a TIL308 display and a Schottky TTL decade counter with a TIL308 display, as just described, and seven TIL306 devices. Part count is minimal, and the complexity of the PC Board is minimized.

Figure 8 is a photo of two counters with identical performance illustrating the difference in component count between a conventional counter consisting of SN7490, SN7475, and SN7447 TTL integrated circuits, resistors, with a display using TIL302 devices, and a counter using TIL306 devices. Both counters are specified to operate up to 15 MHz using a six-digit readout. The counter using TIL306 devices shows a big empty surface in the middle of the board and considerably fewer interconnects to the display. The cost savings resulting from using such a counter are quite obvious.

Figure 9 is a photo of a 100-MHz counter using seven TIL306 devices and two TIL308 devices. A compact assembly technique reduced the total size.

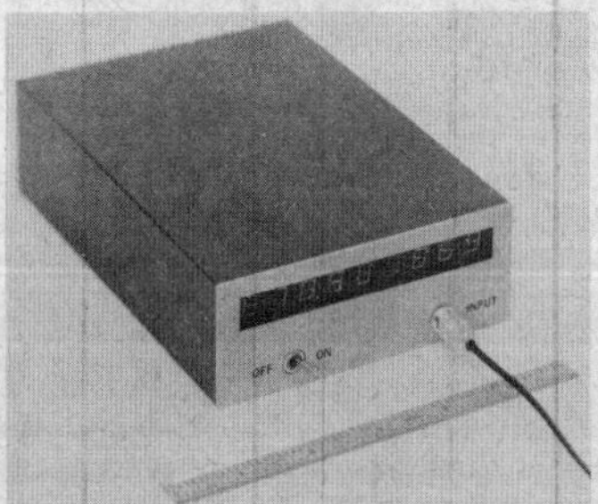

FIGURE 9. A Portable 100-MHz Counter Using Seven TIL306 Devices.

Figure 10 shows all of the basic circuit boards and components used in the counter shown in Figure 9 and shown schematically in Figure 12. The upper board is timebase. The center board is control. The bottom board is counter and display.

FIGURE 10. The Three Basic Circuit Boards of the Portable Counter.

in Figure 5 and seven TIL306 devices. This counter is capable of measuring frequencies up to 100 MHz and time with 10-nanosecond resolution. Again minimum part count and simplicity have been the major objectives. The unit is universal and the counter can be expanded into other functions by adding circuits to the basic building block.

FIGURE 11. The Three Basic Circuit Boards Fastened Together into A Compact, High-Density Unit

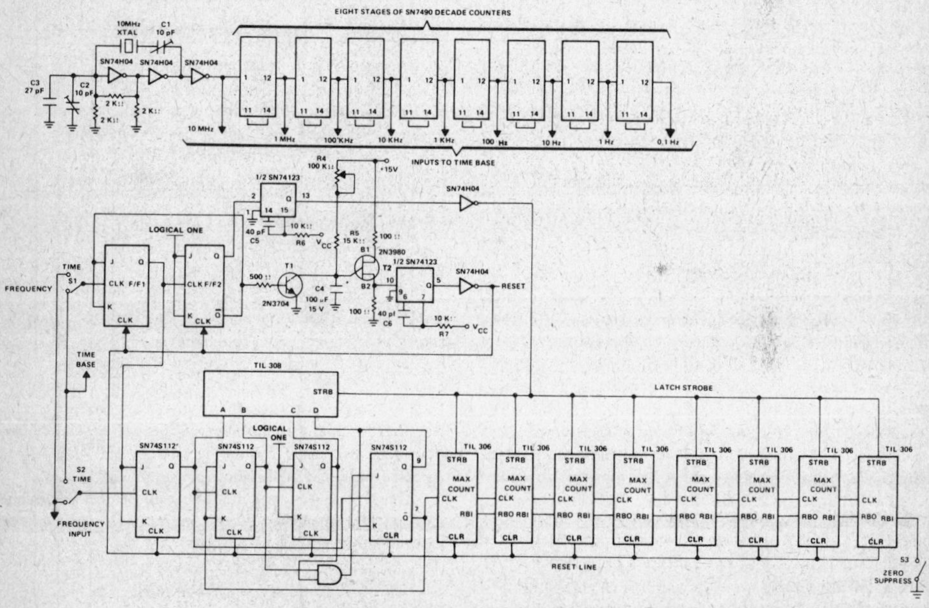

FIGURE 12. Schematic of A Frequency and Time Counter

Figure 11 shows the assembly technique for high density component packing. The total size is 1.2 inches high, 1.2 inches deep and 4.25 inches wide. This counter can be incorporated in a lightweight and portable instrument. Total power dissipation is 9 watts.

Figure 12 shows a complete schematic of a frequency and time counter incorporating the 100-MHz stage shown

The counter has three main functional sections: timebase, control, and counter.

The top part of Figure 12 is the time base. A 10-MHz oscillator is formed using two SN74H04 TTL high-speed inverters. The output is coupled through a third inverter to isolate the oscillator from the rest of the circuit. Capacitor C1 is a coarse adjust and capacitor C2 is a fine adjust. C2 should be a piston capacitor to allow finer resolution during adjustment. For more accurate requirements, a separate oscillator in a temperature-controlled oven with AGC circuitry can replace this circuit. The output of the oscillator is fed into a divider chain consisting of eight SN7490 decade dividers. Timing signals from 10 MHz to 0.1 Hz are generated and switch selectable as the time base. In the middle of the schematic in Figure 10 is the control circuit. The purpose of the control circuit is to gate the counter, and to generate latch strobe, and reset signals.

The input of F/F1 is the time base signal in the frequency measuring mode or the unknown time period in the time measuring mode.

With all circuits reset, the $\overline{Q}$ output of F/F2 holds a high level at the JK inputs of F/F1. With a pulse coming into the F/F1, Q of F/F1 changes from 0 to 1 on the negative-going edge. This 1 is applied to the first stage of the counter, allowing it to count. F/F2 does not change state since it changes only on a negative-going edge. With the next pulse to the clock input of F/F1, F/F1 changes state on the negative-going edge, changing the Q output from logical 1 to logical zero. This negative-going transition sets F/F2 and at the same time stops the counter from counting. With F/F2 set, $\overline{Q}$ of F/F2 is a 0. A 0 at the JK inputs of F/F1 inhibits change with any additional pulses coming into its clock input. The Q output of F/F2 is connected to the input of a monostable multivibrator, 1/2 SN74123. This multivibrator generates a short positive-going pulse at the Q output. The pulse width is determined by the RC combination R6C5 and is set in this application to 150 nanoseconds. The output signal is inverted and applied to the Latch Strobe inputs of the TIL306 and TIL308 devices. This pulse transfers the data from the counters into the latches to be displayed.

The $\overline{Q}$ of F/F2 is connected to the JK inputs of F/F1 and also through a resistor to transistor T1. During counting operation $\overline{Q}2$ is high, turning T1 on and preventing C4 from charging. At the end of the count cycle, the $\overline{Q}2$ is low, turning T1 off. The capacitor C4 begins charging through resistors R4 and R5. R4 is adjustable and allows a variation in the display time. R5 prevents the charging current and the current through T1 from exceeding 1 mA when R4 is turned to zero. Once the charge across C4 reaches the firing potential of the unijunction, T2, the unijunction generates a positive pulse at Base 2, which is coupled into the monostable multivibrator, SN74123. The positive pulse determined by R7C6, 150 nanoseconds wide, is inverted by an inverter, 1/6 of SN74H04, and applied to the reset input of the TIL306 devices, the four F/Fs of the first counter stage, and the two F/Fs in the control section. With F/F1 and F/F2 reset the JK inputs are reset to a high level by F/F2 and the circuit is again ready to handle the incoming signal.

The bottom part of the schematic in Figure 10 shows the counter section. The first stage is made up of two SN74S112, one SN74S11, and one TIL308. The two SN74S112 circuits and one SN74S11 circuit form a decade counter consisting of four flip-flops and one gate. Schottky TTL devices are used because of the speed requirement. If only a 70-MHz counting rate is required, this circuit could be a single SN74196 circuit. The $\overline{Q}$ output of the fourth F/F is connected to the clock input of the first TIL306. The maximum count of the TIL306 is connected to the clock input of the next TIL306. This operation is the asynchronous mode, which is acceptable for counter purposes.

The counter is controlled by the two inputs to the first F/F of the first decade. The clock input is the unknown frequency in the frequency mode, or the known time pulses from the time base in the time-measuring mode. The JK inputs are connected to the Q output of the control F/F. This signal gates the counter. As already explained, a high level to the JK inputs allows the F/F to change state on a negative edge of a pulse applied to the clock input. With the JK inputs low, the clock input does not affect the F/F.

To complete the operation of the counter, the Latch Strobe and the Reset are applied to the circuit as shown. S3 allows choosing between suppression or displaying of zeroes to the left of the most significant digit. With the switch closed, a ground is applied to the ripple blanking input of the most significant digit. If this digit is a zero, the display is blanked and the ripple blanking output goes zero. This output is connected to the next digit and the process repeated until all leading zeroes are suppressed. If switch S3 is opened the high-order zeroes are displayed. All that is necessary for operation of the counter now is to provide a power supply and a signal to be counted.

OPTOCOUPLERS IN CIRCUITS

optocouplers in circuits

There are many situations in which information must be transmitted between switching circuits electrically isolated from each other. This isolation has been commonly provided by relays, isolation transformers, and line drivers and receivers. There is, however, another device that can be used quite effectively to solve these problems. This device is the optocoupler. The need for the optocoupler is most prominent in areas where high voltage and noise isolation, as well as small size, are considered important. By coupling two systems together with the transmission of radiant energy (photons), the necessity for a common ground is eliminated — the main purpose of the optocoupler — and the systems can be effectively isolated.

Four Texas Instruments optocoupler devices, the TIL102, TIL103, TIL120, and TIL121, are discussed in this report. How these devices can be used in various circuits to provide proper isolation in many systems will be a key part of this discussion. There are many circuit applications for optocouplers; however, the ones offered in this report are just several which can be of special use. Complete specifications for these devices are not included here but are available elsewhere in this book.

description of an optocoupler

Basically, a Texas Instruments optocoupler consists of a GaAs (gallium arsenide) infrared-emitting diode (IRED) as the input stage and a silicon n-p-n phototransistor as the output stage. The coupling medium between diode and sensor is an infrared-transmitting (''IR'') glass, as used in the TIL102/TIL103, TIL120/TIL121. Photons emitted from the diode (emitter) have wavelengths of about 900 nanometers. The sensor transistor responds most efficiently to photons having this same wavelength. Consequently, the input and output devices are spectrally matched for optimum transfer characteristics.

Equivalent circuits for the TIL102/TIL103 and TIL120/TIL121 are shown in Figures 1 and 2. For both families of devices, a current source between the collector and base of the sensor is used to represent the virtual base current generated by incident photons striking the base. This base current is proportional to the amount of radiation emitted from the diode. The collector-base and base-emitter junction capacitances are shown for both devices since they are used to determine the rise and fall times of the output current waveform. Because a relatively large transistor base area is necessary for increased sensor efficiency, the collector-base junction capacitance is fairly large.

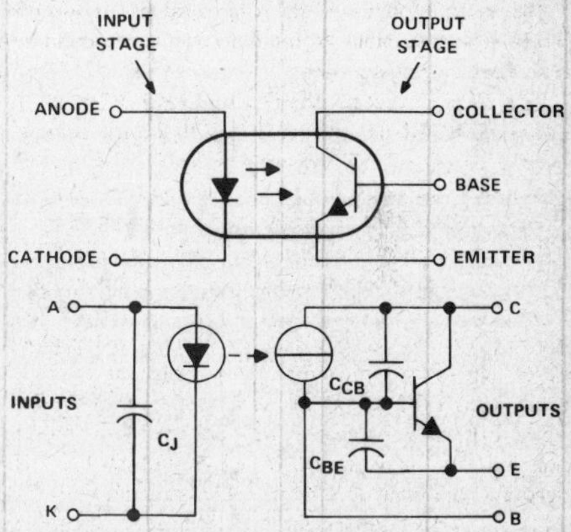

FIGURE 1. Terminal Connections and Equivalent Circuit for the TIL102/TIL103

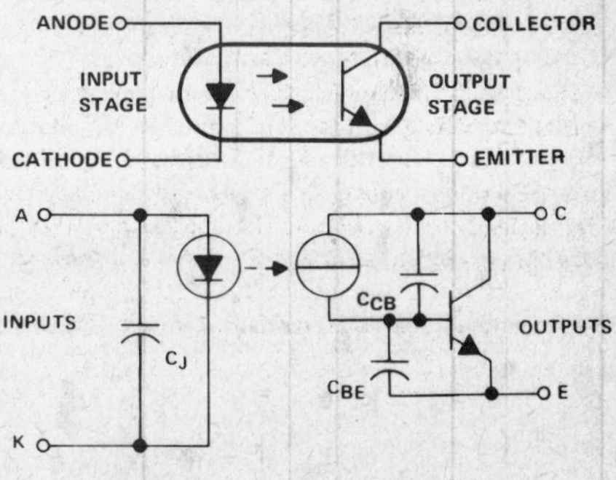

FIGURE 2. Terminal Connections and Equivalent Circuit for the TIL120/TIL121

characteristics of an optocoupler

To fully utilize the advantages offered by an optocoupler, it is necessary that the circuit designer become aware of some of its characteristics. The difference in characteristics between the families is attributed mainly to the difference in construction.

The characteristics most useful to the designer are as follows:

1. High-voltage isolation. High-voltage isolation between the inputs and outputs is obtained by the physical separation between emitter and sensor. This isolation is possibly the most important advantage of the optocoupler. These devices can withstand large potential differences, depending on the type of coupling medium and construction of the package. The IR glass separating the emitter and sensor in the TIL102/TIL103 and TIL120/TIL121 has an isolation capability of 1000 V. The isolation resistance is greater than $10^{12}\ \Omega$.
2. Noise isolation. Electrical noise in digital signals received at the input of the optocoupler is isolated from the output by the coupling medium. Since the input is a diode, common-mode noise is rejected.
3. Current gain. The current gain (output current/input current) of an optocoupler is largely determined by the efficiency of the n-p-n sensor and by the type of transmission medium used. For the TIL103, the current gain is greater than unity, which in many cases eliminates the need for current amplifiers in the output. However, both the TIL102/TIL103 and TIL120/TIL121 have output current levels that are compatible with inputs of digital integrated circuits such as 54/74 TTL. Figures 3 and 4 show typical input-to-output current relationships.
4. Small size. The dimensions of these devices enable them to be used on standard printed-wiring boards. The TIL102/TIL103 and TIL120/TIL121 are built in a metal can similar to a transistor package. The physical dimensions of these packages are shown in Figures 5 and 6.

These are some of the prime characteristics of an optocoupler that can be used effectively to isolate two systems.

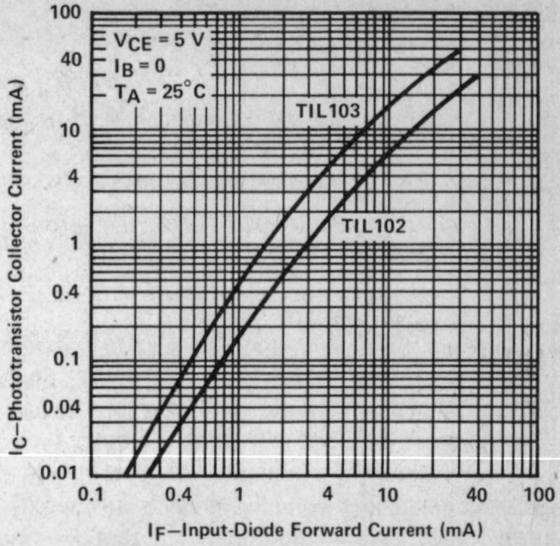

FIGURE 3. Typical Input/Output Current Relationship for the TIL102/TIL103

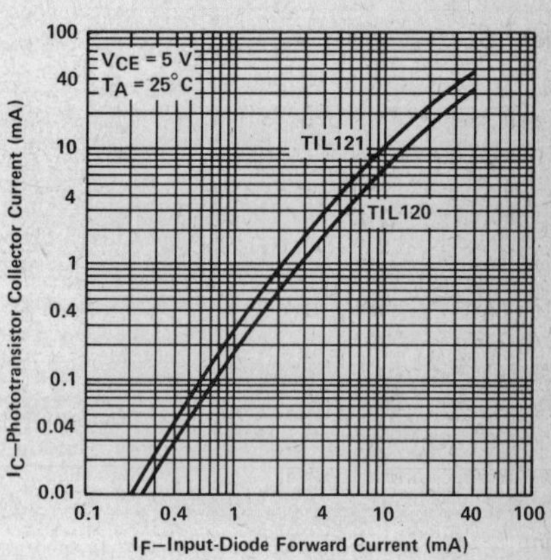

FIGURE 4. Typical Input/Output Current Relationship for the TIL120/TIL121

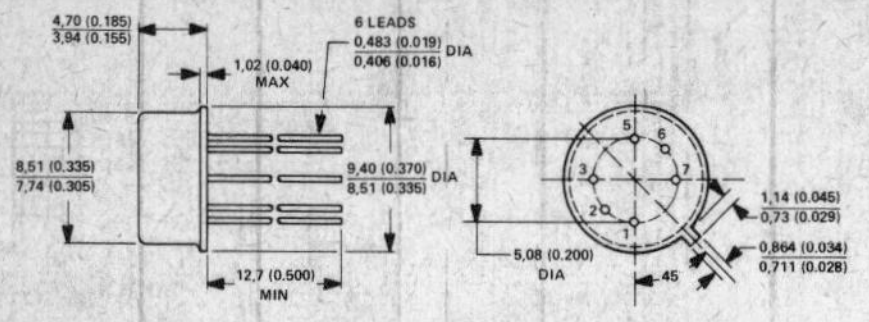

FIGURE 5. Dimensions of the TIL102/TIL103

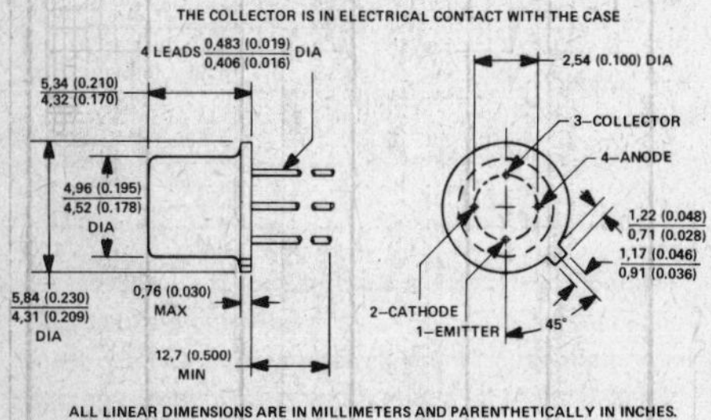

FIGURE 6. Dimensions of the TIL120/TIL121

typical circuit applications

The characteristics and advantages of an optocoupler enable the designer to use it in a wide range of circuit applications. Important among the applications of an optocoupler are those involving 54/74 TTL and similar digital integrated-circuit families. As was mentioned previously, an optocoupler has output currents compatible with TTL inputs. This compatibility enables it to be especially attractive as an interface element between digital systems. The device is particularly beneficial in applications where high voltage differences may exist between systems. However, it is not limited only to digital applications, as shown by the following examples.

driving 54/74 TTL

An effective method of coupling an optocoupler to TTL circuitry is by using a Schmitt trigger that has an output level compatible with standard TTL devices. By coupling any of the Texas Instruments optocouplers to the SN7413, as shown in Figure 7, the isolated signal at the input can be converted to TTL logic levels. Noise immunity is provided by the coupler as well as by the threshold level of the SN7413.

The optocoupler can also be employed as part of a Schmitt trigger circuit that utilizes discrete components. Because the output of the optocoupler is a transistor, it can be used as the input stage to the trigger as shown in Figure 8. For this circuit, regeneration or positive feedback is provided by the coupled emitters of Q1 and Q2. The output of this circuit is noninverting and is compatible with TTL logic.

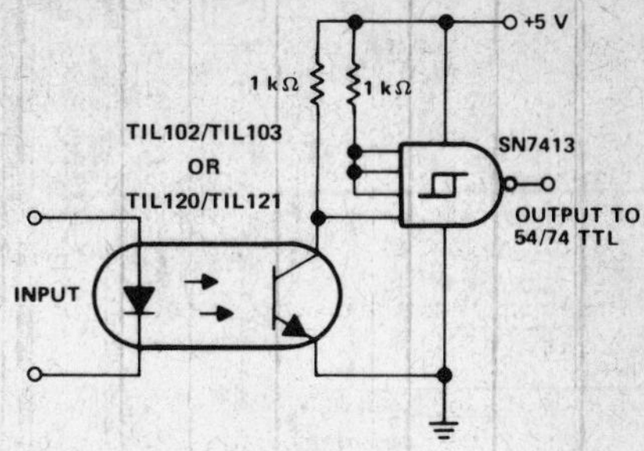

(a) NON-INVERTING FUNCTION

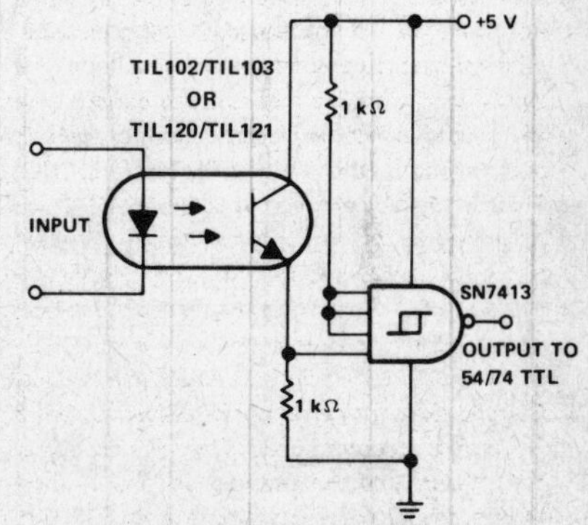

(b) INVERTING FUNCTION

FIGURE 7. Schmitt Trigger Coupling Optocoupler to 54/74 TTL Inputs

Another Schmitt trigger utilizing discrete components that makes use of the base connection of the TIL102/TIL103 is shown in Figure 9. In this circuit, positive feedback is provided from the collector of Q2 to the base of Q1. Resistor R1 limits the base current to Q1 and keeps the device off when there is no signal at the emitter. As with the circuit in Figure 8, the output of this circuit is noninverting and compatible with TTL levels.

transmission-line isolator

By using an optocoupler between two systems coupled by a transmission line, effective line isolation can be achieved. Figure 10 shows a typical interface

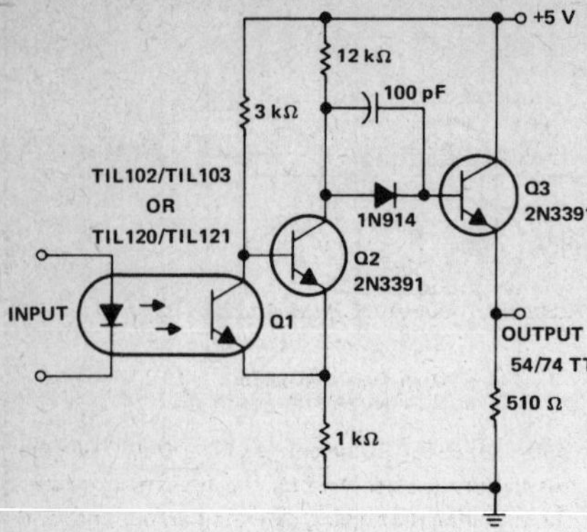

FIGURE 8. Optocoupler with Discrete-Component Schmitt Trigger for Driving 54/74 TTL

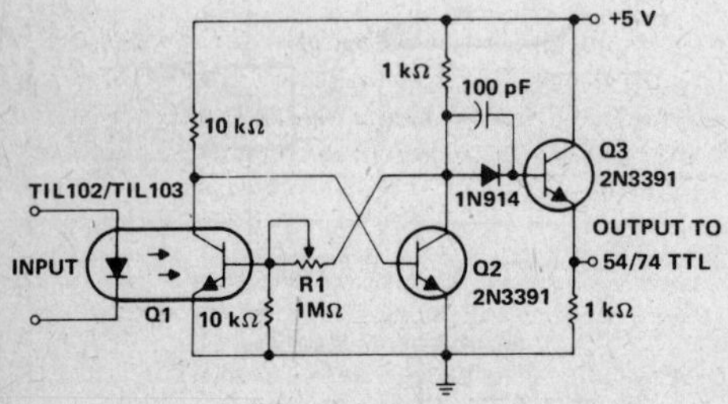

FIGURE 9. TIL102/TIL103 in a Schmitt Trigger for Driving 54/74 TTL

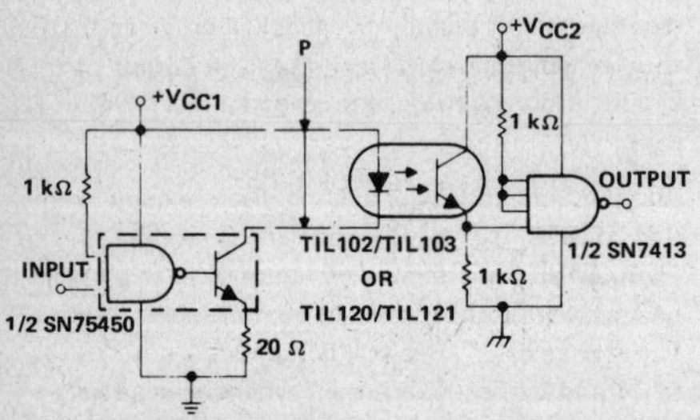

FIGURE 10. Typical Transmission Line Isolator

system using TTL integrated circuitry coupled by a twisted-pair line. The SN75450B is the input stage driving the transmission line and emitter of the optocoupler. The IRED requires about 20 mA during "turn-on," which is well below the maximum current rating of the transistor. At the receiving end of the line, the phototransistor is coupled to an SN7413 for fast pulse generation. The output of this system is a noninverted pulse. However, by rearranging the optocoupler and the SN7413 as shown in Figure 7(a), the output may be inverted.

As simple as it seems, employing an optocoupler this way provides isolation for both noise and high voltage. An isolation transformer or relay could accomplish the task, but it would not be as fast as the optocoupler. Also, a line driver and receiver combination could be used to eliminate the noise and increase the speed, but it would be very ineffective if there were high potential differences between the input and output.

solid-state relay

Through the use of transistor circuits, mechanical relays are slowly being replaced by solid-state relays. In some cases, the solid-state relay (SSR) offers distinct advantages over its mechanical counterpart. For example, an SSR has the advantage that it has neither moving parts nor fragile wires, and it has faster switching speeds and longer operating life. However, one disadvantage of an SSR is that it generally has a lower degree of input/output isolation than a mechanical relay. To overcome this disadvantage in the SSR, an optocoupler can be used as the isolating input stage as shown in the block diagram in Figure 11. The control stage may consist of discrete transistors or integrated circuits, while the output stage consists of high-power switching devices.

A simple isolated latch circuit, which is somewhat of an SSR, is shown in Figure 12. The output of the optocoupler is used to fire the SCR that provides power to the load. To turn off the load current, the supply voltage V_{CC2} must be removed.

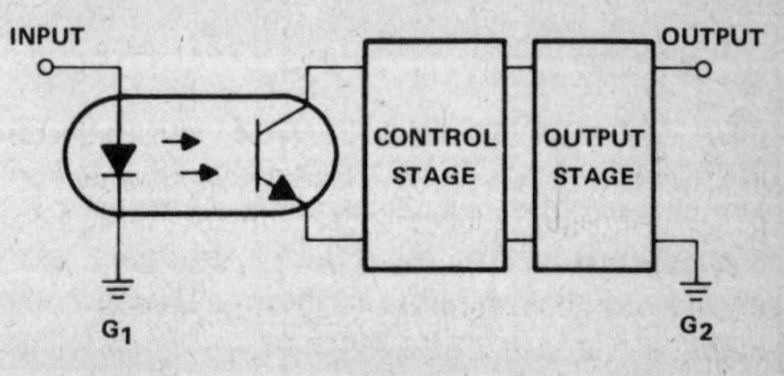

FIGURE 11. Typical Solid-State Relay Using an Optocoupler

isolated chopper circuit

Chopper circuits that use mechanical relays suffer from a speed problem as well as switching transients

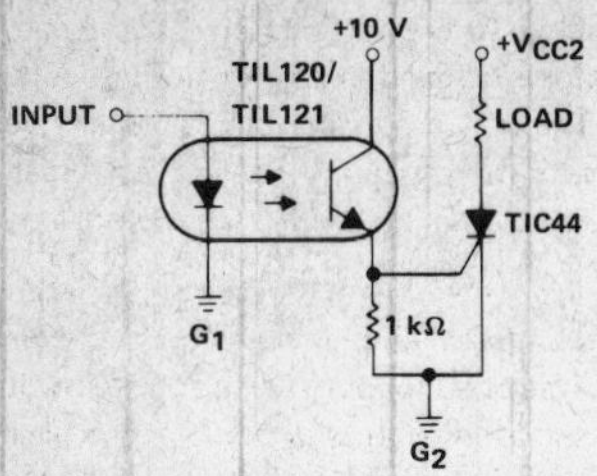

FIGURE 12. Solid-State Latch Using a TIL120/TIL121

at the load. By using bipolar transistors or FETs as series and shunt switching elements, the speed may be improved; but capacitive coupling to the switching circuitry may still produce transient ''spikes'' on the output signal. By using an optocoupler to switch the input signal as shown in Figure 13, the switching circuitry can be isolated from the output, thereby

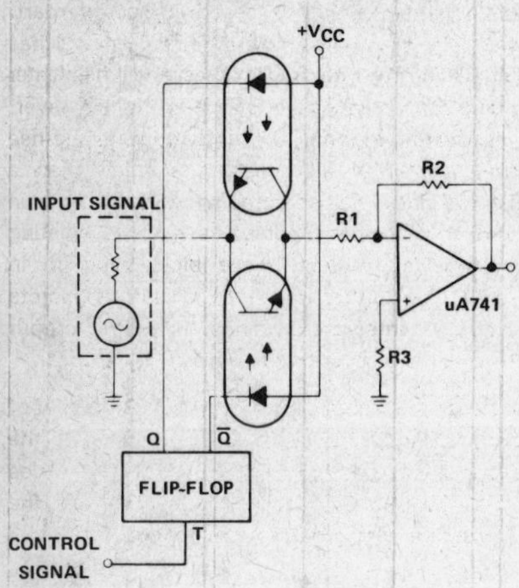

FIGURE 13. Chopper Circuit Using Optocouplers

reducing output ''spikes''. The use of two couplers in the configuration shown allows chopping of either positive or negative input signals with a frequency of one-half that of the input to the flip-flop. The uA741 operational amplifier is used to increase the output signal with a gain of R2/R1.

pulse amplifiers

Pulse amplification, as well as isolation, can be achieved by using an optocoupler with a pulse amplifier. The circuit shown in Figure 14 uses an isolator with a uA741 operational amplifier to amplify

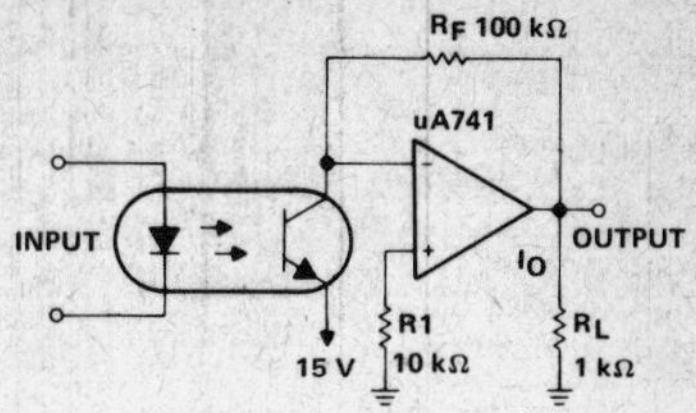

FIGURE 14. Isolated Pulse Amplifier Using Optocoupler uA741 Operational Amplifier

the pulse appearing at the anode of the IRED. The gain of this circuit is controlled by the feedback resistor R_F. An amplifier employing discrete components and that uses the TIL102/TIL103 as part of the current feedback pair is shown in Figure 15. The feedback resistor R1 controls the current gain as well as the output d-c level.

Figure 16 shows an optocoupler with a voltage-feedback amplifier that has a gain of 1 + R2/R1. This type of amplifier offers high input impedance, which will not load the emitter of the sensor transistor.

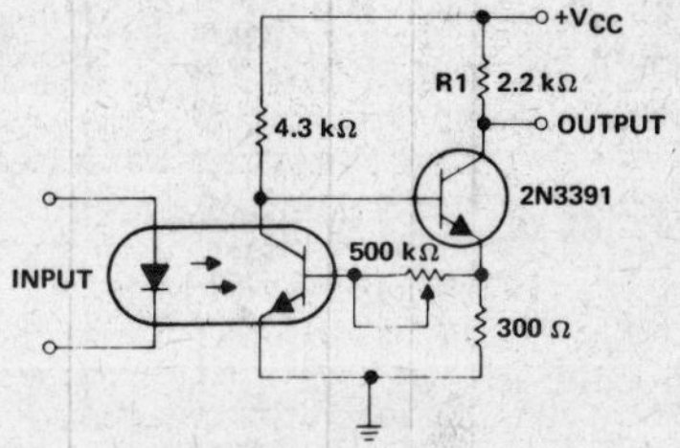

FIGURE 15. Discrete-Component Pulse Amplifier with TIL102/TIL103

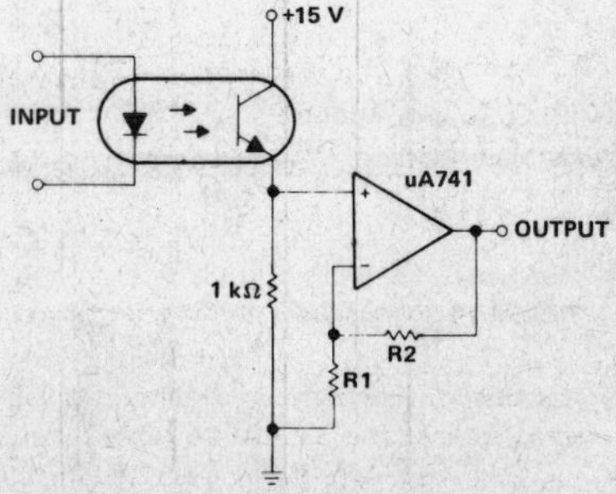

FIGURE 16. Voltage-Feedback Pulse Amplifier with Optocoupler

description

A very useful application of optocouplers is in the interface between different families of digital logic circuits. The worst-case design process should include consideration of data rates, power supply variations, component tolerances, and temperature ranges as well as the characteristics of the digital logic families. Consider the general circuit of Figure 1.

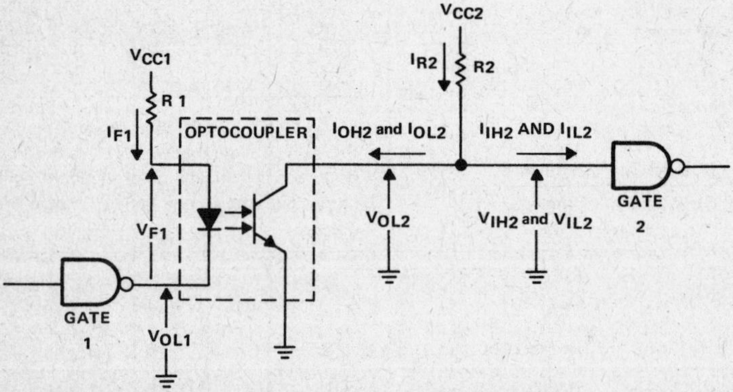

NOTE: V_{OL2} = LOW-LEVEL OUTPUT VOLTAGE OF COUPLER WHEN COUPLER IS ON
V_{IL2} = LOW-LEVEL INPUT VOLTAGE SPECIFIED FOR GATE 2.

Figure 1. Optocoupler Interface Circuit

When the output of logic circuit 1 is low (V_{OL1}), the output of the optocoupler is also low (V_{OL2}). Since V_{OL2} is the input to logic circuit 2, it must be less than the maximum required logic low input voltage (V_{IL2}), in order to hold logic circuit 2 in a stable state. The criteria that must be met at this point is given in equation 1.

$$V_{OL2}\ (\text{coupler}) \leq V_{IL2}\ (\text{max})\ (\text{logic circuit}) \qquad (1)$$

When the coupler output is in the low state it must not only sink the current through R2, I_{R2}, but it must also sink any current required out of the logic circuit 2 input in order to hold logic circuit 2 input to V_{IL2} or less.

Using the current directions specified in Figure 1 and with the conditions of equation 1 satisfied, the conditions required for the coupler current, I_{OL2}, can be expressed as in equation 2.

$$I_{OL2} \geq I_{R2} - I_{L2} \qquad (2)$$

The first step in the design procedure is to select I_{F1}, the forward current through the emitter of the optocoupler. Then using equation 3, R1 is computed:

$$R1 = \frac{V_{CC1} - V_{F1(typ)} - V_{OL1(typ)}}{I_{F1(typ)}} \qquad (3)$$

A standard value resistor for R1 is selected as close to the value computed using equation 3. A tolerance for this resistor is specified from which the maximum and minumum values for R1 are computed using equations 4a and 4b as follows:

$$R1(max) = R1(1 + \frac{tol}{100}) \qquad (4a)$$

$$R1(min) = R1(1 - \frac{tol}{100}) \qquad (4b)$$

"tol" is the percent tolerance of the resistor. With the results of operations 4a and 4b, the maximum and minimum values of I_{F1} can be determined using equation 5a and 5b.

$$I_{F1(max)} = \frac{V_{CC1(max)} - V_{F1(min)} - V_{OL1(min)}}{R1(min)} \qquad (5a)$$

APPLICATIONS 15

$$I_{F1(min)} = \frac{V_{CC1(min)} - V_{F1(max)} - V_{OL1(max)}}{R1_{(max)}} \quad (5b)$$

The output current of the coupler depends on the current transfer ratio (CTR) of the device. CTR is defined by equation 6a as the coupler output current, I_{OL2}, divided by the forward current, I_{F1}, of the coupler diode emitter.

$$CTR = \frac{I_{OL2}}{I_{F1}} \quad (6a)$$

If CTR is not given as a data sheet parameter, it can be calculated from other data sheet specifications (e.g., $I_{C(on)}$ at a certain I_F) or from curves of I_{OL} (sometimes called I_C) vs I_F given in the data sheet. In many cases CTR will be a number less than one, in other cases it will be greater than 1.

Using equation 6a with CTR converted to a percent, the coupler collector current can be computed using equation 6b.

$$I_{OL2(min)} = \frac{(\%\ CTR) \times I_{F1(min)}}{100} \quad (6b)$$

The minimum value for R2 can be calculated using equation 7.

$$R2_{(min)} = \frac{V_{CC2(max)} - V_{OL2(max)}}{I_{OH2(max)} + I_{IH2(max)}} \quad (7)$$

The maximum value of R2 is determined from the condition that exists when the optocoupler output transistor is in the off state. Under these conditions any offstate current, I_{OH}, and any current into the input of gate 2 must not drop the voltage across R2 to the point where the input to gate 2 goes below its required high-level limit value, V_{IH}. These limit conditions are expressed in equation 8, again using Figure 1. I_{OH2} is the current into the output collector and I_{IH2} is the input current to gate 2 when the gate input is at a voltage equal to or greater than the $V_{IH(min)}$ voltage required. $I_{OH2(max)}$, $V_{IH(min)}$, and $I_{IH2(max)}$ are taken from data sheet specifications.

$$R2\ (max) = \frac{V_{CC2(min)} - V_{IH2(min)}}{I_{OH2(max)} - I_{IH2(max)}} \quad (8)$$

R2 is selected between the limits of R2(min) and R2(max). Capacitive effects on response time are less when R2 is closer to R2(min), while maintaining the low-logic-level voltage, V_{IL2}. As the CTR of the optocoupler degrades, correct circuit operation will be maintained longer with R2 closer to R2(max). Final selection depends on which parameter is more important in the application.

example number 1

In Figure 2, a 4N25 optocoupler is to be driven by an SN7404 gate output and will drive the input of an SN7400 gate. The specifications for the logic levels and input and output currents for the Series 74 logic family are given in Table 1.

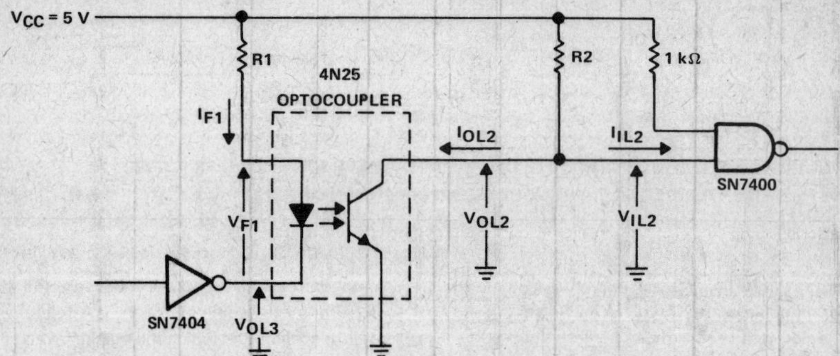

NOTE: V_{OL2} = LOW-LEVEL OUTPUT VOLTAGE OF COUPLER WHEN COUPLER IS ON.
V_{IL2} = LOW-LEVEL INPUT VOLTAGE SPECIFIED FOR SN7400.

Figure 2. Optocoupler Interface Circuit

Table 1. Series 74 Family Data

TTL Family	V_{IL} V	I_{IL} mA	V_{IH} V	I_{IH} μA	V_{OL} V	I_{OL} mA	V_{OH} V	I_{OH} μA
74	0.8	−1.6	2	40	0.4	16	2.4	−400
74H	0.8	−2	2	50	0.4	20	0.24	−500
74LS	0.8	−0.3	2	20	0.5	8	2.7	−400
74L	0.7	−0.18	2	10	0.4	3.6	2.4	−200
74S	0.8	−2	2	50	0.5	20	2.7	−1000

For the particular calculations the values in Table 2 will be used.

Table 2 Calculation Values

TTL	4N25	Power Supply
$V_{IH(min)} = 2\ V$	CTR(min) = 20%	$V_{CC} = 5\ V \pm 5\%$
$V_{IL(max)} = 0.8\ V$	$V_{F(min)} = 1.2\ V$ @ 10 mA	
$I_{IH(max)} = 40\ \mu A$	$V_{F(typ)} = 1.25\ V$ @ 10 mA	
$I_{IL(max)} = -1.6\ mA$	$V_{F(max)} = 1.5\ V$ @ 10 mA	
$I_{OH(max)} = 400\ \mu A$	$I_{OL(max)} = 50\ V$	
$V_{OL(typ)} = 0.2\ V$	$V_{OL(max)} = 0.5\ V$	
$V_{OL(max)} = 0.4\ V$		

calculations

1) Select $I_F = 20\ mA$

2) Check equation 1

$V_{OL}(\text{coupler}) \leq V_{IL2}$ (logic circuit)

0.5 V ≤ 0.8 V It checks.

From equation 3, assuming the V_F at 20 milliamperes is not 0.05 volt greater than the value at 10 milliamperes.

3) $$R1 = \frac{5 - 1.25 - 0.2}{20\ mA}$$

R1 = 178 Ω

Select standard value R1 = 180 Ω ±10%.

Therefore,

4) R1(max) 180 + 18 = 198 Ω

R1(min) 180 − 18 = 172 Ω

5) From equation 5a and 5b, using $V_{OL(typ)} = 0.2\ V$ for $V_{OL(min)}$

$$I_{F1(max)} = \frac{(5.25 - 1.2\ V - 0.2)\ V}{171\ \Omega} = 21.38\ mA$$

$$I_{F1(min)} = \frac{(4.75 - 1.5\ V - 0.4)\ V}{198\ \Omega} = 14.39\ mA$$

6) From equations 6 and 7

$$I_{OL2(min)} = \frac{14.39\ mA \times 20}{100} = 2.878\ mA$$

$$R2(min) = \frac{(5.25 - 0.5)\ V}{2.878\ mA + (-1.6\ mA)} = 3.72\ k\Omega$$

7) $$R2(max) = \frac{4.75 - 2}{400\ \mu A + 40\ \mu A} = 6.25\ k\Omega$$

A choice of 4.7 Ω ±10% for R2 is suitable for this design.

example number 2

A similar approach can be used when interfacing discrete phototransistors to digital logic circuits. Consider a TIL81 connected in the phototransistor mode to an SN7400 as shown in Figure 3. The data for this situation is shown in Table 3.

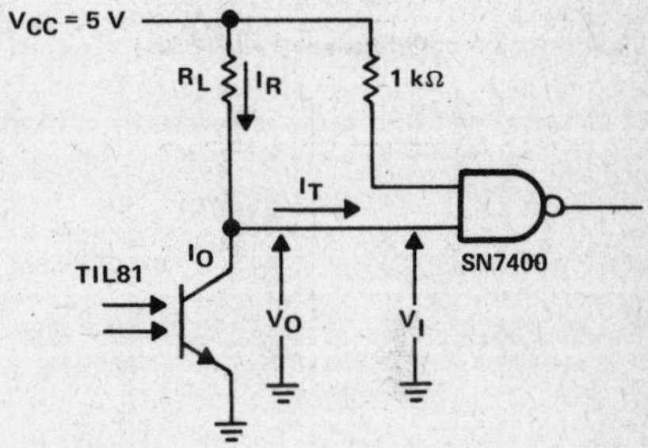

Figure 3. Phototransistor Interface Circuit

Table 3. Calculation Values

SN7400	TIL81	Power Supply
$V_{IH(min)} = 2\ V$	$I_D = 20\ \mu A$ (dark current)	$V_{CC} = 5\ V \pm 5\%$
$V_{IL(max)} = 0.8\ V$	$I_{OH} = I_D + (1 - n/100)\ I_{OL}$	
$I_{IH(max)} = 40\ \mu A$	(where n = % light blocked)	
$I_{IL(max)} = -1.6\ mA$	$V_{OL(max)} = 0.8\ V$	
$I_{OH(max)} = 40\ \mu A$	$I_{OL(min)} = 2\ mA$	
$V_{OL(typ)} = 0.2\ V$		
$V_{OL(max)} = 0.4\ V$		

calculations

In this application the equations before equation 7 are ignored. From equation 7 and 8, the values for R2(min) and R2(max) can be calculated. This application is very sensitive to ambient light. Therefore, care must be taken to shield out ambient light.

Assuming 95% of the ambient light is shielded out,

$$R_{L(min)} = \frac{5.25 - 0.8}{2\text{ mA} + (-1.6\text{ mA})} = \frac{4.45\text{ V}}{0.4\text{ mA}} = 11.1\text{ k}\Omega$$

$$R_{L(max)} = \frac{(4.75 - 2.0)\text{ V}}{I_{OH} + 40\ \mu A}$$

Substituting $I_{OH} = I_D + [1 - (n/100)]\ I_{OL}$, where n = 95%

$$R_{L(max)} = \frac{4.75 - 2.0}{20\ \mu A + (1 - \frac{95}{100})\ 2\text{ mA} + 40\ \mu A}$$

$$= \frac{2.75\text{ V}}{20\ \mu A + 100\ \mu A + 40\ \mu A}$$

$$= \frac{2.75\text{ V}}{160\ \mu A}$$

$$= 17.2\text{ k}\Omega$$

R_L is chosen as a standard value, 14.7 kΩ.

example number 3

If the 74LS series is used with 80% light blocked, from Table 1 $I_{IL(max)}$ = 0.36 mA instead of 1.6 mA and $I_{IH(max)}$ = 20 μA instead of 40 μA.

$$R_{L(max)} = \frac{4.75 - 2}{20\ \mu A + (1 - 80/100)\ 2\text{ mA} + 20\ \mu A} = 6.25\text{ k}\Omega$$

and

$$R_{L(min)} = \frac{5.25 - 0.8}{2\text{ mA} + (-0.36\text{ mA})} = 2.71\text{ k}\Omega$$

Therefore, R_L can be selected between 6.25 kΩ and 2.71 kΩ.

examples number 4 and 5

Substituting appropriate values when the 74L series is used with 80% light blocked, then the values of $R_{L(max)}$ and $R_{L(min)}$ are 6.4 kΩ and 2.5 kΩ.

For the 74H series, $R_{L(max)}$ is 5.85 kΩ and $R_{L(min)}$ is unbounded.

BAR-CODE SCANNING

bar codes and bar code scanners

Many point-of-purchase cash registers (terminals) identify the product that is sold by scanning a code of lines printed on the label or packaging for the product. The printed code is called a bar code. A typical bar code is shown in Figure 1.

Bar codes are usually horizontal with alternating vertical dark bars and light spaces. Data is encoded by varying the width of these bars and spaces. To retrieve data a scanner is moved across the bar code by the operator. The bar-code-pattern must be large enough to allow the operator to easily move the tip of the scanner across the bar code and remain within the space allocated to the code (see Figure 1).

There are a variety of bar codes in use. Some of these are:

MSI
UPC (Universal Product Code)
EAN (European Article Number)
CODABAR
2-of-5
2-of-5-Interleaved
Code 39

Detail specifications and tolerances relating to many of the bar codes do not exist. Lack of detailed specification

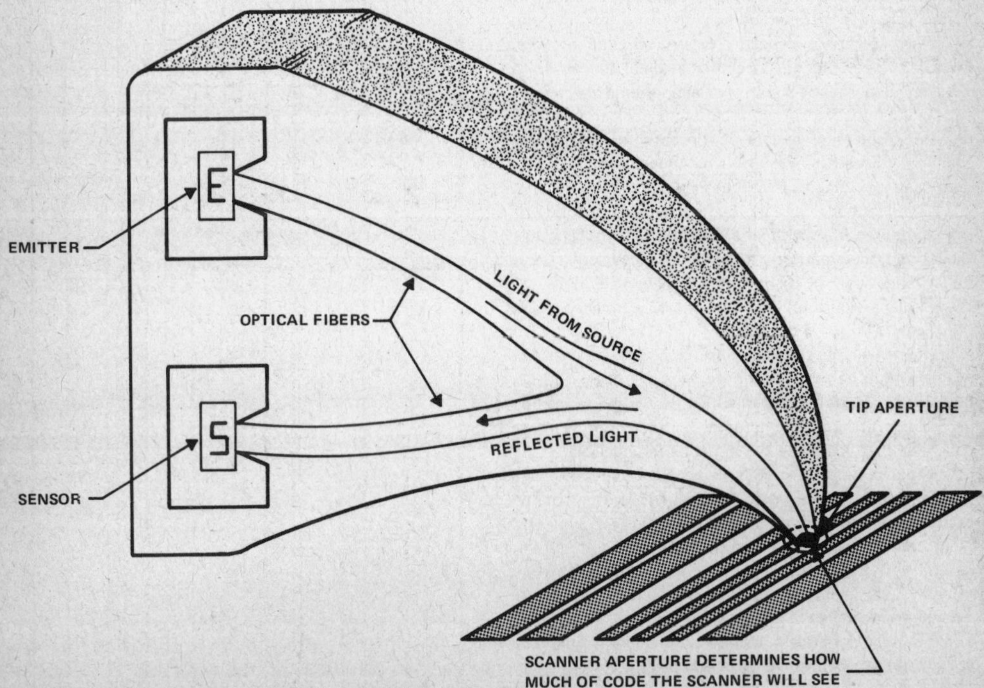

Figure 1. Bar Code Wand

APPLICATIONS
15

allows wide variations in a single bar code by different users. As a result, any bar-code scanner must be designed to handle wide variations.

The TI bar-code reader is a self-contained wand scanner. This scanner has both a light source and a light detector in the same package as shown in Figure 1. In this case, the light emitted from the source is projected through an opening in the tip of the wand. The beam of light strikes the bar code and is reflected back into the wand tip to the light detector. The light source is connected to the tip through an optical fiber, which guides the light to the tip. The reflected light uses the same optical path from the tip back to a ''Y'' junction, and then a portion of the reflected light is directed to the detector

wand scanner aperture

The aperture of a scanner refers to the diameter of the opening through which the reflected light passes. The aperture determines how much of the code the scanner will see. These are shown in Figure 1. Do not mistake the aperture of the scanner with the aperture of the scanner tip. The tip aperture is selected such that all of the wand apertures will work with a single tip. Thus, a scanner with a 6-mil aperture sees an area with a diameter of 0.006 inch.

The amount of light reaching the detector depends on the scanner aperture size. Large scanner apertures allow greater amounts of reflected light to reach the detector, while smaller apertures allow lesser amounts of light. Because the detector has a nominal range of light to which it will respond, scanners with smaller apertures may require more light from the source than larger ones to meet the reflected light requirements of the detector. In general, scanners with small apertures will consume more power than scanners with large apertures.

Some bar codes can be read better by scanners with large apertures while with other bar codes, a small aperture is best. To understand why this is true, a closer look at how the scanner works is in order. Refer to Figure 2. Figure 2a shows a large aperture; Figure 2b shows a small aperture.

If the aperture of the scanner is too large and the bar width too small, the scanner will not recognize the bar. For example, if the bar is 4 mils wide and the aperture is 10 mils, 60% of the aperture will reflect light from the spaces on each side of the bar. The detected light may not decrease to a level that will allow the bar to be recognized.

If the aperture selected is too small, a print flaw such as an ink speck may be incorrectly decoded as a narrow bar, or an ink void recognized as a space.

Contrast or recognition tolerance and power consumption form the selection criteria for scanner aperture selection with respect to a given bar code. The aperture must be small enough to recognize the bars and large enough to tolerate the print flaws. Its power consumption must be acceptable for the desired application.

sample rate and velocity

In order to decode information contained in the bar-code pattern, the relative widths of the light and dark bars have to be determined. If the velocity of the wand moving past the bar code were constant, the distance the wand traveled could be measured linearly and the widths of the bars could be expressed in thousands of inches. Unfortunately the velocity of the wand is not constant. Each person using a wand will scan bar codes differently than another person. Typically scanning velocities are in the range of 76 to 760 millimeters (3 to 30 inches) per second.

Since the movement of the wand is variable, the widths of the bars are determined by measuring the relative time the wand sees them. This is done by sampling the wand output at a constant rate and comparing the rate of change between the light and dark bars to the constant rate. This is controlled by software within the processor that controls the display terminal.

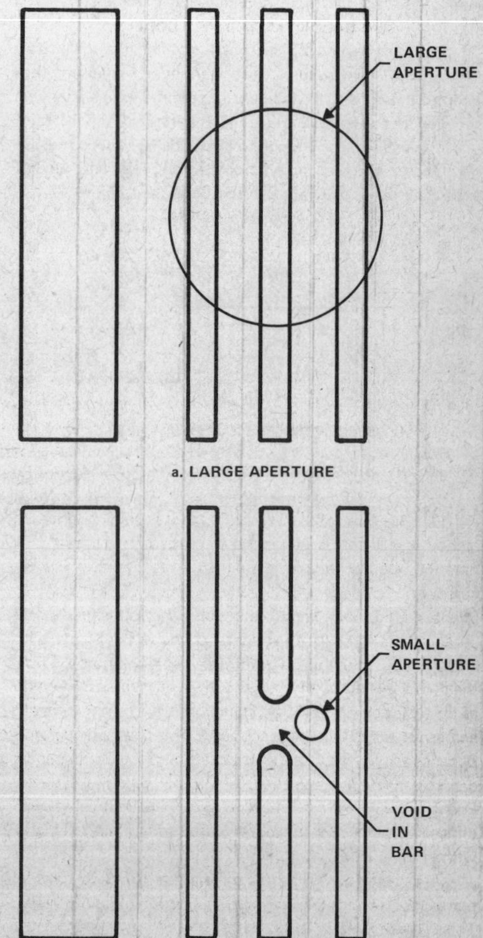

a. LARGE APERTURE

b. SMALL APERTURE

Figure 2. Example of Scanner Apertures

bar-code decoding

When a wand scanner is moved across a bar code as shown in Figure 3a, an electrical signal is produced by the scanner as shown in Figure 3b. It is this signal that is converted to a digital output signal as shown in Figure 3c, and interpreted to determine the proper character represented by the bar code. Each light and dark space in the bar code is equal to or greater than a unit size called a module.

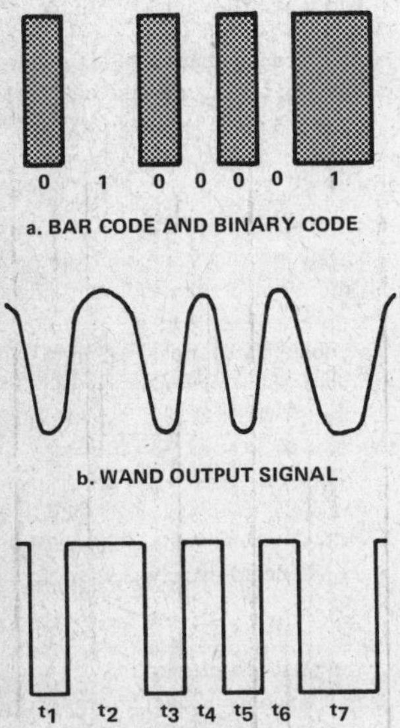

a. BAR CODE AND BINARY CODE

b. WAND OUTPUT SIGNAL

c. DIGITAL OUTPUT SIGNAL

Figure 3. Digital Output Signal From Bar Code

Electrical signal periods are not measured in terms of inches or millimeters, but instead are measured in terms of time. In order to produce the signal, the scanner is moved across the bar code with a certain velocity. If the velocity of the scanner was known and constant, the signal could be accurately translated to inches. The width of the bars and spaces could then be determined and compared. Since the velocity is not known, width is expressed as a function of time. Not only is velocity unknown, it is also not constant across the bar code. This is the result of the human operator. Because the operator cannot maintain precise velocity, some degree of tolerance must be built into the decoding scheme. Even if velocity were constant, dimensional widths between different sizes of bar codes would produce varying signal widths in terms of time.

A velocity of 76 to 760 millimeters (3 to 30 inches) per second is tolerable for most operators and code types. This represents a variable relationship of 1 to 10. Since the module variation in Figure 3 is only 2 to 1, individual module measurement is meaningless. Thus, module width comparison is the only approach.

In the illustrated code, the characters may be represented with seven binary bits. Seven binary bits provide 128 variations. However, in this case, other restraints are imposed on the code to provide such things as self-checking. This limits the variations to 20 possibilities and decoding short cuts may be used.

Instead of arriving at the conclusion that timing widths represented by t_1, t_3, t_4, t_5, and t_6, are approximately the same and are approximately one-half the widths represented by t_2 and t_7, which could yield a binary number of 0100001, a different approach is used.

The odd times representing the bars are compared to each other on a pair-by-pair basis. The possibility for each comparison is that the second time is equal to, less than, or greater than, the first time.

Example:
1. $t_1 : t_3$ equal
2. $t_3 : t_5$ equal
3. $t_5 : t_7$ greater than

Similarly, the even times representing spaces can be compared.

Example:
1. $t_2 : t_4$ less than
2. $t_4 : t_6$ equal

To understand this approach, digits can be assigned to these comparative relationships.

equal is	1
greater is	2
less is	0

The result from the above comparisons would be a number 11201. This number is smaller in size than the number 0100001. The number could then be used as a table pointer to reach the resultant character. The advantage of this short-cut approach is that the table can be divided into five parts. As each comparison is made, the number of possibilities is reduced so that the part of the table to be considered can be reduced until finally the fifth comparison locates the exact number to reach the digit. If the binary approach to interpretation is used, it cannot be achieved until all the times are presented (the complete character is scanned).

MSI bar code characteristics

The MSI Bar Code has these characteristics. The bars and spaces are binary in width. The narrow ones are one module wide and wide ones are two modules wide as shown in Figure 4. Each character is composed of four data bits; each bit is three modules wide and made up of a bar and a space. Thus, each character contains four bars and four spaces. The primary algorithm is binary and applied only to the dark bars. The narrow bars are assigned a binary zero; the wide ones are assigned a binary one. The bar code character set is limited to the digits 0 through 9 by definition. The characters are encoded in the bar code by the secondary algorithm, which is binary coded decimal (BCD). Table 1 lists the character set for MSI Bar Code.

Table 1. MSI Character Set

Decimal	BCD
0	0000
1	0001
2	0010
3	0011
4	0100
5	0101
6	0110
7	0111
8	1000
9	1001
Start Symbol	1
Stop Symbol	00

The code is not self-checking, i.e., the algorithm does not allow each character to be checked independently of the other characters.

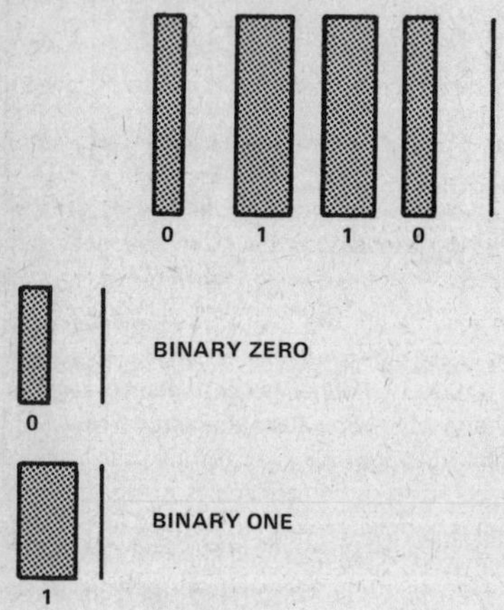

Figure 4. MSI Bar Code

The MSI code can be considered self-clocking, however, with the timing track being an integral part of the code. The complete symbol consists of the following elements: a forward start code, the data characters, one or two check digits, and a reverse start code as shown in Figure 5. The forward start code is a single 'one' bit (wide bar/narrow space) and the reverse start code, sometimes referred to as the stop code, is a single 'zero' bit followed by a narrow bar (narrow bar/wide space/narrow bar). The start symbol and the stop symbol are dissimilar and allow bidirectional scanning. The data field is between the start/stop codes and can extend to 15 characters (including two check digits). The two check digits associated with the MSI bar code are the last two digits (left to right) in the code. Each check digit is the checking digit for all preceding digits. Thus, the second check digit checks the first check digit.

The second check digit is always IBM modulus 10. It is used exclusively by the scanner circuitry and is not transferred to the terminal with the other scanned digits. The check bit is a calculated number based on dividing by the modulus number. The first check digit may be retained with the other digits of the bar code, or it may be discarded. It is, however, used by the checking circuitry and thus, must be valid. The decision to retain it, or drop it, is controlled by a terminal parameter. If the terminal parameter is set to retain this check digit, the check digit may be either IBM modulus 10 or IBM modulus 11. However, if the terminal parameter is set to discard this check digit, it must be IBM modulus 10.

code 39 bar code characteristics

The bars and spaces in Code 39 are binary in width: the narrow bars and spaces represent a binary zero and the wide bars and spaces represent a binary one. Each character is made up of 9 elements, five bars and four spaces. Three of these elements are wide and six are narrow, hence, the name Code 39 (3 or 9). Figure 6 illustrates the character structure. The primary algorithm is binary and is applied to both the bars and the spaces in the code. Narrow bars or spaces represent binary zero and wide bars or spaces represent binary one.

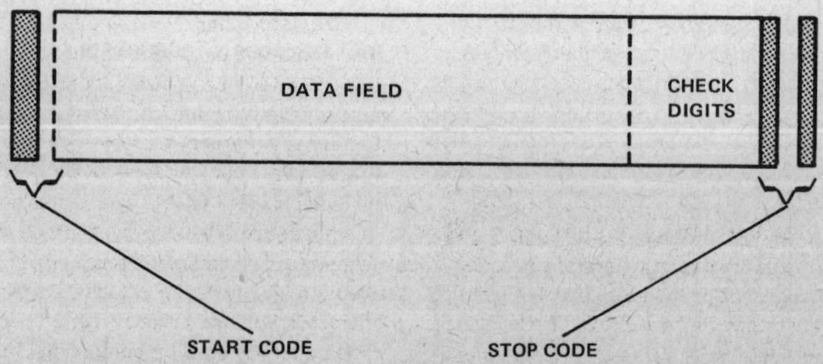

Figure 5. MSI Bar Code

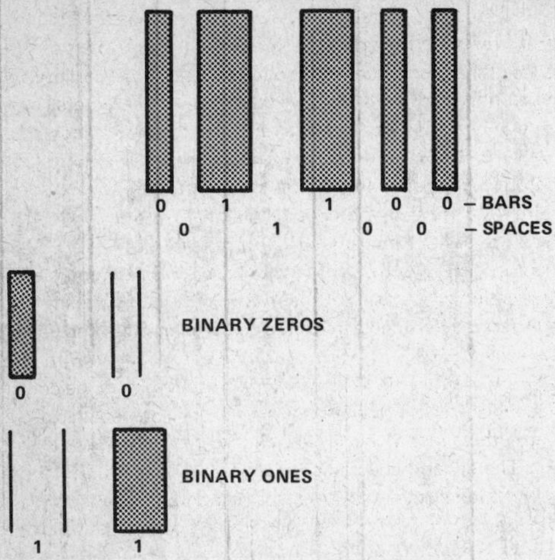

Figure 6. Code 39 Bar Code

Table 2. Code 39 Character Set

Character	Bars	Spaces	Character	Bars	Spaces
1	10001	0100	M	11000	0001
2	01001	0100	N	00101	0001
3	11000	0100	O	10100	0001
4	00101	0100	P	01100	0001
5	10100	0100	Q	00011	0001
6	01100	0100	R	10010	0001
7	00011	0100	S	01010	0001
8	10010	0100	T	00110	0001
9	01010	0100	U	10001	1000
0	00110	0100	V	01001	1000
A	10001	0010	W	11000	1000
B	01001	0010	X	00101	1000
C	11000	0010	Y	10100	1000
D	00101	0010	Z	01100	1000
E	10100	0010	–	00011	1000
F	01100	0010	.	10010	1000
G	00011	0010	Space	01010	1000
H	10010	0010	$	00000	1110
I	01010	0010	/	00000	1101
J	00110	0010	+	00000	1011
K	10001	0001	%	00000	0111
L	01001	0001			
START:	00110	1000			
STOP:	00110	1000			

The character set of Code 39 is alphanumeric being made up of the following 43 characters: 0 through 9 and A through Z, 6 special characters, and a space character. The characters are encoded though the secondary algorithm, which is modified Binary Coded Decimal (BCD). Table 2 lists the complete character set. Code 39 is a variable length code. The maximum length is typically 32 characters. The code is self-checking and utilizes inner character gaps, however, it is not a self-clocking code. The complete symbol comprises a start code, the data field, and a stop code as shown in Figure 7. The asterisk symbol is used for both the start and stop codes, and allows bidirectional scanning.

Numeric values are assigned to each Code 39 character as shown in Table 3.

Table 3 – Code 39 Numeric Values

0	0	F	15	U	30
1	1	G	16	V	31
2	2	H	17	W	32
3	3	I	18	X	33
4	4	J	19	Y	34
5	5	K	20	Z	35
6	6	L	21	–	36
7	7	M	22	.	37
8	8	N	23	Space	38
9	9	O	24	$	39
A	10	P	25	/	40
B	11	Q	26	+	41
C	12	R	27	%	42
D	13	S	28		
E	14	T	29		

Since Code 39 is discrete, i.e., self-checking; it does not require a check digit. An optional check digit may be employed when necessary. The check digit is the modulus 43 sum of all the character values in a given message, and is printed as the last data character. For example, the sum of the values of the following data would be:

Data 12345ABCDE/
Sum of Values = 1 + 2 + 3 + 4 + 5 + 10 + 11 + 12 + 13 + 14 + 40 = 115
115/43 = 2 Remainder 29

The check digit is the character corresponding to the value of the remainder, which in this example is 29 or "T". The data above with its check digit reads:

12345ABCDE/T

2-of-5-interleaved characteristics

The 2-of-5 and 2-of-5-Interleaved bar codes are quite similar and will be discussed together in this section. The differences will be pointed out as they arise. The bars and spaces of both codes are binary in width. The narrow bars and spaces are one module wide and the wide bars and spaces are three modules wide (Refer to Figure 8). A character is made up of 5 bars, 4 spaces that separate them, plus the space following the last bar. In 2-of-5-Interleaved (Figure 9), one character is made up of 5 bars and the second character is made up of the 5 spaces. Thus, the two characters are interleaved. The different spaces of the bar codes can be noted by comparing Figure 8 and Figure 9.

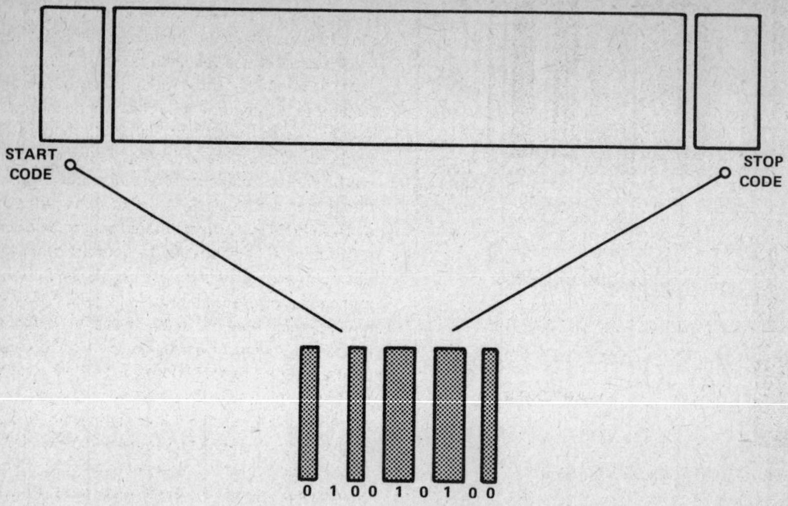

Figure 7. Code 39 Bar Code

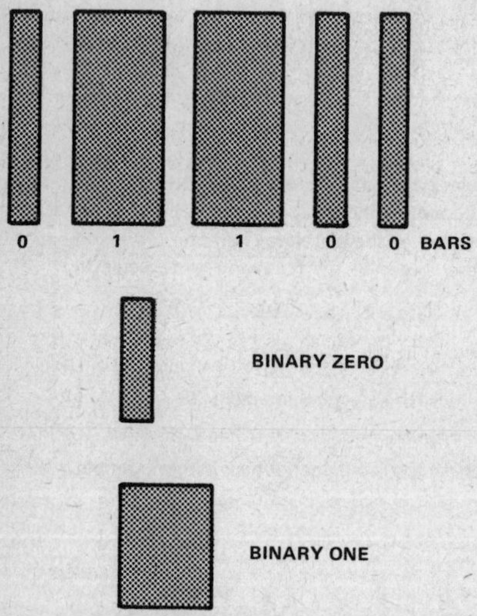

Figure 8. 2-of-5 Bar Code

The primary algorithm is binary for both codes. It is applied only to the dark bars in the 2-of-5 code with all the spaces being one module wide, but in the 2-of-5-Interleaved the algorithm applies to both bars and spaces. A narrow bar or space is assigned a binary zero and a wide bar or space is assigned a binary one.

The character set for both 2-of-5 and 2-of-5-Interleaved is 0 through 9, a start code, and a stop code. The secondary algorithm is modified BCD as shown in Table 4. In the 2-of-5 code, a character is encoded in the dark bars while the spaces have no value. In the 2-of-5-Interleaved code, the first character is encoded in the dark bars and a second character is encoded in the spaces. These codes are both self-checking, however, they are not self-clocking codes.

Figure 10 illustrates the complete symbols for the two codes. The symbol consists of a start code, the data field, and a stop code. The data field is variable in length and may contain any number of characters.

The start and stop codes are unique and provide bidirectional scanning. The start code for 2-of-5 is a "1 1 0" and the stop code is a "1 0 1." The start code for 2-of-5-Interleaved is "0 0" and the stop code is "1 0".

Table 4. 2-of-5, 2-of-5-Interleaved Character Set

Decimal	Modified BCD
0	00110*
1	10001
2	01001
3	11000
4	00101
5	10100
6	01100
7	00011
8	10010
9	01010

	START CODE	STOP CODE
2-of-5	110	101
2-of-5-Interleaved	00	10

*The decimal "0" is an exception to this modified BCD. Its BCD value of eleven must be ignored.

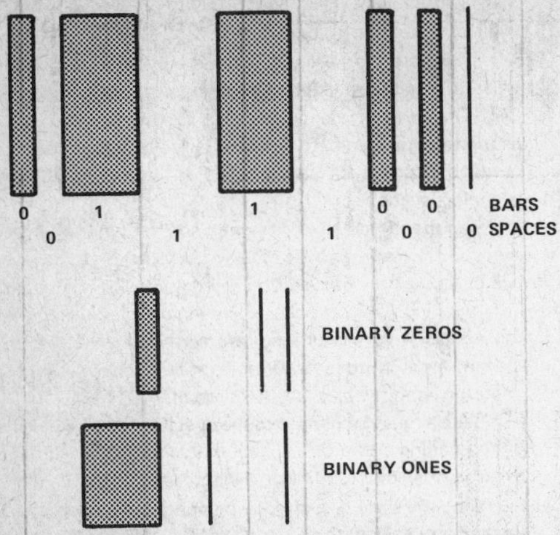

Figure 9. 2-of-5-Interleaved Bar Code

print quality

The process of printing the bar code symbols must be carefully controlled to assure that the printed labels are close to the specifications. Lithography, Gravure, letter press, Offset, and dot-matrix printing techniques are currently being used to print bar-code labels, however, the quality of the print is more important than the type of printer used. If details that effect the quality of the bar code-label are understood, an analysis of any label can be made with little regard for the type of printer used. Dot matrix printers do have special problems because of their nonuniform structure, but the same basic requirements must be met in order to reproduce readable labels.

The items that effect print quality of the labels are: the background substrate, ink reflectance, contrast, voids and specks, edge roughness, and ink spread (or shrink). Background substrate refers to the material that the labels will be printed on. It should be white and have a matte finish rather than a high gloss. A background diffuse reflectance of 70 to 80% in the near infrared spectrum is desirable. This reflectance directly effects the contrast of the label. The ink film color should be black and not exceed 24% reflectance in the near-infrared spectrum. The reflectance value should not vary more than 5% within the same character. A carbon-based ink should be used, not an alcohol-based ink. The contrast refers to the difference of the ink reflectance and the background reflectance and should be 50 to 65%. Voids are a result of missing ink coverage. They usually appear in the form of small light spots within the individual bars. seldom can voids be prevented entirely. However, it is essential to good first-pass read rates that they remain within acceptable tolerances. A speck is the opposite of a void. It is extraneous ink in the space or light bar area. Like voids, specks should be held within acceptable tolerances.

In general, the width of these anomalies is more critical then the height. An anomaly with a large height dimension increases its probbility of detection by the scanner. However, if its width is quite small, the bar code interpretation logic can accommodate the flow. If the width of the anomaly is large, and it is detected by the scanner, the results will be a "no read" of the bar code. Most bar

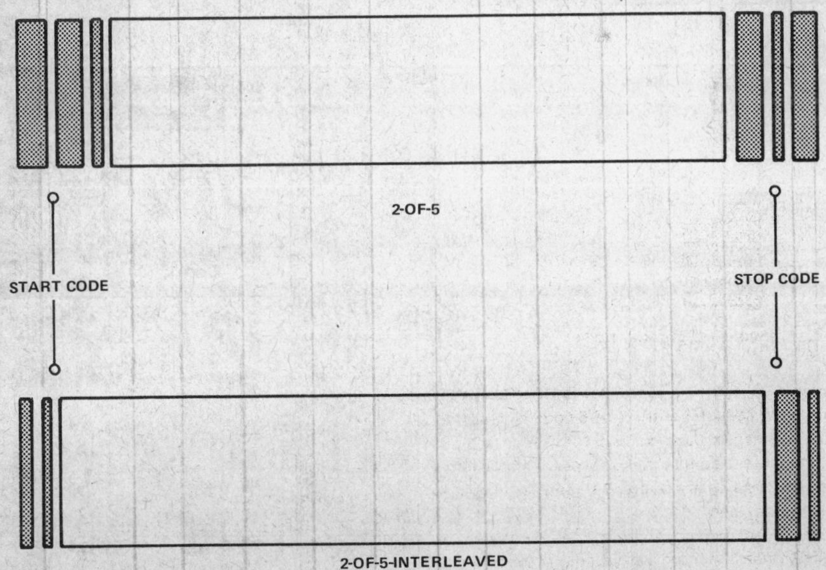

Figure 10. 2-of-5-Interleaved Bar Code

codes have sufficient checks to prevent a misread assuming the vendor makes all of the checks.

Edge roughness pertains to the left and right terminal borders of the dark bars. Edge roughness affects the width of the bars. If the bars are not printed with crisp terminal borders, it is difficult to maintain the bar width tolerances specified by the specific bar code.

Over and under printing is the result of too much or too little ink in printing the bar code. Over printing will make the dark bars wider and the light bars or spaces narrower. Under printing has the opposite effect. Either of these conditions will affect wand scanning if the tolerances of the bars and spaces are exceeded. Figure 11 illustrates the effects of over or under printing.

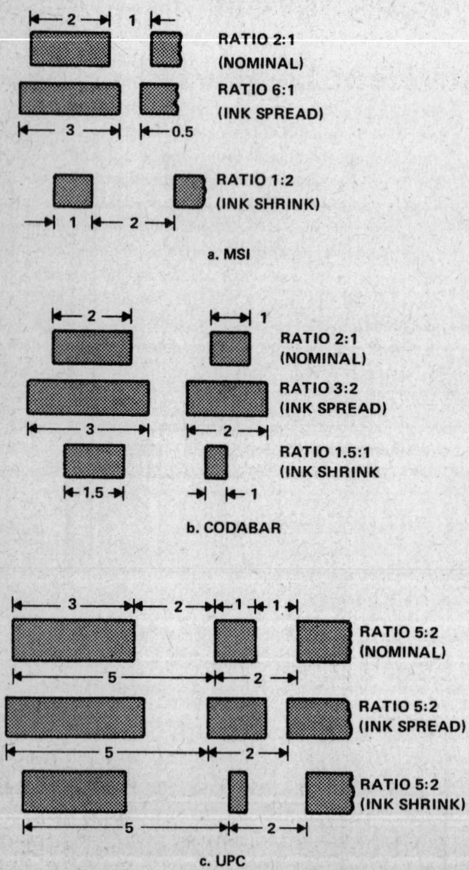

Figure 11. Effects of Uniform Ink Shrinks or Spreads

why codes won't read

There are many reasons some bar codes won't read while others are readable. The first is that the electronics fails. The wand or wand electronics can malfunction as well as the terminal itself. The second is the label itself and any number of problems can degrade an otherwise effective scanning system.

The wand has been designed so that when the tip of the wand is in contact with the symbol, the reflected image will be accurately focused on the detector. If the wand is not touching the symbol, i.e., scanning symbols through thick glass, a very poor read rate will occur. The reason is that the focal length of the wand has been changed and the detector cannot focus on the symbol.

Labels must be printed to the specifications established by the coding authority. A label that is out of specification may or may not read well depending upon how out of tolerance the symbol is after being printed. All labels cannot be read optimally with all wands as shown in Figure 12. You can have a high-resolution wand designed to read complex codes such as a CODABAR and Code 39, and a low-resolution wand for less critical codes. The low-resolution wand is less sensitive to printing anomalis making it more suitable for applications using dot matrix printers. Labels printed using alcohol-based ink may read very poorly or not at all. Alcohol-based ink has a very poor light absorption factor resulting in low contrast. To ensure adequate contrast between the background substrate and the printed bar code, a carbon-based ink should be used.

In summary, even though many bar codes exist, the technology of the scanners and optoelectronic emitters and sensors is such that most variations can be accommodated.

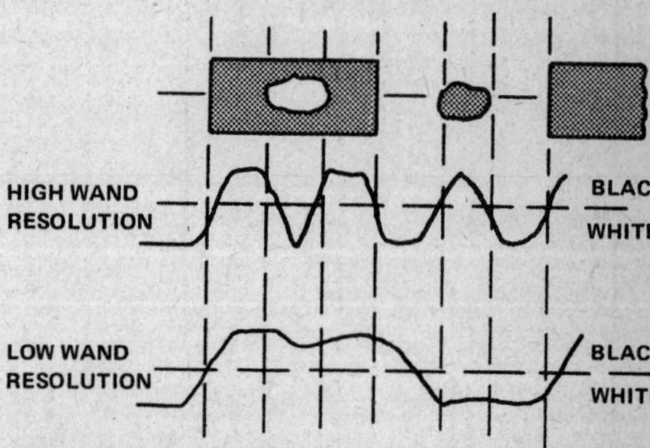

Figure 12. Effects of Ink Voids and Ink Specks

Interchangeability Guide

- Direct Equivalent Devices
- Nearest Equivalent Devices

OPTOELECTRONICS INTERCHANGEABILITY GUIDE

The following interchangeability guide for known optoelectronic devices is intended to serve as a substitution guide for competitive devices to Texas Instruments Optoelectronic Product Line.

Texas Instruments direct replacement devices are believed to be pin-for-pin, mechanically, and electrically interchangeable devices. However, TI does not guarantee that interchangeability in particular application is exact in all respects. Therefore the applicable product sheet should be used to determine product interchangeability. Contact your local TI Sales Office, Authorized Distributor, or Optoelectronic Marketing (Dallas, Texas) for assistance in selecting the appropriate devices for your application.

CODE

A = Direct replacement
B = Electrical or mechanical difference (consult the TI data sheet)

Device	Manufacturer	Equivalent (A) or Nearest (B) TI Device	Code
1N6264	*	TIL31B	B
1N6265	*	TIL33B	B
3N243	*	TIL120	B
3N244	*	TIL120	B
3N245	*	TIL120	B
4N22A	*	4N22	A
4N23A	*	4N23	A
4N24A	*	4N24	A
4N25	*	4N25	A
4N25A	*	TIL154	A
4N26	*	4N26	A
4N27	*	4N27	A
4N28	*	4N28	A
4N29A	*	TIL156	A
4N30	*	TIL113	A
4N31	*	TIL119	A
4N33	*	TIL113	A
4N34	*	TIL113	A
4N35	*	4N35	A
4N36	*	4N36	A
4N37	*	4N37	A
4N51	*	4N41	B
4N54	*	4N56	B
209R	Industrial Electronic Engineers	TIL209A	A
211	Industrial Electronic Engineers	TIL232	B
220R	Industrial Electronic Engineers	TIL220	A
441-0002	Dialight Corp.	TIL111	A
551-0003	Dialight Corp.	TIL112	A
745-0004	Dialight Corp.	TIL304	A
745-0005	Dialight Corp.	TIL305	A
745-0006	Dialight Corp.	TIL302	A
745-0007	Dialight Corp.	TIL311	A
745-0008	Dialight Corp.	TIL308	A
745-0009	Dialight Corp.	TIL306	A
745-0014	Dialight Corp.	TIL312	A
745-0015	Dialight Corp.	TIL327	A
745-0016	Dialight Corp.	TIL313	A
1704R	Industrial Electronic Engineers	TIL305	A
1705R	Industrial Electronic Engineers	TIL306	A
1706R	Industrial Electronic Engineers	TIL308	A
1707R	Industrial Electronic Engineers	TIL311	A
1717R	Industrial Electronic Engineers	TIL309	A
1787R	Industrial Electronic Engineers	TIL321A	A
1788R	Industrial Electronic Engineers	TIL322A	A
5082-4101	Hewlett-Packard	TIL261	B
5082-4150	Hewlett-Packard	TIL281	B
5082-4190	Hewlett-Packard	TIL271	B
5082-4494	Hewlett-Packard	TIL209A	B

5082-4550	Hewlett-Packard	TIL224-1	B
5082-4550	Hewlett-Packard	5082-4550	A
5082-4555	Hewlett-Packard	5082-4555	A
5082-4555	Hewlett-Packard	TIL224-2	B
5082-4650	Hewlett-Packard	5082-4650	A
5082-4655	Hewlett-Packard	5082-4655	A
5082-4855	Hewlett-Packard	TIL220	B
5082-4884	Hewlett-Packard	TIL221	B
5082-4950	Hewlett-Packard	5082-4950	A
5082-4955	Hewlett-Packard	5082-4955	A
5082-7010	Hewlett-Packard	4N57	B
5082-7100	Hewlett-Packard	4N58	B
5082-7100	Hewlett-Packard	TIL305	B
5082-7100	Hewlett-Packard	TIL504	B
5082-7101	Hewlett-Packard	TIL305	B
5082-7300	Hewlett-Packard	TIL307	B
5082-7300	Hewlett-Packard	TIL309	B
5082-7302	Hewlett-Packard	TIL306	B
5082-7302	Hewlett-Packard	TIL308	B
5082-7340	Hewlett-Packard	TIL311	B
5082-7620	Hewlett-Packard	TIL339	A
5082-7621	Hewlett-Packard	TIL339	A
5082-7623	Hewlett-Packard	TIL340	A
5082-7626	Hewlett-Packard	TIL341	B
5082-7730	Hewlett-Packard	5082-7730	A
5082-7731	Hewlett-Packard	5082-7731	A
5082-7740	Hewlett-Packard	5082-7740	A
5082-7751	Hewlett-Packard	TIL321A	B
5082-7756	Hewlett-Packard	TIL330A	B
5082-7760	Hewlett-Packard	TIL322A	B
7620Y	Industrial Electronic Engineers	TIL339	A
7621Y	Industrial Electronic Engineers	TIL339	A
7623Y	Industrial Electronic Engineers	TIL340	A
7630G	Industrial Electronic Engineers	TIL314	A
7631G	Industrial Electronic Engineers	TIL314	A
7633G	Industrial Electronic Engineers	TIL315	A
7730R	Industrial Electronic Engineers	TIL312	A
7731R	Industrial Electronic Engineers	TIL312	A
7740R	Industrial Electronic Engineers	TIL313	A
BPX38	Litronix	TIL99	B
BPX38-2	Litronix	TIL99	B
BPX38-3	Litronix	TIL99	B
BPX38I	Siemens	TIL99	B
BPX38II	Siemens	TIL99	B
BPX38III	Siemens	TIL99	B
BPX43	Litronix	TIL81	B
BPX43-1	Litronix	TIL81	B
BPX43-2	Litronix	TIL81	B
BPX43-3	Litronix	TIL81	B
BPX43-4	Litronix	TIL81	B
BPX43I	Siemens	TIL81	B
BPX43II	Siemens	TIL81	B
BPX43III	Siemens	TIL81	B
BPX43IV	Siemens	TIL81	B
BPX62I	Siemens	TIL602	B
BPX62II	Siemens	TIL602	B
BPX62III	Siemens	TIL603	B
BPX62IV	Siemens	TIL604	B
BPY62-1	Litronix	TIL81	B
BPY62-2	Litronix	TIL81	B
BPY62-3	Litronix	TIL81	B
CLT2035	Clairex	TIL99	B
CLT2150	Clairex	TIL81	B
CLT2160	Clairex	TIL81	B
CLT2164	Clairex	TIL81	B
CLT2165	Clairex	TIL81	B
CLT3160	Clairex	TIL602	B
CLT3170	Clairex	TIL604	B
CL12	Clairex	TIL118	B

CL13	Clairex	4N37	B
CL15	Clairex	TIL116	B
CL16	Clairex	TIL117	B
CL17	Clairex	TIL118	B
CL18	Clairex	TIL116	B
CL19	Clairex	TIL116	B
CL112	Clairex	TIL157	B
CL10506A	Clairex	TIL116	B
CL1210	Clairex	TIL169	B
CL1506	Clairex	TIL118	B
CL1510	Clairex	4N37	B
CL1511	Clairex	4N37	B
CL1800	Clairex	TIL169	B
CL1810	Clairex	TIL143	B
CL1810	Clairex	TIL143/TIL167	B
CL1811	Clairex	TIL143/TIL167	B
CL1835	Clairex	TIL169	B
CL1840	Clairex	TIL145/TIL170	B
CM4-100	Chicago-Miniature	TIL302	A
CM4-101	Chicago-Miniature	TIL304	A
CM4-110	Chicago-Miniature	TIL302	A
CM4-111	Chicago-Miniature	TIL304	A
CM4-5010	Chicago-Miniature	TIL112	A
CM4-5020	Chicago-Miniature	TIL111	A
CNY17-1	Litronix	TIL126	B
CNY17-2	Litronix	TIL126	B
CNY17-3	Litronix	TIL127	B
CNY17-4	Litronix	TIL128	B
CNY18-2	Litronix	TIL120	B
CNY18-3	Litronix	TIL121	B
CQV10-3	Litronix	TIL209A	B
CQV11-4	Litronix	TIL216-1	B
CQV11-5	Litronix	TIL216-1	B
CQV11-6	Litronix	TIL216-2	B
CQV13-5	Litronix	TIL212-2	B
CQV13-6	Litronix	TIL212-2	B
CQV15-3	Litronix	TIL232-2	B
CQV15-4	Litronix	TIL232-2	B
CQV20-4	Litronix	TIL228-1	B
CQV21-4	Litronix	TIL228-1	B
CQV21-5	Litronix	TIL228-1	B
CQV21-6	Litronix	TIL228-2	B
CQV23-4	Litronix	TIL224-1	B
CQV23-5	Litronix	TIL224-1	B
CQV23-6	Litronix	TIL224-2	B
CQV25-3	Litronix	TIL234-1	B
CQV25-4	Litronix	TIL234-1	B
CQV25-5	Litronix	TIL234-2	B
CQV25-6	Litronix	TIL234-2	B
CQY17-4	Litronix	TIL34B	B
CQY17-5	Litronix	TIL31B	B
CQY32	Telefunken	TIL33B/TIL34B	B
CQY34	Telefunken	TIL31B	B
CQY57I	Siemens	TIL25	B
CQY78-3	Litronix	TIL33B	B
CXV20-3	Litronix	TIL220	B
DL-507	Litronix	TIL321A	B
DLO-7010	Litronix	TIL312	B
DLO-7611	Litronix	TIL312	B
DLO-7613	Litronix	TIL313	B
DLO-7614	Litronix	TIL313	B
DL1A	Litronix	TIL302	A
DL10	Litronix	TIL302	A
DL10A	Litronix	TIL302	A
DL57	Litronix	TIL305	A
DL101	Litronix	TIL304	A
DL101A	Litronix	TIL304	A
DL701	Litronix	TIL327	A
DL707	Litronix	TIL312	A

DL1001A	Litronix	TIL304	A
ED123	Sprague	TIL220	A
ED201	Sprague	TIL302	A
ED730	Sprague	TIL113	A
FCD810	Fairchild	TIL111	A
FCD810A	Fairchild	TIL111	A
FCD810B	Fairchild	TIL114	A
FCD810C	Fairchild	TIL124	A
FCD820	Fairchild	TIL111	A
FCD820A	Fairchild	TIL111	A
FCD820B	Fairchild	TIL116	A
FCD820C	Fairchild	TIL125	A
FCD825	Fairchild	TIL116	A
FCD825A	Fairchild	TIL116	A
FCD825B	Fairchild	TIL116	A
FCD825C	Fairchild	TIL126	A
FCD825D	Fairchild	TIL126	B
FCD830	Fairchild	TIL117	B
FCD830A	Fairchild	TIL116	B
FCD830B	Fairchild	TIL116	B
FCD831	Fairchild	TIL118	B
FCD831A	Fairchild	TIL118	B
FCD831B	Fairchild	TIL118	B
FCD836	Fairchild	TIL116	B
FCD850	Fairchild	TIL156	B
FCD850	Fairchild	TIL113	A
FCD850C	Fairchild	TIL125	A
FCD855	Fairchild	TIL113	A
FCD855C	Fairchild	TIL127	A
FCD860	Fairchild	TIL157	B
FCD860	Fairchild	TIL113	A
FCD860C	Fairchild	TIL127	A
FCD865	Fairchild	TIL113	A
FCD865	Fairchild	TIL156	B
FCD865C	Fairchild	TIL127	A
FLV110	Fairchild	TIL228-1	A
FLV111	Fairchild	TIL221	A
FLV112	Fairchild	TIL230-1	A
FLV117	Fairchild	TIL220	A
FLV118	Fairchild	TIL221	A
FLV119	Fairchild	TIL230-1	A
FLV150	Fairchild	TIL228-1	A
FLV160	Fairchild	TIL228-1	A
FLV210	Fairchild	TIL228-2	A
FLV250	Fairchild	TIL228-2	A
FLV260	Fairchild	TIL228-2	A
FLV310	Fairchild	TIL234-2	A
FLV315	Fairchild	TIL234-2	A
FLV350	Fairchild	TIL234-2	A
FLV355	Fairchild	TIL234-1	A
FLV360	Fairchild	TIL234-2	A
FLV365	Fairchild	TIL234-1	A
FLV410	Fairchild	TIL224-2	A
FLV450	Fairchild	5082-4550	A
FLV450	Fairchild	TIL224-2	A
FLV455	Fairchild	TIL224-1	A
FLV460	Fairchild	TIL224-2	A
FLV465	Fairchild	TIL224-1	A
FLV550	Fairchild	TIL228-1	B
FND350	Fairchild	TIL312	B
FND357	Fairchild	TIL313	B
FND500	Fairchild	TIL322A/TIL730	A
FND507	Fairchild	TIL321A/TIL729	A
FND508	Fairchild	TIL330A	A
FND530	Fairchild	TIL324	A
FND537	Fairchild	TIL323	A
FND538	Fairchild	TIL331	A
FND540	Fairchild	TIL346	A
FND547	Fairchild	TIL345	A

FND548	Fairchild	TIL347	B
FND550	Fairchild	TIL326	A
FND557	Fairchild	TIL325	A
FND558	Fairchild	TIL332	A
FND560	Fairchild	TIL322A/TIL730	B
FND567	Fairchild	TIL321A/TIL729	B
FPE30	Fairchild	TIL33B	B
FPE104	Fairchild	TIL39	B
FPE500	Fairchild	TIL34B	B
FPE510	Fairchild	TIL33B	B
FPE520	Fairchild	TIL34B	B
FPE520	Fairchild	TIL31B	B
FPE530	Fairchild	TIL33B	B
FPE700	Fairchild	TIL32	B
FPT100	Fairchild	TIL414	B
FPT100A	Fairchild	TIL414	B
FPT110	Fairchild	TIL99	B
FPT120	Fairchild	TIL414	B
FPT131	Fairchild	TIL414	B
FPT132	Fairchild	TIL414	B
FPT136	Fairchild	TIL99	B
FPT137	Fairchild	TIL99	B
FPT500	Fairchild	TIL81	B
FPT500A	Fairchild	TIL81	B
FPT510	Fairchild	TIL81	B
FPT510	Fairchild	TIL99	B
FPT520	Fairchild	TIL81	B
FPT530	Fairchild	TIL99	B
FPT540	Fairchild	TIL81	B
FPT550	Fairchild	TIL99	B
FPT700	Fairchild	TIL78	B
GL4850	Litronix	TIL322A	A
H11A1	General Electric	TIL117	A
H11A2	General Electric	TIL112	A
H11A3	General Electric	TIL116	A
H11A3	General Electric	TIL114	B
H11A3	General Electric	TIL115	B
H11A3	General Electric	TIL116	A
H11A3	General Electric	TIL114	A
H11A4	General Electric	TIL111	A
H11A5	General Electric	TIL118	B
H11A5	General Electric	TIL116	A
H11A520	General Electric	TIL124/TIL154	B
H11A520	General Electric	TIL125	A
H11A550	General Electric	TIL126/TIL155	B
H11A590	General Electric	TIL126	A
H11B1	General Electric	TIL113	B
H11B2	General Electric	TIL119	B
H11B2	General Electric	TIL113	A
H11B3	General Electric	TIL119	A
H11G2	General Electric	TIL156	B
H13A1	General Electric	TIL143	B
H13A2	General Electric	TIL144	B
H13B1	General Electric	TIL145	B
H13B2	General Electric	TIL146	B
H21A1/A2/A3	General Electric	TIL167-2	B
H21B1/B2/B3	General Electric	TIL168-2	B
H22A1/A2/A3	General Electric	TIL169-2	B
H22B1/B2/B3	General Electric	TIL170	B
HDSP5301	Hewlett-Packard	TIL729	A
HDSP5303	Hewlett-Packard	TIL730	A
HDSP5307	Hewlett-Packard	TIL330A	B
HDSP5701	Hewlett-Packard	TIL345	A
HDSP5703	Hewlett-Packard	TIL346	A
HDSP5707	Hewlett-Packard	TIL347	B
HDSP5801	Hewlett-Packard	TIL323	B
HDSP5803	Hewlett-Packard	TIL324	B
HDSP6504	Hewlett-Packard	HDSP6504	A
HDSP6508	Hewlett-Packard	HDSP6508	A

HD113Y	Litronix	TIL346	A
HD1131Y	Litronix	TIL345	A
IL-1	Litronix	TIL114	A
IL-5	Litronix	TIL117	B
IL-5	Litronix	TIL155	A
IL-12	Litronix	TIL111	B
IL-15	Litronix	TIL112	A
IL-16	Litronix	TIL118	B
IL-30	Litronix	TIL113	B
IL-74	Litronix	TIL111	B
IL-94	Litronix	TIL153	B
IL-203	Litronix	TIL127	B
IL-501	Litronix	TIL125	B
IL-505	Litronix	TIL126	B
IL-512	Litronix	TIL124	B
L14C2	General Electric	TIL99	B
L14G1	General Electric	TIL81	B
L14G2	General Electric	TIL81	B
L14G3	General Electric	TIL81	B
LD56A	Litronix	5082-4550	A
LD56A	Litronix	TIL224-1	A
LD242	Litronix	TIL38	B
LD271	Litronix	TIL39	B
LD561	Litronix	5082-4555	B
LD561	Litronix	TIL224-2	B
LED55B	General Electric	TIL31B	B
LED55BF	General Electric	TIL33B	B
LED55C	General Electric	TIL31B	B
LED55CF	General Electric	TIL33B	B
LED56	General Electric	TIL34B	B
LED56F	General Electric	TIL33B	B
LI5AX601	General Electric	TIL601	A
LI5AX604	General Electric	TIL604	A
LR6504R	Industrial Electronic Engineers	HDSP6504	A
LR6508R	Industrial Electronic Engineers	HDSP6508	A
LT-1188	Liton	TIL393-8	A
LT-1198	Liton	TIL393-9	A
MAN1A	General Instrument	TIL302	A
MAN2A	General Instrument	TIL305	A
MAN10A	General Instrument	TIL302	A
MAN51A	General Instrument	TIL314	B
MAN52A	General Instrument	TIL314	B
MAN71A	General Instrument	TIL312	A
MAN72A	General Instrument	TIL312	A
MAN73A	General Instrument	TIL327	B
MAN74A	General Instrument	TIL313	B
MAN81A	General Instrument	TIL339	A
MAN82A	General Instrument	TIL339	A
MAN83A	General Instrument	TIL341	B
MAN84A	General Instrument	TIL340	B
MAN101A	General Instrument	TIL304	A
MAN1001A	General Instrument	TIL304	A
MAN6760	General Instrument	TIL321A	B
MCA7	General Instrument	TIL149	B
MCA8	General Instrument	TIL145	A
MCA81	General Instrument	TIL145	A
MCA230	General Instrument	TIL113	B
MCA231	General Instrument	TIL113	B
MCT2	General Instrument	MCT2	A
MCT2E	General Instrument	MCT2E	A
MCT8	General Instrument	TIL143	A
MCT26	General Instrument	TIL111	A
MCT81	General Instrument	TIL144	A
MCT271	General Instrument	TIL126	B
MCT272	General Instrument	TIL126	B
MCT273	General Instrument	TIL127	B
MCT274	General Instrument	TIL128	B
MCT275	General Instrument	TIL127	B
MCT277	General Instrument	TIL117	B

MLED650	Motorola	TIL220	A
MLED655	Motorola	TIL220	A
MLED910	Motorola	TIL23	A
MLED930	Motorola	TIL34B	B
MOC119	Motorola	TIL119	B
MOC1000	Motorola	4N26	B
MOC1000	Motorola	TIL116	A
MOC1001	Motorola	4N25	B
MOC1001	Motorola	TIL116	A
MOC1002	Motorola	4N27	B
MOC1002	Motorola	TIL116	A
MOC1003	Motorola	TIL136	A
MOC1003	Motorola	4N28	B
MOC1005	Motorola	TIL118	B
MOC1006	Motorola	TIL118	B
MOC1100	Motorola	TIL113	A
MOC1200	Motorola	TIL113	A
MOC8050	Motorola	TIL113	B
MP52	Fairchild	TILM4	B
MRD450	Motorola	TIL78	B
MRD601	Motorola	TIL601	A
MRD602	Motorola	TIL602	A
MRD603	Motorola	TIL603	A
MRD603	Motorola	TIL604	B
MRD604	Motorola	TIL604	A
MRD3000	Motorola	TIL81	B
MRD3050	Motorola	TIL81	B
MRD3051	Motorola	TIL81	B
MRD3052	Motorola	TIL81	B
MRD3053	Motorola	TIL81	B
MRD3054	Motorola	TIL81	B
MRD3055	Motorola	TIL81	B
MRD3056	Motorola	TIL81	B
MRD3100	Motorola	TIL81	B
MT1	General Instrument	TIL99	B
MT2	General Instrument	TIL81	B
MV5021	General Instrument	5082-4655	A
MV5022	General Instrument	5082-4655	A
MV5023	General Instrument	5082-4655	A
MV5024	General Instrument	TIL228-2	B
MV5026	General Instrument	TIL220	B
MV5050	General Instrument	TIL221	B
MV5051	General Instrument	TIL220	B
MV5052	General Instrument	TIL231-1	B
MV5053	General Instrument	TIL220	B
MV5054-1	General Instrument	TIL228-1	B
MV5054-2	General Instrument	TIL228-2	B
MV5054-3	General Instrument	TIL228-3	B
MV5055	General Instrument	TIL220	B
MV5075B	General Instrument	TIL216-1	A
MV5253	General Instrument	TIL234-1	B
MV5274B	General Instrument	TIL232-1	B
MV5353	General Instrument	TIL224-2	A
MV5353	General Instrument	TIL224-1	B
MV5374B	General Instrument	TIL212-2	B
NSA1188	National Semiconductor	TIL393-8	B
NSA1198	National Semiconductor	TIL393-9	B
NSL5056	National Semiconductor	TIL220	B
NSL5076A	National Semiconductor	TIL209A	B
NSL5086	National Semiconductor	TIL209A	B
NSN71L	National Semiconductor	TIL312	A
NSN71R	National Semiconductor	TIL312	A
NSN373	National Semiconductor	TIL313	B
NSN381	National Semiconductor	TIL313	B
NSN534	National Semiconductor	TIL321A	B
NSN581	National Semiconductor	TIL322A	B
NSN582	National Semiconductor	TIL321A	B

NSN583	National Semiconductor	TIL322A	B
NSN584	National Semiconductor	TIL321A	B
NSN5020	National Semiconductor	TIL221	A
OPB709	TRW Optron	TIL139	A
OPB804	TRW Optron	TIL169	B
OPB806	TRW Optron	TIL138	A
OPB810	TRW Optron	TIL167	B
OPB814	TRW Optron	TIL168	B
OPB815	TRW Optron	TIL168	B
OPI102	TRW Optron	TIL102/4N22	A
OPI103	TRW Optron	TIL103/4N24	A
OPI140	TRW Optron	TIL120	B
OPI2150	TRW Optron	TIL112	A
OPI2252	TRW Optron	TIL116	B
OPI2253	TRW Optron	TIL117	B
OPI3151	TRW Optron	TIL113	A
OP102	TRW Optron	TIL102	A
OP103	TRW Optron	TIL103	A
OP122	TRW Optron	TIL23	A
OP123	TRW Optron	TIL23	A
OP124	TRW Optron	TIL24	A
OP130	TRW Optron	TIL34B	B
OP130W	TRW Optron	TIL31B	B
OP131	TRW Optron	TIL31B	A
OP131W	TRW Optron	TIL33B	B
OP132	TRW Optron	TIL31B	B
OP132W	TRW Optron	TIL33B	B
OP135	TRW Optron	TIL31B	B
OP135W	TRW Optron	TIL33B	B
OP140	TRW Optron	TIL40	B
OP160	TRW Optron	TIL32	B
OP161	TRW Optron	TIL32	B
OP230	TRW Optron	TIL903-1	B
OP231	TRW Optron	TIL903-2	B
OP232	TRW Optron	TIL903-2	B
OP233	TRW Optron	TIL903-2	B
OP260	TRW Optron	TIL902-1	B
OP500	TRW Optron	TIL78	B
OP501	TRW Optron	TIL32	B
OP550	TRW Optron	TIL411/415	B
OP560	TRW Optron	TIL412/416	B
OP600	TRW Optron	LS600	A
OP601	TRW Optron	TIL601	A
OP602	TRW Optron	TIL602	A
OP603	TRW Optron	TIL603	A
OP604	TRW Optron	TIL604	A
OP640	TRW Optron	LS600	A
OP641	TRW Optron	TIL601	A
OP642	TRW Optron	TIL602	A
OP643	TRW Optron	TIL603	A
OP644	TRW Optron	TIL604	A
OP800	TRW Optron	TIL81	A
OP801	TRW Optron	TIL81	A
OP802	TRW Optron	TIL81	A
OP803	TRW Optron	TIL81	A
OP804	TRW Optron	TIL81	A
OP805	TRW Optron	TIL81	B
OP841	TRW Optron	TIL81	B
OP841W	TRW Optron	TIL99	B
OP842	TRW Optron	TIL81	B
OP842W	TRW Optron	TIL99	B
OP843	TRW Optron	TIL81	B
OP843W	TRW Optron	TIL99	B
OP844	TRW Optron	TIL81	B
OP844W	TRW Optron	TIL99	B
OP845	TRW Optron	TIL81	B
OP845W	TRW Optron	TIL99	B
OP905	TRW Optron	TIL100	B

PH302	NEC	TIL100	B
RL2000	Litronix	TIL228-1	A
RL4403	Litronix	TIL220	A
RL4850	Litronix	TIL220	A
RL5054-1	Litronix	TIL228-1	A
RL5054-2	Litronix	TIL228-2	A
RL5054-5	Litronix	TIL228-1	A
SCD11B2	Honeywell	TIL119	B
SCD11B2	Honeywell	TIL113	B
SCD11B2	Honeywell	TIL156	B
SCD11B2	Honeywell	TIL128	B
SCD11B2	Honeywell	TIL127	B
SCD11B2	Honeywell	TIL157	B
SDP8402-1	Honeywell	TIL414	B
SDP8405-3	Honeywell	TIL411	B
SDP8405-4	Honeywell	TIL411	B
SDP8405-5	Honeywell	TIL411	B
SD2440-1	Honeywell	TIL601	A
SD2440-2	Honeywell	TIL602	A
SD2440-3	Honeywell	TIL603	A
SD2440-4	Honeywell	TIL604	A
SD3443-1	Honeywell	TIL99	B
SD3443-1	Honeywell	TIL81	B
SD3443-2	Honeywell	TIL99	B
SD3443-3	Honeywell	TIL99	B
SD3443-3	Honeywell	TIL81	B
SD5443-1	Honeywell	TIL81	B
SD5443-2	Honeywell	TIL81	B
SD5443-3	Honeywell	TIL81	B
SD5443-4	Honeywell	TIL99	B
SD5443-4	Honeywell	TIL81	B
SEP8402	Honeywell	TIL78	B
SEP8403	Honeywell	TIL414	B
SEP8406	Honeywell	TIL411	B
SEP8502	Honeywell	TIL32	B
SEP8502-1	Honeywell	TIL32	B
SEP8503	Honeywell	TIL38	B
SEP8503-1	Honeywell	TIL32	B
SEP8504-1	Honeywell	TIL40	B
SEP8505-1	Honeywell	TIL32	B
SEP8505-2	Honeywell	TIL32	B
SEP8505-3	Honeywell	TIL32	B
SEP8505-4	Honeywell	TIL32	B
SEP8505-5	Honeywell	TIL32	B
SEP8505-6	Honeywell	TIL32	B
SEP8506	Honeywell	TIL40	B
SEP8506-1	Honeywell	TIL32	B
SE2450-1	Honeywell	TIL23	A
SE2450-2	Honeywell	TIL23	A
SE2450-3	Honeywell	TIL25	A
SE2460-1	Honeywell	TIL23	B
SE2460-2	Honeywell	TIL23	B
SE2460-3	Honeywell	TIL24	A
SE2460-4	Honeywell	TIL24	A
SE3455	Honeywell	TIL33B	B
SE5450-1	Honeywell	TIL31B	A
SE5450-2	Honeywell	TIL31B	A
SE5450-3	Honeywell	TIL31B	A
SE5451-1	Honeywell	TIL31B	A
SE5451-2	Honeywell	TIL31B	A
SE5451-3	Honeywell	TIL31B	A
SE5453	Honeywell	TIL31B	B
SE5453-3	Honeywell	TIL31B	B
SE5453-4	Honeywell	TIL31B	B
SE5455	Honeywell	TIL31B	B
SE5455-1	Honeywell	TIL33B	B
SE5455-1	Honeywell	TIL34B	B
SE5455-2	Honeywell	TIL31B	B
SFH309	Litronix	TIL414	B

SFH402-1	Litronix	TIL33B	B
SFH402-2	Litronix	TIL33B	B
SFH409	Litronix	TIL32	B
SFH500	Litronix	TIL99	B
SFH600	Litronix	TIL117	B
SFH600-0	Litronix	TIL117	B
SFH601-1	Litronix	TIL155	B
SPX2E	Honeywell	TIL124	B
SPX2E	Honeywell	TIL125	B
SPX6	Honeywell	TIL126	A
SPX26	Honeywell	TIL153	A
SPX26	Honeywell	TIL118	B
SPX26	Honeywell	TIL115	B
SPX33	Honeywell	TIL153	B
SPX33	Honeywell	TIL154	B
SPX33B2	Honeywell	TIL114	B
SPX33B2	Honeywell	TIL116	B
SPX53	Honeywell	TIL155	A
SPX53	Honeywell	TIL117	B
SPX1402-2	Honeywell	TIL139	B
SPX1404-1	Honeywell	TIL139	B
SPX1404-2	Honeywell	TIL139	B
SPX1404-3	Honeywell	TIL139	B
SPX1872-12	Honeywell	TIL147A/TIL148A	B
SPX1872-13	Honeywell	TIL147/TIL148	B
SPX1872-14	Honeywell	TIL147/TIL148	B
SPX1873-11	Honeywell	TIL144	B
SPX1873-11	Honeywell	TIL143	B
SPX1873-11	Honeywell	TIL159/TIL160	B
SPX1873-12	Honeywell	TIL159/TIL160	B
SPX1873-13	Honeywell	TIL145	B
SPX1873-13	Honeywell	TIL161	B
SPX1873-13	Honeywell	TIL146	B
SPX1878-11	Honeywell	TIL145	B
SPX1878-12	Honeywell	TIL146	B
SPX1878-12	Honeywell	TIL138	B
SPX1878-13	Honeywell	TIL146	B
SPX1879-11	Honeywell	TIL143	B
SPX1879-12	Honeywell	TIL143	B
SPX1879-13	Honeywell	TIL143	B
SPX1879-15	Honeywell	TIL146	B
SPX2862-4	Honeywell	TIL147A	B
SPX2862-4	Honeywell	TIL148	B
SPX18721	Honeywell	TIL158	A
SPX18723	Honeywell	TIL160	A
SPX18731	Honeywell	TIL143	A
SPX18733	Honeywell	TIL145	A
SSOS-700	Silicon Sensor	TIL147A	B
SSOS-800	Silicon Sensor	TIL143	B
STIN-135D1	Sensor Technology	TIL145	B
STIN-135D2	Sensor Technology	TIL146	B
STIN-135T1	Sensor Technology	TIL143	B
STIN-135T2	Sensor Technology	TIL144	B
STPT60	Sensor Technology	LS600	A
XAN71	Xciton	TIL312	A
XAN72	Xciton	TIL312	A
XC-55-A	Xciton	TIL34B	B
XC-55-B	Xciton	TIL34B	B
XC-55-C	Xciton	TIL31B	B
XC55-10	Telefunken	TIL33B/TIL34B	B
XC55PB	Telefunken	TIL33B/TIL34B	B
XC55PC	Telefunken	TIL31B	B
XC66-10	Telefunken	TIL33B/TIL34B	B
XC99-30	Telefunken	TIL906	B
XC209	Xciton	TIL209A	A
XC880-A	Telefunken	TIL902-1	B
XC880-B	Telefunken	TIL902-2	B
XC881-A	Telefunken	TIL905-1	B
XC881-B	Telefunken	TIL905-2	B

XC1209	Telefunken	TIL32	B
YL224-1	Litronix	TIL224-1	B
YL4850	Litronix	TIL224-1	A

CCD IMAGE SENSOR INTERCHANGEABILITY GUIDE

CCD111	Fairchild	TC102	B
CCD121	Fairchild	TC101	B
CCD142	Fairchild	TC103	B
CCD143	Fairchild	TC103	B
CD211	Fairchild	TC201	B
CD221	Fairchild	TC201	B
RL128G	EG&G Reticon	TC102	B
RL1728	EG&G Reticon	TC101	B
RL2048	EG&G Reticon	TC103	B

APPENDIX

- **Glossary**
  - Symbols and Abbreviations
  - Units of Measurements
  - Metric Multipliers
  - Terms and Definitions
- **TI Sales Offices**
- **TI Distributors**
- **TI Worldwide Sales Offices**

GLOSSARY
OPTOELECTRONIC TERMS AND DEFINITIONS

Introduction

This glossary contains letter symbols, abbreviations, terms, and definitions commonly used with optoelectronic devices. Most of the information was obtained from JEDEC Standard No. 77.

Index to Glossary by Symbols and Abbreviations

APD	Avalanche photodiode
B	Demodulation bandwidth
E_e	Irradiance
E_v	Illuminance
f_{mod}	Modulation frequency
H	Irradiance
$I_{C(off)}$	Off-state collector current
$I_{C(on)}$	On-state collector current
I_D	Dark current
I_e	Radiant intensity
I_F	Forward current
I_L	Light current
I_R	Reverse current
IRED	Infrared-emitting diode
I_v	Luminous intensity
L_e	Radiance
L_v	Luminance
LED	Light-emitting diode
M	Photocurrent gain†
NEP	Noise equivalent power (spectral density)
P_n	Noise equivalent power (spectral density)
P_O	Radiant flux or power output
Q_e	Radiant energy
Q_v	Luminous energy
R_e	Radiant responsivity
R_v	Luminous responsivity
sr	Steradian
t_d	Delay time
t_f	Fall time
t_f	Radiant pulse fall time
t_r	Radiant pulse rise time
t_r	Rise time
t_s	Storage time
V_F	Forward voltage
VLED	Visible-light-emitting diode
Δf	Noise equivalent bandwidth
$\Delta\lambda$	Spectral bandwidth
θ_{HI}	Half-intensity beam angle
λp	Wavelength at peak emission
Φ_e	Radiant flux
Φ_v	Luminous flux

†M is also the symbol for luminous or radiant exitance; however, these terms are not used in this publication.

TEXAS INSTRUMENTS
INCORPORATED
POST OFFICE BOX 225012 • DALLAS, TEXAS 75265

Units of Measurement

Unit	Symbol	Note
ampere†	A	
angstrom	Å	1 Å = 10^{-10} m = 10^{-4} μm = 0.1 nm
candela†	cd	1 cd = 1 lm/sr
candela/foot^2	cd/ft^2	1 cd/ft^2 = 10.76391 cd/m^2
candela/meter^2†	cd/m^2	
degree Celsius†	°C	
	°K	See K
farad†	F	
foot	ft	1 ft = 0.3048 m (exactly)
footcandle	fc	1 fc = 1 lm/ft^2 = 10.76391 lx
footlambert	fL	1 fL = (1/π) cd/ft^2 = 3.426259 cd/m^2
hertz†	Hz	
inch	in	1 in = 2.54 cm (exactly)
kelvin†	K	Formerly °K, degree Kelvin
lambert	L	1 L = 3183.099 cd/m^2
lumen†	lm	
lux†	lx	1 lx = 1 lm/m^2
meter†	m	
mho	mho	1 mho = 1 S
micron	μ	The equivalent unit μm is preferred
mil	mil	1 mil = 10^{-3} in = 0.0254 mm (exactly)
nit	nt	1 nt = 1 cd/m^2
ohm†	Ω	
phot	ph	1 ph = 1 lm/cm^2
second†	s	
siemens†	S	
steradian†	sr	
stilb	sb	1 sb = 1 cd/cm^2
volt†	V	
watt†	W	

†International System (SI) units.

Metric Multipliers

Many of the preceding unit symbols can be combined with the metric multipliers that follow.

Symbol	Prefix	Multiple
G	giga	10^9
M	mega	10^6
k	kilo	10^3
h	hecto	10^2
da	deka	10
d	deci	10^{-1}
c	centi	10^{-2}
m	milli	10^{-3}
μ	micro	10^{-6}
n	nano	10^{-9}
p	pico	10^{-12}
f	femto	10^{-15}

Terms and Definitions

Avalanche Photodiode (APD)

A photodiode that is intended to take advantage of avalanche multiplication of photocurrent. As the reverse-bias voltage approaches the breakdown voltage, hole-electron pairs created by absorbed photons acquire sufficient energy to create additional hole-electron pairs when they collide with substrate atoms; thus a multiplication of signal current is achieved.
NOTE: APD's are especially suited for low-noise and/or high-speed applications.

Axis of Measurement

The direction from the source of radiant energy, relative to the mechanical axis, in which the measurement of radiometric and or spectroradiometric characteristics is performed.

Beam-Lead Phototransistor

A phototransistor chip with thick-film leads formed on the chip that project cantilever-style beyond the chip periphery for attachment to a separate substrate.
NOTE: When assembled into arrays and mounted on a ceramic substrate, beam-lead phototransistor arrays offer accurate spacing on centers too close for conventional discrete packages and too far apart for monolithic arrays; see TI Bulletin CB-128 for further information.

Brightness

See Luminance

Color Temperature

The temperature of a blackbody having the same visible color as that of a given non-blackbody radiator.
TYPICAL UNIT: K (formerly °K).

Conversion Efficiency (of a Photon-Emitting Device)

The ratio of maximum available luminous or radiant flux output to total input power.

Dark Current (I_D)

The current that flows through a photosensitive device in the dark condition.
NOTE: The dark condition is attained when the electrical parameter under consideration approaches a value that cannot be altered by further irradiation shielding.

Darlington-Connected Phototransistor

A phototransistor the collector and emitter of which are connected to the collector and base, respectively, of a second transistor. The emitter current of the input transistor is amplified by the second transistor and the device has very high sensitivity to illumination or irradiation.
GRAPHIC SYMBOL:

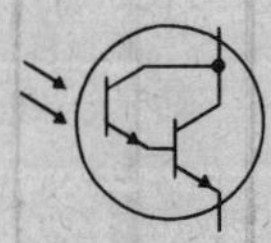

NOTE: The base region(s) may or may not be brought out as (an)electrical terminal(s).

TEXAS INSTRUMENTS
INCORPORATED
POST OFFICE BOX 225012 • DALLAS, TEXAS 75265

D-C Transfer Ratio (of an Opto-coupler)

The ratio of the dc output current to the dc input current.

Delay Time (t_d)

The time interval from the point at which the leading edge of the input pulse has reached 10% of its maximum amplitude to the point at which the leading edge of the output pulse has reached 10% of its maximum amplitude.

Demodulation Bandwidth (B)

The frequency interval in which the demodulated output of a photodetector, or a system including a photodetector, is not more than 3 dB below the midband output. Midband output is the output in the region of flat response or the average output over a specific frequency range.

Electroluminescence

The direct conversion of electrical energy into visible radiation.

Fall Time (t_f)

The time duration during which the trailing edge of a pulse is decreasing from 90% to 10% of its maximum amplitude.

Forward Current (I_F)

The current through a semiconductor diode when the p region (anode) is at a positive potential with respect to the n region (cathode).

Forward Voltage (V_F)

The voltage across a semiconductor diode associated with the flow of forward current. The p-region is at a positive potential with respect to the n-region.

Gain-Bandwidth Product (of an Avalanche Photodiode)

The gain times the frequency of measurement when the device is biased for maximum obtainable gain.

Half-Intensity Beam Angle (θ_{HI})

The angle within which the radiant intensity is not less than half of the maximum intensity.

Hexadecimal Display

A solid-state display capable of exhibiting numbers 0 through 9 and alpha characters A through F.
NOTE: The TIL311 and TIL505 are hexadecimal displays each with an integral TTL circuit that will accept, store, and display 4-bit binary data.

Illuminance (Illumination) (E_v)

The luminous flux density incident on a surface; the quotient of the flux divided by the area of illuminated surface.
TYPICAL UNITS: lm/ft^2, $lx = lm/m^2$. $1\ lm/ft^2 = 10.76391\ lx$.

Infrared Emission

Radiant energy that is characterized by wavelengths longer than visible red, i.e., about 0.78 μm to 100 μm.

Infrared-Emitting Diode (IRED)

A diode capable of emitting radiant energy, in the infrared region of the spectrum, resulting from the recombination of electrons and holes.
NOTE: TI manufactures GaAs and GaAlAs radiant-energy sources that emit in the 0.82-μm to 0.94-μm portion of the near-infrared region. These emitters are spectrally matched with TI silicon photodetectors.
GRAPHIC SYMBOL:

Irradiance (E_e, formerly H)

The radiant flux density incident on a surface; the quotient of the flux divided by the area of irradiated surface.
TYPICAL UNITS: W/ft^2, W/m^2. 1 W/ft^2 = 10.76391 W/m^2.

Light Current (I_L)

The current that flows through a photosensitive device, such as a phototransistor or a photodiode, when it is exposed to radiant energy.

Light-Emitting Diode (LED)

A diode capable of emitting luminous energy resulting from the recombination of electrons and holes.
NOTE: In popular usage, this term is sometimes used for infrared-emitting diodes.
GRAPHIC SYMBOL:

Luminance (L_v) (Photometric Brightness)

The luminous intensity of a surface in a given direction per unit of projected area of the surface as viewed from that direction.
TYPICAL UNITS: fL, cd/ft^2, cd/m^2. 1 fL = $(1/\pi)$ cd/ft^2 = 3.426259 cd/m^2.

Luminous Energy (Q_v)

Energy traveling in the form of visible radiation.
TYPICAL UNITS: lm • s

Luminous Flux (Φ_v)

The time rate of flow of luminous energy.
TYPICAL UNIT: lm
NOTE: Luminous flux is related to radiant flux by the eye-response curve of the International Commission on Illumination (CIE). At the peak response (λ = 555 nm), 1 W = 680 lm.

TEXAS INSTRUMENTS
INCORPORATED
POST OFFICE BOX 225012 • DALLAS, TEXAS 75265

Luminous Intensity (I_V)

Luminous flux per unit solid angle in a given direction.
TYPICAL UNIT: cd. 1 cd = 1 lm/sr.

Luminous Responsivity (R_V)

The quotient of the rms value of the fundamental component of the electrical output divided by the rms value of the fundamental component of the luminous flux of a specified distribution.
TYPICAL UNITS: V/lm, A/lm

Modulation Frequency (f_{mod})

The frequency of modulation of the luminous or radiant flux.

Noise Equivalent Bandwidth (Δf)

The equivalent bandwidth of a flat (or white) sharp-cutoff noise spectrum, having the same maximum value and containing the same noise power as the actual broadband output noise power of the device or circuit.
TYPICAL UNIT: Hz

Noise Equivalent Power (P_n or NEP)

The rms value of the fundamental component of a modulated radiant flux incident on the detector area that will produce a signal (voltage or current) at the detector output that is equal to the broadband rms noise (voltage or current).
TYPICAL UNIT: W
NOTE: The noise equivalent power equals the broadband output noise (voltage or current) divided by the responsivity (in volts/watt or amperes/watt).

Noise Equivalent Power (P_n or NEP) (Spectral Density)

The noise equivalent power in a one-Hertz bandwidth at the detector output.
TYPICAL UNIT: $W/Hz^{1/2}$
NOTE: The noise equivalent power spectral density equals the noise equivalent power divided by the square root of the noise bandwidth.

Off-State Collector Current ($I_{C(off)}$) (of an Opto-coupler)

The output current when the input current is zero.

On-State Collector Current ($I_{C(on)}$) (of an Opto-coupler)

The output current when the input current is above the threshold level.
NOTE: An increase in the input current will usually result in a corresponding increase in the on-state collector current.

Optical Axis

A line about which the radiant-energy pattern is centered.
NOTES: 1. The radiant-energy pattern may be nonsymmetrical.
2. The optical axis may deviate from the mechanical axis.

Opto-coupler (Optically Coupled Isolator, Photo-coupler)

A device designed for the transformation of electrical signals by utilizing optical radiant energy so as to provide coupling with electrical isolation between the input and the output.
NOTE: As manufactured by Texas Instruments, these devices consist of a gallium arsenide infrared-emitting diode and a silicon phototransistor and provide high-voltage isolation between separate pairs of input and output terminals.

Optoelectronic Device

A device that is responsive to or that emits or modifies coherent or noncoherent electromagnetic radiation in the visible, infrared, and/or ultraviolet spectral regions; or a device that utilizes such electromagnetic radiation for its internal operation.

Photocurrent

The difference between light current (I_L) and dark current (I_D) in a photodetector.

Photocurrent Gain (M) (of an Avalanche Photodiode)

The ratio of photocurrent at high bias voltage to that at low bias voltage. (See also avalanche photodiode definition).

Photodetector, Photosensitive Device

A device that is responsive to electromagnetic radiation in the visible, infrared, and/or ultraviolet spectral regions.

Photodiode

A diode that is intended to be responsive to radiant energy.
GRAPHIC SYMBOLS:

NOTE: The photodiode is characterized by linearity between the input radiation and the output current. It has faster switching speeds than a phototransistor.

Photometric Axis

See Axis of Measurement.

Photometric Brightness

See Luminance.

Photon

A quantum (the smallest possible unit) of radiant energy; a photon carries a quantity of energy equal to Planck's constant (6.6262×10^{-34} joule/hertz) times the frequency.

TEXAS INSTRUMENTS
INCORPORATED
POST OFFICE BOX 225012 • DALLAS, TEXAS 75265

Phototransistor

A transistor (bipolar or field-effect) that is intended to be responsive to radiant energy.
NOTE: The base region or gate may or may not be brought out as an external terminal.
GRAPHIC SYMBOLS:

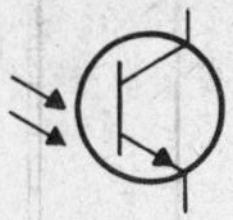

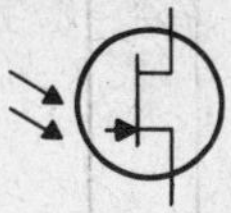

Quantum Efficiency (of a Photosensitive Device)

The fractional number of effective electron-hole pairs produced within the device for each incident photon. For devices that internally amplify or multiply the electron-hole pairs (such as phototransistors or avalanche photodiodes), the effect of the gain is to be excluded from quantum efficiency.

Quantum Efficiency, External (of a Photoemitter)

The number of photons radiated for each electron flowing into the radiant source.

Radiance (L_e)

The radiant intensity of a surface in a given direction per unit of projected area of the surface as viewed from that direction.
TYPICAL UNIT: $W \cdot sr^{-1} m^{-2}$.

Radiant Energy (Q_e)

Energy traveling in the form of electromagnetic waves.
TYPICAL UNITS: W•s, J

Radiant Flux or Power Output (Φ_e or P_O)

The time rate of flow of radiant energy.
TYPICAL UNITS: W

Radiant Intensity (I_e)

Radiant flux per unit solid angle in a given direction.
TYPICAL UNIT: W/sr

Radiant Pulse Fall Time (t_f)

The time required for a radiometric quantity to change from 90% to 10% of its peak value for a step change in electrical input.

Radiant Pulse Rise Time (t_r)

The time required for a radiometric quantity to change from 10% to 90% of its peak value for a step change in electrical input.

Radiant Responsivity (R_e)

The quotient of the rms value of the fundamental component of the electrical output divided by the rms value of the fundamental component of the radiant flux of a specified distribution.
TYPICAL UNITS: V/W, A/W

Reverse Current (I_R)

The current through a semiconductor diode when the n region (cathode) is at a positive potential with respect to the p region (anode).

Reverse Voltage (V_R)

The voltage across a semiconductor diode associated with the flow of reverse current. The n region is at a positive potential with respect to the p region.

Rise Time (t_r)

The time duration during which the leading edge of a pulse is increasing from 10% to 90% of its maximum amplitude.

Series Resistance

The undepleted bulk resistance of the photodiode substrate.
NOTE: This characteristic becomes significant at higher frequencies where the capacitive reactance of the junction is of the same or lower magnitude compared to the series resistance.

Spectral Bandwidth ($\Delta\lambda$)

The wavelength interval in which the spectral concentration of a photometric or radiometric quantity is not less than half of its maximum value.
TYPICAL UNITS: Å, μm, nm

Steradian (sr)

A unit of solid angular measurement equal to the solid angle at the center of a sphere subtended by a portion of the surface area equal to the square of the radius; there are 4π steradians in a complete sphere. The number of steradians in a cone of full angle θ is $2\pi\ (1 - \cos 0.5\theta)$.

Storage Time (t_s)

The time interval from a point at which the trailing edge of the input pulse has dropped to 90% of its maximum amplitude to a point at which the trailing edge of the output pulse has dropped to 90% of its maximum amplitude.

Visible Emission

Radiant energy that is characterized by wavelengths of about 0.38 μm to 0.78 μm.

Visible-Light-Emitting Diode (VLED)

Synonym for Light-Emitting Diode (LED)
NOTE: Strictly speaking, the adjective "visible" is redundant; however, this term is frequently used when there is a likelihood of confusion with infrared-emitting diodes.

Wavelength at Peak Emission (λ_p)

The wavelength at which the spectral radiant intensity is maximum.
TYPICAL UNITS: Å, μm, nm.1 Å = 10^{-4} μm = 0.1 nm.

TEXAS INSTRUMENTS
INCORPORATED
POST OFFICE BOX 225012 • DALLAS, TEXAS 75265

TI Sales Offices

ALABAMA: Huntsville, 500 Wynn Drive, Suite 514, Huntsville, AL 35805, (205) 837-7530.

ARIZONA: Phoenix, P.O. Box 35160, 8102 N. 23rd Ave., Suite A, Phoenix, AZ 85021, (602) 995-1007.

CALIFORNIA: El Segundo, 831 S. Douglas St., El Segundo, CA 90245, (213) 973-2571; **Irvine,** 17891 Cartwright Rd., Irvine, CA 92714, (714) 660-1200; **Sacramento,** 1900 Point West Way, Suite 171, Sacramento, CA 95815, (916) 929-1521; **San Diego,** 4333 View Ridge Ave., Suite B., San Diego, CA 92123, (714) 278-9600; **Santa Clara,** 5353 Betsy Ross Dr., Santa Clara, CA 95054, (408) 980-9000; **Woodland Hills,** 21220 Erwin St., Woodland Hills, CA 91367, (213) 704-7759.

COLORADO: Denver, 9725 E. Hampden St., Suite 301, Denver, CO 80231, (303) 695-2800.

CONNECTICUT: Wallingford, 9 Barnes Industrial Park Rd., Barnes Industrial Park, Wallingford, CT 06492, (203) 269-0074.

FLORIDA: Clearwater, 2280 U.S. Hwy. 19 N., Suite 232, Clearwater, FL 33515, (813) 796-1926; **Ft. Lauderdale,** 2765 N.W. 62nd St., Ft. Lauderdale, FL 33309, (305) 973-8502; **Maitland,** 2601 Maitland Center Parkway, Maitland, FL 32751, (305) 646-9600.

GEORGIA: Atlanta, 3300 Northeast Expy., Building 9, Atlanta, GA 30341, (404) 452-4600.

ILLINOIS: Arlington Heights, 515 W. Algonquin, Arlington Heights, IL 60005, (312) 640-2934.

INDIANA: Ft. Wayne, 2020 Inwood Dr., Ft. Wayne, IN 46805, (219) 424-5174; **Indianapolis,** 2346 S. Lynhurst, Suite J-400, Indianapolis, IN 46241, (317) 248-8555.

IOWA: Cedar Rapids, 373 Collins Rd. NE, Suite 200, Cedar Rapids, IA 52402, (319) 395-9550.

MARYLAND: Baltimore, 1 Rutherford Pl., 7133 Rutherford Rd., Baltimore, MD 21207, (301) 944-8600.

MASSACHUSETTS: Waltham, 504 Totten Pond Rd., Waltham, MA 02154, (617) 890-7400.

MICHIGAN: Farmington Hills, 33737 W. 12 Mile Rd., Farmington Hills, MI 48018, (313) 553-1500.

MINNESOTA: Edina, 7625 Parklawn, Edina, MN 55435, (612) 830-1600.

MISSOURI: Kansas City, 8080 Ward Pkwy., Kansas City, MO 64114, (816) 523-2500; **St. Louis,** 11861 Westline Industrial Drive, St. Louis, MO 63141, (314) 569-7600.

NEW JERSEY: Clark, 292 Terminal Ave. West, Clark, NJ 07066, (201) 574-9800.

NEW MEXICO: Albuquerque, 5907 Alice NSE, Suite E., Albuquerque, NM 87110, (505) 265-8491.

NEW YORK: East Syracuse, 6700 Old Collamer Rd., East Syracuse, NY 13057, (315) 463-9291; **Endicott,** 112 Nanticoke Ave., P.O. Box 618, Endicott, NY 13760, (607) 754-3900; **Melville,** 1 Huntington Quadrangle, Suite 3C10, P.O. Box 2936, Melville, NY 11747, (516) 454-6600; **Poughkeepsie,** 201 South Ave., Poughkeepsie, NY 12601, (914) 473-2900; **Rochester,** 1210 Jefferson Rd., Rochester, NY 14623, (716) 424-5400.

NORTH CAROLINA: Charlotte, 8 Woodlawn Green, Woodlawn Rd., Charlotte, NC 28210, (704) 527-0930; **Raleigh,** 3000 Highwoods Blvd., Suite 118, Raleigh, NC 27625, (919) 876-2725.

OHIO: Beachwood, 23408 Commerce Park Rd., Beachwood, OH 44122, (216) 464-6100; **Dayton,** Kingsley Bldg., 4124 Linden Ave., Dayton, OH 45432, (513) 258-3877.

OKLAHOMA: Tulsa, 7615 East 63rd Place, 3 Memorial Place, Tulsa, OK 74133, (405) 250-0633.

OREGON: Beaverton, 6700 SW 105th St., Suite 110, Beaverton, OR 97005, (503) 643-6758.

PENNSYLVANIA: Ft. Washington, 575 Virginia Dr., Ft. Washington, PA 19034, (215) 643-6450; Coraopolis, PA 15108, 420 Rouser Rd., 3 Airport Office PK, (412) 771-8550.

TEXAS: Austin, 12501 Research Blvd., P.O. Box 2909, Austin, TX 78723, (512) 250-7655; **Dallas,** P. O. Box 1087, Richardson, TX 75080; **Houston,** 9100 Southwest Frwy., Suite 237, Houston, TX 77036, (713) 778-6592; **San Antonio,** 1000 Central Park South, San Antonio, TX 78232, (512) 496-1779.

UTAH: Salt Lake City, 3672 West 2100 South, Salt Lake City, UT 84120, (801) 973-6310.

VIRGINIA: Fairfax, 3001 Prosperity, Fairfax, VA 22031, (703) 849-1400; **Midlothian,** 13711 Sutter's Mill Circle, Midlothian, VA 23113, (804) 744-1007.

WISCONSIN: Brookfield, 205 Bishops Way, Suite 214, Brookfield, WI 53005, (414) 784-3040.

WASHINGTON: Redmond, 2723 152nd Ave., N.E. Bldg. 6, Redmond, WA 98052, (206) 881-3080.

CANADA: Ottawa, 436 Mac Laren St., Ottawa, Canada, K2P0M8, (613) 233-1177; **Richmond Hill,** 280 Centre St. E., Richmond Hill L4C1B1, Ontario, Canada, (416) 884-9181; **St. Laurent,** Ville St. Laurent Quebec, 9460 Trans Canada Hwy., St. Laurent, Quebec, Canada H4S1R7, (514) 334-3635. C

TI Distributors

ALABAMA: Hall-Mark (205) 837-8700.

ARIZONA: Phoenix, Kierulff (602) 243-4101; Marshall (602) 968-6181; Wyle (602) 249-2232; **Tucson,** Kierulff (602) 624-9986.

CALIFORNIA: Los Angeles/Orange County, Arrow (213) 701-7500, (714) 851-8961; Kierulff (213) 725-0325, (714) 731-5711; Marshall (213) 999-5001, (213) 442-7204, (714) 556-6400; R.V. Weatherford (714) 634-9600, (213) 849-3451, (714) 623-1261; Wyle (213) 322-8100, (714) 641-1600, **San Diego,** Arrow (619) 565-4800; Kierulff (619) 278-2112; Marshall (619) 578-9600; R. V. Weatherford (619) 695-1700; Wyle (619) 565-9171; **San Francisco Bay Area,** Arrow (408) 745-6600; Kierulff (415) 968-6292; Marshall (408) 732-1100; Wyle (408) 727-2500; **Santa Barbara,** R. V. Weatherford (805) 965-8551.

COLORADO: Arrow (303) 758-2100; Kierulff (303) 790-4444; Wyle (303) 457-9953.

CONNECTICUT: Arrow (203) 265-7741; Diplomat (203) 797-9674; Kierulff (203) 265-1115; Marshall (203) 265-3822; Milgray (203) 795-0714.

FLORIDA: Ft. Lauderdale, Arrow (305) 776-7790; Diplomat (305) 971-7160; Hall-Mark (305) 971-9280; Kierulff (305) 652-6950; **Orlando,** Arrow (305) 725-1480; Diplomat (305) 725-4520; Hall-Mark (305) 855-4020; Milgray (305) 647-5747; **Tampa,** Diplomat (813) 443-4514; Hall-Mark (813) 576-8691; Kierulff (813) 576-1966.

GEORGIA: Arrow (404) 449-8252; Hall-Mark (404) 447-8000; Kierulff (404) 447-5252; Marshall (404) 923-5750.

ILLINOIS: Arrow (312) 397-3440; Diplomat (312) 595-1000; Hall-Mark (312) 860-3800; Kierulff (312) 640-0200; Newark (312) 638-4411.

INDIANA: Indianapolis, Arrow (317) 243-9353; Graham (317) 634-8202; **Ft. Wayne,** Graham (219) 423-3422.

IOWA: Arrow (319) 395-7230.

KANSAS: Kansas City, Component Specialties (913) 492-3555; Hall-Mark (913) 888-4747; **Wichita,** LCOMP (316) 265-9507.

MARYLAND: Arrow (301) 247-5200; Diplomat (301) 995-1226; Hall-Mark (301) 796-9300; Kierulff (301) 247-5020; Milgray (301) 468-6400.

MASSACHUSETTS: Arrow (617) 933-8130; Diplomat (617) 429-4120; Kierulff (617) 667-8331; Marshall (617) 272-8200; Time (617) 935-8080.

MICHIGAN: Detroit, Arrow (313) 971-8200; Newark (313) 967-0600; **Grand Rapids,** Arrow (616) 243-0912.

MINNESOTA: Arrow (612) 830-1800; Hall-Mark (612) 854-3223; Kierulff (612) 941-7500.

MISSOURI: Kansas City, LCOMP (816) 221-2400; **St. Louis,** Arrow (314) 567-6888; Hall-Mark (314) 291-5350; Kierulff (314) 739-0855.

NEW HAMPSHIRE: Arrow (603) 668-6968.

NEW JERSEY: Arrow (201) 575-5300, (609) 235-1900; Diplomat (201) 785-1830; General Radio (609) 964-8560; Hall-Mark (201) 575-4415, (609) 424-7300, JACO (201) 778-4722, (800) 645-5130; Kierulff (201) 575-6750; Marshall (201) 882-0320; Milgray (609) 983-5010, (800) 645-3956.

NEW MEXICO: Arrow (505) 243-4566; International Electronics (505) 345-8127.

NEW YORK: Long Island, Arrow (516) 231-1000; Diplomat (516) 454-6334; Hall-Mark (516) 737-0600; JACO (516) 273-5500; Marshall (516) 273-2424; Milgray (516) 546-5600, (800) 645-3986; Hall-Mark (516) 737-0600; **Rochester,** Arrow (716) 275-0300; Marshall (716) 235-7620; Rochester Radio Supply (716) 454-7800; **Syracuse,** Arrow (315) 652-1000; Diplomat (315) 652-5000; Marshall (607) 754-1570.

NORTH CAROLINA: Arrow (919) 876-3132, (919) 725-8711; Hall-Mark (919) 872-0712; Kierulff (919) 852-9440.

OHIO: Cincinnati, Graham (513) 772-1661; Hall-Mark (513) 563-5980; **Cleveland,** Arrow (216) 248-3990; Hall-Mark (216) 473-2907; Kierulff (216) 587-6558; **Columbus,** Hall-Mark (614) 891-4555, **Dayton,** Arrow (513) 435-5563; ESCO (513) 226-1133; Marshall (513) 236-8088.

OKLAHOMA: Arrow (918) 665-7700; Component Specialties (918) 664-2820; Hall-Mark (918) 665-3200; Kierulff (918) 252-7537.

OREGON: Kierulff (503) 641-9150; Wyle (503) 640-6000.

PENNSYLVANIA: Arrow (412) 856-7000, (215) 928-1800; General Radio (215) 922-7037; Hall-Mark (215) 355-7300.

TEXAS: Austin, Arrow (512) 835-4180; Component Specialties (512) 837-8922; Hall-Mark (512) 258-8848; Kierulff (512) 835-2090; **Dallas,** Arrow (214) 386-7500; Component Specialties (214) 357-6511; Hall-Mark (214) 341-1147; International Electronics (214) 233-9323; Kierulff (214) 343-2400; **El Paso,** International Electronics (915) 778-9761; **Houston,** Arrow (713) 491-4100; Component Specialties (713) 771-7237; Hall-Mark (713) 781-6100; Harrison Equipment (713) 879-2600; Kierulff (713) 530-7030.

UTAH: Diplomat (801) 486-4134; Kierulff (801) 973-6913; Wyle (801) 974-9953.

VIRGINIA: Arrow (804) 282-0413.

WASHINGTON: Arrow (206) 643-4800; Kierulff (206) 575-4420; Wyle (206) 453-8300.

WISCONSIN: Arrow (414) 764-6600; Hall-Mark (414) 761-3000; Kierulff (414) 784-8160.

CANADA: Calgary, Future (403) 259-6408; Varah (403) 230-1235; **Hamilton,** Varah (416) 561-9311; **Montreal,** CESCO (514) 735-5511; Future (514) 694-7710; **Ottawa,** CESCO (613) 226-6905; Future (613) 820-8313; **Quebec City,** CESCO (418) 687-4231; **Toronto,** CESCO (416) 661-0220; Future (416) 663-5563; **Vancouver,** Future (604) 438-5545; Varah (604) 873-3211; **Winnipeg,** Varah (204) 633-6190. BC

TEXAS INSTRUMENTS

Creating useful products and services for you.

TI Worldwide Sales Offices

ALABAMA: Huntsville, 500 Wynn Drive, Suite 514, Huntsville, AL 35805, (205) 837-7530.

ARIZONA: Phoenix, P.O. Box 35160, 8102 N. 23rd Ave., Suite A, Phoenix, AZ 85021, (602) 995-1007.

CALIFORNIA: El Segundo, 831 S. Douglas St., El Segundo, CA 90245, (213) 973-2571; **Irvine,** 17891 Cartwright Rd., Irvine, CA 92714, (714) 660-1200; **Sacramento,** 1900 Point West Way, Suite 171, Sacramento, CA 95815, (916) 929-1521; **San Diego,** 4333 View Ridge Ave., Suite B., San Diego, CA 92123, (714) 278-9600; **Santa Clara,** 5353 Betsy Ross Dr., Santa Clara, CA 95054, (408) 980-9000; **Woodland Hills,** 21220 Erwin St., Woodland Hills, CA 91367, (213) 704-7759.

COLORADO: Denver, 9725 E. Hampden St., Suite 301, Denver, CO 80231, (303) 695-2800.

CONNECTICUT: Wallingford, 9 Barnes Industrial Park Rd., Barnes Industrial Park, Wallingford, CT 06492, (203) 269-0074.

FLORIDA: Clearwater, 2280 U.S. Hwy. 19 N., Suite 232, Clearwater, FL 33515, (813) 796-1926; **Ft. Lauderdale,** 2765 N.W. 62nd St., Ft. Lauderdale, FL 33309, (305) 973-8502; **Maitland,** 2601 Maitland Center Parkway, Maitland, FL 32751, (305) 646-9600.

GEORGIA: Atlanta, 3300 Northeast Expy., Building 9, Atlanta, GA 30341, (404) 452-4600.

ILLINOIS: Arlington Heights, 515 W. Algonquin, Arlington Heights, IL 60005, (312) 640-2934.

INDIANA: Ft. Wayne, 2020 Inwood Dr., Ft. Wayne, IN 46805, (219) 424-5174; **Indianapolis,** 2346 S. Lynhurst, Suite J-400, Indianapolis, IN 46241, (317) 248-8555.

IOWA: Cedar Rapids, 373 Collins Rd. NE, Suite 200, Cedar Rapids, IA 52402, (319) 395-9550.

MARYLAND: Baltimore, 1 Rutherford Pl., 7133 Rutherford Rd., Baltimore, MD 21207, (301) 944-8600.

MASSACHUSETTS: Waltham, 504 Totten Pond Rd., Waltham, MA 02154, (617) 890-7400.

MICHIGAN: Farmington Hills, 33737 W. 12 Mile Rd., Farmington Hills, MI 48018, (313) 553-1500.

MINNESOTA: Edina, 7625 Parklawn, Edina, MN 55435, (612) 830-1600.

MISSOURI: Kansas City, 8080 Ward Pkwy., Kansas City, MO 64114, (816) 523-2500; **St. Louis,** 11861 Westline Industrial Drive, St. Louis, MO 63141, (314) 569-7600.

NEW JERSEY: Clark, 292 Terminal Ave. West, Clark, NJ 07066, (201) 574-9800.

NEW MEXICO: Albuquerque, 5907 Alice NSE, Suite E., Albuquerque, NM 87110, (505) 265-8491.

NEW YORK: East Syracuse, 6700 Old Collamer Rd., East Syracuse, NY 13057, (315) 463-9291; **Endicott,** 112 Nanticoke Ave., P.O. Box 618, Endicott, NY 13760, (607) 754-3900; **Melville,** 1 Huntington Quadrangle, Suite 3C10, P.O. Box 2936, Melville, NY 11747, (516) 454-6600; **Poughkeepsie,** 201 South Ave., Poughkeepsie, NY 12601, (914) 473-2900; **Rochester,** 1210 Jefferson Rd., Rochester, NY 14623, (716) 424-5400.

NORTH CAROLINA: Charlotte, 8 Woodlawn Green, Woodlawn Rd., Charlotte, NC 28210, (704) 527-0930; **Raleigh,** 3000 Highwoods Blvd., Suite 118, Raleigh, NC 27625, (919) 876-2725.

OHIO: Beachwood, 23408 Commerce Park Rd., Beachwood, OH 44122, (216) 464-6100; **Dayton,** Kingsley Bldg., 4124 Linden Ave., Dayton, OH 45432, (513) 258-3877.

OKLAHOMA: Tulsa, 7615 East 63rd Place, 3 Memorial Place, Tulsa, OK 74133, (405) 250-0633.

OREGON: Beaverton, 6700 SW 105th St., Suite 110, Beaverton, OR 97005, (503) 643-6758.

PENNSYLVANIA: Ft. Washington, 575 Virginia Dr., Ft. Washington, PA 19034, (215) 643-6450; Coraopolis, PA 15108, 420 Rouser Rd., 3 Airport Office PK, (412) 771-8550.

TEXAS: Austin, 12501 Research Blvd., P.O. Box 2909, Austin, TX 78723, (512) 250-7655; **Dallas,** P. O. Box 1087, Richardson, TX 75080; **Houston,** 9100 Southwest Frwy., Suite 237, Houston, TX 77036, (713) 778-6592; **San Antonio,** 1000 Central Park South, San Antonio, TX 78232, (512) 496-1779.

UTAH: Salt Lake City, 3672 West 2100 South, Salt Lake City, UT 84120, (801) 973-6310.

VIRGINIA: Fairfax, 3001 Prosperity, Fairfax, VA 22031, (703) 849-1400; **Midlothian,** 13711 Sutter's Mill Circle, Midlothian, VA 23113, (804) 744-1007.

WISCONSIN: Brookfield, 205 Bishops Way, Suite 214, Brookfield, WI 53005, (414) 784-3040.

WASHINGTON: Redmond, 2723 152nd Ave., N.E. Bldg. 6, Redmond, WA 98052, (206) 881-3080.

CANADA: Ottawa, 436 Mac Laren St., Ottawa, Canada, K2P0M8,(613) 233-1177; **Richmond Hill,** 280 Centre St. E., Richmond Hill L4C1B1, Ontario, Canada, (416) 884-9181; **St. Laurent,** Ville St. Laurent Quebec, 9460 Trans Canada Hwy., St. Laurent, Quebec, Canada H4S1R7, (514) 334-3635. C

ARGENTINA, Texas Instruments Argentina S.A.I.C.F.: Esmeralda 130, 15th Floor, 1035 Buenos Aires, Argentina, 394-2963.

AUSTRALIA (& NEW ZEALAND), Texas Instruments Australia Ltd.: 6-10 Talavera Rd., North Ryde (Sydney), New South Wales, Australia 2113, 02 + 887-1122; 5th Floor, 418 St. Kilda Road, Melbourne, Victoria, Australia 3004, 03 + 267-4677; 171 Philip Highway, Elizabeth, South Australia 5112, 08 + 255-2066.

AUSTRIA, Texas Instruments Ges.m.b.H.: Industriestrabe B/16, A-2345 Brunn/Gebirge, 2236-846210.

BELGIUM, Texas Instruments N.V. Belgium S.A.: Mercure Centre, Raketstraat 100, Rue de la Fusee, 1130 Brussels, Belgium, 02/720.80.00.

BRAZIL, Texas Instruments Electronicos do Brasil Ltda.: Av. Faria Lima, 2003, 20 0 Andar—Pinheiros, Cep-01451 Sao Paulo, Brazil, 815-6166.

DENMARK, Texas Instruments A/S, Marielundvej 46E, DK-2730 Herlev, Denmark, 2 - 91 74 00.

FINLAND, Texas Instruments Finland OY: PL 56, 00510 Helsinki 51, Finland, (90) 7013133.

FRANCE, Texas Instruments France: Headquarters and Prod. Plant, BP 05, 06270 Villeneuve-Loubet, (93) 20-01-01; Paris Office, BP 67 8-10 Avenue Morane-Saulnier, 78141 Velizy-Villacoublay, (3) 946-97-12; Lyon Sales Office, L'Oree D'Ecully, Batiment B, Chemin de la Forestiere, 69130 Ecully, (7) 833-04-40; Strasbourg Sales Office, Le Sebastopol 3, Quai Kleber, 67055 Strasbourg Cedex, (88) 22-12-66; Rennes, 23-25 Rue du Puits Mauger, 35100 Rennes, (99) 79-54-81; Toulouse Sales Office, Le Peripole—2, Chemin du Pigeonnier de la Cepiere, 31100 Toulouse, (61) 44-18-19; Marseille Sales Office, Noilly Paradis—146 Rue Paradis, 13006 Marseille, (91) 37-25-30.

TEXAS INSTRUMENTS

Creating useful products
and services for you

GERMANY, Texas Instruments Deutschland GmbH: Haggerty-strasse 1, D-8050 Freising, 08161-801; Kurfuerstendamm 195/196, D-1000 Berlin 15, 030-8827365; III, Hagen 43/Kibbelstrasse, D-4300 Essen, 0201-24250; Frankfurter Allee 6-8, D-6236 Eschborn 1, 06196-43074; Hamburger Strasse 11, D-2000 Hamburg 76, 040-2201154, Kirchhorsterstrasse 2, D-3000 Hannover 51, 0511-648021; Arabellastrasse 15, D-8000 Muenchen 81, 089-92341; Maybachstrasse 11, D-7302 Ostfildern 2/Nellingen, 0711-34030.

HONG KONG (+ PEOPLES REPUBLIC OF CHINA), Texas Instruments Asia Ltd.: 8th Floor, World Shipping Ctr., Harbour City, 7 Canton Rd., Kowloon, Hong Kong, 3 + 722-1223.

IRELAND, Texas Instruments (Ireland) Limited: 25 St. Stephens Green, Dublin 2, Eire, 01 609222.

ITALY, Texas Instruments Semiconduttori Italia Spa: Viale Delle Scienze, 1, 02015 Cittaducale (Rieti), Italy, 0746 694.1; Via Salaria KM 24 (Palazzo Cosma), Monterotondo Scalo (Rome), Italy, 06 9004395; Viale Europa, 38-44, 20093 Cologno Monzese (Milano), 02 2532541; Corso Svizzera, 185, 10100 Torino, Italy, 011 774545; Via J. Barozzi, 6, 45100 Bologna, Italy, 051 355851.

JAPAN, Texas Instruments Asia Ltd.: 4F Aoyama Fuji Bldg., 6-12, Kita Aoyama 3-Chome, Minato-ku, Tokyo, Japan 107, 03-498-2111; Osaka Branch, 5F, Nissho Iwai Bldg., 30 Imabashi 3-Chome, Higashi-ku, Osaka, Japan 541, 06-204-1881; Nagoya Branch, 7F Daini Toyota West Bldg., 10-27, Meieki 4-Chome, Nakamura-ku, Nagoya, Japan 450, 052-583-8691.

KOREA, Texas Instruments Supply Co.: Room 201, Kwangpoong Bldg., 24-1, Hwayand-Dong, Sung dong-ku, 133 Seoul, Korea, 02 + 464-6274/5.

MEXICO, Texas Instruments de Mexico S.A.: Poniente 116, No. 489, Colonia Vallejo, Mexico, D.F. 02300, 567-9200.

MIDDLE EAST, Texas Instruments: No. 13, 1st Floor Mannai Bldg., Diplomatic Area, Manama, P.O. Box 26335, Bahrain, Arabian Gulf, 973 - 72 46 81.

NETHERLANDS, Texas Instruments Holland B.V., P.O. Box 12995, (Bullewijk) 1100 AZ Amsterdam, Zuid-Oost, Holland (020) 5602911.

NORWAY, Texas Instruments Norway A/S: Kr. Augustsgt. 13, Oslo 1, Norway, (2) 20 60 40.

PHILIPPINES, Texas Instruments Asia Ltd.: 14th Floor, Ba-Lepanto Bldg., 8747 Paseo de Roxas, Makati, Metro Manila, Philippines, 882465.

PORTUGAL, Texas Instruments Equipamento Electronico (Portugal), Lda.: Rua Eng. Frederico Ulrich, 2650 Moreira Da Maia, 4470 Maia, Portugal, 2-9481003.

SINGAPORE (+ INDIA, INDONESIA, MALAYSIA, THAILAND), Texas Instruments Asia Ltd.: P.O. Box 138, Unit #02-08, Block 6, Kolam Ayer Industrial Est., Kallang Sector, Singapore 1334, Republic of Singapore, 747-2255.

SPAIN, Texas Instruments Espana, S.A.: C/Jose Lazaro Galdiano No. 6, Madrid 16, 1/458.14.58. C/Balmes, 89 Barcelona-8, 253 60 00/253 29 02.

SWEDEN, Texas Instruments International Trade Corporation (Sverigefilialen): Box 39103, 10054 Stockholm, Sweden, 08 - 235480.

SWITZERLAND, Texas Instruments, Inc. Riedstrasse 6, CH-8953 Dietikon (Zuerich) Switzerland, 1-740 2220.

TAIWAN, Texas Instruments Supply Co.: 10th Floor, Fu-Shing Bldg., 71 Sung-Kiang Road, Taipei, Taiwan, Republic of China, 02 + 521-9321.

UNITED KINGDOM, Texas Instruments Limited: Manton Lane, Bedford, MK41 7PA, England, 0234 67466; St. James House, Wellington Road North, Stockport, SK4 2RT, England, 061 442-8448. BC

TEXAS INSTRUMENTS
INCORPORATED
POST OFFICE BOX 225012 • DALLAS, TEXAS 75265